Free-radical chemistry

Free-radical chemistry

Structure and mechanism

D. C. NONHEBEL
Senior Lecturer in Chemistry
University of Strathclyde, Glasgow

J. C. WALTON
Lecturer in Chemistry
University of St Andrews

with a Foreword by
Professor Lord Tedder

CAMBRIDGE
AT THE UNIVERSITY PRESS · 1974

CHEMISTRY

Published by the Syndics of the Cambridge University Press
Bentley House, 200 Euston Road, London NW1 2DB
American Branch: 32 East 57th Street, New York, N.Y. 10022

Library of Congress Catalogue Card Number: 72-97887

ISBN: 0 521 20149 7

Printed in Great Britain
at the University Printing House, Cambridge
(Brooke Crutchley, University Printer)

Contents

Contents

Contents

Contents

Foreword

The study of free radicals is the Cinderella of organic chemistry. Students arriving at college in vehicles propelled by internal combustion engines are still taught that alkanes are inert substances. In the majority of text books on organic chemistry free-radical reactions are treated as a special, rather unusual class of reactions despite the fact that in industry and possibly also in nature one-electron transfers are more common than those involving electron pairs. The main reason for this is that the simplest free-radical reactions occur in the gas phase and few organic chemists have the apparatus or the technique to study such processes. The organic chemistry of free radicals in solutions was first fully developed by Kharasch in the United States and by Hey and Waters in Britain. In spite of such a good start free-radical chemistry failed to attract wide-spread interest among organic chemists in general, and very little attempt has been made to correlate the solution-phase work with the extensive kinetic studies in the gas phase. This separation of the 'physical chemistry' of free radicals from the 'organic chemistry' of free radicals has been a major deterrent to advance. The most important feature of this book is that it brings together the physical and organic branches of the subject.

At a time when physical organic chemistry is going through a period of reappraisal the present book emphasizes the important contribution a study of radical reactions can make to the nature of substituent effects. Many of the simple pictures of reactions developed twenty years ago have become clouded by the need to bring the solvent into almost all mechanistic considerations. The present volume shows that radical reactions where solvent effects are minimal, or where the reaction can be carried out in the gas phase, can still be interpreted in the polar and steric terms used by the physical organic chemist. This book arrives at a timely moment and will, I am sure, lead to a wider understanding of the importance of free-radical chemistry.

J. M. TEDDER

Preface

Free-radical chemistry is a subject which has grown enormously in scope and interest since the advent of the first free radical at the turn of the century. Less than twenty years ago it was possible for a book on mechanistic organic chemistry completely to ignore free radicals. Today radical intermediates are recognized and studied in almost every field of chemical activity, including organic, physical, inorganic and biological systems. There are at least two series of books exclusively devoted to recent advances in various specialized areas of the subject. In this volume we have attempted to bring together experimental results and theories from all these different areas and to show how the concept of free-radical intermediates brings an underlying order and unity to an otherwise complicated and confusing situation. In carrying out this aim we have emphasized the role of free radicals as one of the three major reactive intermediates of organic chemistry. We have sought to demonstrate how the structure and reactivity of free radicals is complementary to that of carbonium ions and carbanions.

The book provides a reasonably comprehensive, though not exhaustive, coverage of the general principles governing the behaviour of free radicals. It is directed primarily towards postgraduate students and chemists wishing to obtain an overall knowledge of free-radical chemistry, though it will also be useful to the more advanced undergraduate. We have attempted to integrate the approaches of the organic and the physical chemist by consideration of kinetic and mechanistic aspects of reactions both in solution and in the gas phase. This emphasis on the more physical aspects of the subject marks a departure from the approach found in most books.

The subject matter of the book falls into two main divisions. The first part, consisting of chapters 1 to 6, is concerned with the general principles and technique of radical chemistry, and deals with methods of production and detection of radicals, and also with their shape and stability. The very widespread use of electron spin resonance spectroscopy justifies the fairly extensive treatment of this topic in chapter 3. A comparison of radical reactions in solution and in the gas phase is

Preface

given in chapter 6. The remainder of the book provides a survey, of as wide a scope as could be encompassed in the space, of all the types of radical reactions which are encountered. In the main, each class of reactions is dealt with in a separate chapter, but we have made a slight departure from this practice for addition, abstraction and combination reactions. It was felt that, since these reactions are so widespread and so characteristic of radical species, a reclassification according to the type of radical involved would lead to a clearer and more interesting presentation. Accordingly chapters 7, 8 and 9 are mainly concerned with these three reaction types for atoms, alkyl radicals and heteroradicals respectively. Prominent sub-headings for each reaction type ensure that there is no difficulty in tracing each subject from one chapter to another. There follows an examination of free-radical oxidation and reduction in chapter 10. This is a large and very important area of radical chemistry which is frequently not given the depth of treatment it merits. Thus the important electrochemical oxidations and reductions, as well as metal- and metal-ion-catalysed reactions and other electron-transfer reactions, are all discussed. Autoxidation is also included in this chapter. Homolytic aromatic substitution is the topic of chapter 11. Radical fragmentations, rearrangements and cyclizations are covered in chapters 12, 13 and 14. The final chapter on radical displacement reactions is also to some extent an innovation, and reflects the growing interest in and importance of this type of reaction. Throughout chapters 7 to 15 we have emphasized, wherever appropriate, the synthetic use of radical reactions. We feel that this is important, as ultimately the chemical industry is concerned with the preparation of compounds and not the study of reaction mechanisms, except where this leads to improved synthetic methods. We have reluctantly omitted the treatment of the most important of all industrial radical reactions, polymerization, because we felt that it was impossible to do this large subject justice in the space available. This subject is also very adequately dealt with in other recent books specifically on polymerization.

We should like to express our heartfelt thanks to Professors Lord Tedder and P. L. Pauson, who have read the bulk of the manuscript and made many invaluable comments. We should also like to thank Professor R. O. C. Norman, Dr J. S. Littler, Dr W. Lawrie and Dr W. I. Bengough, who have read sections of the book and given very valuable advice. We are also indebted to Professor R. O. C. Norman for the provision of spectra for Figures 3-2, 3-4 and 3-9, to Professor W. A.

Waters for the spectrum for Figure 3-7, to Professor M. J. Perkins for the spectra for Figure 3-10 and to Dr A. R. Forrester for the spectra for Figure 3-11.

Finally, we should like to express our sincere appreciation to Mrs A. Cumming, Miss G. Forrest and Miss B. Martin, who cheerfully typed and not infrequently retyped the manuscript, to Mrs Jane Walton for excellent work in correcting the manuscript and proofs, and also to the staff of Cambridge University Press for their help in proceeding from typescript to printed book.

Glasgow and St Andrews
December 1972

D. C. NONHEBEL
J. C. WALTON

Note on units

We have written this book at a time when units and the symbols for them are being changed. A new international system is being gradually adopted. Since we are in a period of transition we have used a mixture of SI units and the previously accepted units. The units we have used most frequently, and their relation to other common units, are tabulated below.

Thermodynamic data
Energies in Joules (J) (1 calorie = 4.184 J).
Heats of reaction (ΔH) in kilojoules mole^{-1} (kJ mol^{-1}).
Entropies (S) in joules degree^{-1} (J K^{-1}).
Entropies of reaction (ΔS) in joules degree^{-1} mole^{-1} (J K^{-1} mol^{-1}).
Ionization potentials in electron volts (eV).
Pressure in torr (1 torr = 1 mm Hg).

Kinetic data
Rate constants of first-order reactions in sec^{-1} (s^{-1}).
Rate constants of second-order reactions in litres mole^{-1} sec^{-1} (1 mol^{-1} s^{-1}).
Rate constants of third-order reactions in litres^2 moles^{-2} sec^{-1} (1^2 mol^{-2} s^{-1}).
Activation energies in kilojoules mole^{-1} (kJ mol^{-1}).
A-factors have the same units as the rate constants of the reactions from which they are derived.

Ultraviolet spectra
Wavelength in nanometres (1 nm = 10^{-9} m).

Infrared spectra
Wavenumber in cm^{-1}.
Force constants in mdynes per ångstrom (mdyn Å^{-1}).

Nuclear magnetic resonance spectra
Chemical shifts in parts per million (ppm).
Operating frequency in hertz (Hz).

Note on units

Electron spin resonance spectra

Coupling constants and magnetic field strengths in millitesla (1 mT = 10 gauss).

Bond lengths

Lengths in ångstroms (1 Å = 10^{-1} nm).

1

Introduction

1.1 *THE HISTORY OF FREE-RADICAL CHEMISTRY*

The modern meaning of the term 'radical' has evolved through a series of advances and reverses in the understanding of chemical problems (1). The concept of radicals was introduced by Lavoisier (2) in his theory of acids, which he believed to be compounds of oxygen. He designated the element or group of elements which combined with oxygen in the acid a 'radical'. The oxygen theory of acids soon passed out of favour, but the word 'radical' was still frequently used to signify a group of elements which retained their identity through a series of reactions, e.g. a methyl radical.

The early nineteenth century saw the discovery of numerous 'radicals'. Thus Berzelius and Pontin (3) and Davy prepared the ammonium radical. Cyanogen was separated as an inflammable gas by Gay-Lussac (4), and cacodyl compounds were shown to contain the cacodyl radical C_2H_6As by Bunsen (5). The radical proved a useful organizing concept in both inorganic and organic chemistry. In the 1840s Kolbe obtained gases, by electrolysis of solutions of fatty acids, which he interpreted as free radicals (6). For example, potassium acetate gave 'free methyl' on electrolysis:

$$\text{`}C_2H_3C_2O_3 + HO \longrightarrow C_2H_3 + 2CO_2 + H\text{'}} \qquad (1)$$
'Acetic acid Water   Methyl   Hydrogen'

Frankland (7) heated zinc with ethyl iodide in a sealed tube and obtained a gas which he believed to be 'free ethyl':

$$\text{`}C_4H_5I + Zn \longrightarrow C_4H_5 + ZnI\text{'}} \qquad (2)$$
'Ethyl iodide   Ethyl'

When Cannizzaro placed the measurement of molecular weights on a firm footing by the method of vapour densities it was soon realized that

Introduction

groups such as methyl did not persist in the free state, but combined to form dimers of the radical.

The existence of free organic radicals was also out of harmony with ideas which were developing about structure. The majority of structural formulae could be written by allotting fixed valencies to the elements. Kekulé proved an able supporter of the doctrine of the quadrivalency of carbon which led to rationalization of the formulae of many organic compounds then being studied. This view gained ground until by the beginning of the twentieth century almost all chemists believed carbon to possess the single valency four.

Gomberg's discovery of triphenylmethyl came very unexpectedly in 1900, heralded only by Nef's voice crying in the wilderness (8). Gomberg set out to prepare hexaphenylethane by the reaction of triphenylmethyl chloride with silver in benzene (9). The products of his reaction contained oxygen, and it was only when he carried out the preparation in an atmosphere of carbon dioxide that this was avoided. The product was then a white solid dissolving in benzene to give a yellow solution. The behaviour of the product was very unlike that expected for hexaphenylethane. It rapidly oxidized in air and reacted at once with halogens. On the basis of this evidence Gomberg proposed that the compound was in fact the free triphenylmethyl radical:

$$Ph_3CCl + Ag \longrightarrow Ph_3C\cdot + AgCl \qquad (3)$$

This conclusion was not readily accepted by chemists in general, and the free-radical concept came under heavy fire when cryoscopic determinations showed the compound to have a molecular weight close to that of hexaphenylethane. Undeterred, Gomberg explained this observation by postulating an equilibrium mixture in solution:

$$2Ph_3C\cdot \rightleftharpoons Ph_3C-CPh_3 \qquad (4)$$

The white solid was regarded as hexaphenylethane, and the coloured compound in solution as the free radical. The ability of oxygen and halogens to decolorize solutions of 'triphenylmethyl' was explained as involving scavenging reactions of the radical:

$$Ph_3C\cdot + O_2 \longrightarrow Ph_3COO\cdot \qquad (5)$$

$$Ph_3COO\cdot + Ph_3C\cdot \longrightarrow Ph_3COOCPh_3 \qquad (6)$$

$$Ph_3C\cdot + X_2 \longrightarrow Ph_3CX + X\cdot \qquad (7)$$

$$Ph_3C\cdot + X\cdot \longrightarrow Ph_3CX \qquad (8)$$

Evidence in favour of Gomberg's postulate soon began to accumulate. Piccard (10) showed that the coloured solution did not obey Beer's law, the colour deepening with dilution. Other 'hexa-arylethanes' were prepared and shown to have molecular weights much lower than expected. In 1923 G. N. Lewis deduced that all compounds whose molecules contained an odd number of electrons should be paramagnetic (11). Studies of the magnetic susceptibilities of the solutions showed that the triarylmethyl radicals were indeed present, and enabled the equilibrium dissociation constants to be measured (12).

More and more 'stable' free radicals were discovered in the years following Gomberg's initial work. In 1911 Wieland discovered that tetraphenylhydrazine dissociated in a similar way (13):

$$Ph_2N{-}NPh_2 \rightleftharpoons 2Ph_2N\cdot$$

Radicals with the unpaired electron localized on oxygen were also soon prepared (14).

By the 1920s, transient free radicals were beginning to be proposed as reaction intermediates. Wood and Bonhoeffer obtained atomic hydrogen by passing hydrogen gas through an electrical discharge (15). Oxygen and halogen atoms were later obtained by an extension of this method (16, 17). Taylor proposed that free ethyl radicals were the reactive intermediates in the mercury-photosensitized hydrogenation of ethylene (18):

$$H_2 + Hg \longrightarrow 2H\cdot + Hg \tag{9}$$

$$H\cdot + C_2H_4 \longrightarrow C_2H_5\cdot \tag{10}$$

$$C_2H_5\cdot + H_2 \longrightarrow C_2H_6 + H\cdot \tag{11}$$

Transient free radicals became fairly generally accepted in gas-phase reactions as a result of the work of Paneth and Hofeditz (19) in 1929. They passed a stream of nitrogen or hydrogen saturated with the vapour of tetramethyl lead down a heated tube and showed that a mirror of lead was deposited at the point where the tube was heated:

$$PbMe_4 \longrightarrow Pb + 4Me\cdot \tag{12}$$

They further showed that a second mirror somewhat downstream from the point of heating was gradually removed. This second mirror could be of antimony or zinc as well as lead:

$$4Me\cdot + Pb \longrightarrow PbMe_4 \tag{13}$$

$$2Me\cdot + Zn \longrightarrow ZnMe_2 \tag{14}$$

Introduction

In the absence of this second mirror, ethane was produced when the carrier gas was nitrogen:

$$2CH_3\cdot \longrightarrow C_2H_6 \qquad (15)$$

The importance of free radicals in organic reactions was now rapidly recognized, and radical intermediates were proposed for numerous gas-phase reactions during the 1930s. This was particularly as a result of the study of the kinetics of gas-phase reactions by, amongst others, Dainton, Eyring, Hinshelwood, Kistiakowsky, Laidler, Melville, Norrish, Noyes, Polanyi, Rice, Steacie and Taylor. Particularly important was the publication in 1935 of a book, *The Aliphatic Free Radicals*, by F. O. and K. K. Rice (20).

The next milestone in the development of radical chemistry was the year 1937, when Hey and Waters published a memorable review in which they explained the products and kinetics of a broad range of reactions in solution by free-radical mechanisms (21). In that article they suggested that a number of unusual features encountered in the decompositions of dibenzoyl peroxide, *N*-nitrosoacetanilide, phenylazo-triphenylmethanes and diazonium salts could best be explained by invoking the participation of free phenyl radicals. In particular, they drew attention to the formation of significant quantities of *p*-substituted biphenyls from the phenylation of a variety of monosubstituted benzenes, regardless of the electronic nature of the substituent. They, and independently Kharasch (22), rationalized the ability of peroxides to effect the anti-Markovnikov addition of hydrogen bromide to olefins by a radical-chain process:

$$R\cdot + HBr \longrightarrow RH + Br\cdot \qquad (16)$$

$$Br\cdot + CH_2{=}CHCH_2Br \longrightarrow BrCH_2\dot{C}HCH_2Br \qquad (17)$$

$$BrCH_2\dot{C}HCH_2Br + HBr \longrightarrow BrCH_2CH_2CH_2Br + Br\cdot \qquad (18)$$

These developments gradually led to the acceptance of radical mechanisms for an increasing number of reactions in solution.

In the same year, Flory suggested a radical mechanism for addition polymerization (23). This had first been proposed by Taylor and Bates (24) in 1927, but it was Flory who developed the kinetics of radical polymerization of vinyl monomers and who also introduced the concept of chain transfer.

The subject of radical chemistry has advanced rapidly since those days, both for reactions in the gas phase and in solution. As regards the

4

latter, this has been aided by the ability to make a much more quantitative study of radical reactions as a result of chromatographic and spectroscopic analysis of the reaction products. Even more important has been the ability to demonstrate beyond reasonable doubt the existence of transient free radicals by e.s.r. spectroscopy and other physical techniques. This has opened up a new chapter in the history of radical chemistry, as it is now possible to obtain detailed information about the precise geometry of radical intermediates. In the gas phase ultra-short-lifetime species can now be observed by micro- and nano-second flash photolysis.

It should be said at this stage that the term 'free' radical is now somewhat of a misnomer as it implies that such a species has a finite lifetime enabling it to be studied, or at least intercepted by efficient radical traps. This is by no means always true and there are an increasing number of reactions which, though they proceed by a homolytic pathway, fail to provide evidence of radical intermediates.

1.2 THE CURRENT VIEW OF RADICAL CHEMISTRY

A radical is best defined as an atom or group of atoms with an unpaired electron, and thus includes such species as triphenylmethyl, chlorine atoms, sodium atoms and nitric oxide. A diradical is a species with two unpaired electrons.

Carbon radicals can be considered as being intermediate between carbonium ions and carbanions. Carbonium ions (1) are planar. The central carbon is sp^2-hybridized and there is a vacant p_z-orbital. Carbanions (4) are pyramidal with the carbon centre sp^3-hybridized. Carbon radicals, as will be discussed in chapter 4, are generally planar (2) with the unpaired electron in a p_z-orbital. This may have a certain degree of s-character according to the nature of the substituents attached to the carbon atom (3). In a heteroradical, i.e. a radical in which the unpaired electron is associated with an atom other than carbon, the unpaired electron is in a non-bonding atomic orbital.

    (1)   (2)   (3)   (4)

1.3 CLASSIFICATION OF RADICAL REACTIONS

1.3.1 *Unimolecular reactions*

(i) *Radical fragmentations*

$$\text{e.g. } Me_3CO\cdot \longrightarrow Me\cdot + Me_2CO \qquad (19)$$

Radical fragmentations occur both in solution and in the gas phase, particularly at low pressures. They are the reverse of radical additions. Radical ions also undergo fragmentation:

$$PhCH_3{}^{+\cdot} \longrightarrow PhCH_2\cdot + H^+ \qquad (20)$$

(ii) *Radical rearrangements*

$$\text{e.g. } PhCMe_2CH_2\cdot \longrightarrow PhCH_2\dot{C}Me_2 \qquad (21)$$

Examples of reactions involving rearrangements of radicals are relatively widespread, though much less so than carbonium-ion rearrangements.

(iii) *Radical cyclizations*

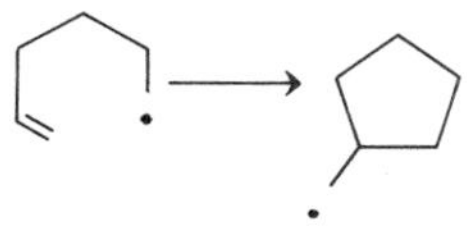

$$(22)$$

Radical cyclizations are perhaps best regarded as intramolecular additions to double bonds. Such reactions are very sensitive to the stereoelectronic requirement of the transition state.

1.3.2 *Bimolecular reactions between two radicals*

(i) *Combinations.* The coupling of like radicals to produce a dimer is perhaps the best known of all radical reactions, and is referred to as dimerization:

$$PhCH_2\cdot + PhCH_2\cdot \longrightarrow PhCH_2CH_2Ph \qquad (23)$$

It is not confined to reactions of like radicals and it is more desirable to refer to this as a combination reaction:

$$PhCH_2\cdot + Me_3C\cdot \longrightarrow PhCH_2CMe_3 \qquad (24)$$

Heteroradicals behave similarly:

$$PhS\cdot + PhS\cdot \longrightarrow PhSSPh \qquad (25)$$

The combination of radicals is a very exothermic process as the heat of formation of the new bond is liberated, and such reactions frequently proceed with little or no activation energy. In solution the excess energy is removed by the solvent. This is not possible in the gas phase when a third body, referred to as a chaperon or surface, is required:

$$R\cdot + R\cdot + M \longrightarrow R_2 + M \qquad (26)$$

Some combination reactions, particularly those involving small radicals, become third-order at low pressures in the gas phase.

(ii) *Disproportionations*. These reactions involve the transfer of a hydrogen atom from one radical to another:

$$Me_3C\cdot + Me_3C\cdot \longrightarrow Me_3CH + CH_2{=}CMe_2 \qquad (27)$$

Like combination, disproportionation is a very exothermic reaction since two bonds are formed and only one broken. It also always competes with combination.

1.3.3 *Bimolecular reactions between radicals and molecules*

(i) *Additions*

$$\text{e.g. } Br\cdot + CH_3CH{=}CH_2 \longrightarrow CH_3\dot{C}HCH_2Br \qquad (28)$$

The unsaturated substrate may also be an alkyne or an arene. In this latter case the overall reaction suggests substitution, rather than addition followed by abstraction:

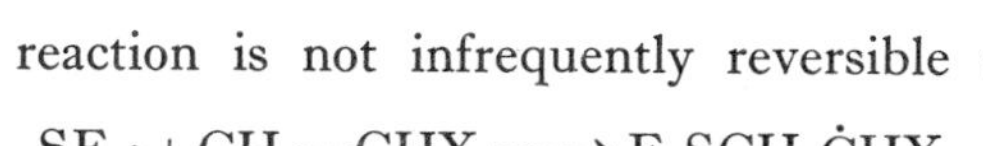

$$Ph\cdot + PhH \longrightarrow \overset{Ph}{\underset{H}{\diagdown}}\!\!\bigcirc\!\cdot \xrightarrow{[-H\cdot]} Ph{-}Ph \qquad (29)$$

Examples are also known in which addition occurs to other multiple bonds including carbonyl and azo groups:

$$Me\cdot + MeN{=}NMe \longrightarrow Me_2N{-}\dot{N}Me \qquad (30)$$

The addition reaction is not infrequently reversible even at room temperature:

$$SF_5\cdot + CH_2{=}CHX \rightleftharpoons F_5SCH_2\dot{C}HX \qquad (31)$$

Radical additions are enormously important, both as the propagation stage in radical polymerization and also in the synthesis of small molecules.

(ii) *Abstractions*. The second stage in the overall process involving addition to an unsaturated system by a radical mechanism involves the

Introduction

abstraction by the intermediate radical of an atom, usually a halogen or hydrogen, from the addend:

$$CH_3\dot{C}HCH_2Br + HBr \longrightarrow CH_3CH_2CH_2Br + Br \cdot \qquad (32)$$

$$C_6H_{13}\dot{C}HCH_2CCl_3 + CCl_4 \longrightarrow C_6H_{13}CHClCH_2CCl_3 + \dot{C}Cl_3 \quad (33)$$

Abstraction reactions are also involved in free-radical halogenation of alkanes and in many other reactions:

$$Br \cdot + CH_3-\underset{\underset{CH_3}{|}}{\overset{\overset{CH_3}{|}}{C}}-CH_3 \longrightarrow HBr + CH_3-\underset{\underset{CH_3}{|}}{\overset{\overset{CH_3}{|}}{C}}-CH_2 \cdot \qquad (34)$$

$$Bu^tO \cdot + PhCH_3 \longrightarrow Bu^tOH + PhCH_2 \cdot \qquad (35)$$

(iii) *Substitutions.* Homolytic substitutions, which are analogous to S_N2 reactions at a saturated carbon atom, constitute a further class of reaction and have been designated as S_H2 reactions. They are very much less common than nucleophilic substitution. Most if not all of the known examples of this type of reaction involve attack at an atom other than carbon:

$$CH_3S \cdot + PhSSPh \longrightarrow CH_3SSPh + PhS \cdot \qquad (36)$$

$$Ph \cdot + Bu^tOOBu^t \longrightarrow PhOBu^t + Bu^tO \cdot \qquad (37)$$

$$CH_3 \cdot + CF_3N{=}NCF_3 \longrightarrow CH_3N{=}NCF_3 + CF_3 \cdot \qquad (38)$$

1.3.4 *Electron-transfer reactions of radicals*

Radicals may undergo oxidation or less commonly reduction by the transfer of an electron. This is usually transferred to or from a transition metal:

$$PhCH_2 \cdot + Cu^{2+} \longrightarrow PhCH_2^+ + Cu^+ \qquad (39)$$

$$PhO^- + Fe(CN)_6{}^{3-} \longrightarrow PhO \cdot + Fe(CN)_6{}^{4-} \qquad (40)$$

REFERENCES

1. A. J. Ihde, IUPAC Symposium, Michigan, *Free Radicals in Solution*, Butterworths, London, 1966, p. 1.
2. A. L. Lavoisier, *Traité Élémentaire de Chimie*, Cuchet, Paris, 1789, vol. 1, p. 293. Translation by R. Kerr reprinted by Dover Press, New York, 1965, p. 66.
3. J. J. Berzelius and M. M. Pontin, *Gilbert's Annalen*, 1810, 6, 247.
4. J. L. Gay-Lussac, *Ann. de Chimie*, 1815, 95, 172.
5. R. Bunsen, *Liebig's Ann. Chem. Pharm.* 1842, 42, 27.

6. H. Kolbe, *Ann. Chim.* 1849, **69**, 257.
7. E. Frankland, *Ann. Chim.* 1849, **71**, 171.
8. J. U. Nef, *Ann. Chim.* 1894, **280**, 302.
9. M. Gomberg, *Ber.* 1900, **33**, 3150; *J. Amer. Chem. Soc.* 1900, **22**, 757.
10. J. Piccard, *Ann. Chim.* 1911, **381**, 347.
11. G. N. Lewis, *Valence and the Structure of Atoms and Molecules*, Chemical Catalog Co., New York, 1923, p. 148.
12. F. L. Allen and S. Sugden, *J. Chem. Soc.* 1936, 440.
13. H. Wieland, *Ann. Chim.* 1911, **381**, 200.
14. S. Goldschmidt and W. Schmidt, *Ber.* 1922, **55**, 3197.
15. K. F. Bonhoeffer, *Z. phys. Chem.* 1924, **113**, 199.
16. P. Harteck, *Z. Physik*, 1929, **54**, 881.
17. W. H. Rodebush and W. C. Klingelhoefer, *J. Amer. Chem. Soc.* 1933, **55**, 130.
18. H. S. Taylor, *Trans. Faraday Soc.* 1925, **21**, 560.
19. F. Paneth and W. Hofeditz, *Ber.* 1929, **62B**, 1335.
20. F. O. Rice and K. K. Rice, *The Aliphatic Free Radicals*, The Johns Hopkins Press, Baltimore, 1933.
21. D. H. Hey and W. A. Waters, *Chem. Rev.* 1937, **21**, 202.
22. M. S. Kharasch, H. Engelmann and F. R. Mayo, *J. Org. Chem.* 1937, **2**, 288; F. R. Mayo and M. S. Kharasch, *Chem. Rev.* 1940, **27**, 351.
23. P. J. Flory, *J. Amer. Chem. Soc.* 1937, **59**, 241.
24. H. S. Taylor and J. R. Bates, *J. Amer. Chem. Soc.* 1927, **49**, 2438.

2

Methods of production of free radicals

2.1 *INTRODUCTION*

Many free-radical reactions are brought about by the addition of radical initiators, such as peroxides or azo compounds, to the reaction system. Details of particular radical sources, e.g. aryl radicals, are given in the appropriate section of the book. In this chapter we shall devote some attention to the principles underlying the breakdown of these compounds into radicals. Broadly speaking these can be divided into three categories. Radicals may be produced by direct homolysis of a covalent bond, i.e. in a unimolecular process as the result of absorption of energy, e.g. in the thermolysis of peroxides:

$$Bu^tOOBu^t \xrightarrow{\Delta} 2Bu^tO\cdot \qquad (1)$$

Alternatively, homolysis may be induced by another reactive species, e.g. radicals can effect the induced decomposition of peroxides:

$$R\cdot + (PhCO_2)_2 \longrightarrow PhCO_2R + PhCO_2\cdot \qquad (2)$$

The third general method of generating radicals is in an electron-transfer reaction or less commonly a ligand-transfer reaction. These reactions are discussed in detail in chapter 10 and will not be further considered here.

Attention is directed primarily towards the use of compounds containing a peroxide linkage and also to azo compounds as radical sources. These classes of compounds are those most commonly employed as radical initiators in addition and polymerization reactions, and thus they merit more detailed treatment. The precise mode of decomposition of these classes of compounds will also be considered. Brief consideration is also given to structural effects, both steric and electronic, which affect the rate of decomposition of a given radical initiator, and also to solvent effects.

2.2 METHODS OF PRIMARY HOMOLYSIS

2.2.1 *Thermolysis*

Thermolysis of all organic compounds at sufficiently high temperatures (~ 800 °C) produces radicals capable of reacting with metallic mirrors. In solution, molecules with relatively weak bonds (bond dissociation energy < 160 kJ mol^{-1}) dissociate at a useful rate at temperatures below 150 °C. This is a widely used and very convenient method of producing radicals. Compounds which are particularly susceptible to homolysis are those containing a peroxide linkage (reactions *3–7*):

$$(PhCO_2)_2 \longrightarrow 2PhCO_2\cdot \tag{3}$$

$$(Bu^tO)_2 \longrightarrow 2Bu^tO\cdot \tag{4}$$

$$PhCO_3Bu^t \longrightarrow PhCO_2\cdot + Bu^tO\cdot \tag{5}$$

$$(Bu^sOCO_2)_2 \longrightarrow 2Bu^sO\cdot + 2CO_2 \tag{6}$$

$$(Bu^tOOCO)_2 \longrightarrow 2Bu^tO\cdot + 2CO_2 \tag{7}$$

Azo compounds, particularly azobisnitriles, are also very valuable radical sources (reaction *8*):

$$Me_2C(CN)N{=}NC(CN)Me_2 \longrightarrow 2Me_2\dot{C}CN + N_2 \tag{8}$$

Simple azoalkanes do not decompose sufficiently rapidly at temperatures below 150 °C to enable them to act as convenient radical sources.

2.2.2 *Photolysis*

Energy may also be provided to effect the homolysis of a bond by absorption of ultraviolet radiation. Absorption of energy results in the excitation of n or π electrons to the excited singlet π^* state in which the electron spins are paired. This excited state normally contains an excess of vibrational energy which is rapidly dissipated, and thereafter one of the following processes takes place: (*a*) radiation is emitted (fluorescence) and the molecule returns to the ground state; (*b*) energy can be lost by collision with solvent; (*c*) the excited singlet may undergo inter-system crossing to an excited triplet state, provided that the particular compound has a triplet state of appropriate energy; (*d*) homolysis may occur, giving a radical pair in which the electron spins are anti-parallel. The triplet state is much longer lived than the singlet because direct transition to the ground singlet state is quantum-mechanically forbidden.

Production of radicals

The process does occur (phosphorescence), though more slowly than the direct emission of radiation from the singlet state. The triplet may also undergo homolysis, in which case radicals with unpaired spins would be produced. The processes are summarized in Fig. 2-1.

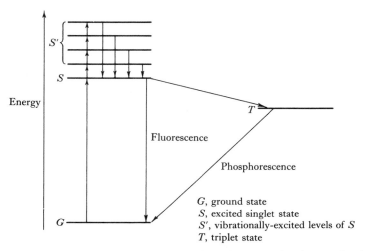

Fig. 2-1. Energy transitions that result from the absorption by a molecule of ultraviolet radiation.

Reactions *9–18* provide some examples in which photolysis has been used to generate radicals:

$$\text{ROOR} \xrightarrow{h\nu} 2\text{RO·} \tag{9}$$

$$\text{(RCO}_2)_2 \xrightarrow{h\nu} 2\text{RCOO·} \tag{10}$$

$$\text{RN}{=}\text{NR} \xrightarrow{h\nu} 2\text{R·} + \text{N}_2 \tag{11}$$

$$\text{RONO} \xrightarrow{h\nu} \text{RO·} + \text{NO} \tag{12}$$

$$\text{ROCl} \xrightarrow{h\nu} \text{RO·} + \text{Cl·} \tag{13}$$

$$\text{ArI} \xrightarrow{h\nu} \text{Ar·} + \text{I·} \tag{14}$$

$$\text{HSiCl}_3 \xrightarrow{h\nu} \text{SiCl}_3\text{·} + \text{H·} \tag{15}$$

$$\text{Ph}_3\text{SnH} \xrightarrow{h\nu} \text{Ph}_3\text{Sn·} + \text{H·} \tag{16}$$

$$\text{CF}_3\text{SCl} \xrightarrow{h\nu} \text{CF}_3\text{S·} + \text{Cl·} \tag{17}$$

$$\text{(EtO)}_2\text{P(O)H} \xrightarrow{h\nu} \text{(EtO)}_2\dot{\text{P}}\text{(O)} + \text{H·} \tag{18}$$

In photolysis as opposed to thermolysis, energy of a particular value is supplied to the molecule, and consequently photolysis is on the whole a more specific method of effecting homolysis. It is also possible to provide sufficient energy to cleave even relatively strong bonds which do not undergo cleavage at moderate temperatures, e.g. azoalkanes. The prerequisite of the method is the ability of the molecule to absorb in the ultraviolet or visible region of the spectrum.

The thermolysis of diacyl peroxides, especially those containing secondary or tertiary alkyl groups, is complicated by heterolytic decomposition, leading mainly to the formation of esters (1). Photolysis of such peroxides proceeds by a radical mechanism and hence is clearly the method of choice for obtaining secondary or tertiary alkyl radicals from the corresponding peroxide. It also has the advantage that radicals can be generated in the cavity of an e.s.r. spectrometer at low temperatures, enabling studies to be carried out on radical shapes and rearrangements (2).

In the flash-photolysis method (3), the radical source is subjected to a very high-intensity short-duration flash of visible or ultraviolet light. The advantage of this technique is that a large proportion of decomposition occurs, producing a momentarily high radical concentration. The radicals thus produced can be detected by physical methods, such as optical absorption spectroscopy, which are incapable of detecting the low stationary radical concentrations produced in continuous photolysis experiments.

In a conventional flash-photolysis set-up, the high-intensity photolysing flash is followed, after a very short time interval of the order of microseconds only, by a second lower-intensity flash which is arranged to traverse the cell at right angles to the photolysing flash. This second flash enters a spectrograph and allows the absorption spectra of the radicals (and any other transient species) to be observed. Alternatively, the photolysing flash can be followed by a series of analysing flashes at timed intervals which enables the rate of reaction of the transient species to be determined.

Ordinary photolysis is used as a method of generation of radicals in preparative organic chemistry, whilst flash photolysis has proved most useful in the determination of radical structure and in following the kinetics of radical reactions. It has been successfully employed in the gas, liquid and solid phases.

Production of radicals

2.2.3 Radiolysis

Energy to bring about the homolysis of covalent bonds can also be supplied by high-energy sources of radiation including γ-rays, which are generated from a ^{60}Co source, X-rays, and high-energy electrons. There are considerable similarities between photochemical and radio-chemical work, though there are also important differences. There is a much greater possibility in radiolysis of effecting quantum-mechanically forbidden absorption of energy with the direct formation of an excited triplet state. The radiation absorbed is also of much higher energy and hence excitation occurs to excited states other than the lowest one. Consequent homolysis of these will produce 'hot radicals'. A second major difference from photolysis is that the very high-energy radiation can cause the extrusion of an electron with the resultant formation of a radical cation (*19*). This can undergo homolysis to a radical and a cation (reaction *20*):

$$H_2O \longrightarrow H_2O^{+\cdot} + e \qquad (19)$$

$$H_2O^{+\cdot} \longrightarrow H^+ + HO\cdot \qquad (20)$$

This is a useful source of hydroxyl radicals. Another important distinction between radiolysis and photolysis is that the radicals in the former are generated in clusters along the path of the radiation, whereas in the latter radicals are produced randomly throughout the system.

The use of radiation in chemical industry is becoming increasingly widespread due to the decrease in cost of radiation sources such as ^{60}Co. γ-Rays are used to induce the radical addition of hydrogen bromide to ethylene.

Pulse radiolysis is essentially a method of producing a high radical concentration in a short time. It is similar to flash photolysis except that the initial decomposition is brought about by a fast pulse of high-energy electrons produced by a microwave linear accelerator, or a Van de Graaf generator (4). Once produced, the decay of the free radicals can be followed by physical methods such as absorption spectroscopy or e.s.r. spectroscopy. The method is fairly wide in application and has been used for studying the reactions of a variety of radicals in solution, particularly the 'simplest free radical', namely the hydrated electron e_{aq}^- (5).

14

2.3 INDUCED DECOMPOSITIONS

2.3.1 *Radical-induced decompositions*

The rate of decomposition of radical initiators, especially peroxides, is strongly dependent on the nature of the solvent, being very much higher in ethers and alcohols than in hydrocarbons (6). Furthermore the rates of decomposition in most solvents show a higher than first-order dependence on peroxide concentration (6). Nozaki and Bartlett demonstrated in the case of dibenzoyl peroxide that these phenomena were due to concurrent unimolecular homolysis and radical-induced homolysis:

$$\frac{-d[\text{Per}]}{dt} = k_1[(\text{PhCO}_2)_2] + k_2[\text{R}\cdot][(\text{PhCO}_2)_2]^n$$

Application of the steady-state approximation to this reaction results in the prediction that the value of n is $\frac{3}{2}$ if termination is between like radicals, and 1 if termination is between unlike radicals. For hydrocarbon and halogenated solvents n is generally $\frac{3}{2}$, whilst for ethers and alcohols it is 1. Results indicate that the rate of the non-induced part of the decomposition is on the whole insensitive to the nature of the solvent.

In aromatic hydrocarbons, Gill and Williams have shown that the induced part of the decomposition is brought about by phenylcyclohexadienyl radicals (7):

$$[\text{PhC}_6\text{H}_6]\cdot + (\text{PhCO}_2)_2 \longrightarrow \text{PhCO}_2\text{H} + \text{PhCO}_2\cdot + \text{Ph—Ph} \qquad (21)$$

This reaction is discussed in greater detail in chapter 11 (p. 429).

The decomposition of dibenzoyl peroxide in diethyl ether gives 0.9 mole of α-ethoxyethyl benzoate per mole of decomposed peroxide, together with substantial quantities of benzoic acid but only small quantities of carbon dioxide (8). These results are rationalized by Scheme 1. Isotopic labelling experiments have shown that radical attack does occur at the peroxidic oxygen (9, 10).

$$(\text{PhCO}_2)_2 \overset{\Delta}{\longrightarrow} 2\text{PhCO}_2\cdot$$

$$\text{PhCO}_2\cdot + \text{CH}_3\text{CH}_2\text{OEt} \longrightarrow \text{PhCO}_2\text{H} + \text{CH}_3\dot{\text{C}}\text{HOEt}$$

$$\text{CH}_3\dot{\text{C}}\text{HOEt} + (\text{PhCO}_2)_2 \longrightarrow \underset{\overset{|}{\text{OCOPh}}}{\text{CH}_3\text{CHOEt}} + \text{PhCO}_2\cdot$$

$$\text{PhCO}_2\cdot + \text{CH}_3\dot{\text{C}}\text{HOEt} \longrightarrow \underset{\overset{|}{\text{OCOPh}}}{\text{CH}_3\text{CHOEt}}$$

Scheme 1

Production of radicals

Similar sequences can be written for the induced decomposition in thioethers, alkenes, and alcohols of diacyl peroxides, in which the chain-propagating species is an alkyl radical, and of dialkyl peroxides (Scheme 2) (11). Of particular interest in this scheme is the reaction between the

$$(Bu^tO)_2 \xrightarrow{\Delta} 2Bu^tO\cdot$$
$$Bu^tO\cdot + R_2CHOH \longrightarrow Bu^tOH + R_2\dot{C}OH$$
$$R_2\dot{C}OH + (Bu^tO)_2 \longrightarrow R_2C{=}O + Bu^tOH + Bu^tO\cdot$$

Scheme 2

α-hydroxyalkyl radical and the peroxide. This could be formulated as a direct displacement reaction with the formation of a hemiacetal which would subsequently decompose to the observed products (reactions *22* and *23*), by analogy with the induced decomposition of dibenzoyl peroxide in ether:

$$R_2\dot{C}OH + (Bu^tO)_2 \longrightarrow R_2\underset{\underset{OBu^t}{|}}{C}OH + Bu^tO\cdot \qquad (22)$$

$$R_2\underset{\underset{OBu^t}{|}}{C}OH \longrightarrow R_2C{=}O + Bu^tOH \qquad (23)$$

Alternatively, it could be a hydrogen-transfer reaction as indicated in Scheme 2. This reaction would be sterically more favourable and would account for the absence of any induced decomposition of di-t-butyl peroxide in ethers, since in this case the α-alkoxyalkyl radicals clearly could not undergo a hydrogen-transfer reaction.

The decomposition of diacyl peroxides is very considerably accelerated by amines. This, however, is a result of nucleophilic attack by the amine on the peroxide linkage (12):

$$Ph\ddot{N}Me_2 + (PhCO_2)_2 \longrightarrow Ph\underset{+}{N}Me_2 + PhCO_2^- \qquad (24)$$

with $OCOPh$ attached above the nitrogen.

The resultant quaternary salt can then decompose by a homolytic or heterolytic pathway (reaction *25* or *26*):

$$Ph\underset{+}{N}Me_2 \longrightarrow Ph\overset{+\cdot}{N}Me_2 + PhCO_2\cdot \qquad (25)$$

with $OCOPh$ attached above the nitrogen.

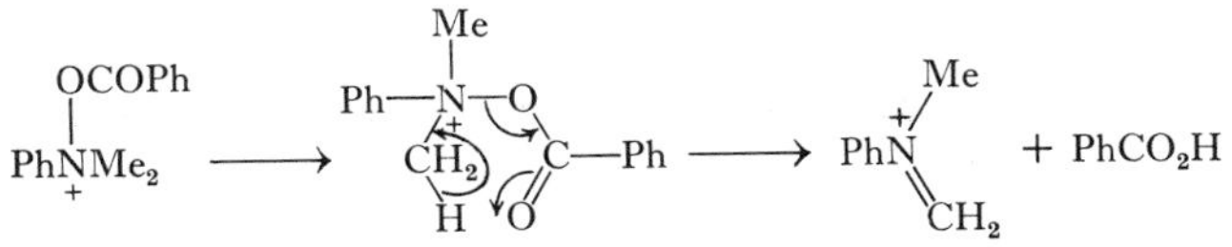

$$\text{PhNMe}_2^+ \quad\longrightarrow\quad \overset{\displaystyle\text{Me}}{\underset{\text{H}}{\text{Ph—N—O}}}\;\text{CH}_2\;\text{C—Ph} \quad\longrightarrow\quad \text{PhN}^{+}\diagdown\!\!\!\diagup^{\text{Me}}_{\text{CH}_2} + \text{PhCO}_2\text{H} \quad (26)$$

Primary and secondary amines similarly react to give quaternary salts, but in general these react further by an ionic mechanism.

In contrast primary and secondary but not tertiary amines induce the decomposition of di-t-butyl peroxide by a homolytic pathway (13):

$$\text{R}_2\text{CHNHR}' + \text{Bu}^t\text{O·} \longrightarrow \text{R}_2\dot{\text{C}}\text{NHR}' + \text{Bu}^t\text{OH}$$

$$\text{R}_2\dot{\text{C}}\text{NHR}' + (\text{Bu}^t\text{O})_2 \longrightarrow \text{R}_2\text{C}{=}\text{NR}' + \text{Bu}^t\text{OH} + \text{Bu}^t\text{O·}$$

The decomposition of di-t-butyl peroxide is considerably faster in the neat liquid than in the gas phase or in dilute solution in an inert solvent. This has indicated that under these conditions an induced decomposition of the peroxide takes place. The isolation of isobutylene oxide has led to the postulation of Scheme 3 (14).

$$\text{Bu}^t\text{O·} \text{ (or Me·)} + (\text{Bu}^t\text{O})_2 \longrightarrow \underset{\underset{\displaystyle\text{CH}_2\text{·}}{|}}{\text{Me}_2\text{COOBu}^t} + \text{Bu}^t\text{OH (or CH}_4)$$

$$\underset{\displaystyle\dot{\text{C}}\text{H}_2}{\overset{\displaystyle}{\text{Me}_2\text{C—O}}}\!\!\diagup\!\text{O—Bu}^t \longrightarrow \text{Me}_2\text{C}\diagdown\!\!\!\diagup\text{CH}_2 + \text{Bu}^t\text{O·}$$

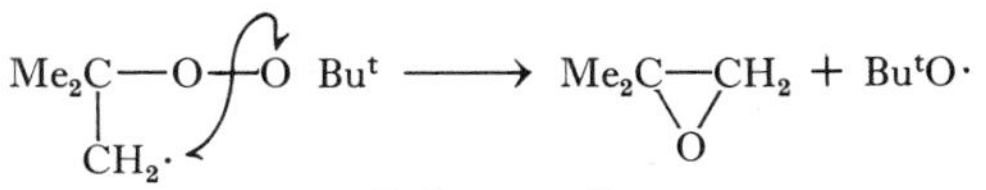

Scheme 3

2.3.2 *Electron-transfer induced decompositions*

The decomposition of diacyl peroxides and peresters can be greatly accelerated by traces of single-electron reductants, especially by copper(I) salts (15). The reaction involves the transfer of an electron to the lowest antibonding orbital of the peroxide oxygen followed by cleavage of the O–O bond:

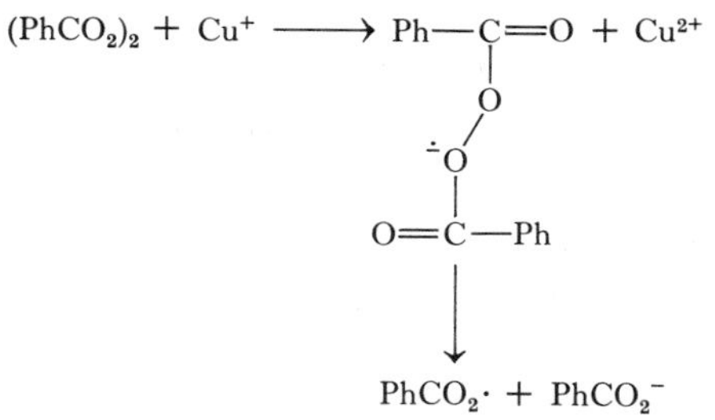

$$(\text{PhCO}_2)_2 + \text{Cu}^+ \longrightarrow \text{Ph—C}{=}\text{O} + \text{Cu}^{2+}$$

$$(27)$$

$$\text{O}{=}\text{C—Ph}$$

$$\downarrow$$

$$\text{PhCO}_2\text{·} + \text{PhCO}_2^-$$

Production of radicals

This technique provides a very useful source of acyloxy radicals and is widely used in the acyloxylation of alkenes (cf. p. 362).

2.3.3 Molecule-induced homolysis

The decomposition of radical initiators can be induced by molecules as well as by radicals and single-electron reductants. It occurs when two molecules react together at an anomalously fast rate to produce radicals. Thus Walling showed that an exceedingly rapid chain reaction took place *in the dark* between styrene and t-butyl hypochlorite giving radical-derived products (16):

$$PhCH{=}CH_2 + Bu^tOCl \longrightarrow PhCHClCH_2OBu^t$$

Similarly, bromination of 1,4,9-trimethylanthracene in acetic acid under nitrogen in the dark, gave exclusively the radical product, 9-bromomethyl-1,4-dimethylanthracene (17). Such reaction conditions would be expected to favour electrophilic substitution and lead to the nuclear brominated product. The driving force involves complexing or bond formation of the incipient radical with the substrate.

A somewhat different type of molecule-induced homolysis occurs in the reaction between diphenylhydroxylamine and dibenzoyl peroxide (18) (Scheme 4). This is essentially an electron-transfer process rather

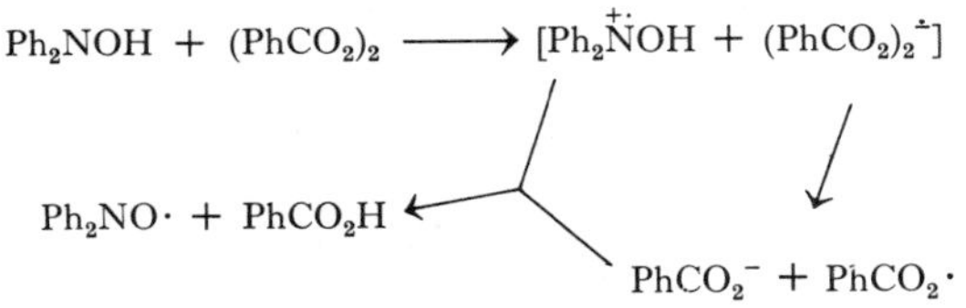

$$Ph_2NOH + (PhCO_2)_2 \longrightarrow [Ph_2\overset{+\cdot}{N}OH + (PhCO_2)_2^{\overset{-}{\cdot}}]$$

$$Ph_2NO\cdot + PhCO_2H \longleftarrow$$

$$PhCO_2^- + PhCO_2\cdot$$

Scheme 4

than a reaction involving nucleophilic attack on the peroxide. This is considerably more probable as the rate of reaction of bis-*o*-chlorobenzoyl peroxide is also high, indicating the absence of appreciable steric hindrance as would have been expected for nucleophilic attack.

2.3.4 Photosensitized decomposition of radical initiators

In the direct photolysis of an initiator, the initiator is excited and thence undergoes homolysis from either the excited singlet or triplet state. A more efficient method of transferring energy to a radical initiator is frequently to add a second compound, known as a photosensitizer, to the system which is itself very easily excited photochemically. The excited molecule can then either transfer energy directly from its excited singlet state to give an excited singlet state of the initiator, or alternatively it may undergo intersystem crossing to an excited triplet state and this can transfer energy giving an excited singlet and an excited triplet of the initiator. The excited triplet states of benzophenone and other triplet photosensitizers are readily quenched by oxygen. This is not true for excited singlet states of aromatic hydrocarbons, e.g. benzene or anthracene. These latter thus provide a means of obtaining an initiator in its excited singlet state.

Scheme 5 indicates the possible pathways by which the excited triplet state of benzophenone may react. The excited triplet of benzophenone

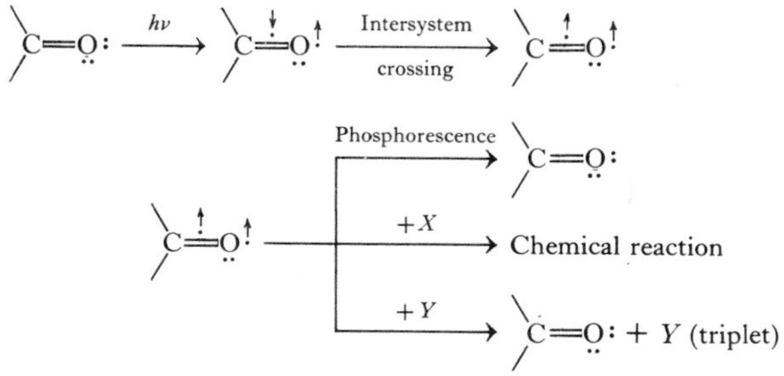

Scheme 5

is capable of abstracting hydrogen from many solvents. This is seen in the photolysis of benzophenone in isopropanol which gives benzopinacol and acetone (20):

$$Ph_2CO^* + Me_2CHOH \longrightarrow Ph_2\dot{C}OH + Me_2\dot{C}OH \qquad (28)$$

$$Ph_2CO^* + Me_2\dot{C}OH \longrightarrow Ph_2\dot{C}OH + Me_2CO \qquad (29)$$

$$2Ph_2\dot{C}OH \longrightarrow \begin{array}{c} Ph_2C-OH \\ | \\ Ph_2C-OH \end{array} \qquad (30)$$

The photolysis of benzophenone in cyclohexane gives benzopinacol and bicyclohexyl, whilst in mixtures of cyclohexane and benzene cyclohexylbenzene is also obtained as a consequence of the cyclohexylation of benzene (21):

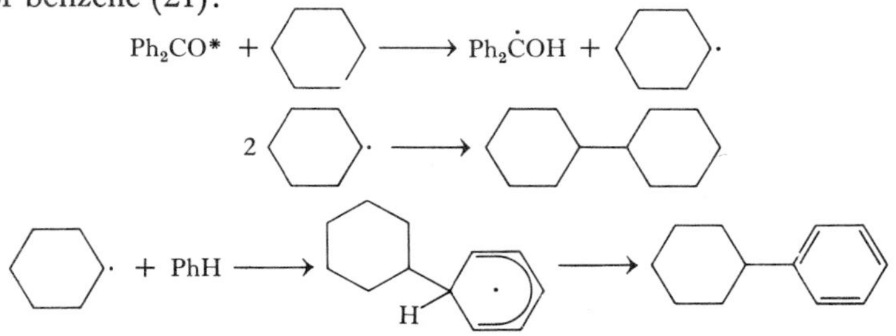

$$(31)$$

$$(32)$$

$$(33)$$

19

Production of radicals

The other way in which the excited triplet of benzophenone behaves is by transferring its energy to another molecule. This is illustrated by the benzophenone-photosensitized decarbonylation of 2-ethylhexanal (22) (reactions *34–37*):

$$Ph_2CO^* + RCHO \longrightarrow Ph_2CO + RCHO^* \qquad (34)$$

$$RCHO^* \longrightarrow R\cdot + \dot{C}HO \qquad (35)$$

$$R\cdot + RCHO \longrightarrow RH + R\dot{C}O \qquad (36)$$

$$R\dot{C}O \longrightarrow R\cdot + CO \qquad (37)$$

The benzophenone causes a very significant speeding up in the rate of decarbonylation. Consistent with the proposed mechanism, no benzopinacol was obtained, indicating that direct energy transfer took place rather than hydrogen abstraction from the aldehyde.

The photosensitization of reactions in the gas phase can be conveniently effected by certain metals, notably mercury (23). The requirements necessary to make a metal suitable as a photosensitizer are that it has a sufficiently high vapour pressure at reasonably low temperatures, that it has an accessible excited state, and that it is chemically inert. These are met by mercury which has a low-lying $6(^3P_1)$ state. This can be quenched by reaction with hydrogen and provides a method of generation of hydrogen atoms:

$$Hg\ 6(^3P_1) + H_2 \longrightarrow Hg\ 6(^1S_0) + 2H\cdot \qquad (38)$$

Alkanes can be similarly decomposed:

$$Hg\ 6(^3P_1) + RH \longrightarrow HgH + R\cdot \qquad (39)$$

$$HgH \longrightarrow Hg\ 6(^1S_0) + H\cdot \qquad (40)$$

Alkenes react by formation of short-lived vibrationally excited triplets which could decompose to alkynes because of this excess vibrational energy. Alternatively, they could be deactivated by collision, with a quencher, Q, or they could undergo spontaneous emission of radiation:

$$Hg\ 6(^3P_1) + C_2H_4 \longrightarrow Hg\ 6(^1S_0) + {}^3C_2H_4^* \qquad (41)$$

$${}^3C_2H_4^* \longrightarrow C_2H_2 + H_2 \qquad (42)$$

$${}^3C_2H_4^* + Q \longrightarrow C_2H_4 + Q \qquad (43)$$

$${}^3C_2H_4^* \longrightarrow C_2H_4 + h\nu \qquad (44)$$

The basic mechanism is supported by the observation that isomerization is observed with *cis-* or *trans*-but-2-enes.

2.4 *MECHANISM OF DECOMPOSITION OF RADICAL INITIATORS*

In this section attention will be devoted to the detailed mechanism of the thermolytic and photolytic decomposition of diacyl peroxides, peresters and azo compounds, as more detailed information is available for these compounds than for other radical initiators. Moreover, the results of studies on these compounds are illustrative of the general behaviour of radical initiators on homolysis.

2.4.1 *Decomposition of dibenzoyl peroxide*

Dibenzoyl peroxide has been successfully used as a source of both phenyl and benzoyloxy radicals. Thus reaction of dibenzoyl peroxide with naphthalene gives both phenylnaphthalenes and benzoyloxy-naphthalenes (24). This indicates that homolysis of dibenzoyl peroxide gives initially benzoyloxy radicals which subsequently undergo decarboxylation outside the solvent cage:

$$(PhCO_2)_2 \longrightarrow 2PhCO_2 \cdot \qquad\qquad (45)$$

$$PhCO_2 \cdot \longrightarrow Ph \cdot + CO_2 \qquad\qquad (46)$$

Decomposition of dibenzoyl peroxide in moist carbon tetrachloride containing iodine results in the quantitative formation of benzoic acid (25):

$$PhCO_2 \cdot + I \cdot \longrightarrow PhCO_2I \xrightarrow{H_2O} PhCO_2H \qquad\qquad (47)$$

This confirms that all benzoyloxy radicals have sufficient lifetime to escape from the solvent cage under these conditions.

Photolysis of dibenzoyl peroxide in benzene gives 13 % of phenyl benzoate whereas thermolysis at 80 °C gives only a trace of this (26). This difference in behaviour results from the more rapid decarboxylation of the benzoyloxy radical at higher temperatures. The amount of phenyl benzoate formed is reduced to 3 % when the photolysis is carried out in the presence of benzophenone as a photosensitizer. An explanation for this is that photolysis in benzene gives the excited peroxide singlet as a result of photosensitization by the excited benzene singlet whereas in the presence of benzophenone the peroxide will be formed in the excited triplet state. The excited peroxide singlet possesses sufficient energy that on homolysis it gives directly or in rapid succession phenyl radicals, carbon dioxide and benzoyloxy radicals, within the solvent cage. Combination of the phenyl and benzoyloxy radicals results in the formation of the ester:

$$(PhCO_2)_2 \xrightarrow[PhH]{h\nu} [Ph\cdot \;\; \cdot CO_2 \;\; PhCO_2\cdot] \begin{cases} \nearrow PhCO_2Ph + CO_2 \\ \\ \searrow Ph\cdot + PhCO_2\cdot + CO_2 \end{cases} \qquad (48)$$

The excited triplet state of the peroxide will undergo homolysis to give benzoyloxy radicals with paired spins. Even if decarboxylation can occur within the solvent cage the radicals will not combine readily as their spins will be parallel:

$$(PhCO_2)_2 \xrightarrow[Ph_2CO]{h\nu} [PhCO_2\cdot^\uparrow \;\; {}^\uparrow\cdot O_2CPh] \longrightarrow [Ph\cdot^\uparrow \;\; CO_2^\uparrow \;\; \cdot O_2CPh]$$
$$\downarrow \qquad\qquad (49)$$
$$Ph\cdot \;\; CO_2 \;\; PhCO_2\cdot$$

2.4.2 Decomposition of diacetyl peroxide

Decomposition of diacetyl peroxide either thermally or photolytically fails to give acetic acid or any products which might be derived from acetoxy radicals. This together with the failure to trap acetoxy radicals with iodine suggests that they have no discrete existence outside the solvent cage. The question then arises as to whether acetoxy radicals are produced initially by O–O bond cleavage, or whether there is concurrent C–CO and O–O bond fission with direct formation of methyl radicals:

$$(CH_3CO_2)_2 \begin{cases} \nearrow [CH_3CO_2\cdot \;\; \cdot O_2CCH_3] \longrightarrow [CH_3\cdot + 2CO_2 + CH_3\cdot] \\ \\ \searrow [CH_3\cdot + 2CO_2 + CH_3\cdot] \end{cases} \qquad (50)$$

The isolation of methyl acetate from decompositions of diacetyl peroxide gives credence to the postulate that acetoxy radicals are formed initially. The methyl acetate arises from coupling of acetoxy radicals and methyl radicals within the solvent cage:

$$[CH_3CO_2\cdot + CH_3\cdot] \longrightarrow CH_3CO_2CH_3 \qquad (51)$$

There is some doubt as to whether this is in fact the correct interpretation of the result, since esters are also formed from diacyl peroxides in a heterolytic process. However, Kochi showed that photolysis of dicyclopropylcarbonyl peroxide, which was thermally stable under the conditions used, gave some cyclopropyl cyclopropanecarboxylate (27):

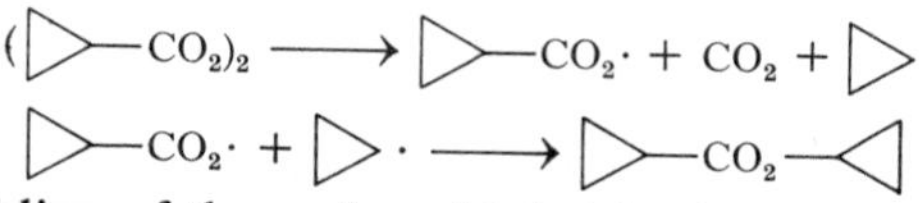

$$\left(\vartriangleright\!\!-CO_2\right)_2 \longrightarrow \vartriangleright\!\!-CO_2\cdot + CO_2 + \vartriangleright\cdot \qquad (52)$$

$$\vartriangleright\!\!-CO_2\cdot + \vartriangleright\cdot \longrightarrow \vartriangleright\!\!-CO_2\!\!-\!\vartriangleleft \qquad (53)$$

Some scrambling of the carbonyl label is observed in the decomposition of diacetyl-^{18}O peroxide (28):

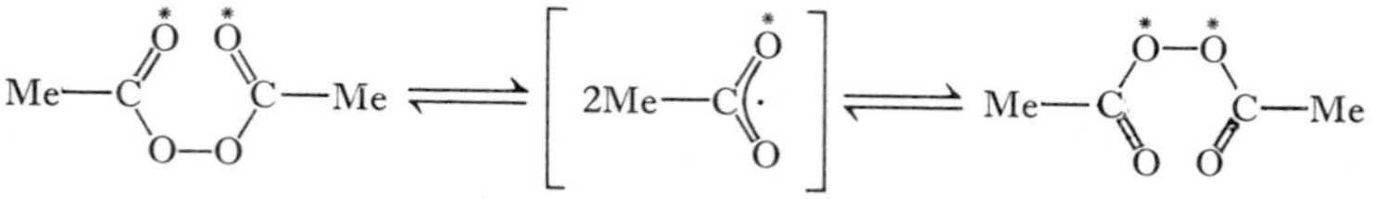

$$Me\!-\!\overset{\overset{*}{O}}{\underset{O-O}{C}}\overset{\overset{*}{O}}{\underset{}{C}}\!-\!Me \rightleftharpoons \left[2Me\!-\!\overset{\overset{*}{O}}{\underset{O}{C\cdot}}\right] \rightleftharpoons Me\!-\!\overset{\overset{*}{O}-\overset{*}{O}}{\underset{O\ O}{C}}\overset{}{\underset{}{C}}\!-\!Me \qquad (54)$$

The extent of scrambling was somewhat greater when the decomposition was carried out in a more viscous solvent (29) in which diffusion from the solvent cage would be more difficult. Recent work has, however, indicated that some but probably not all of this scrambling arises not from the dissociation of the peroxide into radicals followed by their recombination but that it also occurs by a sigmatropic pathway (30).

Further evidence supporting the contention of one-bond homolysis of diacetyl peroxide is the decrease in the rate of decomposition on increasing the viscosity of the solvent (31). This type of behaviour is not observed for azo compounds which decompose by a two-bond process since it is most unlikely that cage recombination of two radicals and nitrogen would occur.

Kochi has studied the decomposition of equimolar mixtures of two peroxides in *n*-pentane and in decalin (32). Considerably more of the unsymmetrical dimer was formed in pentane than in decalin, reflecting the greater degree of reaction occurring outside the solvent cage (Scheme 6). He was able to calculate that approximately 35% of primary alkyl radicals react within the solvent cage in pentane and that this increases to 50% in the more viscous decalin.

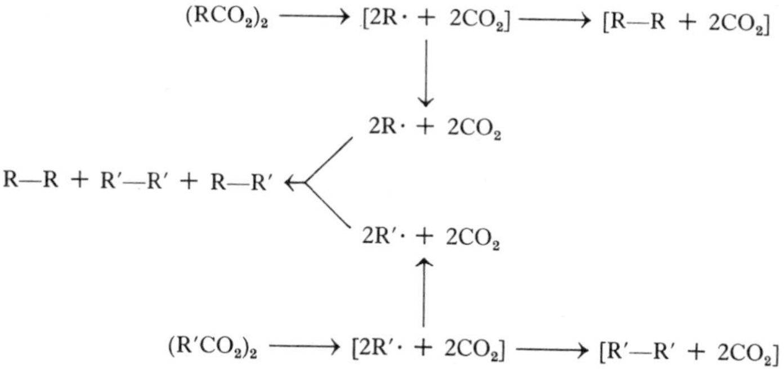

$$(RCO_2)_2 \longrightarrow [2R\cdot + 2CO_2] \longrightarrow [R\!-\!R + 2CO_2]$$

$$\downarrow$$

$$2R\cdot + 2CO_2$$

$$R\!-\!R + R'\!-\!R' + R\!-\!R' \longleftarrow$$

$$2R'\cdot + 2CO_2$$

$$\uparrow$$

$$(R'CO_2)_2 \longrightarrow [2R'\cdot + 2CO_2] \longrightarrow [R'\!-\!R' + 2CO_2]$$

Scheme 6

2.4.3 *Decomposition of peresters*

The question as to whether peresters decompose by a one- or two-bond fission is less clear cut than is the case for peroxides. The existence of a significant secondary deuterium isotope effect in the thermal decomposition of t-butyl phenylperacetate and related peresters is evidence for at least an appreciable degree of stretching of the C–CO bond in the homolysis of the perester (33). The value of this isotope effect (1.03–1.06 per deuterium) is less than that encountered in the thermolyses of related azo compounds which undoubtedly decompose by a two-bond process (34), but greater than that for the thermolysis of diacetyl peroxide. The dependence of the rate of decomposition of t-butyl phenylperacetate on the viscosity of the solvent is also indicative of the decomposition involving two-bond cleavage. A possible explanation of these results is that decomposition proceeds by both the one- and two-bond routes. Scrambling has been observed in the thermolysis of specifically labelled t-butyl peracetate but this could be as a result of a sigmatropic rearrangement (35). Photolysis of t-butyl cyclopropylpercarboxylate failed to give any products containing the cyclopropylcarboxy group even though the cyclopropylcarboxy radical is relatively stable with respect to its decarboxylation. This is further evidence for essentially a two-bond cleavage:

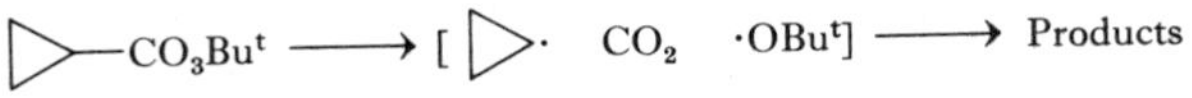

$$\triangleright\!-\!CO_3Bu^t \longrightarrow [\,\triangleright\!\cdot \quad CO_2 \quad \cdot OBu^t\,] \longrightarrow \text{Products} \qquad (55)$$

The e.s.r. spectrum of photolysed t-butyl cyclopropanepercarboxylate only showed the presence of cyclopropyl radicals even at $-100\,°C$ (36).

The situation is different for t-butyl perbenzoate, the decomposition of which gives *inter alia* benzoic acid (36), indicative of a one-bond decomposition. This might be expected because of the greater stability of benzoyloxy radicals:

$$PhCO_3Bu^t \longrightarrow [PhCO_2\cdot \quad \cdot OBu^t] \longrightarrow PhCO_2\cdot + Bu^tO\cdot \qquad (56)$$

t-Butyl peracetate and t-amyl perpropionate gave none of the crossed ethers, t-butyl ethyl ether and t-amyl methyl ether, indicating that ether

$$
\begin{array}{ccc}
MeCO_3Bu^t & & [Me\cdot \quad CO_2 \quad \cdot OBu^t] \longrightarrow MeOBu^t + CO_2 \\
+ & \longrightarrow & + \\
EtCO_3Am^t & & [Et\cdot \quad CO_2 \quad \cdot OAm^t] \longrightarrow EtOAm^t + CO_2
\end{array}
$$

$$EtOBu^t \text{ or } MeOAm^t$$

Scheme 7

formation occurs exclusively within the solvent cage (Scheme 7). On the basis of this evidence Kochi estimated that the extent of cage reactions in the photolysis of t-butyl peracetate at 30 °C rises from 27 % in pentane to 44 % in decalin and 75 % in the much more viscous nujol, whilst the extent of cage reactions in the thermolysis at 115 °C is only 17 % in decalin (36).

2.4.4 *Decomposition of azo compounds*

Symmetrical azo compounds decompose by concurrent homolysis of both C–N bonds, whereas when the two alkyl radicals derived from the azo compound are of very different stability one-bond fission occurs (33):

$$\text{PhCHMeN=NCHMePh} \xrightarrow{h\nu} [2\text{Ph}\dot{\text{C}}\text{HMe} + \text{N}_2] \qquad (57)$$

$$\text{PhCHMeN=NMe} \xrightarrow{h\nu} [\text{Ph}\dot{\text{C}}\text{HMe} + \text{MeN}_2\cdot] \qquad (58)$$

When the two radicals are of similar stability, deuterium isotope effects indicate that in the transition state for the homolysis both C–N bonds are stretched, though to differing extents.

$$\text{PhCHMeN=NCHMe}_2 \longrightarrow [\text{Ph}\dot{\text{C}}\text{HMe} \quad \cdot\text{N=N}\cdots\text{CHMe}_2] \qquad (59)$$

The same conclusions about one- and two-bond fission can be drawn from the dependence of the rate of decomposition on the viscosity of the medium (31).

One of the best studied and most important of azo compounds, from an industrial standpoint, is azobisisobutyronitrile. Thermolysis of this compound occurs readily at moderate temperatures because the energy of activation is lowered as a result of the stability of the cyanoisopropyl radicals. It, together with other azo compounds, has the advantage as a radical initiator that it does not undergo induced decomposition in the same manner as peroxide initiators. The efficiency of it, and other azo compounds, in promoting polymerization of vinyl monomers is low due

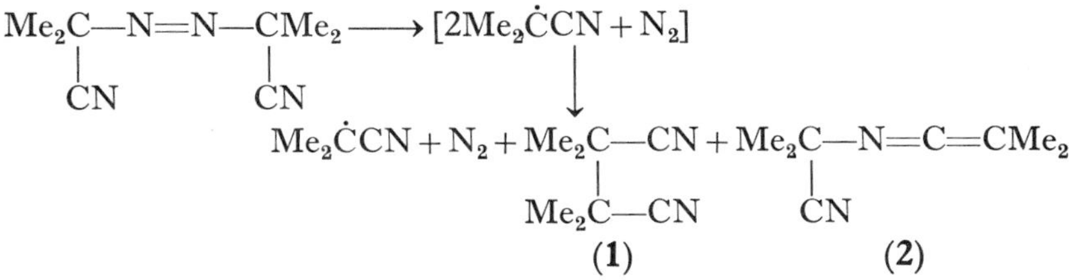

$$
\begin{array}{c}
\underset{\overset{|}{\text{CN}}}{\text{Me}_2\text{C}}\text{—N=N—}\underset{\overset{|}{\text{CN}}}{\text{CMe}_2} \longrightarrow [2\text{Me}_2\dot{\text{C}}\text{CN} + \text{N}_2] \\
\downarrow \\
\text{Me}_2\dot{\text{C}}\text{CN} + \text{N}_2 + \underset{\overset{|}{\text{Me}_2\text{C}}\text{—CN}}{\text{Me}_2\text{C—CN}} + \underset{\overset{|}{\text{CN}}}{\text{Me}_2\text{C—N=C=CMe}_2} \\
\qquad\qquad (1) \qquad\qquad\qquad (2)
\end{array}
$$

Scheme 8

to the importance of cage reactions (37) (Scheme 8). The two products resulting from the cage recombination of cyanoisopropyl radicals are tetramethylsuccinonitrile (**1**) and the ketenimine (**2**). It is impossible to reduce the combined yields of these products to below 20% even when the azo compound is decomposed in the presence of thiols (38).

The importance of azo compounds as radical initiators has resulted in extensive studies of their thermolytic, direct photolytic and photo-sensitized decompositions. In most instances the method of effecting homolysis had remarkably little effect on the products of the reaction. This similarity in products from azo-2-methyl-2-propane arises from the fact that in all instances, even in the presence of triplet sensitizers, radicals are produced with paired spins, i.e. they are derived from an excited singlet state (39). Support for this is that the quantum yield for the photosensitized reaction with triphenylene is unaffected by the addition of piperylene, an effective quencher of triplets. Similarly, aromatic ketones, which undergo rapid intersystem crossing, are very inefficient photosensitizers. Photosensitization with aromatic ketones merely effects *cis–trans* isomerization of acyclic azo compounds. The cyclic azo compound (**3**) undergoes direct photolysis to give predominantly the cyclobutane (**4**) with retention of configuration (> 95%) (40). The photosensitized reaction with the triplet photosensitizer thioxanthone gives both the meso and racemic cyclobutane (**4** and **5**) and also an appreciable amount of 2-methylbut-1-ene (Scheme 9). This is consistent with the much longer lifetime of the triplet diradical.

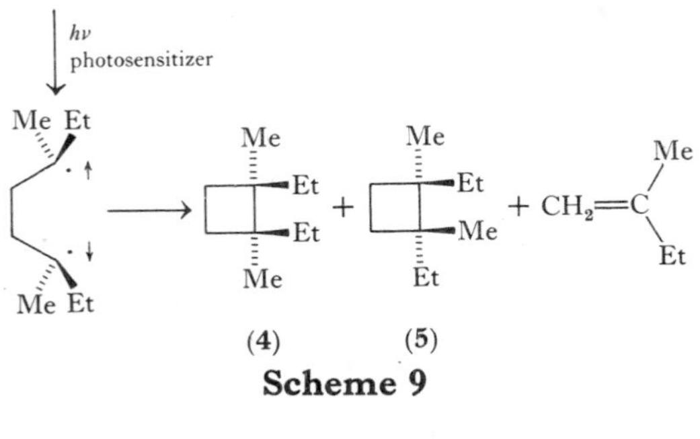

Scheme 9

2.5 STRUCTURE–REACTIVITY RELATIONSHIPS IN THE HOMOLYSIS OF PEROXIDES AND AZO COMPOUNDS

The rates of thermolysis of organic peroxides and azo compounds are dependent on the stability of the resultant radical fragments. This aspect is discussed in greater detail in chapter 5. The dependence of the rate of thermolysis on radical structure is more marked in the case of azo compounds than with diacyl peroxides or peresters since only with azo compounds does one obtain alkyl radicals directly from the homolysis process.

Swain has shown that the rates of thermolysis of a series of *meta-* and *para*-substituted dibenzoyl peroxides are retarded by electron-withdrawing substituents and facilitated by electron-donating substituents (41). The inductive influence of electron-withdrawing groups removes electron density from the peroxidic oxygen, thereby stabilizing it with respect to homolysis. Molecular-orbital calculations by Perkins (42) have shown that the half-life for decomposition of a series of diacyl peroxides, peresters, and dialkyl peroxides is related to both the bond order of the O–O bond, and also to the electronic charges on the two oxygen atoms. Thus electron-withdrawing groups reduce the charge on the oxygen, thereby stabilizing the peroxide. These findings are thus in excellent agreement with the experimental results of Swain (41). *o*-Substituents in dibenzoyl peroxides independent of their electronic nature increase the rate of decomposition (43). This probably results from the relief of steric interaction in the peroxide on decomposition.

REFERENCES

1. C. Walling and M. J. Gibian, *J. Amer. Chem. Soc.* 1965, **87**, 3413; T. Kashiwagi, S. Kozuka and S. Oae, *Tetrahedron*, 1970, **26**, 3619.
2. J. K. Kochi and P. J. Krusic, *J. Amer. Chem. Soc.* 1969, **91**, 3940.
3. G. Porter, *Technique of Organic Chemistry*, ed. G. Weissberger, 2nd edn, vol. VIII, pt II, p. 1055.
4. L. M. Dorfman and M. S. Matheson, *Progr. Reaction Kinetics*, 1965, **3**, 239.
5. F. S. Dainton, IUPAC Symposium, Michigan, *Free Radicals in Solution*, Butterworths, London, 1966, p. 15.
6. K. Nozaki and P. D. Bartlett, *J. Amer. Chem. Soc.* 1946, **68**, 1686.
7. G. B. Gill and G. H. Williams, *J. Chem. Soc.* 1965, 995, 5756, 7127; *J. Chem. Soc.* B, 1966, 880.
8. W. E. Cass, *J. Amer. Chem. Soc.* 1947, **69**, 500.
9. E. H. Drew and J. C. Martin, *Chem. and Ind.* 1959, 925.

10. D. D. Denney and G. Feig, *J. Amer. Chem. Soc.* 1959, **81**, 5322.
11. E. S. Huyser and C. J. Bredeweg, *J. Amer. Chem. Soc.* 1964, **86**, 2401.
12. C. Walling and N. Indictor, *J. Amer. Chem. Soc.* 1958, **80**, 5814.
13. E. S. Huyser, C. J. Bredeweg and R. M. VanScoy, *J. Amer. Chem. Soc.* 1964, **86**, 4148.
14. E. R. Bell, F. F. Rust and W. E. Vaughan, *J. Amer. Chem. Soc.* 1950, **72**, 337.
15. J. K. Kochi, *J. Amer. Chem. Soc.* 1962, **84**, 774.
16. C. Walling, L. Heaton and D. D. Tanner, *J. Amer. Chem. Soc.* 1965, **87**, 1715.
17. J. Flood, A. D. Mosnaim and D. C. Nonhebel, *Chem. Comm.* 1970, 761.
18. G. R. Chalfont and M. J. Perkins, *J. Chem. Soc.* B, 1971, 245.
19. D. C. Neckers, in *Methods in Free-Radical Chemistry*, ed. E. S. Huyser, Dekker, New York, 1969, vol. i, p. 35.
20. A. Schönberg and A. Mustafa, *Chem. Rev.* 1947, **46**, 181.
21. C. Walling and M. J. Gibian, *J. Amer. Chem. Soc.* 1965, **87**, 3361.
22. J. D. Berman, J. H. Stanley, W. V. Sherman and S. G. Cohen, *J. Amer. Chem. Soc.* 1963, **85**, 4010.
23. R. J. Cvetanović, *Progr. Reaction Kinetics*, 1964, **2**, 39.
24. D. I. Davies, D. H. Hey and G. H. Williams, *J. Chem. Soc.* 1958, 1878.
25. G. S. Hammond and L. M. Soffer, *J. Amer. Chem. Soc.* 1950, **72**, 4711.
26. T. Nakata, K. Tokumaru and O. Simamura, *Tetrahedron Letters*, 1967, 3303.
27. R. A. Sheldon and J. K. Kochi, *J. Amer. Chem. Soc.* 1970, **92**, 4395.
28. J. W. Taylor and J. C. Martin, *J. Amer. Chem. Soc.* 1967, **89**, 6904.
29. J. C. Martin and J. H. Hargis, *J. Amer. Chem. Soc.* 1969, **91**, 5399.
30. M. J. Goldstein and H. A. Judson, *J. Amer. Chem. Soc.* 1970, **92**, 4119.
31. W. A. Pryor and K. Smith, *J. Amer. Chem. Soc.* 1970, **92**, 5403.
32. R. A. Sheldon and J. K. Kochi, *J. Amer. Chem. Soc.* 1970, **92**, 4395.
33. T. Koenig, J. Huntington and R. Cruthoff, *J. Amer. Chem. Soc.* 1970, **92**, 5413.
34. S. Seltzer and S. G. Mylonakis, *J. Amer. Chem. Soc.* 1967, **89**, 6586; S. Seltzer, *J. Amer. Chem. Soc.* 1961, **83**, 2625.
35. M. J. Goldstein and H. A. Judson, *J. Amer. Chem. Soc.* 1970, **92**, 4120.
36. R. A. Sheldon and J. K. Kochi, *J. Amer. Chem. Soc.* 1970, **92**, 5175.
37. J. C. Bevington, *Trans. Faraday Soc.* 1955, **51**, 1392.
38. G. S. Hammond, J. N. Sen and C. E. Boozer, *J. Amer. Chem. Soc.* 1955, **77**, 3244.
39. P. D. Bartlett and P. S. Engel, *J. Amer. Chem. Soc.* 1968, **90**, 2960.
40. P. D. Bartlett and N. A. Porter, *J. Amer. Chem. Soc.* 1968, **90**, 5317.
41. C. G. Swain, W. T. Stockmayer and J. T. Clarke, *J. Amer. Chem. Soc.* 1950, **72**, 5426.
42. P. G. Perkins, unpublished results.
43. W. Cooper, *J. Chem. Soc.* 1951, 3106.

3

Physical methods for the detection of free radicals

3.1 INTRODUCTION

In the majority of reaction systems the presence of free radicals can be inferred from the nature of the reactants and products and from the chemical behaviour of the system under study. Free-radical reactions are susceptible to inhibition and retardation, and are accelerated by the presence of initiators. The products of a free-radical reaction may also include dimers, or anti-Markovnikov adducts which enable the reactive species to be characterized. Another sensitive chemical test for the presence of free radicals is the copolymerization of a 50/50 mixture of styrene and methyl methacrylate (1). If the copolymerization is carried out in the presence of cations the product copolymer consists wholly of styrene, whereas if free radicals are present the copolymer consists of equal proportions of styrene and methyl methacrylate. If anions are present then the copolymer consists wholly of methyl methacrylate.

No completely satisfactory physical method has yet been developed for observing free radicals at the low concentrations at which they usually occur in gas-phase reactions. All the methods either depend on specially high radical concentrations or are applicable to only a few select radicals possessing special features. In solution, where higher radical concentrations are normally attained, several methods, in particular electron spin resonance spectroscopy, are of very wide applicability. Stable free radicals can, of course, be detected and characterized by the usual physico-chemical methods of organic chemistry.

3.2 MAGNETIC SUSCEPTIBILITY

Any substance placed in a magnetic field develops an induced magnetic moment and shows the property of diamagnetism. If, in addition, the substance possesses a permanent magnetic moment of its own, then it is also paramagnetic. Free radicals have an unpaired electron which confers a permanent magnetic moment on the molecule by virtue of its spin. Detection of paramagnetism is therefore an excellent diagnostic test for the presence of free radicals in an organic sample.

In a non-uniform magnetic field of gradient $\delta H/\delta z$ in the z direction, the substance experiences a force in this direction, proportional to the product of the field strength and field gradient:

$$f = m\chi H\delta H/\delta z \qquad (1)$$

where m is the mass of the substance and χ is defined as the magnetic susceptibility per unit mass. The molar susceptibility χ_m is the susceptibility per unit mass multiplied by the molecular weight of the substance. Diamagnetic substances have negative susceptibilities, and the force acts to push them out of the magnetic field. Paramagnetic substances have positive susceptibilities, and are drawn into the magnetic field. This is the basis of most methods for measuring susceptibility (1 a). A few milligrams of sample are suspended in a vacuum, from a micro-balance, between the poles of a magnet which provides an inhomogeneous field. The deflection of the balance then gives a measure of the force acting on the sample. $H\delta H/\delta z$ is determined for the magnet by using a standard substance of known susceptibility and hence χ can be determined by use of equation (1).

For a free radical, the gross magnetic susceptibility determined in this way is the sum of the diamagnetic and paramagnetic contributions. The diamagnetic susceptibility is virtually independent of temperature for solids, liquids and gases, provided no chemical change occurs. Pascal (2) has shown that the molar diamagnetism of a compound may be represented by the sum of the diagmagnetisms of the constituent atoms plus a small 'constitutive' correction:

$$\chi_m = \sum_i n_i\chi_i + \lambda \qquad (2)$$

Where n_i is the number of atoms of susceptibility χ_i and λ is the constitutive correction.

Tables of empirical atomic susceptibilities (generally called Pascal Constants), derived from the observed susceptibilities of a large number of compounds, have been constructed (1a, 3). The diamagnetic susceptibility of a molecule can thus be estimated by adding the approximate atomic Pascal constants and constitutive corrections. Agreement with experiment is generally very good. An estimate of the diamagnetic contribution to the gross susceptibility of a free radical can also be obtained in the same way. The procedure is somewhat suspect in this case however, since the correct constitutive constant for the unpaired electron is not known. It is suspected that this correction may be large (1a) because the diamagnetic susceptibilities of aromatic, and other π-conjugated molecules, are unusually large. Radical paramagnetic susceptibilities estimated in this way are therefore subject to a certain degree of doubt, although the total diamagnetism is usually small compared to the paramagnetism of the sample.

The paramagnetic susceptibility, as predicted by quantum mechanics, consists of two parts, one due to the orbital motion of the unpaired electron, and the other due to its spin. In all organic free radicals the spin-orbital coupling is broken down by the fields of the surrounding atoms, and only the spin contribution is left. The molar paramagnetism due to electron spin only is given by:

$$\chi_m = N\beta^2 4S(S+1)/3kT \tag{3}$$

where N is Avogadro's number, β is the Bohr magneton having a value of 0.917×10^{-20} erg gauss^{-1}, and S is the total spin quantum number, which for a radical having one unpaired electron is $\frac{1}{2}$. The calculated molar paramagnetic susceptibility at 20 °C is 1270×10^{-6} mol^{-1}. Experimental values very close to this have been found for many paramagnetic substances with one unpaired electron.

Magnetic susceptibility measurements can be made on all radicals of reasonably long half-life. Radical concentrations can also be measured by comparing the measured paramagnetic susceptibility of a sample containing an unknown concentration of a free radical, with the known susceptibility per mole, i.e. 1270×10^{-6}. This method has been used for calculating the dissociation constants of stable free radicals such as the triarylmethyl radicals (4), some of which are given in Table 5-6. The radical nature of many other organic compounds has also been demonstrated by the susceptibility method (1a), e.g. 1,1-diphenylpicrylhydrazyl, benzophenonepotassium and many other ketyls. The advent of e.s.r. spectroscopy, which is more sensitive, yields much more structural information and by-passes the diamagnetic correction difficulties, has overshadowed this older technique in recent years.

Detection of radicals

3.3 E.S.R. SPECTROSCOPY

Electron spin resonance spectroscopy is by far the most useful method of radical detection. As will be apparent later, detailed analysis of an e.s.r. spectrum frequently makes it possible to deduce not only the gross chemical structure of the radical but also its conformation. Spin densities at various positions in delocalized radicals may also be obtained. The technique can be applied to the measurement of radical concentrations. E.s.r. spectroscopy is of particular value in that, by use of suitable flow systems, short-lived radicals may be detected. The sensitivity of the method allows radical concentrations of 10^{-8} mol to be observed.

This section outlines some of the general principles underlying e.s.r. spectroscopy and is included because of the importance of the method, and also because a knowledge of the principles is essential for deducing shapes of radicals. For a more detailed treatment of the subject, reference should be made to a specialized text (5).

3.3.1 *Principles of e.s.r. spectroscopy*

An unpaired electron has a spin and a magnetic moment. It can thus exist in two spin states, which are of equal energy in the absence of an external magnetic field. In the presence of an applied magnetic field the electron can align itself parallel or antiparallel to this field. The difference in energy of these orientations is given by the equation:

$$\Delta E = g\beta H$$

where g is the Landé factor which is a dimensionless proportionality constant of value 2.0023, β is the Bohr magneton, and H is the strength of the applied magnetic field in gauss. The frequency associated with the change from one energy level to the other is thus given by the equation:

$$\nu = \frac{g\beta H}{h}$$

Initially there are more electrons in the lower energy level. Irradiation with electromagnetic radiation of frequency ν results in transitions from the lower to the higher energy level. Non-emitting relaxation processes maintain the Boltzmann distribution of the two states and normally prevent saturation. The ratio of the numbers of spins in the two energy states is given by the equation:

$$n_1/n_2 = e^{-\Delta E/kT} = e^{-g\beta H/kT}$$

32

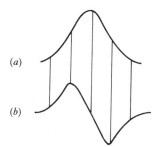

Fig. 3-1. Curves used in the presentation of e.s.r. spectra; (*a*) absorption curve, (*b*) first derivative of absorption curve.

where n_1 and n_2 are the numbers of spins in the higher and lower states respectively. The sensitivity of the method is thus related to the strength of the applied magnetic field. It is imperative that this field be highly homogeneous if the absorption peak is not to be broadened with resultant obscuring of the hyperfine splitting, which is so important in determination of radical structures.

In general, for the study of organic free radicals a magnetic field of about 330 mT (3300 gauss) is employed. Resonance can be achieved either by variation of the irradiation frequency or of the field. Invariably the frequency is kept constant and the magnetic field varied. Further details of the instrumentation are described by Kevan (6). In contrast to n.m.r. spectrometers, e.s.r. spectrometers are arranged to record the first derivative of the absorption curve rather than the absorption curve itself (Fig. 3-1). This gives somewhat greater sensitivity and also better resolution. The area under the absorption curve is proportional to the number of spins in the sample. Integration of the first derivative to give the absorption curve followed by integration of this to obtain its area enables one to determine radical concentrations by comparison of this area with that due to a known concentration of radicals. Fremy's salt, $(KSO_3)_2NO\cdot$, is frequently used as a standard.

3.3.2   *Methods of radical production for e.s.r. studies*

Radicals are generally transient species and hence special experimental techniques have to be devised to allow their observation within the cavity of an e.s.r. spectrometer. There are three principal methods which are used: (*a*) radicals may be generated in a matrix at very low temperatures; (*b*) they can be produced by irradiation of a solution of a suitable radical precursor in the cavity of the spectrometer; or (*c*) they may be formed and continuously introduced into the cavity by use of a flow system.

The lifetime of radicals generated in a matrix is very much greater than would be the case in solution because the slow rate of diffusion prevents or retards reactions of the radical. The spectra of radicals produced in this way are complicated by anisotropic interaction which is absent in the spectra of liquids where the tumbling of the radicals averages out such interactions. The method is thus not of particular interest.

For the study of the e.s.r. spectra of specific radicals the most useful procedure involves irradiation in the cavity of the spectrometer. Fessenden and Schuler (7) accomplished this by irradiating with an electron beam a solution of a hydrocarbon in cyclopropane as an inert solvent. This method usually gives mixtures of radicals but often it is possible to disentangle the spectra of the different radicals. Radiolysis of propane thus gives both propyl and isopropyl radicals. A more convenient procedure for the production of specific alkyl radicals has been devised by Kochi and Krusic (8). They photolysed diacyl peroxides or peresters in inert solvents in the spectrometer cavity. The initially produced acyloxy radicals were not observed, nor in the case of the peresters were t-butoxy radicals. Loss of carbon dioxide from the acyloxy radical led to one and only one alkyl radical whose spectrum could be readily observed:

$$(RCO_2)_2 \longrightarrow 2RCO_2\cdot$$
$$RCO_3Bu^t \longrightarrow RCO_2\cdot + Bu^tO\cdot$$
$$RCO_2\cdot \longrightarrow R\cdot + CO_2$$

Photolysis of di-t-butyl peroxide in presence of a hydrogen donor in an inert solvent also gives rise to alkyl radicals (8) though, in contrast to the photolysis of diacyl peroxides or peresters, more than one alkyl radical may be produced:

$$Bu^tO\cdot + PhCH_3 \longrightarrow PhCH_2\cdot + Bu^tOH$$

Another way of producing specific alkyl radicals has been devised by Hudson and Jackson (9). Photolysis of di-t-butyl peroxide in a trialkyl-silane in the presence of an alkyl bromide gives alkyl radicals. The method is of particular value because of the accessibility of alkyl bromides:

$$Et_3SiH + Bu^tO\cdot \longrightarrow Et_3Si\cdot + Bu^tOH$$
$$Et_3Si\cdot + RBr \longrightarrow R\cdot + Et_3SiBr$$

A flow system has been developed by Norman and Dixon (10) to allow study of radicals produced in chemical reactions. The reactants which give rise to the radicals are mixed just outside the spectrometer cavity. In this way radicals can be observed within 10^{-2} s of their moment of formation. Thus, by the use of this technique, reactions of hydroxyl radicals with a variety of organic substrates may be studied. Hydroxyl radicals are generated by reaction of titanium(III) chloride with hydrogen peroxide. The reaction of hydroxyl radicals with methanol was studied by mixing aqueous methanolic solutions of hydrogen peroxide and titanium(III) chloride just before the cavity. In the cavity $\dot{C}H_2OH$ radicals can be detected:

$$Ti^{III} + H_2O_2 \longrightarrow Ti^{IV} + \cdot OH + OH^-$$

$$HO\cdot + CH_3OH \longrightarrow \dot{C}H_2OH + H_2O$$

A disadvantage of the flow system is the large volume which has to be used.

3.3.3   *Characteristics of e.s.r. spectra*

Electron spin resonance spectra are characterized by three parameters: g-factors, hyperfine splitting constants, and line widths. A close study of these parameters enables much detailed structural information about the particular radical to be gleaned.

(i) *g-Factors.* In a magnetic field an unpaired electron in a free radical possesses, in addition to its spin angular momentum, a small amount of extra orbital angular momentum, with consequent spin-orbit coupling. This results in the electron having a slightly different effective magnetic moment from that which a free electron would possess. As a consequence of this the condition for resonance is somewhat changed, $h\nu = g\beta H$ where g is a function of the particular radical. Hence, for a given frequency, radicals with different g-factors resonate at different field strengths. The difference in the g-factor for a radical and that for the free electron is analogous to the chemical shift in n.m.r. spectroscopy. These differences are small, but they are nevertheless significant and can give valuable information about the structure of a radical.

The g-factor of a methyl radical is 2.00255, i.e. extremely close to that of a free electron. It is, however, modified somewhat by substitution of hydrogen by an alkoxyl (or hydroxyl) group or by a carbonyl-containing group (Table 3-1). It is possible to determine the g-factor of a radical with considerable precision and thus use may be made of it to obtain

TABLE 3-1   *g-Factors of some organic radicals* (11, 12)

Radical	g-Factor
$\dot{C}H_3$	2.00255
$\dot{C}Ph_3$	2.00259
$\dot{C}H_2OH$	2.0033
$CH_3\dot{C}HOH$	2.0033
$CH_3\dot{C}HOEt$	2.0031
$\dot{C}H_2CHO$	2.0045
$\dot{C}H_2COCH_3$	2.0042
$\dot{C}H_2COCH_2OH$	2.0042
$Ph_2N\cdot$	2.0032
$Bu_2{}^sNO\cdot$	2.00585
$PhCMe_2OO\cdot$	2.0155

information about radical structure. Thus, Norman and Pritchett (11) have shown that oxidation of 2,3-butanediol gives rise to the radical **(1)** and not the radical **(2)**, on the basis of its *g*-factor being 2.0039 consistent with a carbonyl-containing radical (cf. Table 3-1). Both of

$$CH_3CO\dot{C}HCH_3 \qquad\qquad CH_3\dot{C}(OH)CH(OH)CH_3$$
$$\textbf{(1)} \qquad\qquad\qquad\qquad \textbf{(2)}$$

these radicals would be expected to show similar hyperfine splitting, namely a large and small quartet splitting and a large doublet splitting. The accurate determination of *g*-factors is particularly important for radicals generated in the solid phase when hyperfine splitting is frequently not observed.

(ii) *Hyperfine splitting*. This is by far the most useful characteristic of e.s.r. spectra both for elucidating the structure and also the shape of the radical under study. It arises from interaction between the unpaired electron and neighbouring magnetic nuclei (^1H, ^{13}C, ^{14}N, ^{18}O, etc.).

Interaction of the unpaired electron with a neighbouring proton causes hyperfine splitting, since the electron experiences not only the applied magnetic field but also the proton magnetic field. Thus, the spectrum of a hydrogen atom is a 1:1 doublet since the proton with a spin $I = \frac{1}{2}$ may be aligned either parallel or antiparallel with the applied magnetic field. The separation of the components of the doublet is known as the hyperfine splitting constant. Interaction of the unpaired electron with three magnetically equivalent protons as in the methyl radical results in a 1:3:3:1 quartet (Fig. 3-2). In general interaction

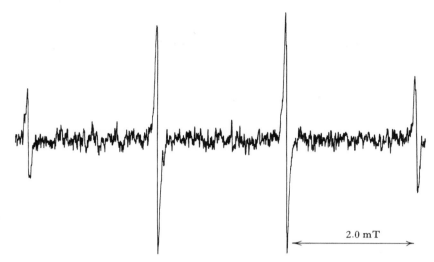

Fig. 3-2. E.s.r. spectrum of the CH_3· radical.

with n equivalent protons gives $n+1$ lines whose relative intensities are given by the coefficients of the binomial expansion. Interaction of the unpaired electron with a nucleus of spin I gives $(2I+1)$ lines. Thus, the e.s.r. spectrum of a nitrogen radical consists of a $1:1:1$ triplet (Fig. 3-3).

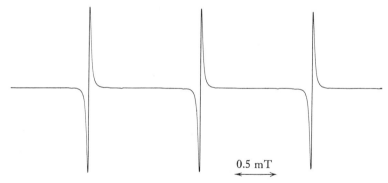

Fig. 3-3. E.s.r. spectrum of Fremy's salt $(KSO_3)_2NO$·

E.s.r. spectra are in general much more complex than n.m.r. spectra because coupling with β-protons is frequently as great as if not greater than that with α-protons. Thus, the spectrum of the ethyl radical consists of 12 lines (Fig. 3-4). The spin of the unpaired electron is split into a quartet by the β-protons of the methyl group and each of these

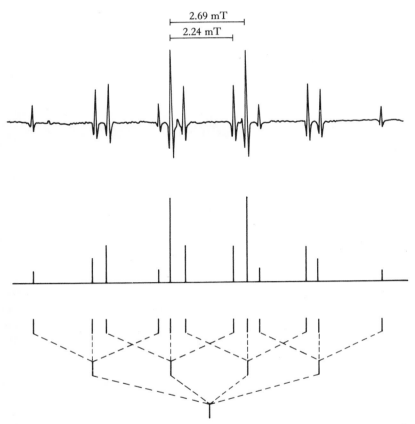

Fig. 3-4. E.s.r. spectrum of the ethyl radical with the stick diagram and interpretation of the spectrum.

lines is split into three lines by the protons of the methylene group. The coupling constants with the α- and β-protons are 2.25 and 2.71 mT respectively.

Careful analysis of the hyperfine splitting enables assignments to be made as to radical structure. The spectrum of the radical resulting from oxidation of ethanol by hydroxyl radicals was a quartet, each component of which was split into a doublet (Fig. 3-5), consistent with the $CH_3\dot{C}HOH$ radical rather than the $\dot{C}H_2CH_2OH$ radical.

In complex situations it is sometimes only possible to obtain the hyperfine splitting constants on the basis of a computer simulation of the spectrum. The technique of 'endor' spectroscopy can also be

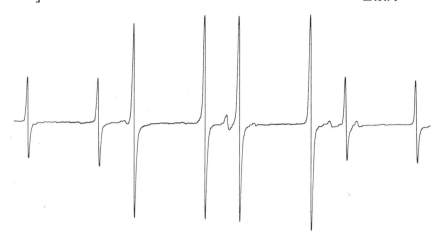

Fig. 3-5. E.s.r. spectrum of the CH₃ĊHOH radical.

successfully applied to the study of radicals which give rise to complex spectra (see p. 53).

3.3.4 *Origin of hyperfine splitting*

The theory of isotropic hyperfine splitting has been discussed in detail by Sales (13). Anisotropic hyperfine splitting, which arises from dipolar–dipolar interactions, is only important in the solid state or in viscous media and will not be considered. In solution such interactions are averaged to zero by rapid tumbling of the radicals.

Isotropic hyperfine splitting results from interaction of the unpaired electron with the magnetic nucleus. This is sometimes called contact interaction. Thus, it might be expected to be observable only when the electron is in an orbital with some s-character, since only then will there be a finite electron density at the nucleus. For π-radicals no splitting would be expected since the unpaired electron is in a p-orbital which has a node at the nucleus. Consequently, it would appear that there is no mechanism for the unpaired electron to interact with the protons attached to the trigonal carbon. Experimentally it is found that though splitting for electrons in orbitals of s-character is very large (it is 50.6 mT for the hydrogen atom), there is nevertheless also some splitting in π-radicals. Thus, for the benzene radical-anion in which the unpaired electron is indubitably in an anti-bonding π-orbital and interacts equally with each of the six protons, the e.s.r. spectrum consists of a symmetrical septet with a coupling constant of 0.375 mT.

Fig. 3-6. Orbital diagram showing the two possible spin polarizations in the C-H σ bond.

The molecular orbital treatment, which leads to the assumption that the spin density at the carbon nucleus is zero, is somewhat crude because it neglects any interaction of the σ- and π-electrons. A more sophisticated approach considers the two possible arrangements of electron spins about the trigonal carbon (Fig. 3-6); the first of these is more probable since there is a somewhat greater probability of the carbon atom being associated with spins of the same type as that in the *p*-orbital (14). This arises as a result of interaction between the σ- and π-systems. The carbon nucleus and the proton are thus associated with spins of the same and opposite signs respectively as that of the unpaired electron. There is thus a net unpaired negative spin density in the *s*-orbital of the proton which gives rise to the hyperfine splitting. The sign of the splitting resulting from this type of interaction is negative. Experimentally the sign of the splitting is not obtained from e.s.r. spectra but it can be obtained from n.m.r. spectra (15). Consideration of the sign of the splitting is important in a few instances, as will be apparent later.

(i) *α-Proton splittings.* McConnell (16) has shown for π-radicals that the coupling constant, a_i^H, of a proton is directly proportional to the spin density, ρ_i^π, on the carbon atom i bearing that proton. This is represented by the McConnell equation:

$$a_i^H = Q^H \rho_i^\pi$$

where Q^H is a proportionality factor with a value between -2.0 and -3.0 mT. The value of Q varies somewhat according to the particular radical. This is seen from examination of radicals for which the value of ρ is accurately known from symmetry considerations. It has a value of -2.303 mT for the methyl radical in which ρ has a value of unity. In the benzene radical-anion, the cycloheptatrienyl and cyclopentadienyl radicals, Q has values of -2.25, -2.77 and -3.00 mT respectively. The negative sign of Q has been confirmed by n.m.r. spectroscopy.

In spite of some uncertainty in the value of Q, the McConnell equa-

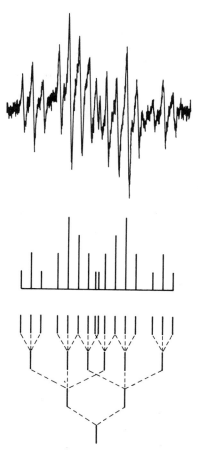

Fig. 3-7. E.s.r. spectrum of the phenoxy radical with the stick diagram and interpretation of the spectrum.

tion is exceedingly valuable in calculating electron densities at various positions in delocalized radicals. Measurement of the three hyperfine splitting constants in the e.s.r. spectrum of the phenoxy radical (Fig. 3-7) enables the relative electron densities at the *ortho-*, *meta-* and *para-*positions to be calculated. If Q is known accurately then the precise spin-densities, ρ_o, ρ_m and ρ_p at these positions can be determined. One snag is that it is not possible to obtain the signs of the coupling constants. It seems reasonable from treatment of the phenoxy radical as an odd-alternant hydrocarbon, to assume that the spin-densities at the *meta-*positions are negative (17). From these values one can obtain by difference the spin density on oxygen,

$$\rho^0 = 1 - (2\rho_o + 2\rho_m + \rho_p)$$

This assumes that the spin-density at C-1 is negligible, whereas it probably has the same sort of value as at C-3 and C-5. To obtain the spin-density at C-1 it is necessary to have ^{13}C data.

Measurement of hyperfine splitting constants can also be used to estimate the spatial configuration of phenyl substituents in, for example, the 2,4,6-triphenylphenoxy radical in which the unpaired electron is delocalized onto the phenyl substituents. Theoretical calculations are made with the phenyl substituents at different angles to the phenoxy ring to obtain the best fit between the experimental and theoretical values. For the 2,4,6-triphenylphenoxy radical (**3**) this obtains when the 2- and 6-phenyl substituents are inclined at an angle of 46° to the

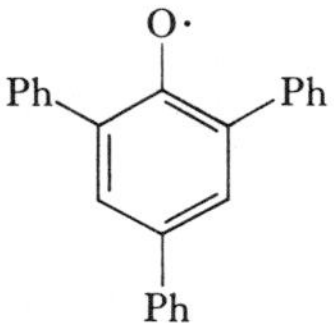

(**3**)

phenoxy ring and the 4-phenyl substituent is coplanar with the phenoxy ring (18).

(ii) *β-Proton splittings.* β-Protons also produce very appreciable hyperfine splitting. This results from a hyperconjugative mechanism which allows some of the unpaired α-spin density to occur at the proton attached to the β-carbon atom producing a positive coupling constant as is observed in n.m.r. spectroscopy (cf. the resonance hybrid **4**).

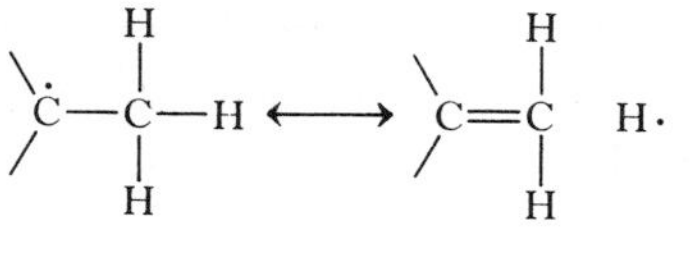

(**4**)

Fessenden and Schuler (7) established that the hyperfine splitting constant for the protons of the β-methyl group, $a_H^{CH_3}$, in radicals of the type $CH_3\dot{C}XY$ is proportional to the spin density at the α-carbon atom

$$a_H^{CH_3} = \rho_\alpha Q_H^{CH_3}$$

42

They further showed that for the ethyl, isopropyl and t-butyl radicals there was a unique value for the proportionality constant $Q_H^{CH_3}$ equal to 2.93 mT. This value of $Q_H^{CH_3}$ corresponds to a loss of 8.1 % of spin density at the carbon atom to each methyl group. Fischer (19) extended these studies to a series of radicals of the type $CH_3\dot{C}HX$ and showed that there was a general relationship:

$$\rho_\alpha = \prod_{i=1}^{3} [1 - \Delta X_i]$$

where ΔX_i is a measure of the spin-withdrawing efficiency of the group X. Table 3-2 shows some typical values of ΔX. The values for the hydroxy and ethoxy groups may well turn out to be of little use since there is evidence that these groups 'bend' carbon radicals towards a pyramidal structure and that this affects the value of a_H as well as the degree of delocalization (20). Use may be made of this type of information in assigning radical structures.

TABLE 3-2   *Values of ΔX in radicals of the type $CH_3\dot{C}HX$ (19)*

X	H	Me	CH$_2$OH	CO$_2$R	CN	COEt	OH	OEt
ΔX	0	0.081	0.079	0.072	0.145	0.162	0.160	0.172

More interesting information is obtainable in cases where one or two of the β-protons are replaced by other substituents when the latter have a fixed conformation. The hyperfine splitting constant, a_H, for such radicals has been shown to depend on θ, the dihedral angle between the p-orbital containing the unpaired electron and the β carbon–hydrogen bond (see Fig. 3-8)

$$a_H = (B_0 + B \cos^2\theta)\rho_\alpha$$

where ρ_α is the electron-density at the α-carbon atom and B_0 and B are constants. B_0 is small (~ 0.3–0.4 mT) relative to B (~ 5.0 mT). The coupling constant with a β-proton is thus at a maximum when it is in the same plane as that of the p-orbital.

For freely rotating C–H bonds as in the isopropyl radical, θ will have an average value of 45° and $\cos^2\theta$ will be $\frac{1}{2}$. In the conformationally fixed cyclobutyl radical in which the β–C–H bonds make an angle of 30° with the plane of the orbital containing the unpaired electron, $\cos^2\theta$ will be equal to $\frac{3}{4}$. If B_0 is small relative to B, the ratio of the hyperfine splitting constants for protons attached to the β-carbon atom should be

Detection of radicals

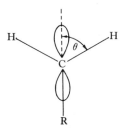

Fig. 3-8. Isotropic coupling for β-protons looking along the C_α–C_β bond.

2:3 for the isopropyl and cyclobutyl radicals (21). This is very close to the ratio of the experimental values observed. Use has been made of this method in determining conformations of radicals (22). Thus, Sevilla and Vincow (23) observed that the β-proton splitting constants for the 9-methyl-, 9-ethyl- and 9-isopropylxanthyl radicals (**5**, R = Me, Et and Pr^i) were 1.22, 0.62 and 0.08 mT respectively. These values are consistent with the 9-ethylxanthyl radical being in the conformation in which the p-orbital bisects the HCH angle of the ethyl group, i.e. $\theta = 60°$, and is orthogonal to the plane of the β-C–H bond of the isopropyl group in the 9-isopropylxanthyl radical. For the ethyl derivative the coupling constant should be half that for the methyl derivative in which the methyl group is freely rotating. These conformations are the least hindered for these radicals.

(5)

(iii) *Other proton splittings.* Splittings by γ-protons in alkyl radicals are generally very small, and are not always resolved. The γ-proton splitting in the propyl radical is 0.038 mT. Such splittings are thought to arise by a hyperconjugative mechanism involving alkyl-hyperconjugation. This is represented for the propyl radical by the resonance hybrid (**6**).

$$\underset{CH_2-CH_2\cdot}{\overset{CH_3}{\diagdown}} \longleftrightarrow \underset{CH_2=CH_2}{CH_3\cdot}$$

(6)

44

Splittings are greater when the spin density is transmitted across oxygen rather than a methylene group. Thus the γ-proton splitting in the radical $CH_3\dot{C}HOCH_2CH_3$ is 0.14 mT. This is due to the existence of significant spin density on the oxygen, e.g. (7).

$$CH_3\dot{C}H\!-\!\ddot{O}\!-\!CH_2CH_3 \longleftrightarrow CH_3\dot{C}H\!-\!\overset{=}{\overset{+\cdot}{O}}\!-\!CH_2CH_3$$

(7)

Small long-range splittings are also observed in certain bicyclic radicals. Thus, Russell has shown that the semidione (8) has the

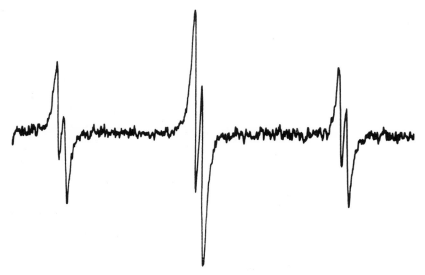

(8)

splittings shown (23 *a*). These are greatest when the bonds between the radical centre and the proton are in a **W**-shaped arrangement. This is well exemplified in (8) in which the coupling constant with the *anti*-hydrogen at C-7 is much greater than that with the *syn*-hydrogen

Fig. 3-9. E.s.r. spectrum of the $\dot{C}H_2OH$ radical.

attached to this carbon atom. A similar type of long-range splitting is also observed in certain σ-radicals (p. 48) and also in n.m.r. spectroscopy.

Splittings due to the hydroxyl proton in hydroxyalkyl radicals are very small (0.02–0.18 mT) (cf. Fig. 3-9), and are frequently not observed because of too rapid an exchange in hydroxylic solvents.

(iv) *Hyperfine splitting with other nuclei.* Hyperfine splitting also occurs with other magnetic nuclei. Use can be made of such splittings in elucidation of radical structures.

Replacement of hydrogen by deuterium markedly alters e.s.r. spectra. The magnitude of coupling with deuterium is about one-sixth that with hydrogen and hence substitution of hydrogen by deuterium at positions of low spin density results in simplification of the spectrum. The spectrum of the 2,4,6-tri-t-butylphenoxy radical (9) is a 1:2:1 triplet (24). On replacement of the hydrogen at the 3-position by deuterium a

O·

Bu^t Bu^t

Bu^t

(9)

1:1 doublet is observed in place of the triplet. Replacement of hydrogen by deuterium at a site of high electron density results in an increase in the number of lines observed since coupling with deuterium results in a 1:1:1 triplet (the spin of deuterium is 1).

Coupling with ^{13}C has been used by Fessenden to obtain information about the shapes of alkyl radicals (25). The ^{13}C splitting is very much larger if the unpaired electron is in an *s*-containing orbital. ^{13}C Splitting constants can thus be used to give an indication of the *s*-character of the orbital of the unpaired electron and hence the shape of the radical (see chapter 4). ^{13}C Coupling is the only way of obtaining information on the spin density at a fully substituted carbon atom, e.g. at the 1-position in phenoxy radicals. Hyperfine splitting with ^{17}O has similarly been used to obtain the spin density on oxygen. Otherwise it is necessary to estimate this, as has been previously mentioned, from a knowledge of the hyperfine splittings with the *o-*, *m-* and *p*-protons.

The simple McConnell equation for correlating coupling constants with spin densities for protons does not apply for ^{13}C coupling (nor for coupling with ^{14}N, ^{17}O and ^{19}F) though more complex modified equations have been used successfully (26). This is because it is necessary to consider interactions of the unpaired spin with several nuclei and also with lone-pair electrons.

The value of the coupling constant with ^{14}N in a series of nitroxide radicals, RR′NO·, has been found to be strongly dependent on the nature of the groups R and R′ (12). The value of the coupling constant is lower for alkylarylnitroxides than for dialkylnitroxides as a consequence of increased delocalization of the unpaired electron (**10–12**). Steric factors also markedly influence the magnitude of the splitting in the spectra of nitroxides.

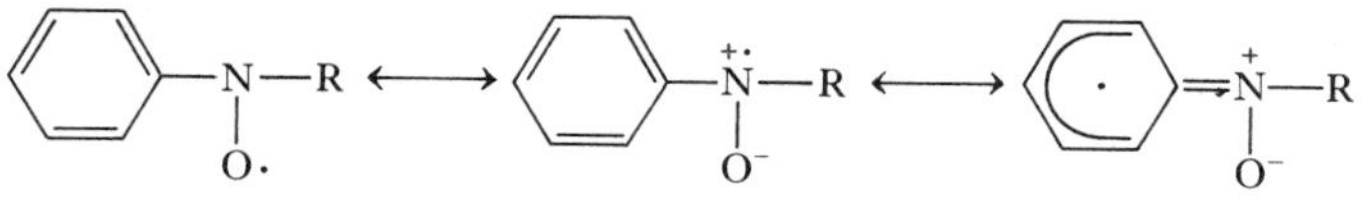

(**10**)   (**11**)   (**12**)

Halogen atoms, particularly fluorine, can also interact with an unpaired electron with resultant hyperfine splitting. The splitting constants are in the order:

$$a_F > a_{Cl} > a_{Br}$$

(v) *Hyperfine splitting in σ-radicals.* The discussion of hyperfine splitting thus far has been confined to π-radicals. The pattern of splitting is significantly different in σ-radicals in which the unpaired electron lies in the nodal plane of the π-system. A study of e.s.r. parameters can show unambiguously whether a given radical is a σ- or π-radical. Thus, the ^{13}C splitting of the formyl radical is large (13 mT), indicative of the unpaired electron being in an orbital in which the s-character is appreciable.

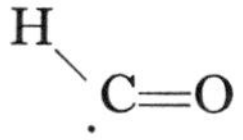

The e.s.r. spectrum of the phenyl radical is notable in that the coupling constants with the *ortho-*, *meta-* and *para-*protons are about 1.8, 0.6 and 0.2 mT (28) respectively. This contrasts with the situation for π-radicals where delocalization leads to the splitting constants being the order: *para > ortho ≫ meta*. The mechanism of interaction in the phenyl radical must involve the σ-bonds of the aromatic nucleus.

Another example of a σ-radical is the iminoxy radical. Iminoxy radicals are best represented by the canonical structures (**13** and **14**) in

47

which the unpaired electron is in an orbital orthogonal to the π-system. The large ^{14}N splitting constant (3 mT) confirms that these radicals are σ-radicals and that the unpaired electron is in an orbital with considerable s-character. An unusual feature of the e.s.r. spectra of iminoxy radicals is the observation in certain cases of a long-range coupling (10).

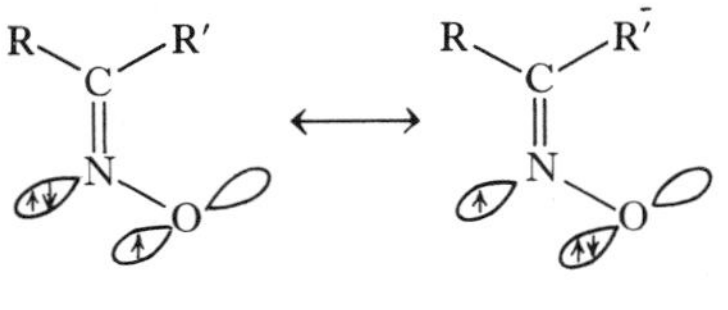

(13)   (14)

This is well exemplified in the radicals derived from camphor oxime (**15**, R = H) and 1-methylcamphor oxime (**15**, R = Me) when coupling with the bridgehead proton is as strong as that with the protons of the α-methylene group. Such interaction is strong when the proton, the

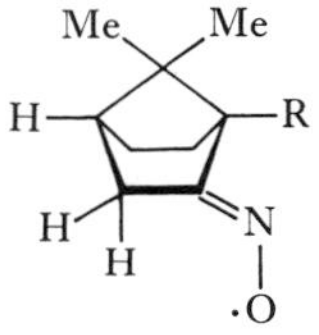

(15)

iminoxy group, and the intervening carbon atoms are in the same plane, and particularly so when the bonds between the nitrogen and the proton form a W-shaped configuration. It is also notable that coupling through space is observed. Thus coupling occurs with the 1-proton in the radical derived from 1-halogenofluorenone oxime (**16**), and with the

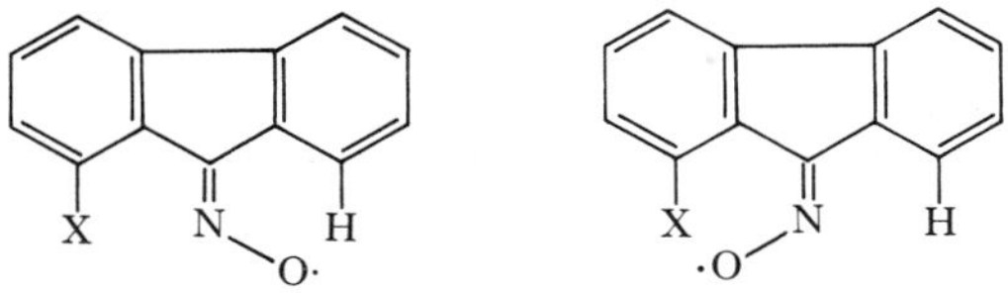

(16)   (17)

halogen atom if X is a large halogen. In the isomeric radical (**17**) this interaction is not observable. Instead there is interaction with the halogen. The magnitude of this interaction is in the order: iodine >

48

bromine > chlorine > fluorine. In consequence of this, it is suggested that these splittings arise from interaction of the unpaired electron with an orbital of the halogen atom. Such interaction would be greater for the larger halogens.

3.3.5 Line-widths

The line-width of an e.s.r. absorption is related to the lifetime of a particular spin state. This lifetime is reduced if there is rapid interchange of interacting magnetic nuclei. The e.s.r. spectrum of the cyclohexane-1,2-semidione radical anion has been examined at different temperatures (29). At room temperature the spectrum consists of a quintet indicating rapid ring flipping of the two half-chair conformers (18 and 19) which results in the equivalence of the four α-protons. As

(18)   (19)

the temperature is decreased the rate of ring flipping slows down and consequently the quintet broadens and at $-96\,°C$ a poorly resolved septet is observed. This arises because the protons H_1 and H_2 are no longer magnetically equivalent. The dihedral angles that the C–H_1 and C–H_2 bonds make with the plane of the unpaired electron are different, so that the coupling constants will be different. In this way it is possible to estimate the energy barrier and rate of ring flipping (29).

3.3.6 Detection of free radicals by the spin-trap technique

Transient free radicals may be detected, as has been discussed, by e.s.r. spectral studies of flow systems. This procedure requires the use of relatively large amounts of material and hence it is limited in scope. The problem then arises as to how radicals in other reactions may be detected. Several groups of workers (30–33) have solved this problem by carrying out the particular reaction in the presence of a suitable diamagnetic compound, which can scavenge a transient radical giving rise to a stable radical. The e.s.r. spectrum of the latter may be obtained without recourse to special techniques.

Suitable scavengers include nitroso compounds (20) and nitrones (22), both of which on reaction with radicals give rise to nitroxides (21 and

23). Analysis of the e.s.r. spectra of the resultant stable nitroxide radicals can give information about the nature of the transient radical. The spectra of nitroxides are characterized by a $1:1:1$ triplet due to splitting by ^{14}N, the value of the splitting constant depending on the nature of the groups attached to nitrogen. This triplet is further split as a result of coupling with magnetic nuclei on the α- and β-carbon atoms.

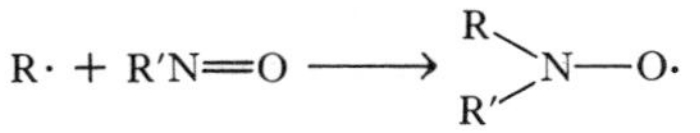

$$R\cdot + R'N{=}O \longrightarrow \begin{array}{c} R \\ {\diagdown} \\ R' {\diagup} \end{array} N{-}O\cdot$$

$$\qquad (20) \qquad\qquad (21)$$

$$R\cdot + R^1CH{=}\overset{+}{N}\overset{O^-}{\diagup}_{R^2} \longrightarrow \begin{array}{c} R^1 \\ | \\ RCH \\ {\diagdown} \\ R^2 {\diagup} \end{array} N{-}O\cdot$$

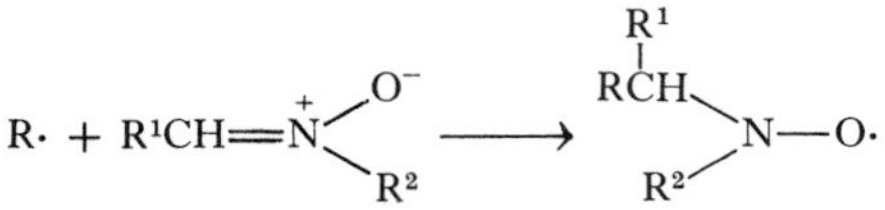

$$\qquad (22) \qquad\qquad\qquad (23)$$

The ideal scavenger is one in which the hyperfine splitting of the derived nitroxide radical is due entirely to magnetic nuclei in the transient radical. For this reason 2-methyl-2-nitrosopropane (nitroso-butane) **(20:** $R' = Bu^t$) has been extensively used by Perkins (31). Interaction with the protons of the t-butyl group does cause some line-broadening. Perdeuterionitrosobutane is much superior to its undeuteriated analogue since the spectral lines of the resultant nitroxide radical are much sharper, because of the much reduced magnitude of the splittings of the nine equivalent deuterium nuclei (34). Fig. 3-10 gives comparison spectra of t-butyl methoxycarbonyl nitroxide obtained by trapping methoxycarbonyl radicals with nitrosobutane and its deuteriated counterpart.

Janzen and his group have used the nitrone **(24)**. The e.s.r. spectra of the derived nitroxide radicals show little evidence of line-broadening

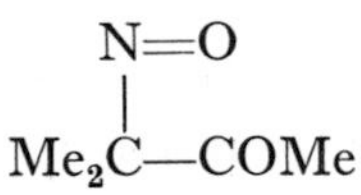

$$\begin{array}{c} N{=}O \\ | \\ Me_2C{-}COMe \end{array}$$

$$(24)$$

as a result of coupling with the protons in the methyl groups. Hence this compound is rather more useful than nitrosobutane as a radical trap, though probably less so that perdeuterionitrosobutane.

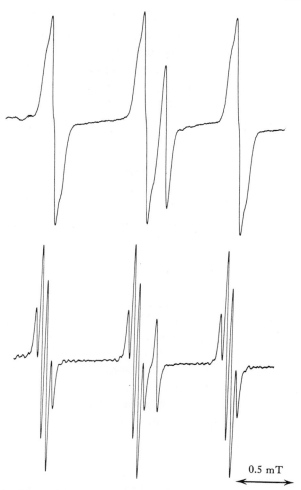

Fig. 3-10. E.s.r. spectra obtained by trapping $\dot{C}O_2Me$ radicals with (*a*) t–C_4H_9NO and (*b*) t–C_4D_9NO.

The technique of spin trapping has been used to obtain evidence for the formation of alkoxyl radicals in the lead tetra-acetate oxidation of alcohols (32):

$$ROH \xrightarrow[h\nu]{Pb(OAc)_4} RO\cdot$$

The e.s.r. spectra of the derived alkoxynitroxides are characterized by large nitrogen splittings (2.7–2.9 mT) and very small proton splittings (0.1–0.15 mT). The oxidation of certain alcohols including 2-phenylethanol proceeded differently giving rise to alkyl- rather than alkoxy-nitroxides. Thus oxidation of 2-phenylethanol in the presence of the nitroso compound (24) gave a nitroxide whose e.s.r. spectrum consisted of a triplet of triplets with nitrogen and proton splittings of 1.47 and 0.82 mT respectively. This was consistent with the formation of the benzyl radical, which was produced by loss of formaldehyde from the 2-phenylethoxy radical:

$$PhCH_2CH_2O\cdot \xrightarrow{\;-CH_2O\;} PhCH_2\cdot$$

Nitrones, though possessing greater thermal and photochemical stability than nitroso compounds, are less attractive as scavengers because, in the derived nitroxide (23) the transient radical fragment is fairly remote from the radical centre and thus contributes little to the splitting pattern. In such cases identification of the radical can only be made by very precise measurements of the nitrogen and proton splittings. The method is really only satisfactory if the suspected nitroxide is made by an independent procedure for direct spectral comparison. N-t-Butyl methylene nitrone (22; R^1 = H, R^2 = Bu^t) is superior to N-t-butyl benzylidene nitrone (22; R^1 = Ph, R^2 = Bu^t) as a much wider range of proton splittings are observed with nitroxides derived from the former than from the latter.

The *aci*-anion $[CH_2NO_2]^-$ has also been used as a spin trap for radicals generated in aqueous solution. This reacts with radicals to give stable radical anions which can be independently obtained by one-electron reduction of the appropriate nitro compound (33):

$$R\cdot + CH_2NO_2^- \longrightarrow RCH_2NO_2^{-\cdot}$$
$$RCH_2NO_2 + Me_2\dot{C}OH \longrightarrow RCH_2NO_2^{-\cdot} + Me_2\overset{+}{C}OH$$

Of especial value is the fact that splitting is observed from the γ- and in some cases δ-protons in the resultant radical anion. The values of the β-proton and nitrogen coupling constants are also markedly dependent on the nature of the radical trapped.

The technique of spin-trapping has to be used with caution because nitrosoalkanes are neither photochemically nor thermally stable. 2-Methyl-2-nitrosopropane readily loses nitric oxide, giving t-butyl radicals, which combine with the nitroso compound to give di-t-butylnitroxide.

$$Bu^tN=O \longrightarrow Bu^t\cdot + NO$$
$$Bu^t\cdot + Bu^tN=O \longrightarrow Bu_2^tNO\cdot$$

Nitroso compounds have also been found to react readily with nucleophiles to give intermediates which on oxidation with oxygen or the
nitroso compound can give nitroxides:

$$R^- + Bu^tN{=}O \longrightarrow \underset{Bu^t}{\overset{R}{>}}N{-}O^- \xrightarrow{[O]} \underset{Bu^t}{\overset{R}{>}}N{-}O\cdot$$

Care must thus be used in interpretation of the results.

A further problem is that the detection of a particular radical does not
necessarily mean that it is involved in the major reaction pathway in the
system under study. Thus, even though *N*-bromosuccinimide has been
shown to give succinimidyl radicals, it is almost certain that these are
not involved to a major extent in the allylic bromination of alkenes (31)
(see chapter 7, p. 190).

3.4 *ENDOR SPECTROSCOPY*

The e.s.r. spectra of radicals in which the unpaired electron can be
delocalized onto a large number of sites are exceedingly complex and
frequently defy complete analysis. Thus, the 2,4,6-triphenylphenoxy
radical, in which there are seven different hyperfine splittings, gives rise
to 4050 lines. Endor or electron nuclear double resonance spectroscopy
can be applied to such radicals. Each class of equivalent nuclei gives rise
to two endor lines, the separation of which corresponds to the appropriate
electron-nuclear (i.e. e.s.r.) hyperfine splitting constant. The endor
spectrum of a radical is consequently enormously simplified in comparison with the e.s.r. spectrum of the same radical.

The technique of endor spectroscopy involves placing the sample in a
magnetic field and subjecting it to two radiofrequency fields, one
corresponding to the precession frequency of the electrons and the
other to that of the protons. The instrument is maintained continuously
on an e.s.r. line and the frequency of a radio source is swept. The endor
spectrum of a radical containing only protons will consist of pairs of
lines centred on the proton frequency and separated by the hyperfine
splitting. The underlying theory of endor spectroscopy has been recently
reviewed by Kwiram (35) and Atherton (35*a*).

In order to observe the endor spectrum of a radical it is necessary that
the nuclear transitions occur very rapidly at a rate which will compete
with electron relaxation times. For free radicals in solution these times

are of the order of 10^{-5} to 10^{-6} s. Very intense nuclear radiofrequency magnetic fields (10–100 Gauss) are thus necessary (36). It is thus not a simple procedure to obtain endor spectra of radicals in solution. Endor presents fewer problems in the solid phase because very low temperatures may be used at which the electron spin lattice relaxation times are much greater.

As has been mentioned the endor spectrum of a radical is very much simpler than its e.s.r. spectrum, thus enabling coupling constants of the various protons to be obtained. The endor spectrum does not, however, indicate the number of equivalent nuclei responsible for a particular coupling. It is thus necessary to obtain both the endor and e.s.r. spectra of a radical, and then to simulate the e.s.r. spectrum using the endor coupling constants until a fit with the experimental spectrum is achieved.

The high resolution of endor makes it possible to obtain even small coupling constants, which may be unobservable in the e.s.r. spectrum. The value of endor in determining small coupling constants can be illustrated by reference to a study of the radical **(25)** (36 *a*). It was noted that at -80 °C, two coupling constants with the methylene protons were observable, but only one could be seen at -20 °C. These results can be interpreted on the basis that at -80 °C the radical exists in two equally populated conformations corresponding to the left-hand and right-hand propeller forms of **(25)**, which at -20 °C interconvert rapidly.

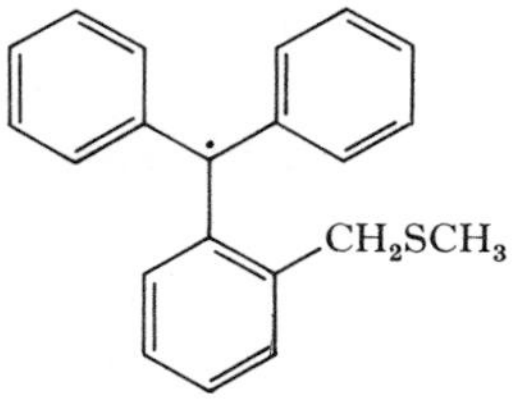

(**25**)

Replacement of hydrogen by deuterium results in the elimination of coupling at that site in the endor spectrum. This is in contrast to the effect of deuteriation in e.s.r. spectroscopy, which smears out and complicates the hyperfine splitting pattern (see p. 46). Use of deuteriation was made in the study of the 2,4,6-triphenylphenoxy radical (37). Analysis of the endor spectrum of this radical was not possible because of the near-degeneracy of certain proton splittings. This was removed by deuteriation.

54

There are several examples in the literature of the use of endor to determine coupling constants and hence conformations of phenyl-substituted phenoxyls and nitroxides (cf. ref. 35 a). Endor has also been used to sort out overlapping signals from two different radicals (37 a).

The resolution of the endor spectrum of a radical in the solid state is much greater than that of its e.s.r. spectrum. The reviews of Kwiram and Atherton survey the use of endor in determining coupling constants on trapped radicals in single crystals. Thus, the radical $Me_2\dot{C}CO_2^-$ is observed in one conformation at room temperature but in three conformations by endor when the crystal is cooled to 4 °K (38). A further use of endor is the study of triplet states when information about spin populations can be obtained.

3.5   N.M.R. SPECTROSCOPY

The magnitudes of e.s.r. coupling constants can also be obtained by n.m.r. spectroscopy. This is applicable to the determination of coupling constants of a smaller magnitude than can be obtained by e.s.r. or endor (39). In addition the sign of the coupling constant can be obtained. The n.m.r. spectrum is also easy to interpret in that each group of equivalent nuclei gives rise to a single line.

The n.m.r. spectrum of a radical is most readily seen when the electron has a short spin-lattice relaxation time, T_1, and/or short spin-exchange time, T_e. These conditions exist in concentrated solutions of radicals.

The position of the absorption peak in the n.m.r. spectrum of a radical is shifted relative to its position in a very closely related diamagnetic compound by an amount given in equation (4):

$$\Delta_i = \frac{\nu_D - \nu_P}{\nu_P} = -a_i h \frac{\gamma_e}{\gamma_i} \frac{S(S+1)}{3kT} \tag{4}$$

where ν_D and ν_P are the resonant frequencies of a nucleus i in a diamagnetic and paramagnetic compound respectively; γ_e and γ_i are the gyromagnetic ratios of the electron and the nucleus i, and S is the total electron spin. Hence by substituting in this equation, the value of the coupling constant for protons, a_H, at 295 °K is:

$$a_H = -3.73\Delta H \times 10^4\ \text{Hz} \tag{5}$$

Thus, the value of an e.s.r. coupling constant can be obtained from the shift in the n.m.r. signal, and the sign of the coupling constant from the direction of the shift. Signals of protons with the positive coupling constants are moved downfield, and those with negative values upfield.

The width of an absorption line at half height, h_i, in the n.m.r. spectrum is also related to its coupling constant, a_i, and the spin-lattice relaxation time, T_1:

$$h_i = T_1 a_i^2/8 \qquad (6)$$

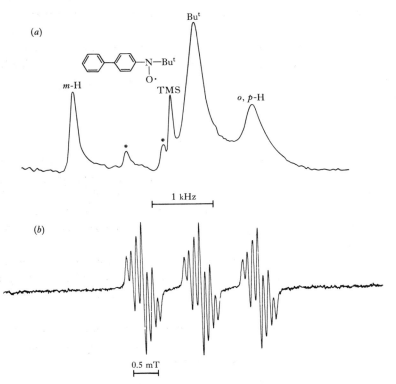

Fig. 3-11. (a) N.m.r. spectrum, and (b) e.s.r. spectrum of t-butyl *p*-biphenylyl nitroxide.

The dependence of the width of the signal on the square of the coupling constant means that it is not possible to measure large coupling constants with n.m.r. spectroscopy because the signals of such protons would be excessively broad. The dependence of the signal width on T_1 emphasizes the need to use concentrated solutions of radicals.

N.m.r. spectroscopy has been fairly widely used to determine coupling constants in nitroxides (e.g. 39a). Fig. 3-11 shows the e.s.r. and n.m.r. spectra of t-butyl *p*-biphenylyl nitroxide to illustrate the complementary use of the two techniques (39*b*). The nitrogen coupling constant and the couplings of the ring protons nearest to the nitrogen are easily measured by e.s.r., but the smaller splittings of the other ring protons can only be measured by n.m.r. The peaks marked with an asterisk in the n.m.r. spectrum are due to diamagnetic impurities, in this case the corresponding hydroxylamine. These are used to obtain ν_D in equation (4).

The values of coupling constants obtained by n.m.r. and e.s.r. are in close agreement with each other even though the experimental conditions used in the two experiments are necessarily very different. E.s.r. measurements are carried out on very dilute solutions ($< 10^{-5}$ M) to increase times between radical collisions, whereas n.m.r. studies can only be made with concentrated solutions.

One of the chief limitations of the method is the need to measure the spectra on concentrated solutions of the radical. Kreilick *et al.* (39*c*) have partially overcome this by using solutions of the radical under study in a stable radical as solvent such as di-t-butylnitroxide.

3.6 *CHEMICALLY INDUCED DYNAMIC NUCLEAR POLARIZATION (CIDNP)*

N.m.r. spectra recorded during the occurrence of certain reactions show anomalous behaviour. Some of the n.m.r. absorption peaks are enlarged, i.e. enhanced absorption is occurring, and some negative peaks are observed, i.e. emission is occurring (40, 41). For example, when dibenzoyl peroxide was decomposed at 100 °C in cyclohexanone the n.m.r. spectrum at the moment when the sample was transferred into the probe ($t = 0$) showed only the normal absorption lines of dibenzoyl peroxide in the aromatic proton region (42). During the reaction these lines disappeared, and a negative emission line at the position corresponding to benzene appeared, which reached its maximum at $t = 4$ min. The intensity of this line then decreased to zero and after 7 min it reappeared in absorption reaching a constant maximum at the end of the reaction. This line corresponds to the protons of benzene, which is formed through hydrogen abstraction from the solvent by phenyl radicals:

$$(PhCO_2)_2 \longrightarrow 2PhCOO\cdot \longrightarrow 2Ph\cdot + 2CO_2$$
$$Ph\cdot + S \longrightarrow PhH + S(-H)$$

The product benzene is formed with non-equilibrium nuclear spin state populations, and emission from the upper states occurs until equilibrium populations are attained. The normal absorption n.m.r. spectrum of benzene is then observed.

Enhanced absorption and/or emission lines are observed in the n.m.r. spectra of solutions which react rapidly by a radical pathway. Observation of this effect is often useful in diagnosing the presence of a radical intermediate in an organic reaction (see chapter 10, p. 392). A qualitative explanation of why the nuclear spin state populations of the products deviate strongly from the thermal equilibrium value, so that emission or absorption occurs, has been given by Fischer and Bargon (42), and others. If the electron spin states of the free radical are *equally* populated when it is first formed, then magnetic interaction between the unpaired electron and magnetic nuclei of the radical causes the population of the nuclear spin states to change from the equilibrium value. The corresponding n.m.r. transitions of the product show emission or enhanced absorption as the nuclear spin state populations relax to their equilibrium values; provided that the nuclei conserve their nuclear spin states during product formation from the radical. The electron spin states of the radicals are initially equally populated, if in the chemical reaction two radicals with antiparallel spins are formed simultaneously. The mechanism whereby the nuclear spin state populations are then changed is known as dynamic nuclear polarization (DNP) and the observable phenomenon is named chemically induced dynamic nuclear polarization (CIDNP).

The emission behaviour of the products can be put on a quantitative scale in terms of the enhancement factor V_{exp}:

$$V_{exp} = (I - I^0)/I^0 \qquad (7)$$

Here I is the negative maximum emission intensity, and I^0 is the absorption intensity which the normal non-emitting product would show at the time when the emission reached its maximum. I^0 is usually measured in a separate experiment where paramagnetic ions are added to the solution. These cause relaxation of the radical electron spins, so that DNP does not occur and no emission from the products is observed. The enhancement factor depends on the operating frequency of the n.m.r. spectrometer, and generally increases as the frequency decreases. Thus in the case of the dibenzoyl peroxide decomposition described above, Fischer and Bargon (42) found enhancement factors for the benzene emission line of -2.9 at 100 MHz, -8.5 at 56.4 MHz and -20 at 40 MHz.

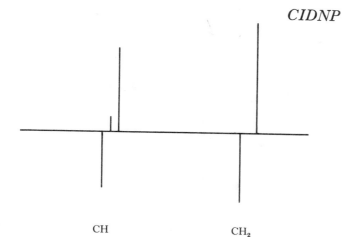

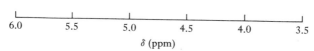

Fig. 3-12. CIDNP spectrum of $CBr_3CH_2CHBrPh$.

Another interesting feature of the CIDNP spectrum is the occurrence, within a single multiplet due to one set of protons, of both emission and enhanced absorption. This effect can be illustrated by reference to the decomposition reaction of dibenzoyl peroxide in the presence of carbon tetrabromide and styrene (42). The phenyl radicals generate tribromo-methyl radicals by bromine abstraction from CBr_4.

$$Ph\cdot + CBr_4 \longrightarrow PhBr + \dot{C}Br_3$$
$$\dot{C}Br_3 + CH_2 {=} CHPh \longrightarrow CBr_3CH_2\dot{C}HPh$$
$$CBr_3CH_2\dot{C}HPh + CBr_4 \longrightarrow CBr_3CH_2CHBrPh + \dot{C}Br_3$$

The product of the chain addition of CBr_4 to styrene is $CBr_3CH_2CHBrPh$. The transient n.m.r. spectrum of this product shows a triplet for the methine proton and a doublet for the methylene protons. In the doublet the low field component emits and the high field component absorbs. In the triplet the low field component emits and the high field component absorbs and the intermediate component does not show appreciable CIDNP; see Fig. 3-12.

This effect, known as the multiplet effect, has been observed in radical abstraction reactions and radical combination and disproportion-ation reactions (43). Kaptein and Oosterhoff have observed CIDNP

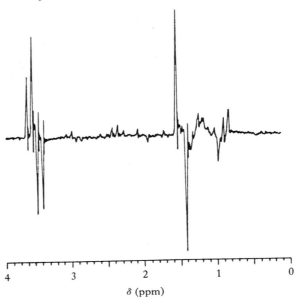

Fig. 3-13. CIDNP spectrum of CH_3CH_2Cl. From Kaptein and Oosterhoff (44). (Reproduced by permission of *Chemical Physics Letters*.)

effects in the reactions of methyl, ethyl, n-propyl, isopropyl and t-butyl radicals (44). They generated ethyl radicals by the thermal decomposition of dipropionyl peroxide at 110 °C in hexachloroacetone. The transient n.m.r. spectrum at 100 MHz of the product ethyl chloride is shown in Fig. 3-13.

The low field lines of the CH_2 quartet show enhanced absorption, and the high field lines emission. The outer lines of the CH_2 quartet are more enhanced than the inner. Similarly, the low field line of the CH_3 triplet shows absorption and the high field line emission. The central line shows virtually no enhancement. Multiplet effects were also observed in the CIDNP spectra of products of the reactions of the other radicals.

Several research groups have given explanations of the multiplet effect (45), the most comprehensive being that of Closs and Trifunac (46) for radical combination reactions. A radical pair is generated from a source in a single step. A fraction of the radicals react by combination or

$$R^1R^2 \longrightarrow [R^1 ------ R^2] \longrightarrow R^1R^2 + R^1H + R^2(-H)$$
$$\downarrow$$
$$R^{1\cdot} + R^{2\cdot}$$

disproportionation inside the solvent cage, and the remainder diffuse apart. The initial radical pair in which there is still weak electronic interaction may be a singlet or triplet, depending on its mode of formation. The nuclear spins of the protons are coupled to the electron spin, and the populations of the nuclear spin states depend therefore on the amount of mixing of the singlet and triplet electron spin states. The enhancement factor is therefore a function of the singlet/triplet mixing coefficient M which itself depends on the electron-nuclear coupling a, and the difference in g-values, Δg, of the two radicals forming the pair. For instance, if the electron spin is coupled to only a single proton in the radical pair, the mixing coefficient is:

$$M = \beta H_0 \Delta g / 2 \pm \tfrac{1}{4} a \qquad (8)$$

where β is the Bohr magneton and H_0 is the applied magnetic field strength. The sign of the enhancement, i.e. emission or absorption, and its intensity depend therefore, on the initial spin state of the radical pair, and the relative values of the g-factors of the two radicals. The CIDNP effect offers a method of studying the spin states of radical pairs, and the influence of spin multiplicity on the radical coupling and disproportionation reaction.

Kaptein *et al.* have used this method for studying the disproportionation of isopropyl radicals and trichloromethyl radicals (47). They decomposed di-isobutyryl peroxide, which gave isopropyl radicals in the presence of bromotrichloromethane. Chloroform could then be formed either from a pair of radicals which reacted within the solvent cage and which had a singlet multiplicity (S), or from a pair of freely diffusing radicals (F) which gave both singlet and triplet states:

$$(\text{Pr}^{\text{i}}\text{CO}_2)_2 \xrightarrow{-2\text{CO}_2} [2\text{CH}_3\dot{\text{C}}\text{HCH}_3] \xrightarrow{\text{CCl}_3\text{Br}} \text{CH}_3\text{CHBrCH}_3$$
$$+ [\text{CCl}_3\cdot + \text{CH}_3\dot{\text{C}}\text{HCH}_3]$$
$$(\text{S})$$
$$\downarrow$$
$$\text{CHCl}_3 + \dot{\text{C}}\text{H}_3\text{CH}{=}\text{CH}_2$$

$$\text{CCl}_3\cdot + \text{CH}_3\dot{\text{C}}\text{HCH}_3 \longrightarrow \text{CHCl}_3 + \text{CH}_3\text{CH}{=}\text{CH}_2$$
$$(\text{F})$$

It was found that at higher CCl_3Br concentrations enhanced absorption occurred, so that disproportionation came predominantly from the singlet radical pairs (S). At 0.11 M CCl_3Br the signal changed sign and at lower concentrations emission of the chloroform signal was observed.

At these lower concentrations there is insufficient $CClBr_3$ for the cage process to occur efficiently, and the reaction proceeds mainly by encounter of freely diffusing radicals (F).

3.7 ELECTRONIC ABSORPTION SPECTRA OF FREE RADICALS

Transitions by the unpaired electron between the energy levels of the radical require less energy than transitions of the paired electrons of the stable parent molecule. A free radical tends to absorb light of longer wavelength than the fully bonded parent molecule, and consequently many anilino and phenoxy radicals are coloured.

The electronic spectra of many diatomic radicals have been known for a long time from spectroscopic observations on high-temperature gases (48). The flash photolysis method (49) has enabled these observations to be carried out under more readily controlled conditions, and has facilitated the identification of radical spectra in both the gas and solution phases.

The spectrum of the methyl radical was observed in the vacuum-ultraviolet by Herzberg and Shoosmith (51) by the flash photolysis of dimethylmercury and its perdeuteriated derivative. Long-wavelength bands were absent, so they concluded that the radical was planar, and this conclusion has been confirmed by more recent e.s.r. and infrared evidence (cf. chapter 4, p. 79). The only other alkyl radical whose spectrum has so far been fully identified is trifluoromethyl (50). The radical concentrations can be determined from the intensities of the absorption bands and this enables reaction rates to be studied by kinetic spectroscopy. The rates of combination of methyl radicals (52, 53) and trifluoromethyl radicals (50) have been determined in this way.

The absorption spectra of a large number of aromatic radicals have been observed in solution (54). Many of the radicals so far studied are related to the isoelectronic benzyl, anilino and phenoxy radicals. The absorption spectra of all three radicals exhibit a sharp band in the vicinity of 300 nm and another band in the region 400–500 nm (see Fig. 3-14). Kinetic and mechanistic studies have been carried out with alkyl, aryl and halogen-substituted radicals (55–58).

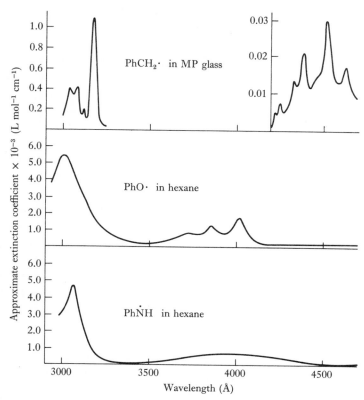

Fig. 3-14. Comparison of the ultraviolet absorption spectra of benzyl, phenoxy and anilino radicals. Benzyl in hydrocarbon glass at 77 °K. Phenoxy and anilino in hexane solvent at 300 °K. From Land (54). (Reproduced by permission of *Progress in Reaction Kinetics*, © 1965, Pergamon Press Ltd.)

3.8 *INFRARED SPECTROMETRY OF FREE RADICALS*

3.8.1 *Experimental methods*

The infrared spectrometry of free radicals is a relatively new application of the well established infrared method. The spectra of a number of free radicals have been measured by stabilization of the reactive species in solid argon matrices at liquid hydrogen or liquid helium temperatures. The technique, originally developed for infrared studies by Pimentel and Herr (59), consisted of irradiating a methyl halide, trapped in solid argon, with a beam of lithium atoms. Some of the free-radical products were sufficiently long-lived at the temperatures employed for their infrared spectra to be recorded. More recently, radicals have been generated by flash photolysis (in the vacuum-ultraviolet) of various radical sources such as alkyl halides, ketones and azo-compounds. The products were again trapped in solid argon. An interesting development in the matrix isolation technique is the rotating cryostat (60) which enables reactive species to be trapped on a preparative scale for study of their infrared, ultraviolet and e.s.r. spectra. It has also been possible in a few cases to measure the gas-phase infrared spectrum of a radical by flash photolysis generation and rapid-scan infrared spectrometry (61).

Identification of the radical species responsible for the infrared absorption bands presents formidable difficulties. It is usually accomplished by generation of the same radical from a number of different sources. Absorption bands are then observed at the same frequencies in each experiment, so that the bands due to a particular radical can be identified. The behaviour of the absorbing species on warm-up of the matrix is also helpful since it gives some idea of the reactivity of the species. Experiments with isotopically different radicals can also aid in the correct assignment of the infrared bands to a particular radical.

3.8.2 *Structural studies*

The number and intensity of infrared absorption bands observed for a particular radical gives information about its structure. Thus, depending on the geometry of the radical, certain vibrational transitions are symmetry-forbidden in the infrared. Vibrational analysis of the spectrum can give information about the symmetry and hence the structure of the radical. For a tetratomic radical $\cdot CX_3$ there are four possible fundamental frequencies: ν_1 symmetric stretch, ν_2 symmetric bend, ν_3 asymmetric stretch and ν_4 asymmetric bend. If the radical is planar, ν_1 is not excited in the infrared and no corresponding absorption should be observed. The infrared spectrum of methyl itself has been observed in solid argon (62), and no absorption corresponding to ν_1 has been found (see Fig. 3-15). The infrared evidence, therefore, favours a planar structure for the methyl radical and confirms Herzberg's deductions from the electronic spectrum (51). Infrared spectral measurements were also carried out with a series of deuterium- and carbon-13-substituted methyl radicals. The normal co-ordinate analysis, made assuming various out-of-plane structures for $CH_3\cdot$, did not correctly predict the positions of the absorptions for the isotopically labelled species. Agree-

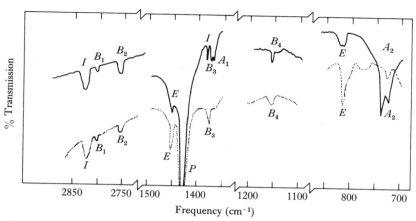

Fig. 3-15. Infrared spectrum of methyl radical recorded by Andrews and Pimentel (62). Spectra recorded for CH_3Br and ^6Li deposited in solid argon at 15 °C. A_1 and A_2 bands at 1383 and 730 cm^{-1} are attributed to $CH_3 \cdot$. E bands are attributed to ethane. B bands are attributed to $LiCH_3$. P band is the parent CH_3Br and I bands are from the CsI cell windows. Dotted line is the spectrum after warming the sample to 53 °K and recooling to 15 °K. Note the decrease in intensity of the methyl bands and the increase in intensity of the ethane bands.

ment could only be obtained for structures planar, or within two or three degrees of planarity (62).

The infrared spectrum of the trifluoromethyl radical has been obtained in both gas (61) and solid (64) phases. In both phases the symmetric stretch mode ν_1 was observed, and $CF_3 \cdot$ was found to be non-planar, in agreement with e.s.r. and other evidence (see chapter 4, p. 79). The structures of other methyl and silyl radicals which have been investigated are given in Table 3-3 (see also chapter 4, p. 79).

Isotopic substitution experiments have enabled detailed deductions about bond angles and lengths to be drawn. In contrast to the trifluoro-methyl radical, the trichloromethyl radical is found to be planar. It was originally proposed on the basis of the infrared spectra that this radical and the tribromomethyl radical were non-planar (68, 69). However, from the most recent experiments (65, 67), it appears that the original assignment of frequencies as ν_1 in the spectra may be in error. The infrared evidence now points to a planar structure for both radicals. All the chlorine- and bromine-containing methyl radicals which have been examined appear to be planar. Force constants and bond orders for the C–Cl or C–Br bonds in these radicals were deduced from the normal co-ordinate analysis. They were found to be greater than the force

TABLE 3-3   *Free radicals detected by infrared spectroscopy in solid argon matrices*

Radical	Structure[a]	Reference
$CH_3\cdot$	Planar	(62)
$CH_2F\cdot$	Nearly planar[b]	(63)
$CF_3\cdot$	Non-planar[c]	(64, 61)
$CH_2Cl\cdot$	Planar	(65, 66)
$CHCl_2\cdot$	Planar	(66)
$CCl_3\cdot$	Planar	(67, 68)
$CBr_3\cdot$	Planar	(67, 69)
$SiH_3\cdot$	Pyramidal	(70)
$SiF_3\cdot$	Pyramidal	(71)
$SiCl_3\cdot$	Pyramidal	(72)

[a] Structure determined from the infrared spectrum in the solid phase.
[b] Definite structure assignment was not possible.
[c] E.s.r. evidence indicates a pyramidal structure for $CF_3\cdot$.

constants or bond orders of C–Cl or C–Br bonds in the fully saturated parent molecules. The reason may be that the C–X bond has some π-character, because of resonance stabilization of the radical, i.e. structures such as:

$$X_2C{=}\dot{X} \longleftrightarrow X_2\dot{C}{-}\ddot{X}$$

may make a significant contribution to the bonding (69). Analysis of the infrared spectra of the three silyl radicals $\cdot SiH_3$, $\cdot SiF_3$ and $\cdot SiCl_3$ showed that for all three the angle between the Si–X bond and the threefold axis of symmetry was about 72°, which differs insignificantly from that of the pyramidal structure.

The infrared spectra of a series of carbonyl-containing radicals $X\dot{C}{=}O$, have also been investigated (73, 74). The carbonyl stretching frequencies, and the corresponding force constants K_{CO} which are shown in the table, are higher than those of the fully bonded molecules from which the radicals are derived. The conclusion is again that there is some p–π^* or s–π^* interaction between the carbonyl function and the attached group X.

	K_{CO}(mdyn Å^{-1})		K_{CO}(mdyn Å^{-1})
$H{-}\dot{C}{=}O$	13.7	$H_2C{=}O$	12.6
$F{-}\dot{C}{=}O$	14.3	$F_2C{=}O$	12.8
$Cl{-}\dot{C}{=}O$	14.5	$Cl_2C{=}O$	12.6

3.9 *MASS SPECTROMETRY OF FREE RADICALS*

3.9.1 *General principles*

The mass spectrometer provides a detector of high sensitivity for both free radicals and other chemical constituents in a reaction system. This is a distinct advantage of the method, since it permits the analysis of all the components simultaneously. It can be a handicap, however, in determining small concentrations of radicals, because it is necessary to distinguish the ions derived from the radical, from those in the background derived from other molecules. The theoretical threshold for detection and identification of a radical by this method is less than one radical in a million background molecules (75). In practice, however, few radicals at these concentrations have been identified, and several per cent radical concentration is usually needed for unequivocal results.

The mass spectrometer does not possess any intrinsic ability to distinguish between stable molecules and free radicals. If a peak in the mass spectrum is observed with the right mass-to-charge ratio for a particular radical, this is only the first step towards its identification. Other contributions to this peak from other molecules present in the ion source must first be subtracted. The ion current remaining after this can then be attributed to the radical. Suppose a radical $R\cdot$ is generated by thermolysis or photolysis of the parent molecule RX. Then the radicals are normally present in the ion source of the mass spectrometer accompanied by an excess of the parent RX, together with product molecules such as the radical dimer R_2. The radical ion R^+ can be generated by at least two processes:

$$R\cdot + e \longrightarrow R^+ + 2e \quad I(R\cdot) \tag{9}$$

$$RX + e \longrightarrow R^+ + X\cdot + 2e \quad A(R^+) \tag{10}$$

Additional processes are:

$$R_2 + e \longrightarrow R^+ + R\cdot + 2e \tag{11}$$

$$RX + e \longrightarrow R^+ + X^- + e \tag{12}$$

Measurements can usually be carried out under conditions which minimize or eliminate steps (*11*), (*12*), etc. The energy required for process (*9*) is simply the radical ionization potential $I(R\cdot)$. The energy required for process (*10*) is the appearance potential of R^+ from RX, i.e. $A(R^+)$. This appearance potential equals the radical ionization potential, plus the energy required to break the R–X bond, plus any kinetic energy of the fragments, i.e.

$$A(R^+) = I(R\cdot) + D(R-X) + K.E. \qquad (13)$$

The energy required to form the ion R^+ from the parent molecule is therefore greater than that required to form the same ion from the free radical. This provides another way of detecting radicals in the mass spectrometer. The energy of the ionizing electron beam is set to less than the appearance potential $A(R^+)$, so that parent molecules are not ionized, but greater than $I(R\cdot)$ so that any ions at the mass-to-charge ratio corresponding to R^+ come wholly from the free radical.

3.9.2 *Experimental method*

The mass spectrometer has proved most useful in the study of gas-phase reactions. The pioneering work was carried out by Eltenton (76). Numerous studies of the formation of free radicals on heated metal surfaces have been made by Robertson (77) and LeGoff (78).

The chief experimental difficulty is to admit free radicals from the gas-phase reaction system into the mass spectrometer source without them colliding and recombining at metal surfaces on the way. The usual arrangement is to connect the reaction system to the mass spectrometer by a quartz or glass thimble having a pin-hole at the end. Molecules can then effuse straight through the leak without colliding with any surfaces, and pass immediately into the ionizing region of the mass spectrometer (79, 80). The narrowness of the pin-hole also enables the radical generation system to be operated at a much higher pressure than the mass spectrometer source. Foner and collaborators (75) have overcome the problem of background interference by modulating the inlet molecular beam with a chopper device. The mass spectrometer operates with a phase-sensitive detector which is set in phase with the modulation of the molecular beam. Only the ions formed from the modulated beam are detected, and hence the d.c. background signal makes a negligible contribution.

It is desirable to have as high a concentration of radicals as possible, so the radicals are usually generated in a flow system. Conventional photolysis methods seldom provide a high enough radical concentration, but thermal generation at high temperatures, is found to be one of the most useful methods (80). Lossing has also developed a mercury photosensitization technique, which provides sufficient radicals for mass spectrometric study. His apparatus and inlet system are shown in Fig. 3-16.

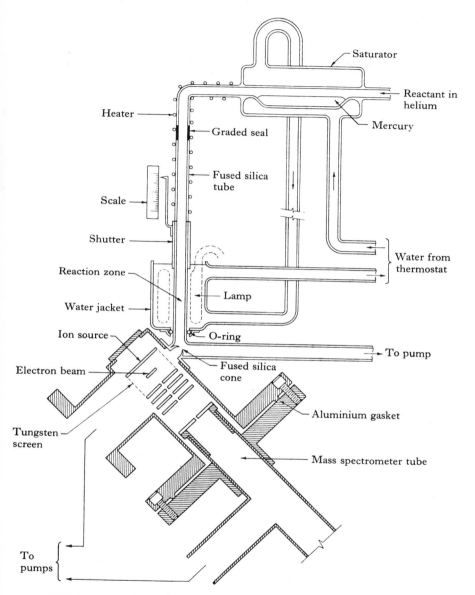

Fig. 3-16. Apparatus for the study of free-radical mass spectra. From Lossing and Kebarle (see ref. 80). (Reproduced by permission of the National Research Council of Canada.)

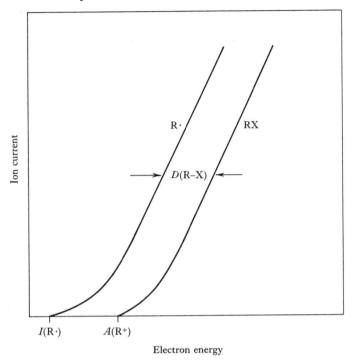

Fig. 3-17. Ionization efficiency curves for a free radical R· and molecule RX.

The method of ionization most commonly used is electron-impact, since the ionization cross-section of molecules is high by this technique. Photoionization gives a sharper ionization onset, which is obviously desirable when attempting to distinguish a free radical from the parent molecule by its lower appearance potential. The electron bombardment source of a mass spectrometer is a heated filament, which emits electrons having a spread of energies. Consequently, the ionization efficiency curves of the radical and molecule do not show a sharp onset of ionization, but become curved at lower energies (see Fig. 3-17). If the curvature on the parent-molecule ionization efficiency curve extends to low energies, detection of the radical becomes difficult or impossible. With photoionization, the energy spread of the photons can be made extremely small, and the energy scale is also known precisely. Unfortunately, because of the low photoionization cross-sections of most molecules, this method has as yet had only very limited application to free radicals (81).

3.9.3   *Measurement of radical concentrations*

Radicals are most clearly distinguished from the parent molecules when the mass spectrometer is operated at a low ionizing electron voltage set to a value between the appearance potential of the ion R^+ from the radical and that from the parent molecule. At this low voltage, however, the ion current which is due wholly to ions derived from the radical is small (see Fig. 3-17) and fluctuations due to filament temperature etc. are greatest. It is generally necessary to work at much higher electron voltages; of the order of 50 eV. The contributions of all stable reactants and products to the total ion current at the mass to charge ratio corresponding to R^+, are subtracted, and the remaining ion current is then proportional to the radical concentration. Relative radical concentrations can be determined in a straight-forward manner by this method, provided the radical concentration is sufficiently high (a few per cent or more).

Absolute radical concentrations are much more difficult to obtain. The sensitivity of the instrument to the particular radical must first be determined. Lossing and Tickner (82) were able to make absolute concentration measurements on the methyl radical generated from dimethylmercury. A known concentration of dimethylmercury was decomposed in the reactor, and all the products except the methyl radical, i.e. methane, ethane and mercury, were quantitatively analysed in the usual way, after calibration of the instrument with known amounts of methane, ethane, etc. By assuming 100% carbon balance, the sensitivity for methyl radicals relative to that for methane, $S(CH_3 \cdot)/S(CH_4)$, was then found. The relative sensitivity to the radicals is a function of instrumental factors such as pumping speed and other unknown parameters (75, 80), so that figures obtained by different research groups for $S(CH_3 \cdot)/S(CH_4)$ are not the same. It appears that calibration for a radical must be carried out separately for the particular instrument and experimental conditions being used. Absolute radical concentration measurements have only been made for certain reaction systems, such as this, where all the products are known, and the concentration of radicals is high.

3.9.4   *Ionization potentials of free radicals*

One of the most interesting achievements of the mass spectrometric method has been the measurement of radical ionization potentials. The technique is similar to that used for measuring ionization potentials of stable molecules. The free radical passes into the ionization chamber mixed with a gas of known ionization potential; usually argon, krypton or xenon. The ionization efficiency curves for the radical and standard are then obtained. The radical ionization potential is found by comparing the two curves, using one of the standard extrapolation methods available for this purpose (75). The electron-impact ionization potentials of a number of radicals are given in Table 3-4.

TABLE 3-4 *Ionization potentials of free radicals measured by electron impact*

Radical	$I(R\cdot)(eV)^a$	$I(R\cdot)$ calc. $(eV)^b$
OH·	13.2	
NH·	13.1	
HO_2·	11.5	
NH_2·	11.4	
NF_2·	12.0	
CH_3·	9.9 (9.84)[c]	9.95
CD_3·	9.9 (9.83)[c]	
C_2H_5·	8.7	8.71
n-C_3H_7·	8.7	
iso-C_3H_7·	7.9	7.84
n-C_4H_9·	8.6	
iso-C_4H_9·	8.4	
sec-C_4H_9·	7.9	
t-C_4H_9·	7.4	7.40
CH_2F·	9.4	
CH_2Cl·	9.3–9.7	9.42
CH_2Br·	8.3–9.3	
CHF_2·	9.5	
$CHCl_2$·	9.3–9.5	9.04
$CHBr_2$·	8.1	
CF_3·	10.2	
CCl_3·	8.8	8.64

[a] See Lossing (80) and Foner (75) for original sources.
[b] Molecular orbital calculations of radical ionization potentials by A. Streitwieser (83).
[c] Spectroscopic value from Herzberg (84).

Ionization potentials of free radicals measured by electron impact methods are generally referred to as 'vertical' ionization potentials. The electronic transition takes place from the ground vibrational state of the radical, vertically to the upper ionized state. If the potential energy curves of the ground state and the upper ionized state are displaced relative to each other, probability favours formation of the ion in a high vibrational level. The adiabatic ionization potentials measured spectro-scopically, from Rydberg series, refer to ionization from the ground vibrational level of the radical to the ground vibrational level of the ion. In the case of free radicals the displacement, if any, of the upper state relative to the lower is not known, but it is to be anticipated that the electron impact ionization potential will be greater than the spectroscopic value by a small fraction of an electron volt, representing the excess of vibrational energy with which it is formed.

Spectroscopic ionization potentials have only been determined for methyl radicals (84, 51) and for a few other diatomic and triatomic species. For these radicals good agreement is found between the results of the two methods. For the vast majority of radicals, the electron impact ionization potential is the only one available. A few radical ionization potentials have also been determined by photoionization (81). For methyl radicals, the value agrees well with the electron impact datum, but for ethyl and isopropyl radicals the photoionization values are considerably lower. The cause of this, whether it is experimental or whether it indicates some inherent flaw in either method, is not known at present.

The electron impact ionization potentials of the alkyl radicals show certain regularities. A steady decrease is observed from methyl to ethyl to isopropyl to t-butyl radicals. The ionization potentials of the primary radicals, ethyl, n-propyl, and n-butyl, are all very similar, and those of the two secondary radicals isopropyl and *sec*-butyl, are also almost identical. The decrease in ionization potential with increasing alkyl substitution of the radical, probably reflects the greater ease of formation of the ions whose order of stability is:

$$\text{tertiary} > \text{secondary} > \text{primary}.$$

The alkyl substituents stabilize the carbonium ion character of the atom to which they are attached. A linear correlation of ionization potential with Taft σ^* constants has been found for these alkyl radicals (85).

The ionization potentials of various substituted benzyl radicals have been found to correlate with the known electron-releasing or electron-attracting power of the substituents. A plot of the ionization potential against the σ^+ substituent constants showed a linear relationship between the two quantities (86). A similar linear correlation has been found for substituted phenoxy radicals (87).

The radical ionization potentials have been very useful in helping to establish good values for bond dissociation energies of various molecules, and have contributed towards the calculation of radical heats of formation. From equation (*13*) we obtain:

$$D(\text{R--X}) = A(\text{R}^+) - I(\text{R·}) - \text{K.E.} \qquad (14)$$

The dissociation energy of the bond R–X is given in terms of the appearance potential of the ion R^+ from RX, the radical ionization potential and the kinetic energy. Stevenson's rule (88) states that the kinetic energy is negligible provided the ionization potential of the non-ionized fragment (X in the above example) is greater than that of the ionized fragment, i.e. K.E. $= 0$ if $I(\text{X·}) > I(\text{R·})$. This is usually the case. Once the radical ionization potential is known, bond dissociation energies for various bonds can be found simply by measuring the appearance potential of R^+ from the appropriate molecule. Bond dissociation energies are related to radical heats of formation by the equation below:

$$D(\text{R--X}) = \Delta H_f^0(\text{R·}) + \Delta H_f^0(\text{X·}) - \Delta H_f^0(\text{RX}) \qquad (15)$$

The heats of formation of many organic molecules, $\Delta H_f^0(\text{RX})$, are known, and atom heats of formation are known accurately from spectroscopic data (89); so that if X is an atom $\Delta H_f^0(\text{R·})$ can be found. Thermodynamic data derived from electron impact data in this way are seldom the most accurate, but the results are very useful as a cross-check on other methods.

Radical ionization potentials can be calculated by the molecular orbital method. The ionization potential is given as the difference in the computed electronic energies for the radical and corresponding ion. Calculations have been carried out by Hush and Pople (90) and Streitwieser (83). Good agreement with the experimental results is obtained when electron correlation terms are included in the calculations. The calculated values for various radicals are included in Table 3-4.

REFERENCES

1. C. Walling, E. R. Briggs, W. Cumming and F. R. Mayo, *J. Amer. Chem. Soc.* 1950, **72**, 48.
1a. L. N. Mulay in *Techniques of Chemistry*, ed. A. Weissberger, and B. W. Rossiter, 4th edn, Interscience, New York, 1972, vol. I, pt IV, p. 431.
2. P. Pascal, *Ann. Chim. Phys.* 1910, **19**, 5.
3. A. Pacault, *Rev. Sci.* 1948, **86**, 38.
4. M. F. Roy and C. S. Marvel, *J. Amer. Chem. Soc.* 1937, **59**, 2622; P. W. Selwood and R. M. Dobres, *J. Amer. Chem. Soc.* 1950, **72**, 3860.

5. A. Carrington and A. D. McLachlan, *Introduction to Magnetic Resonance*, Harper and Row, New York, 1967.
6. L. Kevan in *Methods in Free-Radical Chemistry*, vol. 1, ch. 1, ed. E. S. Huyser, Marcel Dekker, New York, 1969.
7. R. W. Fessenden and R. H. Schuler, *J. Chem. Phys.* 1963, **39**, 2147.
8. J. K. Kochi and P. J. Krusic, *Essays in Free-Radical Chemistry, Chem. Soc. Special Publ.* No. 24, 1970, 147.
9. A. Hudson and R. A. Jackson, *Chem. Comm.* 1969, 1323.
10. R. O. C. Norman in *Essays in Free-Radical Chemistry, Chem. Soc. Special Publ.* No. 24, 1970, 117.
11. R. O. C. Norman and R. J. Pritchett, *Chem. and Ind.* 1965, 2040.
12. A. R. Forrester, J. M. Hay and R. H. Thompson, *Organic Chemistry of Stable Free Radicals*, Academic Press, New York and London, 1968.
13. K. D. Sales, *Adv. Free-Radical Chem.* 1969, **3**, 169.
14. A. Carrington, *Quart. Rev.* 1963, **17**, 67.
15. E. de Boer and H. van Willigen, *Progr. N.M.R. Spectroscopy*, 1967, **2**, 111.
16. H. M. McConnell, *J. Chem. Phys.* 1956, **24**, 632, 764.
17. T. J. Stone and W. A. Waters, *J. Chem. Soc.* 1964, 213.
18. K. Dimroth, A. Berndt, F. Bar, A. Schweig and R. Volland, *Angew. Chem. Internat. Edn*, 1967, **6**, 34.
19. H. Fischer, *Z. Naturforsch.* 1964, **19a**, 866; 1965, **20a**, 428.
20. A. J. Dobbs, B. C. Gilbert and R. O. C. Norman, *J. Chem. Soc. A*, 1971, 124.
21. R. O. C. Norman and B. C. Gilbert, *Adv. Phys. Org. Chem.* 1967, **5**, 53.
22. D. H. Geske, *Progr. Phys. Org. Chem.* 1967, **4**, 125.
23. M. D. Sevilla and G. Vincow, *J. Phys. Chem.* 1968, **72**, 3647.
23a. G. A. Russell, G. Holland, K.-Y. Chang and L. H. Zalkow, *Tetrahedron Letters*, 1967, 1955.
24. E. Müller, A. Rieker, K. Scheffler and A. Moosmayer, *Angew. Chem. Internat. Edn*, 1966, **5**, 6.
25. R. W. Fessenden, *J. Phys. Chem.* 1967, **71**, 74.
26. G. Fraenkel, M. Kaplan and J. Bolton, *J. Chem. Phys.* 1965, **42**, 955.
27. H. Lemaire, Y. Marechal, R. Ramasseul and A. Rassat, *Bull. Soc. chim. France*, 1965, 372.
28. P. H. Kasai, E. Hedaya and E. B. Whipple, *J. Amer. Chem. Soc.* 1969, **91**, 4364.
29. G. A. Russell, *Radical Ions*, ed. E. T. Kaiser and L. Kevan, Interscience, New York, 1968, p. 108.
30. S. Forshult, C. Lagercrantz and K. Torsell, *Acta Chem. Scand.* 1969, **23**, 522.
31. M. J. Perkins, *Essays in Free-Radical Chemistry, Chem. Soc. Special Publ.* No. 24, 1970, 97.
32. E. G. Janzen and B. J. Blackburn, *J. Amer. Chem. Soc.* 1968, **90**, 5909.
33. B. C. Gilbert, J. P. Larkin and R. O. C. Norman, *J. C. S. Perkin II*, 1972, 1272.
34. R. J. Holman and M. J. Perkins, *Chem. Comm.* 1971, 244; *J. Chem. Soc. C*, 1971, 2324.
35. A. L. Kwiram, *Ann. Rev. Phys. Chem.* 1971, **22**, 133.

35*a*. N. M. Atherton, *Chem. Soc. Specialist Report in E.S.R. Spectroscopy*, 1973, **1**, 32.
36. J. S. Hyde, *J. Chem. Phys.* 1965, **43**, 1806.
36*a*. J. S. Hyde, R. Breslow and C. DeBoer, *J. Amer. Chem. Soc.*, 1966, **88**, 4763.
37. J. S. Hyde, *J. Phys. Chem.*, 1967, **71**, 68.
37*a*. N. M. Atherton and A. J. Blackhurst, *J. C. S. Faraday II*, 1972, **68**, 470.
38. J. W. Wells and H. C. Box, *J. Chem. Phys.* 1967, **47**, 2935.
39. K. H. Hausser, H. Brunner and J. C. Jochims, *Mol. Phys.* 1966, **10**, 253.
39*a*. A. Calder, A. R. Forrester, J. W. Emsley, G. R. Luckhurst and R. A. Storey, *Mol. Phys.* 1970, **4**, 481.
39*b*. A. R. Forrester, unpublished results.
39*c*. W. Esperen and R. W. Kreilick, *J. Phys. Chem.* 1969, **73**, 3370; F. Yamauchi and R. W. Kreilick, *J. Amer. Chem. Soc.* 1969, **91**, 3429.
40. J. Bargon, H. Fischer and U. Johnsen, *Z. Naturforsch.* 1967, **22a**, 1551, 1556.
41. H. R. Ward and R. G. Lawler, *J. Amer. Chem. Soc.* 1967, **89**, 5518; R. G. Lawler, *ibid.* 1967, **89**, 5519.
42. H. Fischer and J. Bargon, *Accounts Chem. Res.* 1969, **2**, 110.
43. A. R. Lepley, *J. Amer. Chem. Soc.* 1968, **90**, 2710.
44. R. Kaptein and L. J. Oosterhoff, *Chem. Phys. Letters*, 1968, **2**, 261.
45. R. Kaptein and L. J. Oosterhoff, *Chem. Phys. Letters*, 1969, **4**, 195, 214; A. R. Lepley, *J. Amer. Chem. Soc.* 1968, **90**, 2710; 1969, **91**, 1237.
46. G. L. Closs and A. D. Trifunac, *J. Amer. Chem. Soc.* 1970, **92**, 2183, 2186.
47. R. Kaptein, F. W. Verheus and L. J. Oosterhoff, *Chem. Comm.* 1971, 877.
48. G. Herzberg, *Spectra of Diatomic Molecules*, Van Nostrand, Princeton, 1950.
49. G. Porter in *Technique of Organic Chemistry*, ed. A. Weissberger, Interscience, New York, 1960, vol. VIII, pt II, p. 1055.
50. N. Basco and F. G. M. Hawthorn, *Chem. Phys. Letters.* 1971, **8**, 291.
51. G. Herzberg and J. Shoosmith, *Canad. J. Phys.* 1956, **34**, 523.
52. H. E. Van den Berg, A. B. Callear and R. J. Norstrom, *Chem. Phys. Letters*, 1969, **4**, 101.
53. N. Basco, D. G. L. James and R. D. Suart, *Internat. J. Chem. Kinetics*, 1970, **2**, 215.
54. E. J. Land, *Progr. Reaction Kinetics*, 1965, **3**, 371.
55. G. Porter and F. J. Wright, *Trans. Faraday Soc.* 1955, **51**, 1469.
56. R. L. McCarthy and A. MacLachlan, *Trans. Faraday Soc.* 1960, **56**, 1187.
57. G. Porter and M. W. Windsor, *Nature*, 1957, **180**, 187; *see also* ref. 54.
58. E. J. Land and G. Porter, *Trans. Faraday Soc.* 1963, **59**, 2016, 2027.
59. A. J. Barnes and H. E. Hallam, *Quart. Rev.* 1969, **23**, 392; G. C. Pimentel, *J. Amer. Chem. Soc.* 1958, **80**, 62; *Pure Appl. Chem.* 1965, **11**, 563.
60. J. E. Bennett, B. Mile, A. Thomas and B. Ward, *Adv. Phys. Org. Chem.* 1970, **8**, 1.
61. G. A. Carlson and G. C. Pimentel, *J. Chem. Phys.* 1966, **44**, 4053.
62. W. L. S. Andrews and G. C. Pimentel, *J. Chem. Phys.* 1966, **44**, 2527; 1967, **47**, 3637.
63. M. E. Jacox and D. E. Milligan, *J. Chem. Phys.* 1969, **50**, 3252; J. I. Raymond and L. Andrews, *J. Phys. Chem.* 1971, **75**, 3239.

64. D. E. Milligan, M. E. Jacox and J. J. Comeford, *J. Chem. Phys.* 1966, **44**, 4058.
65. L. Andrews and D. W. Smith, *J. Chem. Phys.* 1970, **53**, 2956.
66. M. E. Jacox and D. E. Milligan, *J. Chem. Phys.* 1970, **53**, 2688.
67. E. E. Rogers, S. Abramowitz, M. E. Jacox and D. E. Milligan, *J. Chem. Phys.* 1970, **52**, 2198.
68. W. L. S. Andrews, *J. Phys. Chem.* 1967, **71**, 2761.
69. L. Andrews and T. G. Carver, *J. Chem. Phys.* 1968, **49**, 896.
70. D. E. Milligan and M. E. Jacox, *J. Chem. Phys.* 1970, **52**, 2594.
71. D. E. Milligan and M. E. Jacox, *J. Chem. Phys.* 1968, **49**, 5330.
72. M. E. Jacox and D. E. Milligan, *J. Chem. Phys.* 1968, **49**, 3130.
73. D. E. Milligan, M. E. Jacox, A. M. Bass, J. J. Comeford and D. E. Mann, *J. Chem. Phys.* 1965, **42**, 3187.
74. J. S. Shirk and G. C. Pimentel, *J. Amer. Chem. Soc.* 1968, **90**, 3349.
75. S. N. Foner, *Adv. Atomic and Molecular Phys.* 1966, **2**, 385.
76. G. C. Eltenton, *J. Chem. Phys.* 1942, **10**, 403; 1947, **15**, 455.
77. A. J. B. Robertson, *Proc. Roy. Soc. (Lond.)* A, 1949, **199**, 394; A. J. B. Robertson, *Mass Spectrometry*, Methuen, London, 1954.
78. P. LeGoff, *Applied Mass Spectrometry*, Institute of Petroleum, London, 1954; *J. Chim. phys.* 1953, **50**, 423; 1956, **53**, 269; L. P. Blanchard and P. LeGoff, *Adv. Mass Spectrometry*, 1959, 570.
79. J. Cuthbert, *Quart. Rev.* 1959, **13**, 215.
80. F. P. Lossing in *Mass Spectrometry*, ed. C. A. McDowell, McGraw-Hill, New York, 1963, p. 442.
81. F. P. Lossing and I. Tanaka, *J. Chem. Phys.* 1956, **25**, 1031; F. A. Elder, C. Giese, B. Steiner and M. Inghram, *ibid.* 1962, **36**, 3292.
82. F. P. Lossing and A. W. Tickner, *J. Chem. Phys.* 1952, **20**, 907.
83. A. Streitwieser, *J. Amer. Chem. Soc.* 1960, **82**, 4123; *Tetrahedron*, 1959, **5**, 149.
84. G. Herzberg, *Canad. J. Phys.* 1961, **39**, 1511.
85. A. Streitwieser, *Progr. Phys. Org. Chem.* 1963, **1**, 1.
86. A. G. Harrison, P. Kebarle and F. P. Lossing, *J. Amer. Chem. Soc.* 1961, **83**, 777.
87. J. M. S. Tait, T. W. Shannon and A. G. Harrison, *J. Amer. Chem. Soc.* 1962, **84**, 4.
88. D. P. Stevenson, *Discuss. Faraday Soc.* 1951, **10**, 35.
89. J. A. Kerr, *Chem. Rev.* 1966, **66**, 494.
90. N. S. Hush and J. A. Pople, *Trans. Faraday Soc.* 1955, **51**, 600.

4

Shapes of free radicals

4.1 *INTRODUCTION*

The shape of carbon free radicals has been a topic of considerable interest and not a little controversy over a period of time. Alkyl radicals have been variously postulated to be planar, as is the case for the corresponding carbonium ions, in which the carbon atom is sp^2-hybridized and the unpaired electron is in a p-orbital perpendicular to the plane of the sp^2-orbitals (Fig. 4-1 a). Alternatively, they have been postulated to be pyramidal with the carbon atom sp^3-hybridized (Fig. 4-1 b). In the latter case there could be rapid inversion of the two enantiomeric configurations of the pyramidal radical.

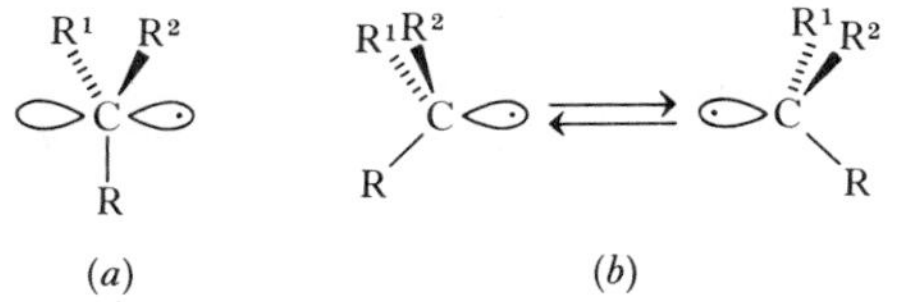

Fig. 4-1. Possible conformations of the radical $R\dot{C}R^1R^2$: (a) planar conformation; (b) pyramidal conformation

The shape of a radical may be deduced by use of either physical or chemical methods. Physical methods, which in the main involve e.s.r. studies, are undoubtedly more satisfactory than chemical methods since one is dealing with the ground state of the radical. Chemical methods generally involve the generation of a radical at an asymmetric carbon atom and observation of whether or not racemization takes place. Retention (or inversion) of configuration would argue strongly in favour of a pyramidal radical. Racemization, however, is consistent with either a planar or a rapidly inverting pyramidal radical in which inversion proceeds more rapidly than reaction.

4.2 ALKYL RADICALS

Pertinent information about the shape of alkyl radicals can be obtained from the ^{13}C splittings in their e.s.r. spectra. ^{13}C Splitting is a particularly sensitive measure of the degree of non-planarity of this type of radical (2), increasing with increasing s-character of the orbital carrying the unpaired electron. Table 4-1 shows the values for the methyl radical and the fluorinated methyl radicals. The value steadily increases with the degree of fluorine substitution. That for the trifluoromethyl radical is approximately one quarter of the calculated ^{13}C splitting for an electron in a pure $2s$-orbital. This is thus consistent with the unpaired electron in this radical being in an sp^3-orbital, i.e. with the trifluoromethyl radical being pyramidal. The ^{13}C splitting for the methyl radical is very much smaller. The low value can be considered to arise because the unpaired electron is in an orbital containing some s-character or as a result of spin polarization.

TABLE 4-1   *Hyperfine splitting constants of methyl and fluorinated methyl radicals* (1)

| | Coupling constants (mT) | | |
Radical	a (H)	a (^{13}C)	a (F)
$\dot{C}H_3$	2.30	3.85	—
$\dot{C}H_2F$	2.11	5.48	6.43
$\dot{C}HF_2$	2.22	14.88	8.42
$\dot{C}F_3$	—	27.16	14.24

The proton splitting in the methyl radical (2.3 mT) is about six times the value in the benzene radical anion in which the electron density at each carbon atom is one sixth. The benzene radical anion is undoubtedly a π-radical and hence the proton splitting for the methyl radical is consistent with it being a π-radical, i.e. a planar radical. The ^{13}C splittings for the fluoromethyl and difluoromethyl radicals are consistent with their having the unpaired electron in an orbital containing a degree of s-character, i.e. there is some bending in these radicals. It is also apparent in the fluorinated radicals that the value of ^{19}F splitting (Table 4-1) varies with the degree of bending of the radical.

The proton splitting of a planar radical is negative as a result of spin polarization, which induces a negative spin density at the trigonal carbon atom. Pyramidal radicals, in which the unpaired electron is in an orbital possessing some *s*-character, have positive proton splittings since there is a positive spin density associated with the radical centre. One can thus envisage the proton splittings of radicals numerically decreasing in value and then increasing as the degree of bending increases, since it is not possible to measure the signs of splittings in e.s.r. spectra. On this basis it seems possible that the proton splitting of $\dot{C}H_2F$ is negative and that of $\dot{C}HF_2$ is positive when the proton splittings are considered along with the ^{13}C splittings (Table 4-1).

Infrared (31) and ultraviolet (32) studies on the methyl radical like-wise support the contention that it adopts a planar conformation (see chapter 3, pp. 62 and 64).

A study of ^{13}C and particularly proton splitting constants of oxygen-containing radicals has shown that they are bent (3). ^{13}C Satellite lines in these radicals frequently cannot be observed, and perforce one has to rely on the values of the proton splittings to infer the geometry of the radical. This is complicated by not knowing the sign of the splitting constant, though by consideration of a series of related radicals one can frequently infer the sign of the coupling constants. This is exemplified in the series of radicals (1–3) which have the proton splitting constants

H (2.15)   H (1.24)   H (2.15)

(1)   (2)   (3)

shown. The smaller numerical value for (2) as compared with (1) is to be expected, partly because of the ability of the oxygen to reduce the spin density on carbon by resonance interaction,

$$\overset{\backslash}{\underset{/}{\ddot{C}}} - \ddot{O} \overset{\backslash}{\underset{\backslash}{}} \longleftrightarrow \overset{\backslash}{\underset{/}{\bar{C}}} - \overset{+}{\ddot{O}} \overset{\backslash}{\underset{\backslash}{}}$$

and partly because of bending at the radical centre. It seems almost certain that the splitting constant for (3) must be positive, indicative of a degree of bending at the radical centre. There is also evidence of bending in the acyclic radicals $\dot{C}H_2OH$ and $\dot{C}Me_2OH$ as judged from their ^{13}C and proton splitting constants.

The *g*-factor of a radical has also been shown to be sensitive to the geometry of the radical centre. Thus, whilst substitution of one hydro-

gen by a hydroxyl or an alkoxyl group results in an increase of the
g-factor, further substitution causes a decrease in the g-factor. The
g-factor for $\dot{C}(OEt)_3$ is significantly lower than that for $\dot{C}H_2OH$ as a
consequence of bending at the radical centre. This is consistent with the
observation that g-factors of σ-radicals are lower than those for π-
radicals.

Chemical evidence for the configuration of carbon radicals is very
much less satisfactory than the physical evidence discussed above. Thus,
although chlorination of 1-chloro-2-methylbutane (4) results in the
formation of racemic 1,2-dichloro-2-methylbutane (5), consistent with
the reaction proceeding via a planar radical (Scheme 1) or a rapidly
inverting pyramidal radical, bromination leads to the optically active
product in which inversion at the radical centre has occurred (reaction *1*) (4, 5). This could be attributed to bromine being a more

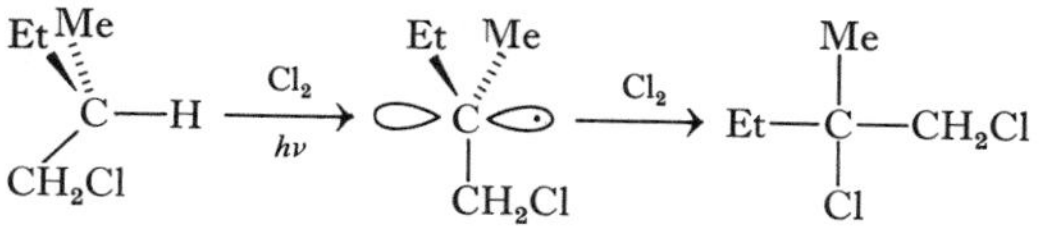

(4)   Racemic
                                         (5)

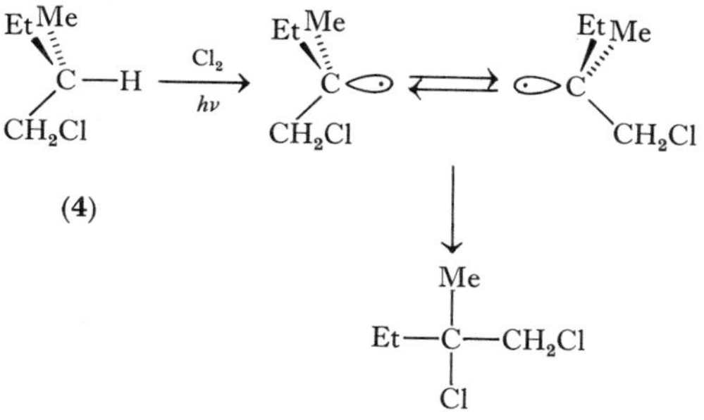

(4)

Me
|
Et—C—CH$_2$Cl
|
Cl

Racemic
(5)

Scheme 1

$4\cdot f\,114$

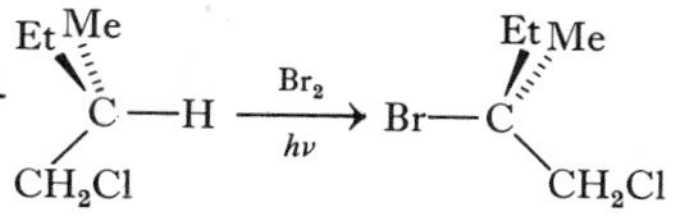

(4)

(*1*)

efficient transfer agent than chlorine, reaction occurring before inversion of the pyramidal radical. In the case of a planar radical, attack could take place from the side opposite to that from which hydrogen abstraction took place. Alternatively, one can postulate that the intermediate radical is a bridged species somewhat analogous to a bromonium ion which thereby ensures attack from the other side. Racemization does not occur as inversion cannot take place. It is less easy to see why, if this is the case, chlorination leads to racemic products unless the bridged radical in this case is converted into the planar radical, whereas in bromination the bridged radical undergoes attack before this second step can occur. Maximum optical purity in the bromination of (4) only occurs at high bromine concentrations, suggesting a competition between the reaction of the bridged radical (6) with bromine and its isomerization to the acyclic radical (Scheme 2). These results illustrate the care that must be used in the interpretation of chemical results for inferring radical conformations.

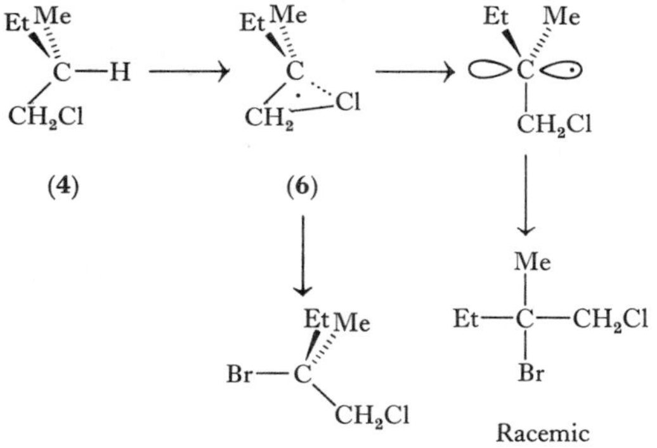

Scheme 2

Another experiment involved the study of the radical-induced decarbonylation of the optically active aldehyde (7) which resulted in the formation of the racemic hydrocarbon (8) (Scheme 3) (6). This is somewhat more satisfactory than the preceding example in that the possibility of bridged radicals is absent. It is still not possible to distinguish between rapidly inverting pyramidal and planar radicals.

Decomposition of optically active diacyl peroxides has also been studied to give information regarding radical configurations (7). Thus bis(2-methyl-3-phenylpropionyl) peroxide (9) on decomposition in

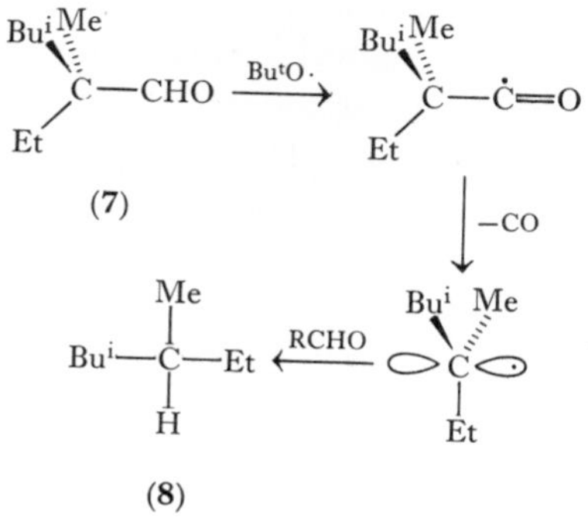

Scheme 3

carbon tetrachloride gave the ester **(10)** in which the acyl group had completely retained its activity and the alkyl group had undergone slight racemization (Scheme 4). In addition optically inactive products (**11** and **12**) arose from reaction with the solvent and trichloromethyl radicals respectively. The predominant retention of configuration in the ester is considered to arise as a result of combination of acyl and alkyl radicals within the solvent cage, though the ester could also be formed in a cyclic process.

$$(R*CO_2)_2 \longrightarrow R*CO_2\cdot + R*\cdot + CO_2$$
$$\text{(9)} \qquad\qquad \downarrow$$
$$R*CO_2R*$$
$$\text{(10)}$$

$$R*\cdot \longrightarrow R\cdot$$

$$R\cdot + CCl_4 \longrightarrow RCl + \dot{C}Cl_3$$
$$\text{(11)}$$

$$R\cdot + \dot{C}Cl_3 \longrightarrow RCCl_3$$
$$\text{(12)}$$

$$\overset{\displaystyle Me}{\underset{\displaystyle}{(R = PhCH_2\overset{|}{C}H{-})}}$$

Scheme 4

4.3 CYCLOALKYL RADICALS

The e.s.r. spectra of cyclohexyl radicals are consistent with rapidly interconverting chair conformers with a planar radical centre. The coupling constants for the 4-t-butylcyclohexyl radical are very similar to those of the cyclohexyl radical, again indicative of it being in the chair conformation with a planar radical centre. Chemical studies on the 4-t-butylcyclohexyl radical generated from *cis-* and *trans-*4-t-butyl-cyclohexanecarboxylic acids in the Hunsdiecker reaction support this contention in that both isomers lead to the same product mixture (9) (Scheme 5).

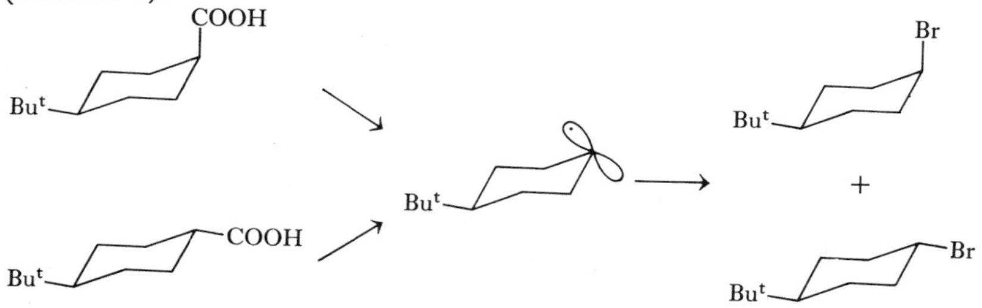

Scheme 5

The e.s.r. spectra of cyclobutyl and cyclopentyl radicals are interpretable on the basis of their having a planar radical centre. In the case of cyclobutyl radicals (13) there is no evidence whatsoever for their interconversion with cyclopropylmethyl (14) or 3-butenyl radicals (15) which might have been expected if there was any tendency towards a non-classical structure for the cyclobutyl radical analogous to the non-classical cyclobutyl carbonium ion.

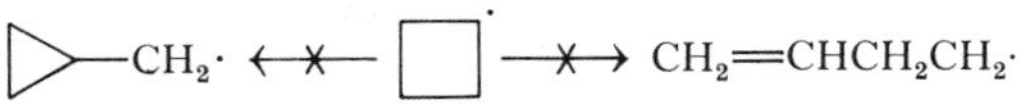

$$\text{(14)} \qquad \text{(13)} \qquad \text{(15)} \qquad\qquad (2)$$

The α-proton splitting constant of the cyclopropyl radical is only 0.65 mT, indicative of a significant degree of bending at the radical centre. This is supported by chemical studies (10–12) on optically active cyclopropyl compounds which indicate that cyclopropyl radicals react with partial retention of configuration at the radical centre, indicative of a pyramidal structure (reactions *3–5*).

4.4 9-DECALYL RADICALS

Physical measurements clearly indicate that alkyl and cycloalkyl radicals (with the exception of the cyclopropyl radical) adopt a planar configuration at the radical centre. The question arises whether a non-

84

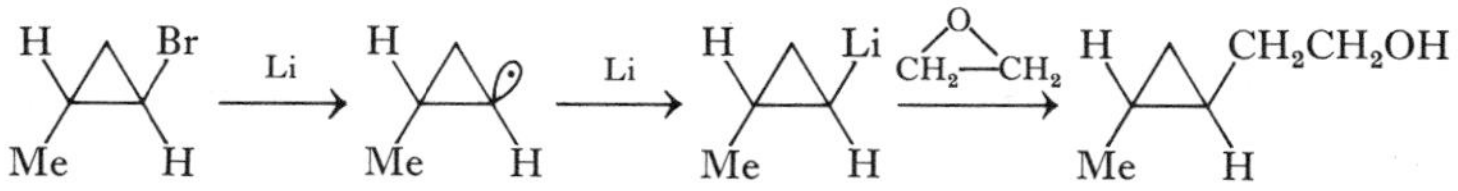

$$(3)$$

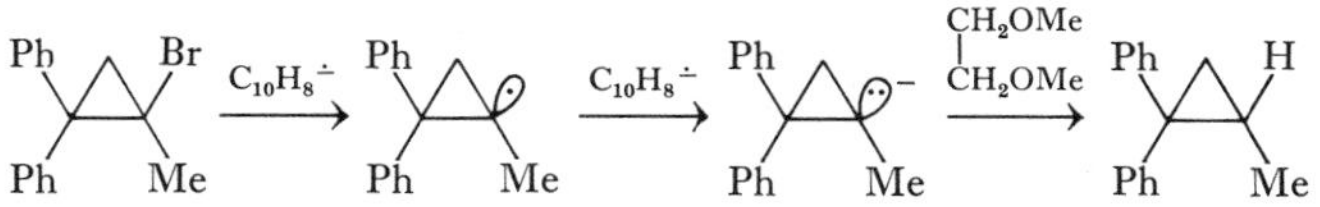

$$(4)$$

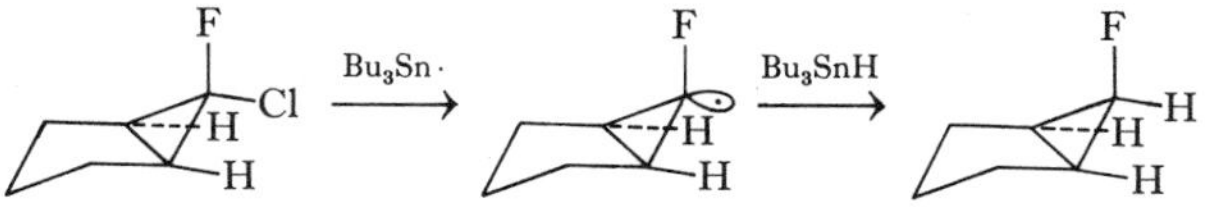

$$(5)$$

planar pyramidal radical is initially formed in some systems and this subsequently rearranges to a planar radical. Information about this possibility has been obtained from a study of the radicals generated in the decomposition of the t-butyl peresters of *cis-* and *trans*-9-decalin-carboxylic acids in the presence of oxygen followed by reduction of the hydroperoxides with lithium aluminium hydride to give *cis-* and *trans*-9-decalols (13):

$$RCO_3Bu^t \xrightarrow{\Delta} R\cdot + CO_2 + Bu^tO\cdot$$

$$R\cdot + O_2 \longrightarrow RO_2\cdot \xrightarrow{SH} RO_2H$$

$$RO_2H \xrightarrow{LiAlH_4} ROH$$

At ordinary pressures of oxygen both peresters give identical mixtures of the decalols, indicative of a common intermediate planar radical. At very high oxygen pressures the *cis*-perester gives more *cis*-9-decalol than the *trans*-perester. This implies that the *cis*-perester gives initially a pyramidal radical (**16**) which at high oxygen pressures is scavenged by oxygen at a rate comparable to its rate of rearrangement to the planar radical (**17**), as indicated in Scheme 6. This is reasonable since the *cis*-pyramidal radical can only attain a planar configuration by ring flipping whereas the *trans*-pyramidal radical (**18**) has a structure very similar to that of the planar radical. Hence, interconversion of the *cis*-pyramidal radical to the planar radical is a more energetic process than for the *trans*-pyramidal radical.

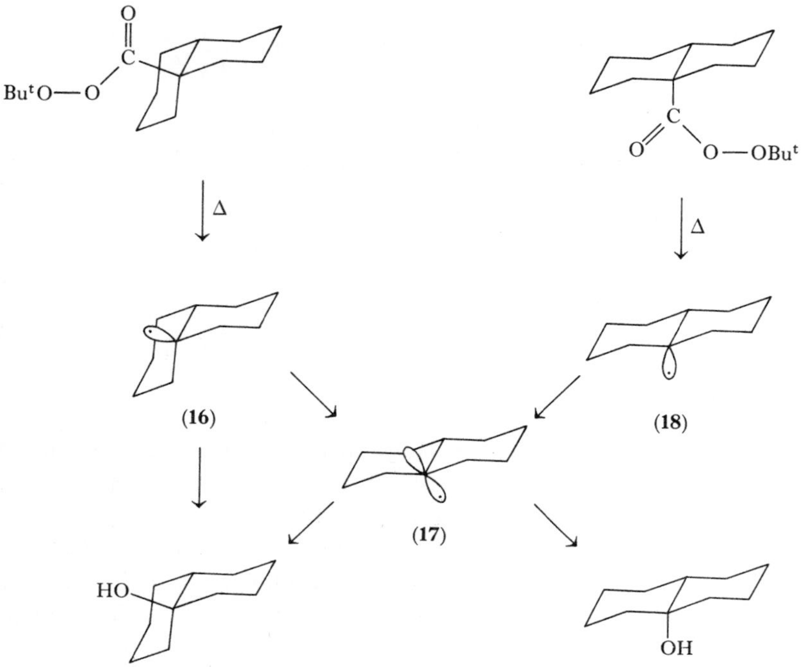

Scheme 6

4.5 *BRIDGEHEAD RADICALS*

Information about the favoured geometry at a radical centre can also be obtained from consideration of the ease of formation and reactivity of bridgehead radicals. Were the preferred radical conformation to be rigorously planar, then it would be expected that bridgehead radicals would be formed only with great difficulty, analogous to the situation for bridgehead carbonium ions. It would also be expected that the strained bridgehead radicals would be very highly reactive and hence unselective in their reactions (14). The results, which will be discussed in greater detail in the following chapter, indicate that bridgehead radicals are considerably strained, though much less so than the corresponding carbonium ions. From this it is inferred that though the preferred shape of radicals is planar, deviation from planarity is less costly energetically than is the case for carbonium ions.

4.6 *THE 7-NORBORNENYL RADICAL*

The similarity in conformations of radicals and carbonium ions stimulated an interest in the possibility of the 7-norbornenyl radical (**19**) being non-classical, analogous to the well-authenticated non-classical

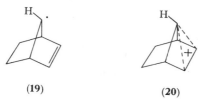

(19)   (20)

7-norbornenyl carbonium ion (**20**). The same e.s.r. spectrum was obtained for the 7-norbornenyl radical derived from photolysis of either *syn*- or *anti*-t-butylperoxy-2-norbornene-7-carboxylate (15). The assignments for the hyperfine splitting constants (Fig. 4-2) are based on

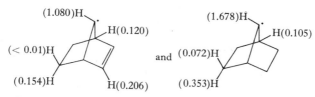

Fig. 4-2. Hyperfine splitting constants for the 7-norbornenyl and 7-norbornyl radicals.

analogy with the spectrum of the norbornyl radical. If the radical were significantly non-classical in structure this should manifest itself in removal of electron density from the 7-position to the 2- and 3-positions. One would thus expect that the hyperfine splittings associated with these protons would be much greater in the non-classical norbornenyl radical than in the norbornyl radical. In fact the hyperfine splitting for the 7-norbornenyl radical is close to the mean of the values for the splittings associated with the *endo*- and *exo*-protons in the norbornyl radical. This can be taken as support for a classical structure for the radical. A detailed analysis of the spectrum together with comparisons with simulated spectra indicates that the radical centre is bent and that the 7-carbon atom is displaced somewhat towards the double bond as shown (Fig. 4-3).

Chemical studies on the 7-norbornenyl and related radicals correlate well with the e.s.r. study (16). Thus reaction of both *syn*- and *anti*-7-

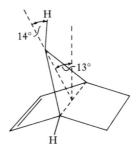

Fig. 4-3. Structure of the 7-norbornenyl radical.

bromonorbornene (**21** and **22**) with tri-n-butyltin deuteride give a 1:4.9 mixture of *syn-* and *anti*-7-deuterionorbornene (**23** and **24**) (reaction 6).

$$\text{(21)} \quad \xrightarrow{\qquad} \quad \xrightarrow{\qquad} \quad \text{(23)} \atop + \quad \text{(24)} \qquad (6)$$

The stereoselectivity arises as a result of the 7-hydrogen lying closer to the C_2–C_3 bond than to the C_5–C_6 bond in consequence of non-bonded interactions, as is indicated from the e.s.r. results. If the radical were non-classical one might expect exclusive formation of the *anti*-7-deuterionorbornene. The selectivity is even less in the case of reaction of *syn*-7-bromo-*anti*-7-methylnorbornene with tri-n-butyltin hydride which gives a nearly 1:1 mixture of 7-*syn*- and 7-*anti*-methylnorbornenes.

4.7 *ALLYLIC RADICALS*

Allylic radicals like allylic carbonium ions show configurational stability. This has been shown by Kochi and Krusic (15) from a study of the e.s.r. spectra of *cis-* and *trans*-1-methylallyl radicals (**25** and **26**) generated from *cis*- and *trans* but-2-enes. The spectra of these radicals were quite distinct and there was no evidence for any isomerization of the radicals over a temperature range of -130 to $0\,°C$. From these results it was possible to conclude that the rate of isomerization of these allylic radicals was lower than $10^2\ \mathrm{s}^{-1}$.

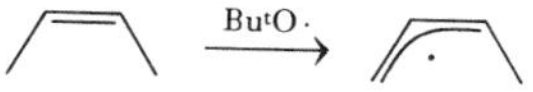

(25)

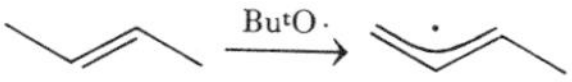

(26)

The delocalized nature of the allylic radical is disrupted by extreme substitution. Thus the e.s.r. spectrum of the 1,1,3,3-tetramethylallyl radical (27) consists of two septets ($a = 2.31\ \mathrm{mT}$). The large value of the proton splitting ($3.49\ \mathrm{mT}$) indicates that the central hydrogen must

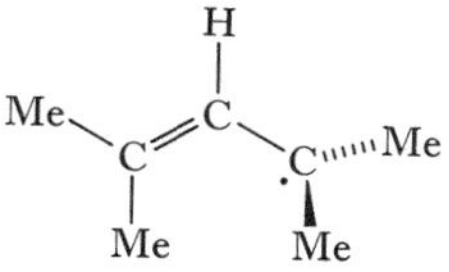

(27)

be close to the anti-node of the orbital containing the unpaired electron. Models indicate that steric interaction of the methyl groups precludes the carbon atoms being coplanar. The same situation also arises with the analogous carbonium ions. It is particularly interesting that 2,4-dimethyl-pent-2-ene gives none of this radical but only the isomeric *cis*- and *trans*-2,4-dimethylpent-2-en-4-yl radicals (29 and 30) on reaction with t-butoxy radicals. This is consistent with the absence of allylic stabiliza-tion in the radical (28).

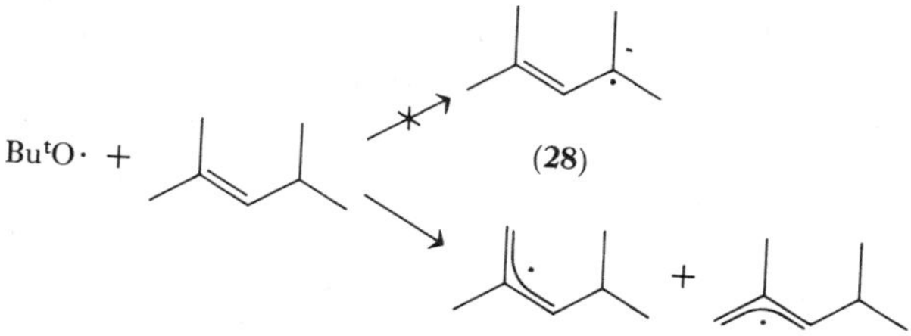

(28)

(29)   (30)

4.8 *VINYLIC RADICALS*

Vinylic radicals could adopt either a bent or linear structure (**31** and **32**) according to whether the carbon atom carrying the unpaired electron

(31)   (32)

is sp^2- or sp-hybridized, i.e. the unpaired electron is in an sp^2- or p-orbital. In the case of the bent configuration there is also the possibility of inversion of the two isomeric radical species (**33** and **34**).

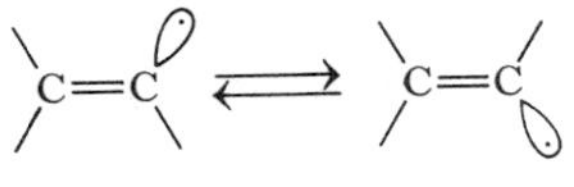

(33)   (34)

Detailed analysis of the e.s.r. spectra of vinyl radicals is consistent with a bent formulation with rapid inversion of its configuration even at $-180°$ (18). The barrier to inversion has been estimated to be about 8 kJ mol^{-1} and the life-time of the radical between 3×10^{-10} and 3×10^{-8} s. The e.s.r. spectrum of the 1-methylvinyl radical (**35**) revealed two different β-proton splittings, indicative of two non-equivalent

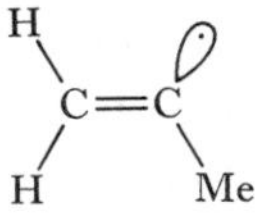

(35)

protons. This supports the formulation of a non-linear structure which is configurationally stable at the temperature of the study ($-172°$C). The greater size of the methyl group with respect to hydrogen is credited with causing this decreased rate of inversion.

Chemical studies on transfer reactions of substituted vinyl radicals have also been used to give evidence about their structure (19). Decomposition of both *cis*- and *trans*-dicinnamoyl peroxide in carbon tetrachloride gave the same mixture of *cis*- and *trans*-β-chlorostyrenes (20) (reactions 7 and 8). These results are interpretable on the basis of either a linear vinyl radical or a rapidly inverting bent radical. It is possible to distinguish between these processes by studying the decompositions of the isomeric peroxides in a more efficient transfer agent such as bromo-trichloromethane. In this solvent the ratio of *cis-* and *trans-α-bromo-*

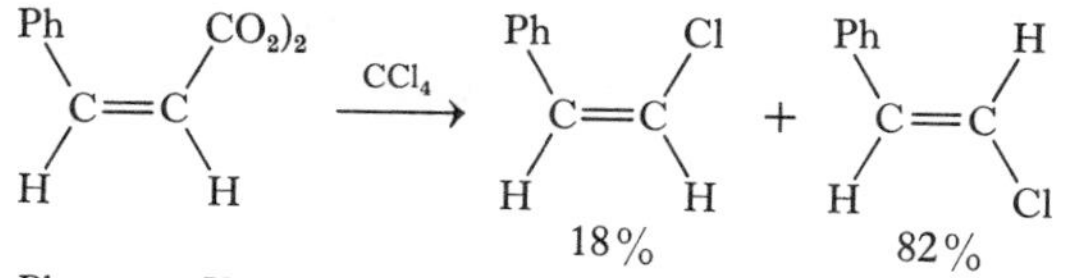

$$(7)$$

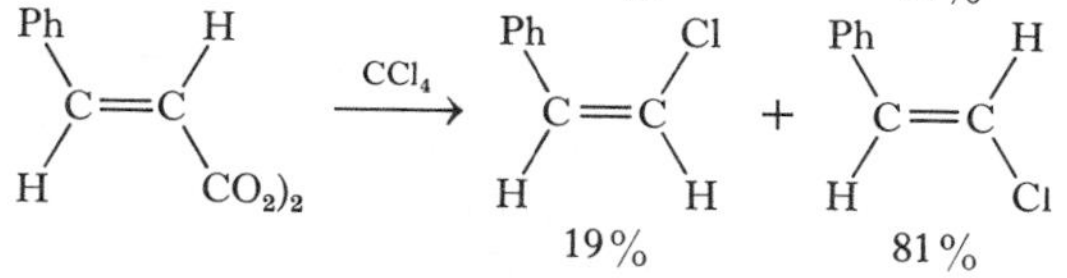

$$(8)$$

styrenes was dependent on the stereochemistry of the initial peroxide (reactions *9* and *10*). This can only be explained by postulating that reaction of the bent radical with bromotrichloromethane competes with inversion of the radical (reactions *11* and *12*).

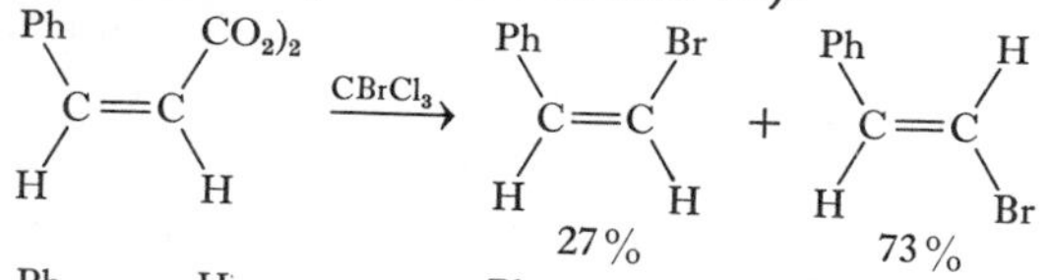

$$(9)$$

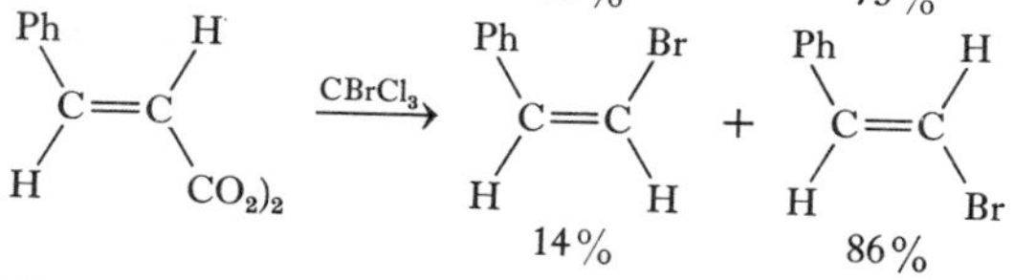

$$(10)$$

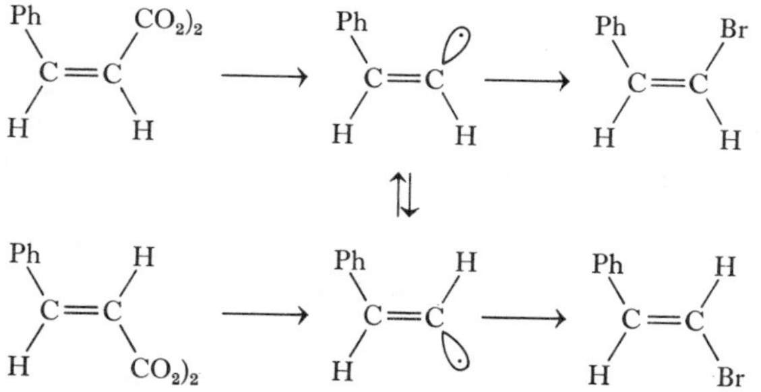

$$(11)$$

$$(12)$$

The size of the group attached to the α-carbon markedly affects the rate of inversion of vinylic radicals. Thus, the decomposition of t-butyl *cis-* and *trans-α-bromopercinnamates* in cumene give different ratios of *cis-* and *trans-α-bromostyrenes*, whereas under the same conditions both isomeric t-butyl α-methylpercinnamates give the same mixture of the isomeric α-methylstyrenes.

Shapes of radicals

The possibility of vinylic radicals adopting a linear configuration is clearly greatest in α-phenylvinyl radicals (36), in which the unpaired

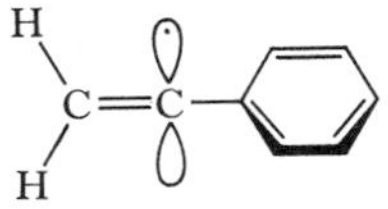

(36)

electron would be in an orbital capable of overlapping with the π-electron system of the phenyl group. Molecular orbital calculations indicate that for the unsubstituted vinyl radical the H–C–C bond angle is 140–150° in the most stable configuration, whilst for the 1-vinylvinyl radical the most stable arrangement is a linear structure. The 1-vinylvinyl radical can be considered as a reasonable model for the 1-phenylvinyl radical. From a study of the influence of the bulk of the solvent on the *cis/trans* ratio of propenylbenzenes derived from β-methyl-α-phenylvinyl radicals, it has been claimed that these radicals are linear (22). 1,2-Diphenylvinyl radicals are also said to be planar on the same type of evidence. Studies on the decomposition of α-phenylpercinnamates in carbon tetrachloride, however, indicate that the radical is non-linear, since different ratios of the *cis-* and *trans*-chlorostilbenes are obtained from the two isomers.

4.9   SILYL RADICALS

Table 4-2 records the hyperfine splitting constants for a series of silyl radicals (15, 24). The high ^{29}Si splitting in the silyl radical, $\dot{\text{SiH}}_3$, indicates that the unpaired electron is in an orbital of appreciable s-character, i.e. the radical is non-planar. The degree of s-character has

TABLE 4-2   *Hyperfine splitting constants (mT) of some silyl radicals*
(16, 24)

Radical	Coupling constants (mT)		
	a (α-H)	a (β-H)	a (^{29}Si)
$\text{Me}_3\text{Si}\cdot$	—	0.634	18.1
$\text{Me}_2\dot{\text{Si}}\text{H}$	1.729	0.730	18.3
$\text{Me}\dot{\text{Si}}\text{H}_2$	1.211	0.821	—
$\dot{\text{Si}}\text{H}_3$	0.784	—	26.6
$\text{Me}_3\text{Si}\dot{\text{Si}}\text{Me}_2$	—	0.821	13.7

been estimated to be 22%. As hydrogens are successively replaced by methyl groups, the ^{29}Si splitting falls, indicative of the radicals becoming more planar. In a planar radical the α-proton splitting is negative in sign, and this increases with increasing bending of the radical. The trend of the α-proton splittings observed for the series of radicals is reasonable if the coupling constants for the dimethylsilyl and methylsilyl radicals are negative, though that of the parent silyl radical may be positive. The small β-proton splittings of the silyl radicals, as compared with their carbon analogues, is attributed to the decrease in planarity of the radical and to decreased hyperconjugation with the larger silicon.

The ^{29}Si splitting for $Me_3SiSiMe_2$ is substantially less than that for $Me_3Si\cdot$, consistent with it being more planar. This may be due to conjugative interaction of the unpaired electron with the β-silicon, or electron withdrawal by the silicon. Electronegative groups would reduce the charge density on the silicon, which would result in an increased tendency for the unpaired electron to go into a lower energy orbital, i.e. one with more s-character. A similar explanation can be invoked for difference in conformation of the methyl and trifluoromethyl radicals.

Chemical studies (25, 26) on silyl radicals generated in asymmetric silicon compounds indicate that such radicals react with retention of configuration (reactions *13* and *14*).

$$\underset{Np^\alpha}{\overset{Ph\,Me}{Si}}\!\!-H \xrightarrow{R\cdot} \underset{Np^\alpha}{\overset{Ph\,Me}{Si}}\cdot \xrightarrow{CCl_4} \underset{Np^\alpha}{\overset{Ph\,Me}{Si}}\!\!-Cl \qquad (13)$$

$$\underset{Np^\alpha}{\overset{Me\,Ph}{Si}}\!\!-COCH_3 \xrightarrow{h\nu} \underset{Np^\alpha}{\overset{Me\,Ph}{Si}}\cdot \xrightarrow{CCl_4} \underset{Np^\alpha}{\overset{Me\,Ph}{Si}}\!\!-Cl \qquad (14)$$

4.10 β-SUBSTITUTED ALKYL RADICALS

The e.s.r. spectra of alkyl radicals substituted in the β-position by Group IV elements, e.g. $H_3SiCH_2CH_2\cdot$, are characterized by anomalously low β-proton splittings and exceptionally large γ-proton splittings (see Table 4-3) (15). The β-proton splittings are also temperature dependent, indicative of a barrier to rotation about the C_α–C_β bond. The reduced β-proton splittings are consistent with the most stable conformation (**37**) being that in which the group IV element is eclipsed with the

93

TABLE 4-3   *Hyperfine splitting constants for β-substituted ethyl radicals, $R_3MCH_2CH_2\cdot$ (16)*

Radical	Coupling constants (mT)		
	$a\ (\alpha\text{-H})$	$a\ (\beta\text{-H})$	$a\ (\gamma\text{-H})$
$H_3C—CH_2CH_2\cdot$	2.214	3.033	0.027
$H_3Si—CH_2CH_2\cdot$	2.139	1.768	0.277
$H_3Ge—CH_2CH_2\cdot$	2.096	1.584	0.432

p-orbital carrying the unpaired electrons. This is probably as a consequence of interaction between the *p*-orbital and the *d*-orbital of the group IV element.

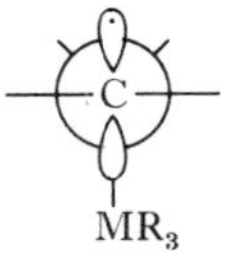

(37)

β-Thioalkyl radicals similarly have relatively small *β*-proton splittings which are temperature dependent (15). This is again explained on the basis of interaction of the partially filled *p*-orbital and the unfilled 3*d*-orbital of the sulphur. This results in the most stable conformation of the radical being that in which the *β*-sulphur atom is eclipsed with the *p*-orbital of the radical centre (**38**). This interaction, while not being sufficient to cause bridging of the sulphur rendering the *α*- and *β*-protons completely equivalent, could lead to stereoselective control of

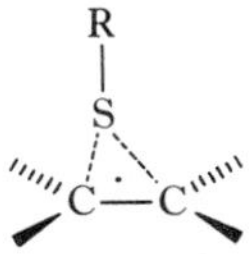

(**38**)

reactions involving this type of radical. This is commonly encountered in the *trans*-addition of thiols to alkenes.

Chemical evidence (26*a*) supporting the concept of bridged *β*-bromo and *β*-thioalkyl radicals has been obtained from a study of the peroxide-induced decomposition of bromohydrins (X = Br) and *β*-hydroxysulphides (X = RS) (Scheme 7). The relative rates of hydrogen abstraction

94

from *threo-* and *erythro*-3-bromo-2-butanols and -3-methylthio-2-butanols should be dependent on the relative stabilities of the radicals

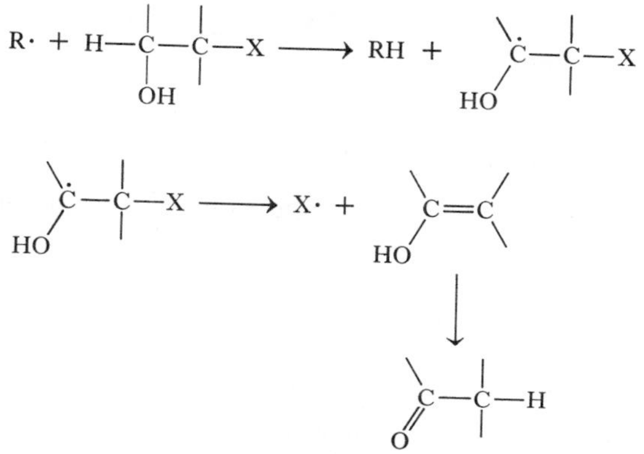

Scheme 7

obtained from the two isomers. If bridging occurs in the transition state leading to the intermediate radical, then one would predict that the *threo*-isomer (**39**) would react more rapidly than the *erythro*-isomer (**41**), on the grounds that in the bridged radical from the latter (**42**), but not from the former (**40**), there is a gauche interaction of two methyl groups.

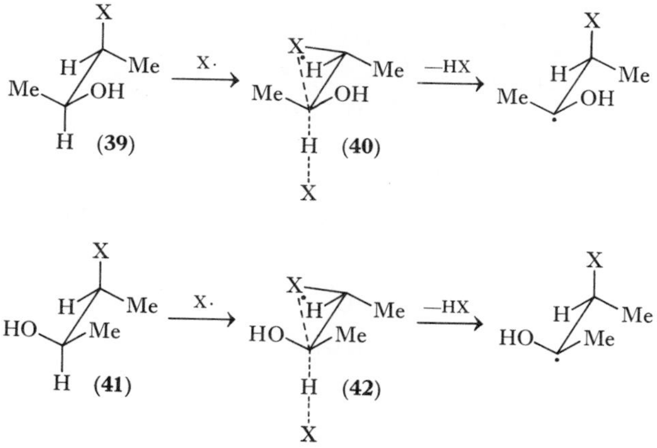

The relative rates of these reactions indicate that bridging is important when the bridging atom is bromine or sulphur but is barely significant in the case of chlorine. Studies on the bromination of (+)-1-bromo-2-methylbutane and (+)-1-chloro-2-methylbutane indicate that in both cases reaction proceeds through a bridged radical and that this is more important in the case of a bridging bromine (5).

Substitution of oxygen or nitrogen at the β-position of an alkyl radical does not produce the same type of effect as is found with sulphur and second-period elements.

4.11   NITROGEN RADICALS

The nitrogen splitting in dialkylamino radicals is relatively small, and is consistent with the unpaired electron being in a p-orbital rather than an s-containing orbital (27).

4.12   PHOSPHORUS RADICALS

Attempts to observe the e.s.r. spectra of phosphorus radicals have not yet been successful. Chemical studies have been used, and these point to the radicals being non-planar. Reaction of the optically active hydrogen phosphinate (43) with thiophenyl radicals proceeds with retention of configuration at the phosphorus (Scheme 8) (28). Addition of the phosphinyl radical to 1-heptene similarly proceeds with retention of configuration. These results indicate that the phosphinyl radical is configurationally stable.

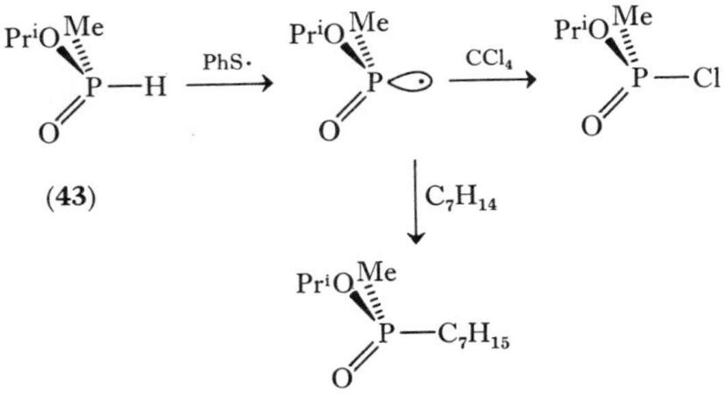

Scheme 8

4.13   CONCLUSIONS

There is ample evidence in the foregoing discussion of the structures of a range of radicals which shows that, in general, free radicals have the same shapes as the analogous carbonium ions. Thus, alkyl radicals preferentially adopt a planar configuration in which there is sp^2-hybridization at the radical centre. Radicals do, however, differ from carbonium ions in that distortion from a planar configuration is very much more readily accommodated as shown by the relative ease of formation of bridgehead radicals (p. 107).

The biggest difference between radicals and carbonium ions is that the latter, but not the former, can be non-classical. There is no indication of any non-classical behaviour of 7-norbornenyl or cyclopropylmethyl radicals, as might have been expected by analogy with the corresponding carbonium ions. The same is true of 3,5-cyclocholestan-6-yl radicals (44) (29) and of 2-phenylethyl radicals (45) (30). If the 7-norbornenyl radial were non-classical it would be

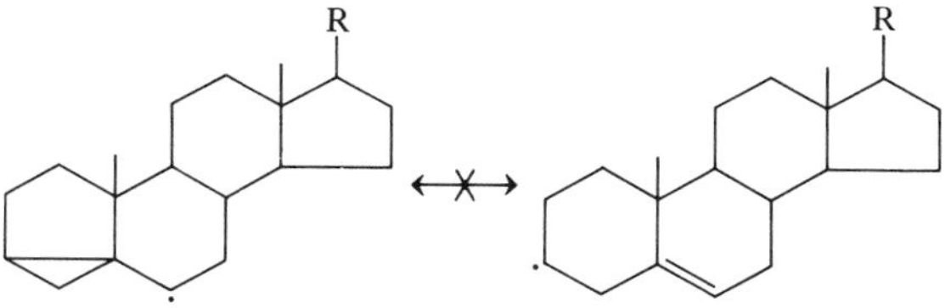

(44)

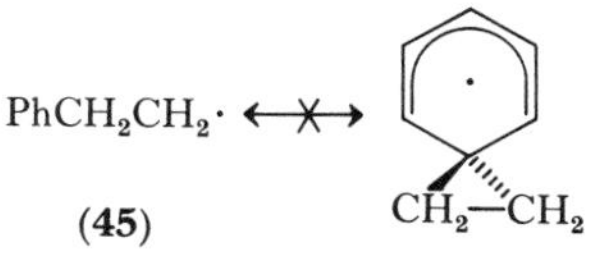

$$PhCH_2CH_2\cdot \longleftarrow X \longrightarrow$$

(45) CH_2-CH_2

necessary to place three electrons in the available molecular orbitals. The extra electron in the radical compared to the carbonium ion has perforce to go into a higher-energy orbital than is the case for the carbonium ion. The resultant situation is thus of higher energy than the corresponding carbonium ion and apparently of higher energy than the classical radical in which the orbital containing the unpaired electron does not overlap with the π-electron system.

REFERENCES

1. R. W. Fessenden and R. H. Schuler, *J. Chem. Phys.* 1965, **43**, 2704.
2. M. C. R. Symons, *Nature*, 1969, **222**, 1123.
3. R. O. C. Norman, *Chem. in Britain*, 1970, **6**, 66; A. J. Dobbs, B. C. Gilbert and R. O. C. Norman, *J. Chem. Soc.* A, 1971, 124.
4. H. C. Brown, M. S. Kharasch and T. H. Chao, *J. Amer. Chem. Soc.* 1940, **62**, 3439.
5. P. S. Skell, D. L. Tuleen and P. D. Readio, *J. Amer. Chem. Soc.* 1963, **85**, 2849.

6. W. von Doering, M. Farber, M. Sprecher and K. B. Wiberg, *J. Amer. Chem. Soc.* 1952, **74**, 3000.
7. D. F. DeTar and C. Weis, *J. Amer. Chem. Soc.* 1957, **79**, 3045.
8. S. Ogawa and R. W. Fessenden, *J. Chem. Phys.* 1964, **41**, 994; R. W. Fessenden, *J. Phys. Chem.* 1967, **71**, 74.
9. F. R. Jensen, L. H. Gale and J. E. Rodgers, *J. Amer. Chem. Soc.* 1968, **90**, 5793.
10. M. J. S. Dewar and T. M. Harris, *J. Amer. Chem. Soc.* 1969, **91**, 3652.
11. J. Jacobus and D. Pensak, *Chem. Comm.* 1969, 400.
12. T. Ando, F. Namingata, H. Yamanaka and W. Funasaka, *J. Amer. Chem. Soc.* 1967, **89**, 5719.
13. P. D. Bartlett, R. E. Pincock, J. H. Rolston, W. G. Schindel and L. A. Singer, *J. Amer. Chem. Soc.* 1965, **87**, 2590.
14. R. C. Fort and P. von R. Schleyer, *Adv. Alicyclic Chem.* 1966, **1**, 337.
15. J. K. Kochi and P. J. Krusic, *Essays in Free-Radical Chemistry, Chem. Soc. Special Publ.* No. 24, 1970, 147.
16. G. A. Russell and G. W. Holland, *J. Amer. Chem. Soc.* 1969, **91**, 3968.
17. G. A. Russell and G. W. Holland, *J. Amer. Chem. Soc.* 1969, **91**, 3969.
18. R. W. Fessenden and R. H. Schuler, *J. Chem. Phys.* 1963, **39**, 2147.
19. O. Simamura in *Topics in Stereochemistry*, ed. N. L. Allinger and E. L. Eliel, Wiley, New York, 1969, vol. 4, p. 1.
20. O. Simamura, K. Tokumaru and H. Yui, *Tetrahedron Letters*, 1966, 5141.
21. L. A. Singer and N. P. Kong, *Tetrahedron Letters*, 1966, 2089; 1967, 643.
22. L. A. Singer and J. Chen, *Tetrahedron Letters*, 1969, 4849.
23. R. M. Kopchik and J. A. Kampmeier, *J. Amer. Chem. Soc.* 1968, **90**, 6733.
24. S. W. Bennett, C. Eaborn, A. Hudson, R. A. Jackson and K. D. J. Root, *J. Chem. Soc.* A, 1970, 348.
25. A. G. Brook and J. M. Duff, *J. Amer. Chem. Soc.* 1969, **91**, 2118.
26. H. Sakurai, M. Murakami and M. Kumada, *J. Amer. Chem. Soc.* 1969, **91**, 511.
26a. E. S. Huyser and R. H. C. Feng, *J. Org. Chem.* 1971, **36**, 731.
27. W. C. Danen and T. T. Kensler, *J. Amer. Chem. Soc.* 1970, **92**, 5235.
28. L. P. Rieff and H. S. Aaron, *J. Amer. Chem. Soc.* 1970, **92**, 5275.
29. S. J. Cristol and R. V. Barbour, *J. Amer. Chem. Soc.* 1968, **90**, 2832.
30. J. K. Kochi and P. J. Krusic, *J. Amer. Chem. Soc.* 1969, **91**, 3938.
31. D. E. Milligan and M. E. Jacox, *J. Chem. Phys.* 1967, **47**, 5146.
32. G. Herzberg, *Proc. Roy. Soc.* A, 1961, **262**, 291.

5

Stabilities of free radicals

5.1 *INTRODUCTION*

A variety of methods have been employed to obtain information about the relative stabilities of free radicals. This is relatively easily accomplished for stable radicals, i.e. those with half-lives sufficiently long for them to be studied by normal spectroscopic methods. Their stability can be determined by measurement of their half-lives or, with very stable radicals, by the extent of dissociation of the radical dimer.

The relative stabilities of transient radicals can be adduced from consideration of the relative ease of formation of radicals in a variety of ways, some of which we will consider. One of the most widely used procedures is to consider the relative rates of decomposition of radical precursors, such as diacyl peroxides, peresters and azo compounds:

$$(RCO_2)_2 \longrightarrow 2R\cdot + 2CO_2 \tag{1}$$

$$RCO_3R' \longrightarrow R\cdot + CO_2 + R'O\cdot \tag{2}$$

$$RN{=}NR \longrightarrow 2R\cdot + N_2 \tag{3}$$

The rate of decomposition of a radical precursor can only give a valid indication of radical stability if the transition state for the reaction resembles the incipient radical. This will only be a valid assumption in the case of diacyl peroxides and peresters if the C–CO bond is significantly stretched in the transition state. This is not the case for diaroyl peroxides and peresters (cf. chapter 2). A second objection to the use of the rate of decomposition of diacyl peroxides and peresters as a measure of radical stability is that the transition state has significant polar character (cf. **1**) (1). Consequently the rate of decomposition will reflect not

$$\overset{\text{O}}{\underset{\overset{\|}{\underset{\delta+}{\text{R}----\text{C}--\text{O}----\text{OR}'}}}{}} \overset{}{\underset{\delta-}{}}$$

(1)

only the stability of the radical but also the carbonium ion. These

Stability of radicals

objections do not apply to the same extent in the decomposition of azoalkanes, and consequently their relative rates of decomposition provide a much truer assessment of the stability of the incipient radical (2).

The orientation of radical addition to alkenes has frequently been used as a measure of radical stability:

$$RCH{=}CH_2 + X\cdot \underset{\searrow}{\overset{\nearrow}{\times}} \begin{array}{l} RCHXCH_2\cdot \\ \\ R\dot{C}HCH_2X \end{array} \qquad (4)$$

As discussed in chapter 7, this is in fact more dependent on steric and polar factors than on the stability of the adduct radical.

That free-radical substitution occurs at a tertiary hydrogen rather than at a secondary or primary hydrogen gives support to the concept that a tertiary radical is more stable than a secondary or primary radical. This is an over-simplified view, since the ease of reaction is also dependent on the polar nature of the transition state.

A further method which has been extensively used is based on the fact that some radicals, e.g. acyl and α-alkoxybenzyl radicals, can undergo fragmentation to alkyl radicals as well as reaction with solvent or dimerization (reactions 5 and 6) (3, 4) (cf. chapter 12, pp. 483 and 486):

$$R\dot{C}O \underset{k_2}{\overset{k_1}{\diagdown}}\, \begin{array}{l} R\cdot\ +\ CO \\ \\ \overset{CCl_4}{} RCOCl\ +\ \dot{C}Cl_3 \end{array} \qquad (5)$$

$$Ph\dot{C}HOR \underset{k_2}{\overset{k_1}{\diagdown}}\, \begin{array}{l} R\cdot\ +\ PhCHO \\ \\ PhCHOR \\ | \\ Ph\dot{C}HOR \end{array} \qquad (6)$$

The relative importance of the competing pathways, i.e. the value of k_1/k_2, has been related to radical stability. Once again it is necessary to consider the importance of polar and steric factors in the transition state.

t-Alkoxy radicals undergo fragmentation by competing pathways, the relative importance of which has been used as a measure of radical stability (5, 6) (cf. chapter 12, p. 482):

$$R-\underset{\underset{R^2}{|}}{\overset{\overset{R^1}{|}}{C}}-O\cdot \quad
\begin{cases}
\longrightarrow R\cdot \;+\; R^1COR^2 \\
\longrightarrow R^1\cdot \;+\; RCOR^2 \\
\longrightarrow R^2\cdot \;+\; RCOR^1
\end{cases} \qquad (7)$$

Closely related studies on the fragmentation of α,α-dialkoxyalkyl radicals (reaction *8*) indicate qualitatively the same relative order of stabilities. Quantitatively a much greater selectivity in the mode of fragmentation is observed with the t-alkoxy radical because of the more polar nature of the transition state.

$$CH_3\overset{\overset{\displaystyle OR}{\diagup}}{\underset{\underset{\displaystyle OR'}{\diagdown}}{C}}
\begin{array}{l}
\nearrow R\cdot \;+\; CH_3CO_2R' \\[2ex]
\searrow R'\cdot \;+\; CH_3CO_2R
\end{array} \qquad (8)$$

The other procedure used for determination of stabilities of transient free radicals is based on the measurement of bond dissociation energies. The bond dissociation energy is the energy required for the homolysis of a bond. Bond dissociation does not proceed via a transition state with significant energy maxima, since the reverse reaction involves the dimerization of free radicals, consequently any stabilization of a radical should be reflected in a lower bond dissociation energy. It is, however, necessary to consider that on homolysis of a C–X bond, the hybridization of the carbon atom will be changed from sp^3 to sp^2. The steric situation in the radical is thus different from that in the ground state. Release of steric strain will thus accompany homolysis of the bond, and the bond dissociation energy will thus include a component due to this release of strain (2).

In spite of the above objections to the methods used for assessing stabilities of radicals, the order of stability is almost invariably independent of the method though there are quantitative differences.

5.2 *FACTORS CONTROLLING THE STABILITY OF RADICALS*

The stability of free radicals is influenced by the ability of the group attached to the radical centre to delocalize the unpaired electron as in the benzyl radical. Such effects diminish the tendency of the radical to undergo reactions by decreasing the spin density at the radical centre.

Steric factors are also very important in influencing the stability of radicals because of hindrance in bimolecular reactions of the radical which would lead to disproportionation or dimerization. This is well exemplified in the triphenylmethyl radical, which is very much more stable than the benzyl radical even though there is not a great difference in the spin density at the radical centre. The stability of this radical is due largely to steric interaction in the dimerization reaction. The tri(*o*-tolyl)methyl radical is much more stable because of the even greater steric hindrance to dimerization (10). The stability of a radical is also dependent on its environment. This is particularly important in considering the role of steric factors, since these will be much more important for the dimerization or disproportionation of radicals than for their reaction with small molecules, e.g. oxygen.

A third factor, which is important with carbon radicals, is the ability of the radical centre to attain a planar conformation. Deviations from planarity usually decrease the stability of a radical. These considerations will be amplified in the discussion of particular radicals, both for transient and stable radicals. No attempt has been made to give a complete coverage of all types of stable radical, and the examples which are discussed have been chosen to illustrate the general principles governing radical stabilities.

5.3 *STABILITY OF TRANSIENT RADICALS*

5.3.1 *Alkyl and substituted alkyl radicals*

The order of stability of alkyl radicals is:

tertiary > secondary > primary.

This is indicated from the bond dissociation energies of primary, secondary and tertiary C–H bonds (Table 5-1), from the relative rates of decomposition of azoalkanes (Table 5-2) and from the relative reactivities of primary, secondary and tertiary hydrogens towards radicals (Table 7-7). That all these methods and others (cf. Tables 12-1, 12-2 and 12-3) indicate the same relative order of radical stability gives one confidence in the results despite doubts which may be felt about any one of the methods used. The ability of alkyl groups to stabilize radicals is due to hyperconjugation. This is supported by e.s.r. studies which show that alkyl groups withdraw spin from a radical centre (cf. chapter 3, p. 43). The difference in the relative stabilities of primary, secondary and tertiary radicals is, however, very much less than that of the corresponding carbonium ions.

TABLE 5-1　*Bond dissociation energies of C–H bonds in different environments* (8)

R—H	Bond dissociation energy (kJ mol^{-1})
Ph—H	431
CH$_3$—H	434
C$_2$H$_5$—H	410
Me$_2$CH—H	395
Me$_3$C—H	381
CH$_2$=CHCH$_2$—H	356
PhCH$_2$—H	356
Ph$_3$C—H	314
HOCH$_2$—H	389
F$_3$C—H	443
Cl$_3$C—H	402

TABLE 5-2　*Relative rates of decomposition of azoalkanes, RN=NR, at 300 °C* (10)

R	Relative rate	ΔE (kJ mol^{-1})
Me	1	220
Et	5	209
Pr^i	60	199
Bu^t	130	184
PhCH$_2$	700	157

Replacement of a hydrogen by a phenyl group results in a great increase in the stability of the radical arising from delocalization of the unpaired electron. E.s.r. measurements indicate that the spin density on the methyl carbon has decreased to about 0.7 in the benzyl radical (2) (11).

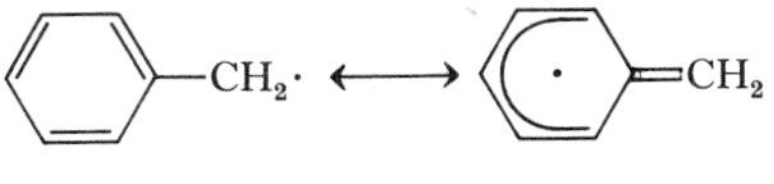

(2)

Cyclopropyl groups have also been shown to exert a considerable stabilizing effect by a resonance interaction between the cyclopropyl group and the radical centres. Thus, the cyclopropylmethyl radical is much more stable than the methyl radical. Evidence for this is based on the accelerating effect that cyclopropyl groups have on the rate of decomposition of azomethanes. The 9:1 ratio of the relative rates of decomposition of the azo compounds (**3** and **4**) shows that this is not a steric effect (12). Cyclopropyl and isopropyl groups should exert similar

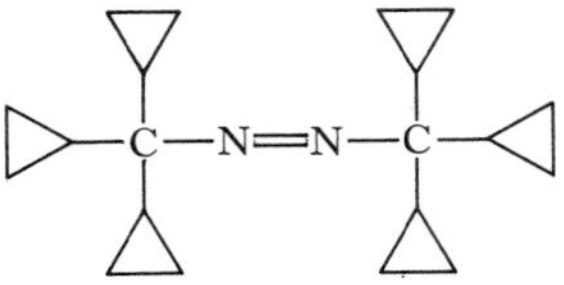

(3) 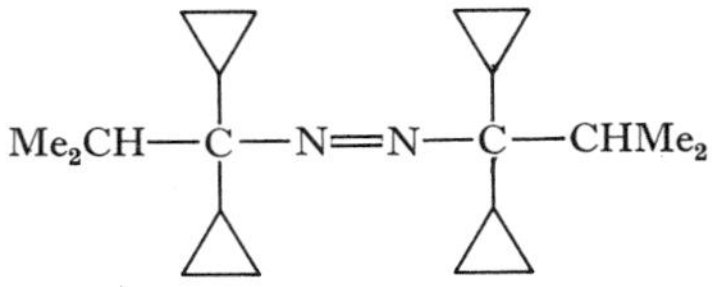   (4)

steric effects. There is ample evidence for the stabilizing effects of cyclopropyl groups from other experiments to support these conclusions (7, 13).

The apparent relative stabilities of allyl and propargyl radicals vary according to the particular reaction studied. Thus, methylacetylene undergoes hydrogen abstraction by methyl radicals more readily than does propene (14), whereas the reverse is true for abstraction by t-butoxy radicals (15). This difference in behaviour can probably be related to the different polar character of methyl and t-butoxy radicals. Measurements of bond dissociation energies (24) and of the rates of decomposition of peresters (Table 5-3) indicate that the allyl radical is somewhat more stable than the propargyl radical (17). In the former the two canonical structures have identical energies whereas this is not true for the latter:

$$CH_2=CH-CH_2 \cdot \longleftrightarrow \dot{C}H_2-CH=CH_2$$
$$HC\equiv C-CH_2 \cdot \longleftrightarrow H\dot{C}=C=CH_2$$

TABLE 5-3 *Relative rates of decomposition of allylic and propargylic peresters, RCO_3Bu^t, at 60 °C (17)*

R	Relative rate	ΔE^* (kJ mol^{-1})	ΔS^* (J K^{-1} mol^{-1})
$CH_3C\equiv CCH_2$	1.0	125	25
$CH_3CH=CHCH_2$	7.8	108	-10
$PhC\equiv CCH_2$	2.1	121	19
$PhCH=CHCH_2$	30.7	98	-25

Formation of the allyl radical involves a negative entropy of activation because of the restriction in rotation about the C–C bond imposed on the radical. This effect is not felt by the cylindrically symmetrical propargyl radical. Consonant with this a γ-phenyl group has a greater stabilizing effect on an allyl than a propargyl radical. A similar difference operates in the case of carbonium ions (18).

E.s.r. measurements have indicated that both $-R$ and $+R$ substituents are effective in delocalizing spin:

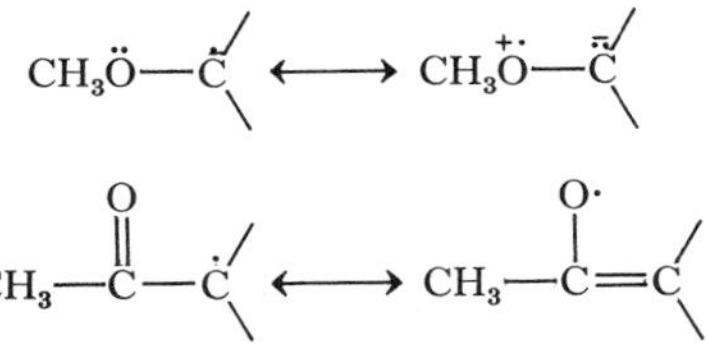

That these results have some bearing on radical stability is indicated from the greatly increased rate of decomposition of α-alkoxyalkylpercarboxylates compared with simple alkylpercarboxylates (Table 5-4), and from the ease of fragmentation of t-alkoxy radicals containing methoxymethyl groups into methoxymethyl radicals. The transition states in both these processes have significant polar character. This overemphasizes the stabilizing effect of α-alkoxy groups on the stability of the radical. That this is so is indicated from the small accelerating effect of α-alkoxy groups on the rate of decomposition of azomethane (19).

TABLE 5-4	*Relative rates of decomposition of peresters and azoalkanes leading to phenoxymethyl and thiophenoxymethyl radicals (21, 20)*

| | $XCH_2CO_3Bu^t$ | | $XCMe_2N{=}NCMe_2X$ | |
| | Relative rate at 70.5 °C | ΔE^* (kJ mol^{-1}) | Relative rate at 100 °C | ΔE^* (kJ mol^{-1}) |
X				
H	1.0	154	—	—
Me	—	—	1.0	151
PhO	4.5×10^3	113	50.0	139
PhS	4.14×10^3	105	6.6×10^3	124

α-Thioalkoxy groups have a much greater accelerating effect than alkoxy groups on the rates of decomposition of azo compounds (Table 5-4) on account of their having a greater stabilizing effect on the radical centre (20). This is because the unpaired electron can now be delocalized

Stability of radicals

in two ways, the second arises because of the ability of sulphur to undergo octet expansion:

$$CH_3\ddot{\ddot{S}}-\overset{/}{\underset{\backslash}{C}} \longleftrightarrow CH_3\overset{+:}{\ddot{S}}-\overset{\overset{\cdot\cdot}{}}{\underset{\backslash}{C}} \longleftrightarrow CH_3\dot{S}=\overset{/}{\underset{\backslash}{C}}$$

In a similar manner α-chlorine atoms are much more effective in causing stabilization of a radical than α-fluorine atoms:

$$\ddot{\ddot{C}}l-\overset{/}{\underset{\backslash}{C}} \longleftrightarrow \overset{+:}{\ddot{C}}l-\overset{\overset{\cdot\cdot}{}}{\underset{\backslash}{C}} \longleftrightarrow \dot{C}l=\overset{/}{\underset{\backslash}{C}}$$

$$\ddot{\ddot{F}}-\overset{/}{\underset{\backslash}{C}} \longleftrightarrow \overset{+:}{\ddot{F}}-\overset{\overset{\cdot\cdot}{}}{\underset{\backslash}{C}}$$

Polar effects have little, if any, effect in determining radical stabilities. This was demonstrated by Cadogan in a study of the addition of bromotrichloromethane to substituted stilbenes (22). Substituents in the 3-position, which could only exert an inductive effect, have very little influence on the position of addition of the trichloromethyl radical. The 4-nitro group has a rather larger effect. This favours addition of the radical to the carbon remote from the 4-nitrophenyl group since the unpaired electron in this radical can be delocalized by resonance interaction with the nitro group (cf. **5**).

(**5**)

5.3.2 Cycloalkyl radicals

The stability of cycloalkyl radicals is dependent on the strain in the radical compared to that in the reactant or the product, depending on the reaction being studied. Factors which influence strain include distortion of bond angles, bond opposition forces, and compression of Van der Waals radii. The relative stabilities of cycloalkyl radicals are thus determined by the same factors which control the stabilities of cycloalkyl carbonium ions. The expected order of stability of cycloalkyl radicals is thus:

cyclobutyl < cyclohexyl < cyclopentyl < cycloheptyl < cyclo-octyl.

The relative rates of decomposition of azocycloalkanes indicate that the relative stabilities of cycloalkyl radicals lie in this order (2). The rates of decomposition of t-butyl cycloalkylpercarboxylates are in a different order because there is little stretching of the C–CO bond in the transition state and hence different factors control their rates of thermolysis. A variety of other chemical studies, such as the photochlorination of cycloalkanes and the reaction of cycloalkanes with methyl radicals, also give the same order of stability (23, 24). The polar characteristics of methyl radicals and chlorine atoms are very different and the fact that both these reactions lead to the same order of radical stability gives confidence in the order of radical stability obtained.

5.3.3 Bridgehead radicals

Bridgehead radicals have been extensively studied because of the interest in the preferred geometry of radicals. Considerable work has thus been focused on the relative reactivity and ease of formation of bridgehead radicals. If these radicals demand a rigidly planar structure as is the case with carbonium ions then it is reasonable to argue that reactions proceeding via bridgehead radicals would proceed with considerable difficulty if at all. This arises as a result of the increase in strain as the bridgehead carbon changes from sp^3- to sp^2-hybridization (25).

There is considerable experimental evidence for the relative ease of formation of bridgehead radicals compared with the analogous bridgehead carbonium ions. This is based on the number of reactions which proceed through a bridgehead radical (reactions *9–13*) (26, 3, 5, 27, 28).

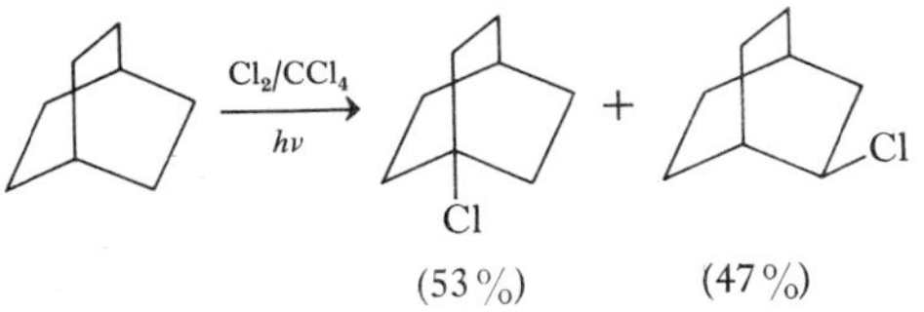

$$(9)$$

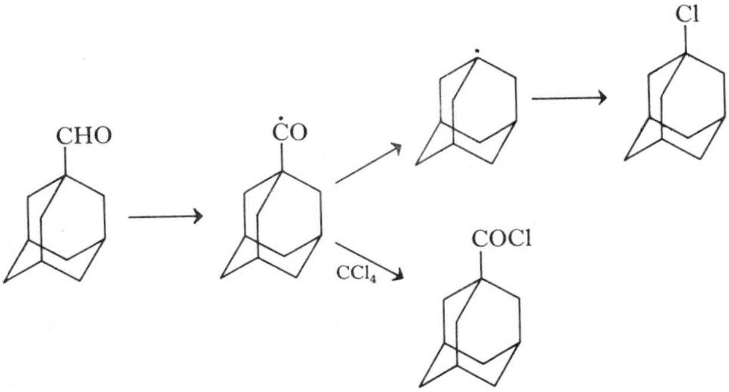

$$(10)$$

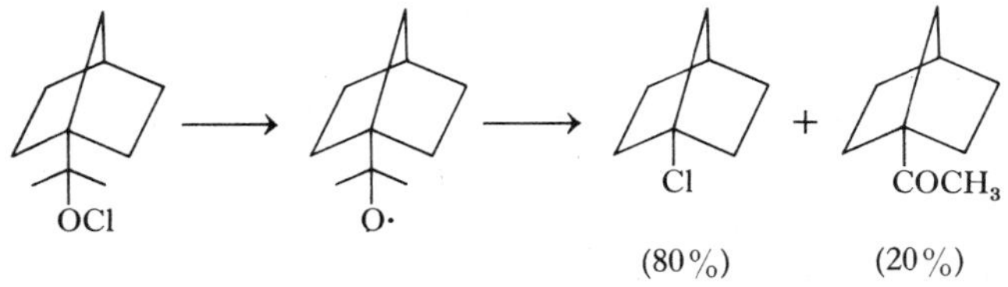

Stability of radicals

(11)

Structures showing OCl → O· → Cl (80%) + COCH₃ (20%)

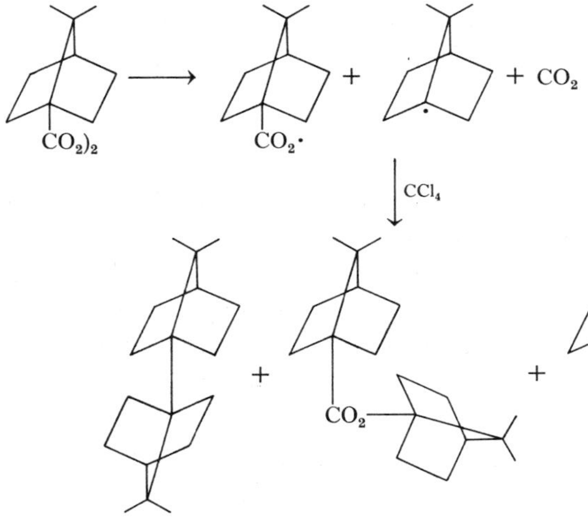

(12)

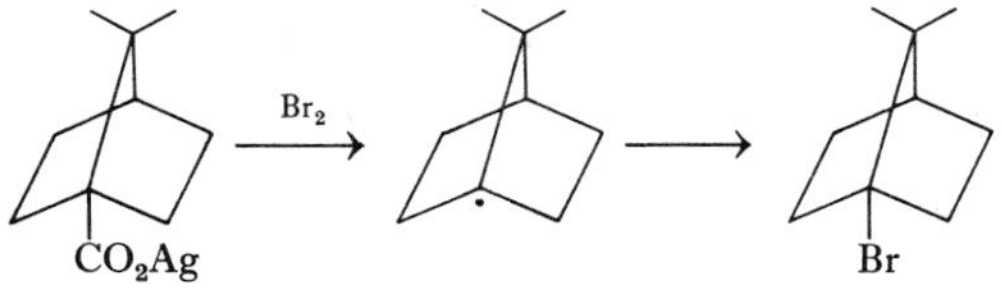

(13)

An idea of how strong a preference bridgehead radicals have for a planar structure can be obtained by comparing their stabilities with the stability of an acyclic tertiary radical such as the t-butyl radical. Table 5-5 indicates that the t-butyl radical is more stable than the 1-adamantyl radical and very much more stable than the more highly strained 1-norbornyl radical. The relative stabilities of t-butyl and bridgehead radicals depends markedly on the method used for their assessment, and undoubtedly reflect differences in the nature of the transition states leading to their formation. The results show quite definitely that bridgehead radicals are less unstable relative to the t-butyl radical than is the case with the analogous carbonium ions. This suggests that bridgehead

TABLE 5-5   *Relative reactivities of t-butyl and bridgehead radicals*

Radical	Relative rate of decomposition of RCO_3Bu^t (ref. 29)	Relative rate of decomposition of $RN{=}NR$ (ref. 2)	Ease of fragmentation of $R\dot{C}O$ k_1/k_2[b]	Ease of fragmentation of $Ph\dot{C}HOR$ [PhCHO][c]/ [(PhCHOR)$_2$]	Relative rate of solvolysis of RBr (ref. 25)
Bu^t	1.0	1.0	12.3	3.0	1.0
1-Adamantyl	1.48 (0.3)[a]	4.02×10^{-4}	30.5	0.42	10^{-3}
1-Bicyclo[2,2,2]-octyl	0.103 (0.03)[a]	5.06×10^{-5}	15.2	—	10^{-6}
1-Norbornyl	1.57×10^{-3}	1.98×10^{-6}	0.08	—	10^{-14}

[a] Values in parentheses are corrected values after consideration of differences in inductive effect.
[b] See reaction 5 (p. 100).
[c] See reaction 6 (p. 100).

radicals can tolerate a much greater degree of non-planarity than the corresponding carbonium ions. It is reasonable to expect the energy levels of sp^2- and sp^3-hybridized carbon to be much closer for radicals which have one non-bonding electron than for carbonium ions which have none. Consequently a non-planar radical centre can be formed relatively readily.

Information on the relative stability of bridgehead radicals can also be obtained by comparing their selectivity in competing reactions. Evidence here suggests that they are particularly indiscriminate in their reactions, indicative of strain in the radical which is relieved on undergoing reaction. This was observed in the reactions of bridgehead radicals with mixtures of carbon tetrachloride and bromotrichloromethane (2). It was also shown that the 1-bicyclo[2,2,2]octyl radical gives a mixture of 1-bromo- and 1-chlorobicyclo[2,2,2]octanes when the bridgehead carboxylic acid is subjected to Cristol's modification of the Hunsdiecker reaction (30) (reaction *14*):

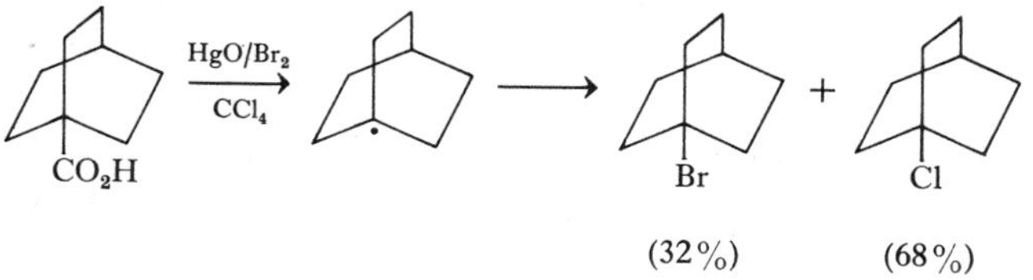

$$\text{(14)}$$

(32%)   (68%)

5.4 STABILITY OF STABLE RADICALS

5.4.1 *Triarylmethyl and related carbon radicals*

The best known of all stable radicals is undoubtedly the triphenylmethyl radical, which was discovered accidentally by Gomberg in 1900 whilst attempting to prepare hexaphenylethane (31). He found that the colourless hydrocarbon (6) which he believed to be hexaphenylethane (7) gave

$$Ph_3C \diagdown_H \diagup \bigcirc = CPh_2 \qquad\qquad Ph_3C\text{—}CPh_3$$

(6)   (7)

yellow solutions in a variety of solvents. The intensity of the colour increased on heating and was discharged by bubbling oxygen or nitric oxide into the solution. These solutions failed to obey Beer's law. This evidence indicated that free triphenylmethyl radicals were in equilibrium with their dimers. This was confirmed initially by molecular weight determinations and later by e.s.r. spectroscopy.

The e.s.r. spectra of triphenylmethyl radicals indicate that there is considerable delocalization of the unpaired electron onto the *ortho-* and *para*-positions of the phenyl rings (8). The extent of delocalization of the unpaired electron was not, however, very much greater for triphenylmethyl radicals than for diphenylmethyl or benzyl radicals (32).

$$\bigcirc\!\!-\dot{C}Ph_2 \longleftrightarrow \bigcirc\!\!=CPh_2 \longleftrightarrow \dot{\bigcirc}\!\!=CPh_2$$

(8)

The stability of triphenylmethyl radicals must thus be due more to steric influences which prevent dimerization rather than delocalization of the unpaired electron. That the unpaired electron is not that much more extensively delocalized in the triphenylmethyl radical than the benzyl radical arises from the lack of coplanarity of the phenyl groups in the former. X-Ray crystallographic studies have shown that the benzene rings are tilted at an angle of 30° in the stable tris(*p*-nitrophenyl)methyl radical (33), whilst e.s.r. studies indicate that the aryl rings are tilted at an angle of 50° in the tris(2,6-dimethoxyphenyl)methyl radical (34).

The structure of the dimer from triphenylmethyl radicals has recently

TABLE 5-6   *Degree of association of triarylmethyl radicals (0.2 M*
             solutions in benzene) (36)

Ar	ArPh$_2$C·	Ar$_2$PhC·	Ar$_3$C·
o-MeC$_6$H$_4$	75	18	13
m-MeC$_6$H$_4$	93.5	93	60
p-MeC$_6$H$_4$	95	94.5	84

been shown to be 1-diphenylmethylene-4-triphenyl-methylcyclohexa-
2,5-diene (6) rather than hexaphenylethane (7) (35). The correct structure of the dimer was temporarily advocated by Gomberg but was soon
rejected in favour of the hexaphenylethane structure. The dimer arises
from attack by a triphenylmethyl radical at the *para*-position of a
phenyl group in a second radical. This is preferred to dimerization to
hexaphenylethane because of the great frontal strain between the
triphenylmethyl groups which would be present in this.

TABLE 5-7   *Degree of association of some tris(o-halogenophenyl)-*
             methyl radicals (0.2 M solutions in benzene) (37)

	X=F	Cl	Br
% Association of (o-XC$_6$H$_4$)$_3$C·	92.5	88	83

The importance of steric effects upon the stability of triarylmethyl
radicals with respect to their dimers, is clear from a study of the effect
of substituents on the degree of dissociation of the dimer (Tables 5-6,
5-7, and 5-8). The results enable the following conclusions to be
drawn:

(*a*) The stability of the radical is increased more by substitution at the
ortho- than at the *meta*- and *para*-positions.

TABLE 5-8   *Degree of association of some arylbis(p-t-butylphenyl)-*
             methyl radicals (0.1 M solutions in benzene) (38, 39)

	Ar=Ph	o-MeC$_6$H$_4$	o-BrC$_6$H$_4$	p-Bu^{t}C$_6$H$_4$
% Association of ArĊ(C$_6$H$_4$—Bu^t-p)$_2$	92.5	35	6	0

Stability of radicals

(b) The stability of the radical increases with the number and size of the *ortho*-substituents.

(c) The stability of the radical is very high if all three rings are substituted in the *para*-position by bulky substituents.

An increasing amount of *ortho*-substitution increases the frontal strain in the radical, thereby inhibiting attack on a second radical. Substitution in all three *para*-positions of the triphenylmethyl radical by bulky groups renders attack by a second radical impossible. If, however, there is one free *para*-position the degree of dissociation is markedly reduced. These results were not explicable with the hexa-arylethane formulation for the dimer.

Similar considerations apply to the stability of diarylmethyl radicals. These are only stable when there is appreciable steric hindrance to dimerization, as in the case of the *bis*(2,5-di-t-butylphenyl)methyl radical (9) which is completely non-associated (40). The bulky *ortho*-

(9)

groups result in a large frontal strain to attack to form a tetra-arylethane, whilst the *meta*-t-butyl groups inhibit attack at the *para*-positions.

In recent years Ballester has prepared a number of perchloro radicals such as perchlorotriphenylmethyl and perchlorodiphenylmethyl (41). These are very much more stable than even the most stable of the radicals mentioned previously, as is witnessed by their inertness to attack by oxygen and halogens. Perchlorotriphenylmethyl can be recovered unchanged from boiling toluene and does not decompose appreciably at 300 °C.

A model of this radical indicates that the aromatic rings are tilted at 60° to each other and thus there is little or no delocalization of the unpaired electron. That this is so is supported by its e.s.r. spectrum consisting of a single line. The reason for its complete inertness is evident from an examination of a Stuart–Briegleb model of the radical, which indicates that the methyl carbon is almost perfectly shielded by the perchlorophenyl rings.

The above results indicate that steric effects are much more important than delocalization effects in determining the stability of triarylmethyl radicals. This is confirmed by consideration of the 9-phenylfluorenyl (10) (42) and the sesquixanthydryl radicals (11) (34), both of which are

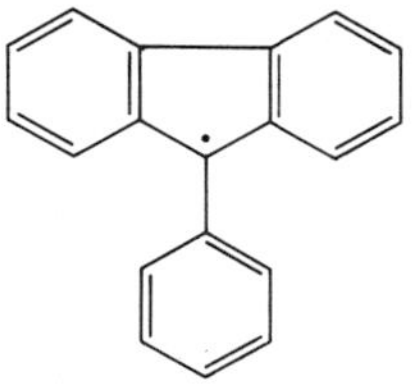

(10) 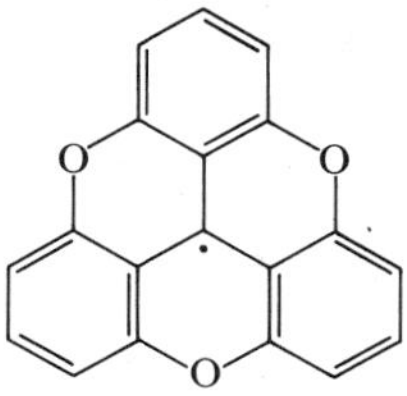   (11)

almost completely associated. In these radicals the degree of delocalization of the unpaired electron is considerably enhanced on account of their planarity, but at the same time there is a marked decrease in steric effects which allows dimerization to occur. In contrast, the related 9-mesitylfluorenyl (12) and 9-(o-t-butylphenyl)fluorenyl radicals (13) are completely disassociated (43): steric effects prevent dimerization.

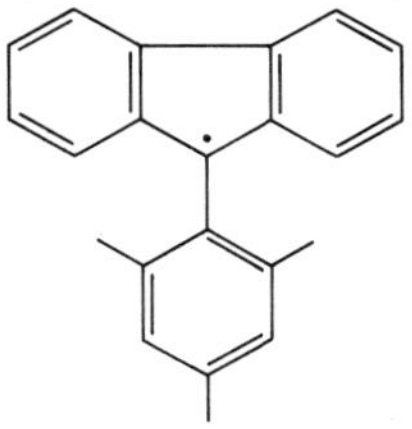

(12) 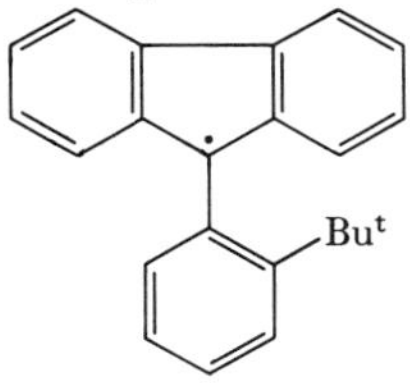   (13)

The stability of triarylmethyl and other radicals is not limited to their reluctance or otherwise to undergo dimerization or reaction with oxygen or nitric oxide. There is also the possibility of their undergoing disproportionation. A solution of the tri-*p*-tolylmethyl radical which is initially yellow becomes colourless in a few hours at room temperature as a result of disproportionation. One radical attacks the benzylic methyl group of a second radical giving tri-*p*-tolylmethane and 1,1-di-*p*-methylphenyl xylylidene, which polymerizes (reaction *15*) (44):

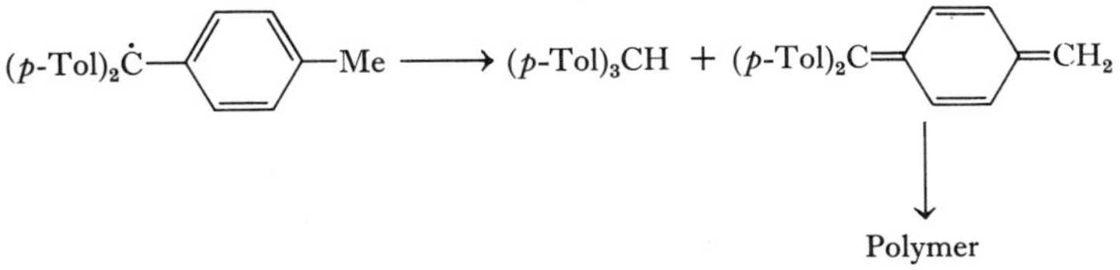

$$(15)$$

Polymer

Stability of radicals

The stability of triarylmethyl radicals is generally considered to be enhanced by both $-R$ and $+R$ substituents (45). It is difficult to get an accurate measure of this because of the greater importance of steric effects in influencing radical stabilities.

The importance of electron delocalization in enhancing radical stabilities is seen much more clearly if the conjugated system is extended, as in the case of the tetraphenylallyl (14) and pentaphenylbiallyl radicals (15). The radical (14) is partly disassociated (54), whilst (15) is completely disassociated (55). These radicals show increasing inertness to reaction with oxygen. This stability towards oxygen is even more marked with Koelsch's radical (16). The paper dealing with this work (48) was submitted for publication in 1932 but was rejected by a referee

$$Ph_2C\!\!=\!\!CH\dot{C}Ph_2$$

(14)

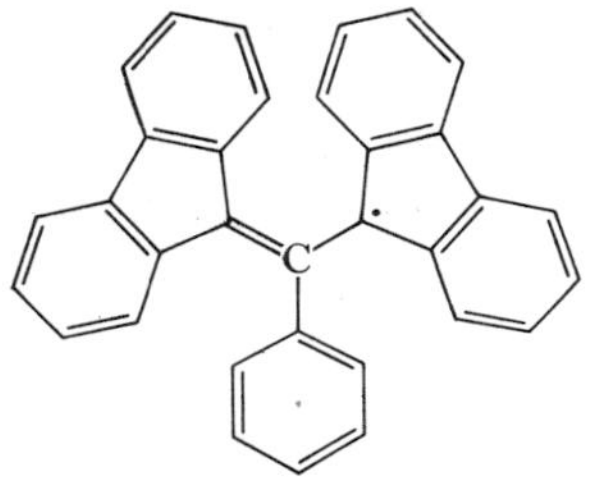

(16)

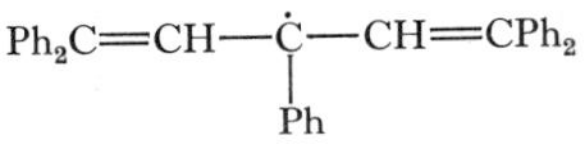

$$Ph_2C\!\!=\!\!CH\!\!-\!\!\underset{\underset{Ph}{|}}{\dot{C}}\!\!-\!\!CH\!\!=\!\!CPh_2$$

(15)

on the grounds that free radicals react readily with oxygen. The work was successfully published twenty-five years later!

Biradicals related to triarylmethyl radicals are well known. Two different types of radicals can be distinguished according to whether the radical centres are insulated from each other by a group which prevents electronic interaction between the two unpaired electrons, e.g. (17), or the radical centres may be in conjugation with each other, e.g. (18). In this latter case the compounds may exist as an equilibrium mixture of diamagnetic and paramagnetic forms.

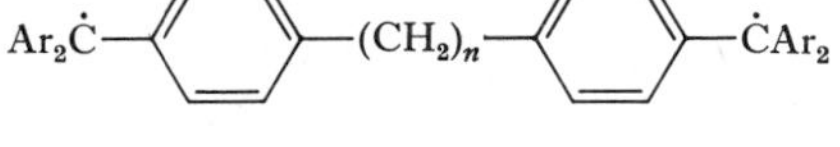

(17)

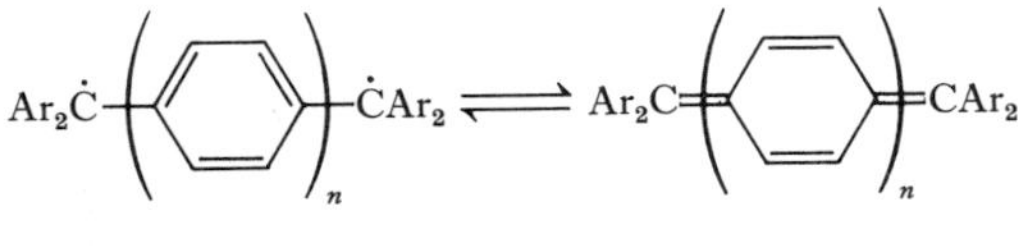

(18)

5.4.2 *Stable phenoxy radicals*

The phenoxy radical (19) is a much more stable radical than an alkoxy radical, on account of the delocalization of the unpaired electron to the *ortho-* and *para*-positions of the benzene ring. Coupling of the mesomeric

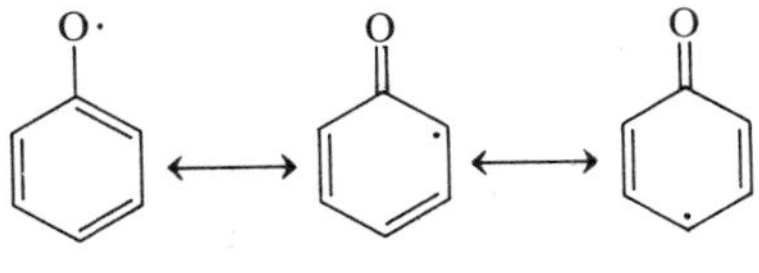

(19)

phenoxy radical produces dimers which are formed by C–O or C–C but not O–O combination. This latter does not take place on account of the instability of the resultant peroxide. In the case of C–O and C–C dimerization this may take place at either the *ortho-* or *para*-positions. Thus, when the *ortho-* and *para*-positions are substituted by bulky groups there will be steric hindrance to dimerization, and consequently 2,4,6-trisubstituted phenoxy radicals are much more stable than the phenoxy radical. The 2,4,6-tri-t-butylphenoxy radical (20) is completely monomeric both in solution and in the solid state (49, 50). The 2,4,6-

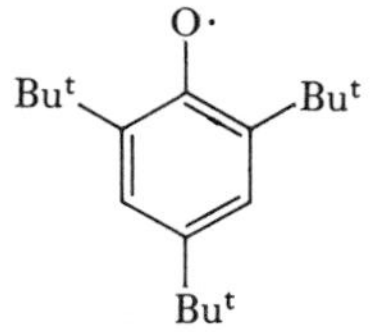

(20)

triphenylphenoxy radical (21) is likewise completely disassociated in solution (51) though it exists as its dimer (22) in the solid state (52).

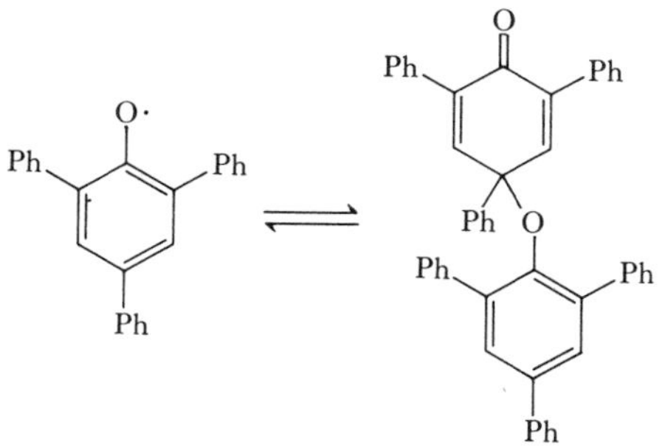

(21)   (22)

Stability of radicals

The greater tendency of the 2,4,6-triphenylphenoxy radical to exist as the dimer is attributed to the smaller steric requirement of the phenyl group which allows C–O coupling to occur at the *para*-position.

The stability of 2,6-di-t-butyl-4-substituted phenoxy radicals is markedly reduced if the substituent at the 4-position is a primary or secondary alkyl group, owing to the ease with which disproportionation can occur to give the phenol and quinone methide (reaction *16*) (53):

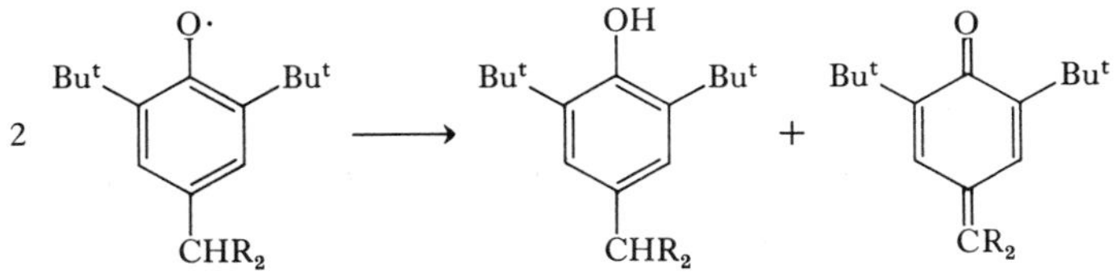

$$(16)$$

The importance of electron delocalization in the stability of phenoxy radicals is illustrated by the relatively long-lived radical derived from ethyl 2,6-di-t-butyl-4-hydroxycinnamate (23) (54). This has a half-life

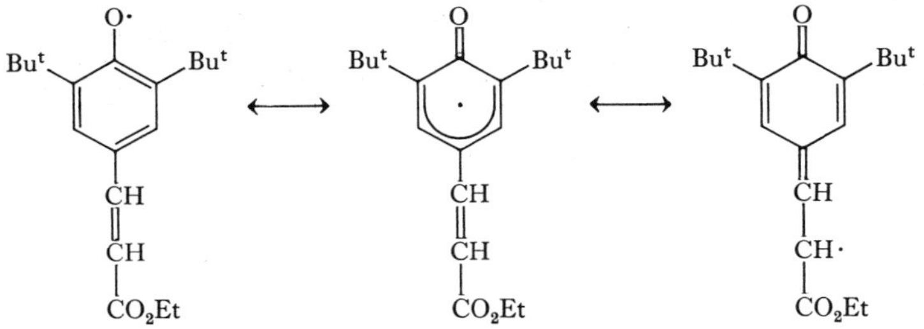

(23)

of about two hours in dilute solution. The extra stability of this radical compared to the 2,6-di-t-butyl-4-methylphenoxy radical arises from the

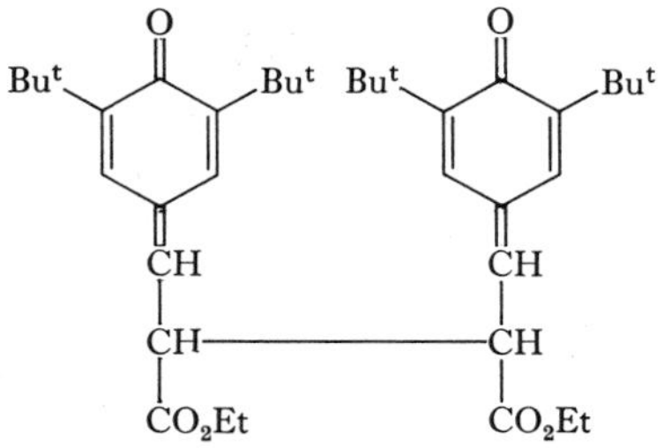

(24)

116

possibility of electron delocalization onto the side chain. That this occurs is indicated by the structure of its dimer (24).

Galvinoxyl (25), in which the degree of electron delocalization is greater, is even more stable than the 2,4,6-tri-t-butylphenoxy radical (55). The e.s.r. spectrum of this consists of a doublet, from splitting by

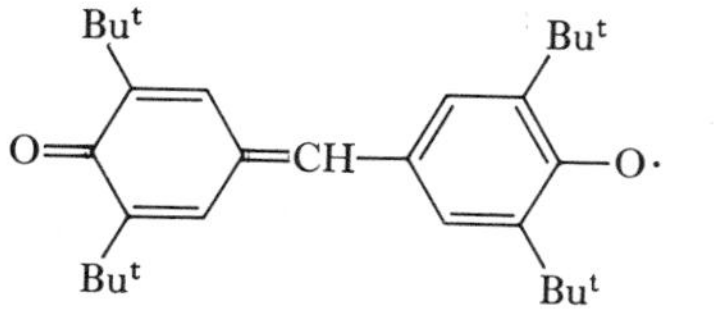

(25)

the methine proton, each component of which is split into a quintet by the four ring protons. These are equivalent, indicative of complete delocalization of the unpaired electron through the radical.

In the foregoing discussion phenoxy radicals have been considered solely in the light of their stability with respect to their dimers in an inert atmosphere. A significantly different order of stability is obtained if one considers their inertness towards reaction with oxygen. Thus, the 2,4,6-tri-t-butylphenoxy radical is decolorized in thirty minutes (56), whereas the 2,6-di-t-butyl-4-phenylphenoxy radical requires eight hours (57), and solutions of the 2,4,6-triphenylphenoxy radical are completely stable in air (51). Reaction of radicals with oxygen leads to peroxides. In general, reaction occurs at the 4-position to give peroxides such as (26).

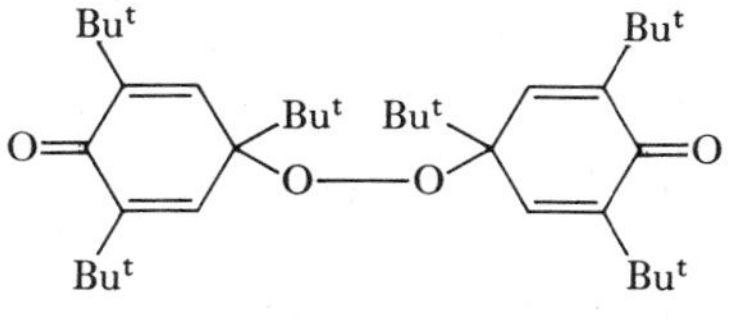

(26)

Electron-withdrawing substituents in the phenoxy radical reduce its tendency to react with oxygen (56). This is readily explained by postulating that the transition state in the reaction of the radical with oxygen has appreciable polar character (58):

$$R\cdot + O_2 \longrightarrow [R^{\delta+}\ {}^{\delta-}O\!-\!O\cdot] \longrightarrow R\!-\!OO\cdot \qquad (17)$$

5.4.3 *Nitroxides*

The first completely monomeric organic radical to be isolated was the nitroxide (27), which was prepared in 1901 by Piloty and Schwerin (59).

Stability of radicals

Even earlier, in 1845, the inorganic nitroxide (**28**), Fremy's salt, had been obtained (60). Interest in nitroxides has increased greatly in recent

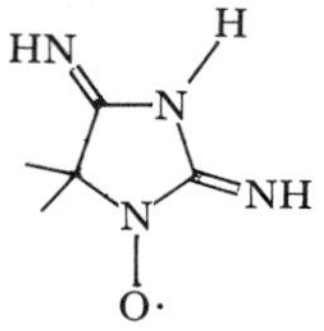

(27) 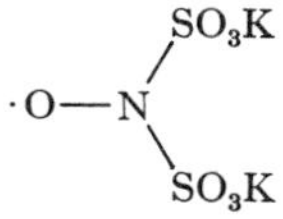   (28)

years with the advent of e.s.r. spectroscopy. This has permitted study of less stable nitroxides as well as stable ones.

The stability of nitroxides is due to an inherently stable electronic configuration about the nitrogen and oxygen rather than to steric and electronic influences of the groups attached to the nitrogen. They are conveniently represented as a resonance hybrid of the forms (**29** and **30**), though they are perhaps more accurately represented in the form (**31**) in

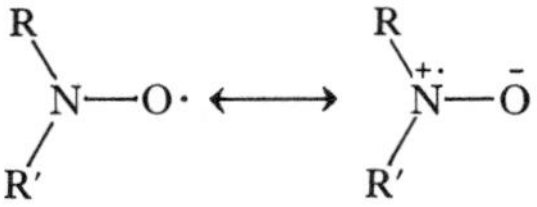

(29)   (30) 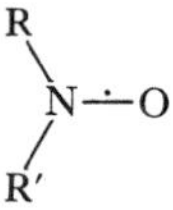   (31)

which there is a three-electron bond between the nitrogen and the oxygen. As a consequence of this stable electronic arrangement they do not undergo dimerization at the nitrogen or oxygen atoms.

The stability of nitroxides is largely determined by their ability to react with themselves by disproportionation. Nitroxides are stable when the groups attached to nitrogen do not allow the radical to undergo reaction with itself. Thus, whilst dialkylnitroxides (**32**) in which either

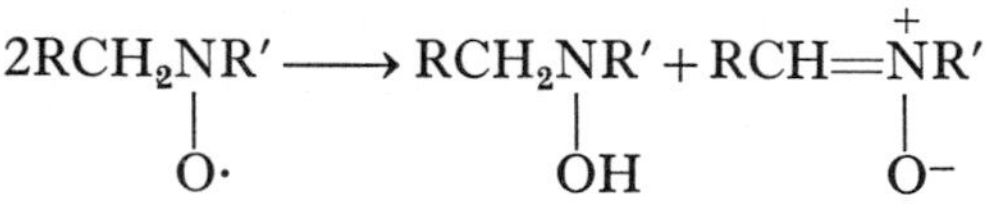

$$2RCH_2NR' \longrightarrow RCH_2NR' + RCH{=}\overset{+}{N}R'$$

$$\underset{\underset{(32)}{O\cdot}}{|} \qquad \underset{\underset{(33)}{OH}}{|} \qquad \underset{\underset{(34)}{O^-}}{|}$$

alkyl group is primary or secondary are extremely short-lived because of the ease with which they undergo disproportionation to hydroxylamine (**33**) and nitrone (**34**), di-t-alkylnitroxides are quite stable (61). The stability of di-t-alkylnitroxides is due more to their inability to undergo reaction because of the lack of a mechanism whereby they may react, rather than to steric shielding of the $\diagdown$N—O· group. This is supported by the stability of the much less sterically hindered bis(trifluoromethyl)-nitroxide (**35**) (62).

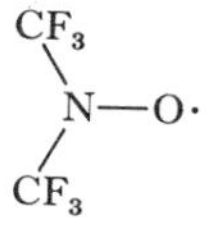

$$
\begin{array}{c}
CF_3 \\
\diagdown \\
N\!-\!O\cdot \\
\diagup \\
CF_3
\end{array}
$$

(**35**)

The groups attached to the nitrogen may make the radical more susceptible to dimerization at some other part of the radical by increasing the amount of electron delocalization, and hence the number of reactive sites. Thus, t-butylphenylnitroxide is much less stable than di-t-butylnitroxide, because the unpaired electron may be delocalized onto the aromatic nucleus. As a result of this extra delocalization of the unpaired electron, the nitrogen coupling constant is lower for t-butyl-phenylnitroxide (1.34 mT) than for di-t-butylnitroxide (1.62 mT) (63). Chemically this delocalization makes the aromatic nucleus susceptible to attack by a second radical at the *para*-position (Scheme 1) (64). The resulting dimer immediately breaks down to *N*-t-butylaniline and *N*-t-butyl-*p*-benzoquinonimine-*N*-oxide. Carbon–carbon dimerization

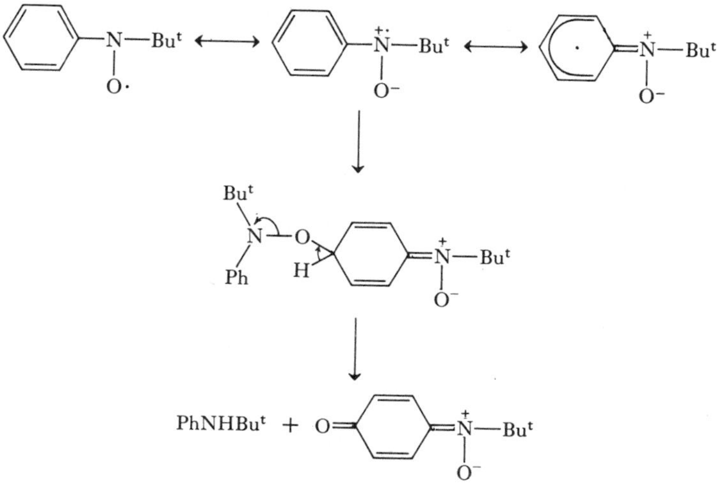

Scheme 1

Stability of radicals

does not occur as is the case for phenoxy radicals because most of the electron density is associated with the nitrogen and oxygen atoms.

Carbon–oxygen dimerization in t-alkylarylnitroxides can be prevented by the same sort of steric factors that increase the stability of other classes of stable radicals. Thus, substitution of the phenyl ring in the *para*-position by a tertiary alkyl or phenyl group results in a great increase in radical stability. Thus, t-butyl-*p*-t-butylphenylnitroxide (**36**)

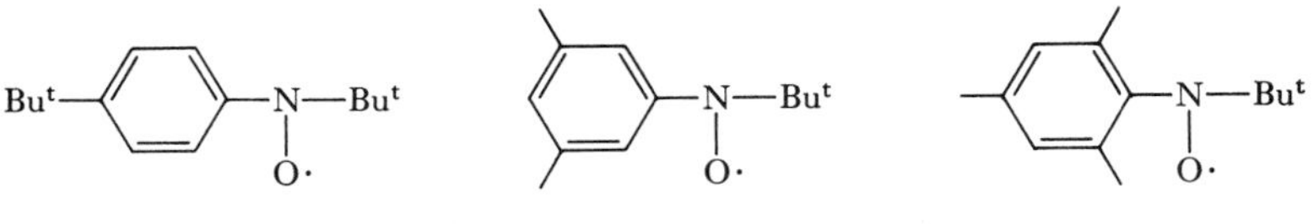

(36)   (37)   (38)

is stable in the solid state for several months. Stabilization is also increased by groups in the *meta*-positions which likewise hinder attack at the *para*-position. t-Butyl-3,5-dimethylphenylnitroxide (**37**) is very much more stable than t-butylphenylnitroxide (65). Substitution in the *ortho*-positions has the same effect but for different reasons. t-Butylmesitylnitroxide (**38**) is stable in the solid state for several months (66). In this case the aromatic ring is prevented for steric reasons from lying in the

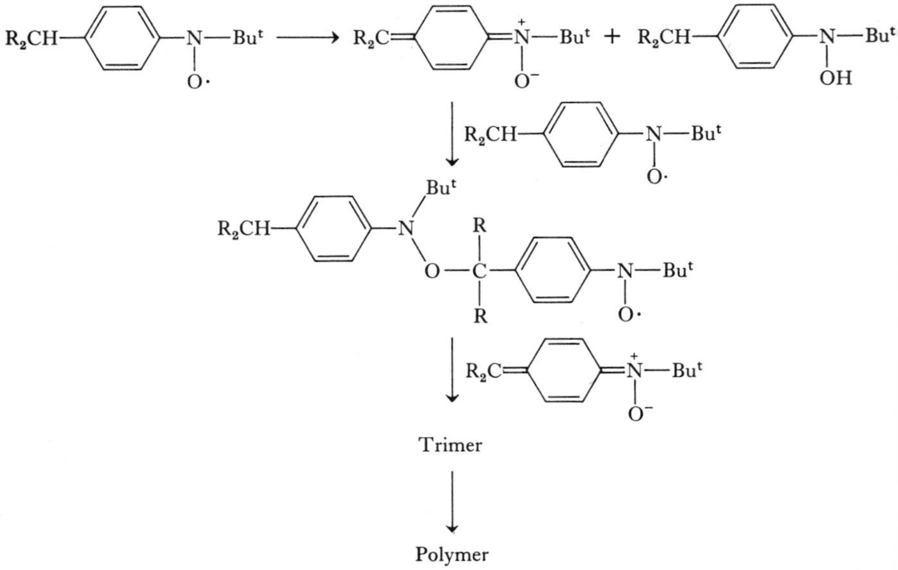

Scheme 2

same plane as the nitroxide group with a consequent decrease in delocalization onto the aromatic ring.

Aryl-t-butylnitroxides with primary or secondary alkyl groups in the *para*-position of the aromatic ring are much less stable than t-butyl-phenylnitroxide as disproportionation may occur (Scheme 2). The initially produced quinone methide imine *N*-oxide can react further with the nitroxide to give dimers, trimers or polymers (65).

5.4.4 *Diarylamino radicals*

Diarylamino radicals (39) are formed reversibly by heating tetra-arylhydrazines. They owe their stability to the delocalization of the unpaired electron onto the aromatic nuclei.

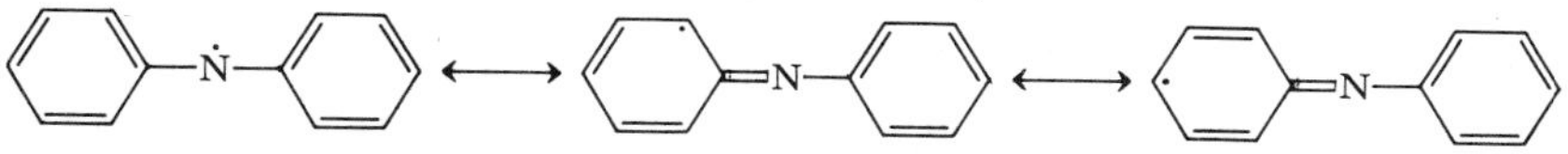

(39)

The stability of diarylamino radicals has been found to be increased by electron-donating substituents and decreased by electron-withdraw-ing substituents (67). This contrasts with the behaviour of triarylmethyl radicals whose stability is increased by both electron-donating and electron-withdrawing substituents. There are two possible electronic arrangements (40 and 41) about the nitrogen atom in the diarylamino radical. The former in which the higher-energy *p*-orbital is in the same plane as that of the aromatic nuclei will normally be preferred. Intro-duction of an electron-withdrawing substituent would favour the form (41) in which conjugation of the electron-pair with the electron-with-drawing group will be possible. The unpaired electron density on the

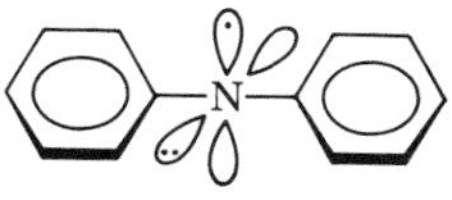

(40) 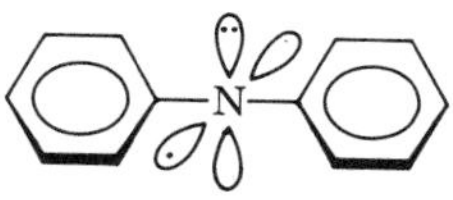   (41)

nitrogen will thus be increased, rendering dimerization to the hydrazine more probable (45).

Stability of radicals

5.4.5 Hydrazyl radicals

Probably the best known stable organic radical is the 2,2-diphenyl-1-picrylhydrazyl radical (42) which was first obtained by Goldschmidt (68). It is quite stable in the solid state and also in dilute solution. Its

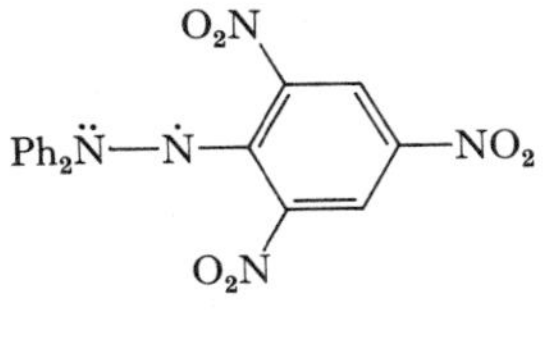

(42)

importance lies in its extensive use in polymer chemistry as a radical scavenger, and to a lesser extent as a standard in e.s.r. spectroscopy for measuring radical concentrations.

The e.s.r. spectrum of the 2,2-diphenyl-1-picrylhydrazyl radical indicates that the unpaired electron is associated to an almost equal extent with both nitrogens; the nitrogen coupling constants with N_1 and N_2 being 0.935 and 0.785 mT respectively (69). There is also appreciable delocalization of the unpaired electron onto the aromatic residues. Six canonical forms can be considered (43–48).

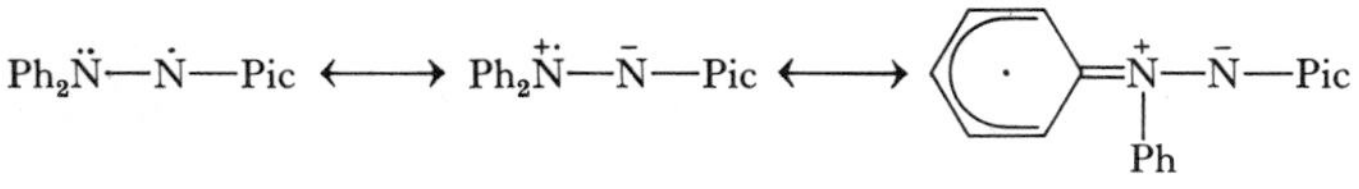

(43)   (44)   (45)

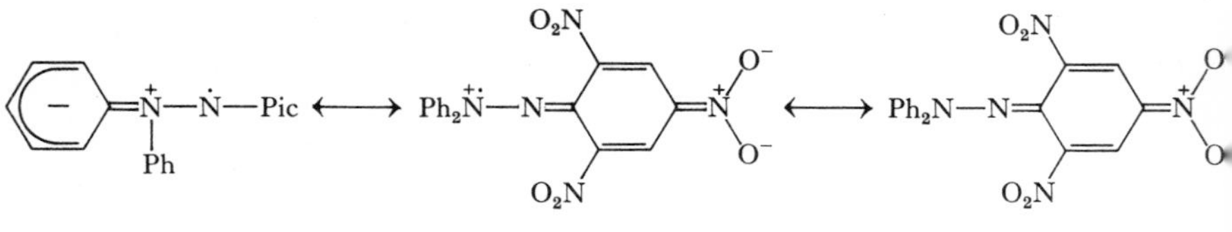

(46)   (47)   (48)

The tendency for triarylhydrazyl radicals to dimerize can be related to the relative importance of the canonical forms (49 and 50) to the resonance hybrid. Dimerization will be favoured, the greater the

$$Ar_2\ddot{N}-\dot{N}Ar' \longrightarrow Ar_2\overset{+\cdot}{N}-\bar{N}Ar'$$

(49)   (50)

contribution of (49). Electron-withdrawing substituents in the Ar′
group and electron-releasing substituents in the Ar group will favour
the canonical form (50) and hence increase the stability of the radical.
This is based on Walter's theory that unshared electron-pair delocalization is favoured over unpaired-electron delocalization (45). It is found
that steric considerations are also important, in that whereas the radical
(51) is a stable solid, the radical (52) only exists in solution (70). E.s.r.

5.f187

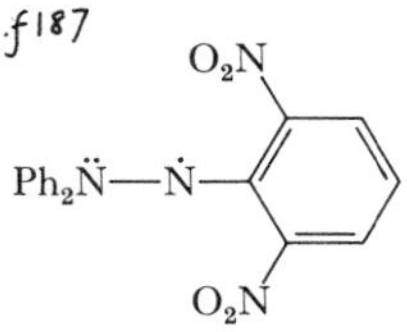

(51) 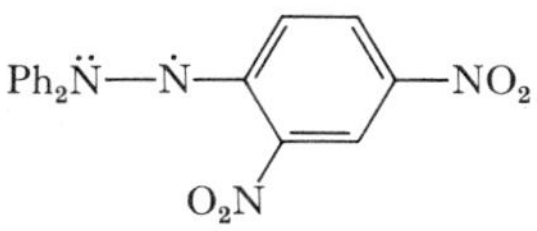   (52)

measurements indicate that the radicals have almost identical $a_{N(2)}/a_{N(1)}$
ratios.

Goldschmidt showed in a series of diarylbenzoylhydrazyls (53) that
electron-releasing substituents increased the stability of the radical,

$$Ar_2\ddot{N}-\dot{N}COPh \longleftrightarrow Ar_2\overset{+\cdot}{N}-\overset{-}{N}COPh$$
$$(53) \qquad\qquad\qquad (53a)$$

whereas electron-withdrawing substituents reduced its stability (68).
Electron-releasing substituents would favour the canonical form (53a)
and hence decrease the tendency for dimerization to occur.

5.4.6   *Radical ions*

Although formally outside the scope of this book, brief reference will
be made to the stability of both radical anions and radical cations. The
characteristic feature of these species is that quite apart from structural
considerations they are stable only over a fairly sharply defined pH
range.

Semiquinone radical anions (54). These are produced by mild
oxidation of hydroquinone in alkaline solution. They owe their stability

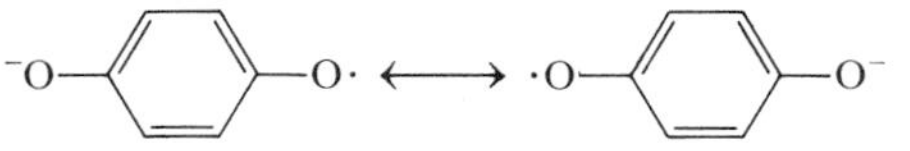

(54)

Stability of radicals

to the delocalization of the unpaired electron between the two oxygens. The radical anion is only stable in basic solutions. In acid conditions protonation occurs to give the *p*-hydroxyphenoxy radical which disproportionates to quinhydrone.

Triarylaminium radical cations (**55**). These are isoelectronic with triarylmethyl radicals, and are stabilized by delocalization of the un-

$$Ar_3\overset{+\cdot}{N}$$

(55)

paired electron. The unpaired electron density is greatest at the nitrogen and at the *ortho-* and *para*-positions of the aromatic rings. Electron-donating substituents, e.g. *p*-MeO, markedly increase the stability of these radical cations, whilst electron-withdrawing groups, e.g. *p*-NO$_2$, greatly reduce their stability (71). The reason for this is not clear, since both + R and − R substituents would enhance the degree of delocalization of the unpaired electron. As in the case of triarylmethyl radicals, dimerization involves *para–para* C–C coupling, giving eventually the Wurster salt (**56**) (Scheme 3).

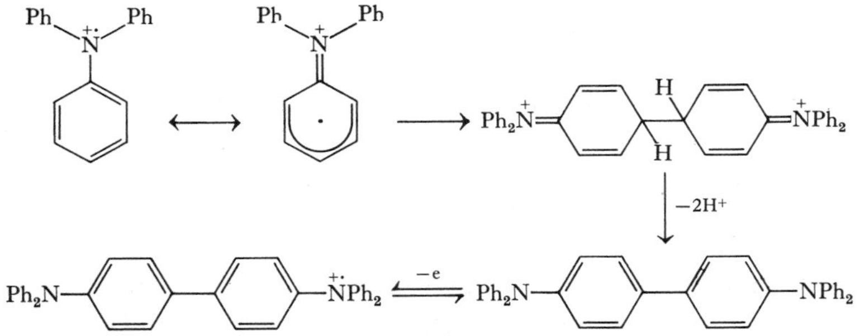

(56)

Scheme 3

Wurster's salts (**57**). These are very closely related to triarylaminium radical cations and are derived from aromatic *para*-diamines. The

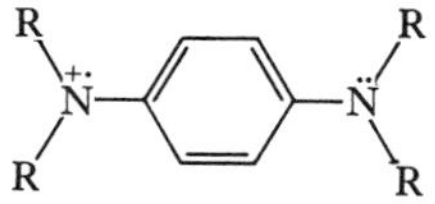

(57)

unpaired electron is delocalized onto both nitrogens and to a lesser extent onto all four ring positions (**58–61**). This is indicated by e.s.r. studies. They are only stable over a limited pH range. In basic solution

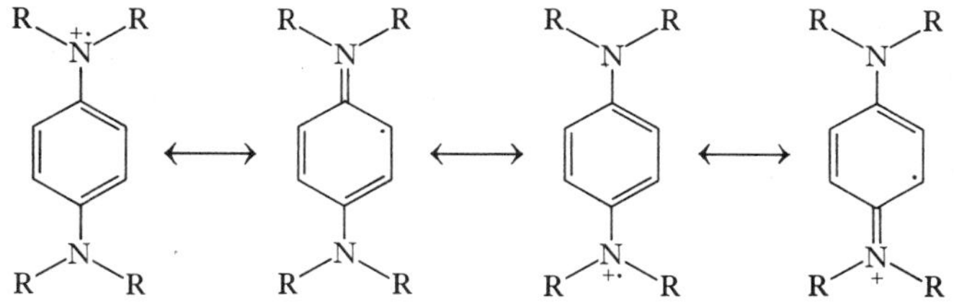

(58)   (59)   (60)   (61)

the neutral radical (**62**) is formed, whilst in acid solution the dipositive ion (**63**) is obtained. In neither case can the unpaired electron be as extensively delocalized. The stability of Wurster's salt is sensitive to the nature of the groups R attached to the nitrogen atoms. Both alkyl and aryl groups increase the stability of the radical cation, enhancing the degree of delocalization of the unpaired electron.

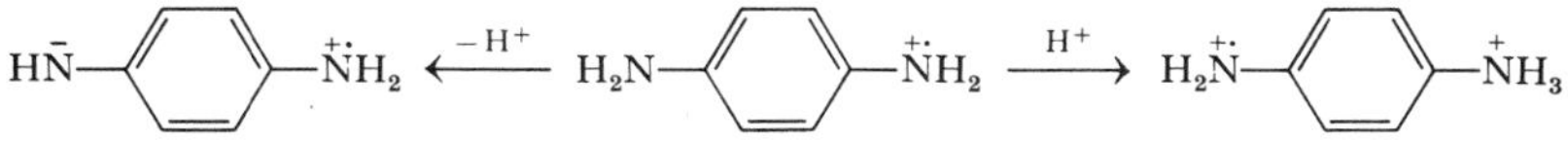

(62)   (63)

REFERENCES

1. J. P. Lorand and P. D. Bartlett, *J. Amer. Chem. Soc.* 1966, **88**, 3294, and earlier papers.
2. C. Rüchardt, *Angew. Chem. Internat. Edn*, 1970, **9**, 830.
3. D. E. Applequist and L. Kaplan, *J. Amer. Chem. Soc.* 1965, **87**, 2194.
4. W. H. Chick and S. H. Ong, *Chem. Comm.* 1969, 216.
5. F. D. Greene, M. L. Savitz, F. D. Osterholtz, H. H. Lau, W. N. Smith and P. M. Zanet, *J. Org. Chem.* 1963, **28**, 55.
6. D. G. Hoare and W. A. Waters, *J. Chem. Soc.* 1964, 2552.
7. E. S. Huyser and D. T. Wang, *J. Org. Chem.* 1964, **29**, 2720.
8. S. W. Benson, *J. Chem. Ed.* 1965, **42**, 502.
9. W. Theilacker and M.-L. Wessel-Ewald, *Annalen*, 1955, **594**, 214.
10. S. W. Benson and H. E. O'Neal, *Kinetic Data on Gas Phase Unimolecular Reactions*, National Bureau of Standards, Washington, D.C., 1970.
11. W. T. Dixon and R. O. C. Norman, *J. Chem. Soc.* 1964, 4857.
12. J. C. Martin and J. W. Timberlake, *J. Amer. Chem. Soc.* 1970, **92**, 978.
13. R. C. P. Cubbon, *Progr. Reaction Kinetics*, 1970, **5**, 29.
14. M. Gazith and M. Szwarc, *J. Amer. Chem. Soc.* 1957, **79**, 3339.
15. C. Walling, L. Heaton and D. Tanner, *J. Amer. Chem. Soc.* 1965, **87**, 1715.

16. J. Collin and F. P. Lossing, *J. Amer. Chem. Soc.* 1957, **79**, 5848.
17. M. M. Martin and E. B. Sanders, *J. Amer. Chem. Soc.* 1967, **89**, 3777.
18. J. K. Kochi and G. S. Hammond, *J. Amer. Chem. Soc.* 1953, **75**, 3452.
19. J. W. Timberlake and M. L. Hodges, *Tetrahedron Letters*, 1970, 4147.
20. A. Ohno and Y. Ohnishi, *Tetrahedron Letters*, 1969, 4405.
21. C. Rüchardt and H. Bock, *Chem. Ber.* 1971, **104**, 577.
22. J. I. G. Cadogan, E. G. Duell and P. W. Inward, *J. Chem. Soc.* 1962, 4164.
23. H. C. Brown and G. A. Russell, *J. Amer. Chem. Soc.* 1952, **74**, 3995.
24. A. F. Trotman-Dickenson, *Adv. Free-Radical Chem.* 1965, **1**, 1.
25. R. C. Fort and P. von R. Schleyer, *Adv. Alicyclic Chem.* 1966, **1**, 283.
26. A. F. Bickel, J. Knotnerus, E. C. Kooyman and G. C. Vegter, *Tetrahedron*, 1960, **9**, 230.
27. M. Kharasch, F. Engelman and W. H. Urry, *J. Amer. Chem. Soc.* 1943, **65**, 2428.
28. P. Wilder and A. Winston, *J. Amer. Chem. Soc.* 1953, **75**, 5370.
29. J. P. Lorand, S. D. Chodroff and R. W. Wallace, *J. Amer. Chem. Soc.* 1968, **90**, 5266; R. C. Fort and R. E. Franklin, *ibid.* 5267; L. B. Humphrey, B. Hodgson and R. E. Pincock, *Canad. J. Chem.* 1968, **46**, 3099.
30. F. W. Baker, H. D. Holtz and L. M. Stock, *J. Org. Chem.* 1963, **28**, 514.
31. M. Gomberg, *Chem. Rev.* 1925, **1**, 91.
32. H. van Willigen and S. I. Weissman, *J. Chem. Phys.* 1966, **44**, 420.
33. P. Anderson and B. Klewe, *Acta Chem. Scand.* 1962, **16**, 1817.
34. M. J. Sabacky, C. S. Johnson, R. G. Smith, H. S. Gutowsky and J. C. Martin, *J. Amer. Chem. Soc.* 1967, **89**, 2054.
35. H. Laukamp, W. Th. Nanta and C. MacLean, *Tetrahedron Letters*, 1968, 249.
36. W. Theilacker and M.-L. Wessel-Ewald, *Annalen*, 1955, **594**, 214.
37. C. S. Marvel, F. C. Dietz and Ch. M. Himel, *J. Org. Chem.* 1942, **7**, 392.
38. C. S. Marvel, H. W. Johnston, J. W. Meier, T. W. Martin, J. Whitson and Ch. M. Himel, *J. Amer. Chem. Soc.* 1944, **66**, 914.
39. P. W. Selwood and R. M. Dobres, *J. Amer. Chem. Soc.* 1950, **72**, 3860.
40. W. Theilacker and F. Koch, *Angew. Chem. Internat. Edn*, 1966, **5**, 246.
41. M. Ballester, *Pure Appl. Chem.* 1967, **15**, 123.
42. H. E. Bent and J. E. Cline, *J. Amer. Chem. Soc.* 1936, **58**, 1624.
43. W. Theilacker, H. Schulz, U. Baumgarte, H.-G. Drössler, W. Rohde, F. Thater and H. Uffman, *Angew. Chem.* 1957, **69**, 322.
44. C. S. Marvel, W. H. Rieger and M. B. Mueller, *J. Amer. Chem. Soc.* 1939, **61**, 2769.
45. A. I. Walter, *J. Amer. Chem. Soc.* 1966, **88**, 1923.
46. K. Ziegler, G. Bremer, F. Thiel and F. Thielmann, *Annalen*, 1923, **434**, 34.
47. G. Wittig and H. Kosack, *Annalen*, 1937, **529**, 167.
48. C. F. Koelsch, *J. Amer. Chem. Soc.* 1957, **79**, 4439.
49. C. D. Cook, *J. Org. Chem.* 1953, **18**, 261.
50. E. Müller and K. Ley, *Z. Naturforsch.* 1953, **8b**, 694.
51. K. Dimroth, F. Kalk and G. Neubauer, *Chem. Ber.* 1957, **90**, 2058.
52. K. Dimroth and A. Berndt, *Angew. Chem. Internat. Edn*, 1964, **3**, 385.
53. C. D. Cook and B. E. Norcross, *J. Amer. Chem. Soc.* 1959, **81**, 1176.
54. K. Scheffler, *Z. Electrochem.* 1961, **65**, 439.

References

55. G. M. Coppinger, *J. Amer. Chem. Soc.* 1957, **79**, 501.
56. C. D. Cook and N. D. Gilmour, *J. Org. Chem.* 1960, **25**, 1429.
57. E. Müller, A. Schick and K. Scheffler, *Chem. Ber.* 1959, **92**, 474.
58. G. A. Russell and R. F. Bridger, *J. Amer. Chem. Soc.* 1963, **85**, 3765.
59. O. Piloty and B. G. Schwerin, *Ber.* 1901, **34**, 1870, 2354.
60. E. Fremy, *Ann. Chim. Phys.* 1845, **15**, 459.
61. A. K. Hoffmann and A. T. Henderson, *J. Amer. Chem. Soc.* 1961, **83**, 4671.
62. W. D. Blackley and R. R. Reinhard, *J. Amer. Chem. Soc.* 1965, **87**, 802.
63. H. Lemaire, Y. Marechal, R. Ramasseul and A. Rassat, *Bull. Soc. chim. France*, 1965, 372.
64. A. R. Forrester and R. H. Thomson, *Nature*, 1964, **203**, 74.
65. A. Calder and A. R. Forrester, *J. Chem. Soc.* C, 1969, 1459.
66. A. R. Forrester and S. P. Hepburn, *J. Chem. Soc.* C, 1970, 1277.
67. H. Wieland and A. H. Lecher, *Annalen*, 1912, **392**, 186, and earlier papers.
68. S. Goldschmidt and J. Bader, *Annalen*, 1929, **473**, 137, and earlier papers.
69. N. W. Lord and S. M. Blinder, *J. Chem. Phys.* 1961, **34**, 1693.
70. A. T. Balaban, P. T. Frangopol, M. Marculescu and J. Bally, *Tetrahedron*, 1961, **13**, 258.
71. E. T. Seo, R. F. Nelson, J. M. Fritsch, L. S. Marcoux, D. W. Leedy and R. N. Adams, *J. Amer. Chem. Soc.* 1966, **88**, 3498.

6

Comparison of free-radical reactions in the gas phase and in solution

6.1 *INTRODUCTION*

In many free-radical reactions in solution the solvent actually takes part in the process as a reactant, and becomes wholly or partly incorporated in the products. In this chapter we are concerned only with the behaviour of the solvent as a medium, i.e. those reactions in which the solvent is not a reactant.

A single molecule of a gas experiences relatively very small interactions with its neighbours, and its situation approximates far more closely to that of an 'isolated' molecule than is the case of a molecule in solution. Theoretical treatment of 'isolated' molecules is much simplified, so that gas-phase experimental studies are generally more useful for comparison with theoretical predictions of direction and rate, or for comparison with structural parameters of the molecules themselves. In solution, the concentration of reactants is normally far greater than in the gas phase: typically a solution concentration would be 10 mol l^{-1}, compared to a gas-phase concentration of 0.05 mol l^{-1}; consequently the amount of reactants and the gross reaction rate are greater in solution, so that preparation or synthesis is frequently easier and more convenient in this medium.

The effects which the solvent medium can have on the course and rate of an electroneutral free-radical reaction can be divided into two classes:

(i) Internal volume effects which arise because of restrictions imposed on the motion of the solute molecules and the space available for them, by the solvent molecules, and

(ii) Solvation effects which arise from specific interaction of solvent molecules with reactants, intermediates or products of the free-radical process.

6.2 INTERNAL VOLUME EFFECTS

Solutions which have no specific solvation effects, and in which the distributions and orientations of the molecules are random, are called 'regular' solutions (1). Mayo (2) has suggested a method of estimating the size of the internal volume effect from the Van der Waals equation. The rate of a reaction is proportional to the concentration of the reactants. Under ideal conditions the concentrations of the reactants are proportional to their vapour pressures in both gas and solution phases. The vapour pressures are given approximately by the expression:

$$P(V-b) = nRT \tag{1}$$

where $V-b$ is the corrected volume in the Van der Waals equation, and corresponds roughly to the free volume of the container not occupied by molecules. If we consider the reactants at a concentration of 0.01 mol l^{-1} in the gas phase, and at the same concentration in benzene solution, then $V-b$ decreases by about a factor of four from the gas to the solution. The vapour pressure given by (1) increases by the same amount. For a first-order reaction the reactant pressure is increased fourfold in solution but the volume is decreased fourfold, so that internal volume effects in the phase change are slight. For a second-order reaction, however, the product of the reactant vapour pressures is increased sixteen-fold in solution, whilst the reactants are present in one-fourth the volume; the rate of reaction should therefore be four times as great in solution. Similarly, the rate of a third-order reaction is predicted to be about sixteen times as great in solution. The free volume factor $V-b$ varies from solvent to solvent, so that this can cause changes in the rate from one solvent to another as great as, or greater than, the change from gas phase to solution.

Benson (3), from a consideration of transition-state theory, derives the expression:

$$R_n = \frac{k_n(\text{solution})}{k_n(\text{gas})} \simeq \frac{10^{2n-2}}{ne^{n-1}} \tag{2}$$

which gives a comparison of the rate constant of the reaction in solution with the rate constant in the gas phase, where n is the order of the reaction. For a first-order reaction $n = 1$ and $R_1 \simeq 1$, i.e. the rate constants should be about the same in the two phases. For bimolecular reactions $n = 2$ and $R_2 \simeq 20$, so that reaction would be somewhat faster in solution. The best chance of observing a solvent effect comes with third-order reactions ($n = 3$) for which $R_3 \simeq 450$.

Solvent effects

Alternatively the 'internal volume effect' may be discussed in terms of the internal pressures of the reactants and solvent (4). For a bimolecular reaction between A and B in solvent S, the rate constant is given by:

$$RT \ln k = V_A \Delta_A + V_B \Delta_B + V_\# \Delta_\# + \text{const.} \qquad (3)$$

provided solvation effects can be ignored. V_A, V_B and $V_\#$ represent the molal volumes of A, B and the transition state, and Δ_A represents the difference between internal pressure of reactant A and the solvent, etc.

$$\Delta_A = (P_A^{\frac{1}{2}} - P_S^{\frac{1}{2}})^2, \ \Delta_B = (P_B^{\frac{1}{2}} - P_S^{\frac{1}{2}})^2, \ \Delta_\# = (P_\#^{\frac{1}{2}} - P_S^{\frac{1}{2}})^2$$

Since molal volumes do not vary to any great extent, the rate constant is determined largely by the difference between the internal pressure of the solvent and the internal pressures of the reactants and transition state.

The rates of free-radical reactions are normally measured relative to the rate of some standard reaction, so that only ratios of rate constants are available in most cases. It is clear from the foregoing discussion, that changing the phase can have only a very small effect on the ratio of rate constants, especially if the standard reaction and the reaction being measured are of the same order. This conclusion only applies of course to those solvents, such as saturated hydrocarbons, in which specific solvation effects are negligible.

6.2.1 Diffusion in solution

In the solution phase each molecule of solute is closely surrounded by solvent molecules, the number of which tends to be a maximum, so that the distance between the centres of the molecules is generally not much greater than the sum of the molecular radii. For a reaction of the type:

$$A + B \longrightarrow \text{Products}$$

the reactants A and B must diffuse together through the solvent, before they can react. Once they have encountered each other, however, they will be constrained by the caging effect of the solvent molecules to undergo several collisions before they diffuse apart. For electroneutral species in solution, therefore, the rate at which A and B diffuse together represents the maximum possible reaction rate, analogous to the collision rate in the gas phase. Initially both A and B are randomly distributed in the solution: however, as reaction occurs there is a net flux of B molecules toward A molecules. The situation can be treated in terms of Fick's Laws (5). Solution of the appropriate differential equation leads to the expression:

$$k(\text{Diff.}) = 4\pi\sigma D_{AB}(N/1000)/[1 + 4\pi\sigma D_{AB}N/1000k] \qquad (4)$$

where σ is the sum of the radii of molecules A and B, D_{AB} is the relative diffusion coefficient of A and B, and k is the rate constant that would describe the reaction if equilibrium distributions were maintained; it is usually equated to the gas-phase rate constant. For normal molecules having σ of the order of a few ångstrom units and $D_{AB} \approx 10^{-5}$ cm^2 s^{-1}, $k(\text{Diff.})$ is about 7×10^9 l mol^{-1} s^{-1}. This may be compared with the collision number of gases $Z_{AB} = N_A B_B \sigma^2 (8\pi kT/\mu)^{\frac{1}{2}}$, which predicts a maximum rate about thirty times greater. It is clear that for fast reactions, such as radical combination and disproportionation, which occur at virtually every collision in the gas phase, the reaction rate in solution will be dependent on the rate at which the species A and B can diffuse together, i.e. these reactions are diffusion-controlled.

The diffusion coefficient D_{AB} for a free radical is generally estimated by considering molecules of similar shape and size. Noyes, however, has measured the diffusion coefficient of iodine atoms by the space intermittency method (6), in which a solution is illuminated with a pattern of light and dark areas. The average steady-state concentration of the atoms is dependent upon the size of the light and dark areas, and can be calculated theoretically in terms of the diffusion coefficient. The theory is somewhat similar to that of the time intermittent illumination method, (cf. chapter 8, p. 196) and D_{AB} is obtained by comparing the theoretical average radical concentrations with the experimentally measured ones for various light and dark pattern sizes. Burkhart and co-workers (28) have used this method for measuring diffusion coefficients of some alkyl radicals in cyclohexane solution.

The combination rates of atoms such as iodine and bromine, and of alkyl radicals, are undoubtedly diffusion-controlled in solution. The experimental combination rates do not agree exactly with the predictions of equation (4) (see chapter 8, p. 207). The solution-phase combinations show many of the characteristics to be expected for diffusion control, e.g. very small temperature coefficient, little difference in the combination rate from one radical to another, and a dependence of the combination rate constant on solution viscosity. The diffusion coefficient D is given approximately by the Stokes–Einstein equation:

$$D = kT/6\pi\eta\sigma$$

Solvent effects

where η is the solvent viscosity, so that the rate of combination is predicted to be approximately proportional to $\eta^{-1 \cdot 0}$. Viscosity dependence of termination rate is well known from polymerization studies, and Carlsson and Ingold (7) have recently shown that the rate of combination of t-butyl radicals depends on $\eta^{-0 \cdot 8}$, in rough agreement with the prediction.

6.2.2 Cage effects

When two radicals A and B encounter each other in solution, they are surrounded by solvent molecules so that they undergo many collisions before diffusing apart. This 'caging' effect of the solvent was first appreciated by Rabinowitch and Wood (8). It is particularly obvious when two radicals are generated together in solution as in the photolysis of an azo compound, e.g.

$$\mathrm{EtN_2Me} + h\nu \longrightarrow \mathrm{Et\cdot} + \mathrm{N_2} + \mathrm{Me\cdot}$$

There is a high probability that the methyl and ethyl radicals will combine to give propane, since they are generated together and constrained to undergo several collisions within the solvent cage. Only if they manage to separate by diffusion can the symmetrical dimers n-butane and ethane be formed. Hence the cage effect leads to a preponderance of the unsymmetrical combination product propane.

Lyon and Levy (9) decomposed mixtures of azomethane and hexadeuterioazomethane in both gas and liquid phases. In the gas phase there was statistical combination of $\mathrm{CH_3\cdot}$ and $\mathrm{CD_3\cdot}$ to form ethane, showing the absence of any cage effect. In solution, all the ethane found was either $\mathrm{C_2H_6}$ or $\mathrm{C_2D_6}$, so that virtually no radicals were able to escape from the solvent cage in which they were formed. Some caging can also occur in the gas phase at high pressures, as has been demonstrated for the photolysis of azomethane in propane at 50 atmospheres (10).

The cage effect has now been observed in the decomposition of numerous azo compounds, peroxides, etc., which generate two radicals simultaneously. Greene and co-workers (12) and Kopecky and Gillan (13) have shown that in spite of the high probability of cage combination, the radicals can *rotate* many times within the solvent cage prior to combination. Optically active azo compounds were decomposed in solution, with the expectation that combination within the cage might be so fast that the radicals would retain their configuration. However, a great deal of randomization of the configuration was found, so that inversion or rotation of the radicals is fast in comparison to cage combination (see chapter 8, p. 209).

Szwarc and collaborators have studied cage combination reactions by photolysis of azo compounds in the presence of a suitable radical scavenger, so that all radicals escaping from the cage are trapped (14). All the radical dimer then comes from cage combination. In hydrocarbon olefinic or aromatic solvents all the radicals escaping from the cages are removed by reaction with the solvent. In other solvents α-methylstyrene was added as a scavenger:

$$RN_2R' + h\nu \longrightarrow [R\cdot + N_2 + R'\cdot] \longrightarrow RR' + N_2$$

The probability of cage combination is thus given by the yield of combination product RR' divided by the nitrogen yield, i.e. $p(RR')$ = [RR']/[N$_2$]. For CF$_3\cdot$ and for CH$_3\cdot$ radicals the probability of cage combination was determined in a wide variety of solvents, and over a range of temperatures (15). Szwarc *et al.* found that $p(RR')$ increased with increasing viscosity of the solvent and decreased with increasing temperature for both radicals.

Noyes (5) has developed a model of cage reactions in which the radicals are treated as spheres of radius b, their centres being initially separated by a distance a. The reaction takes place in a liquid treated as a continuum of viscosity η, and diffusion is random. The probability of combination on each collision is α. The treatment (15) leads to the expression (5) where A is a constant:

$$\frac{1}{p(RR')} = \frac{a}{2b} + \left(\frac{Aa}{2b\alpha}\right)\left(\frac{T^{\frac{1}{2}}}{\eta}\right) \tag{5}$$

$1/p(CH_3\cdot CF_3)$ should therefore be a linear function of $T^{\frac{1}{2}}/\eta$. For individual solvents Szwarc found excellent correlation of the experimental data with $T^{\frac{1}{2}}/\eta$. Combining all the $p(CH_3\cdot CF_3)$ values for all the solvents also gave a reasonable correlation, with deviations only for the most fluid solvents. Noyes' simplified model accounts satisfactorily for the main features of the cage combinations, although for fluid solvents some specific interactions, not included in the model, are obviously important.

That the proportion of cage combination depends on the type of solvent used, was also shown by Kochi (11), who decomposed a series of diacyl peroxides in pentane and in decalin. In decalin, the more viscous solvent, photolysis of two diacyl peroxides gave a high yield of the symmetrical dimers and only a minor yield of cross-dimer, showing the strong cage effect. In pentane much more of the cross-products were formed, the cage effect being less important in the more fluid solvent.

Kiefer and Traylor (16) have studied the cage combination of t-butoxy radicals generated by decomposition of di-tertiary-butyl peroxide and other precursors, in a series of hydrocarbon solvents. The probability of $Bu^tO\cdot$ cage combination was determined in each solvent at a single temperature. It was found, however, that no simple relationship of $1/p(Bu^tO\cdot Bu^tO\cdot)$ with viscosity existed for these radicals. The assumptions inherent in the simple cage model, such as spherical radicals, absence of specific solvation effects, random diffusion, and treatment of the solvent as a continuum, are not justified in this case.

6.2.3 *Disproportionation in solution*

Disproportionation in solution has been the subject of study by several groups of workers, particularly Szwarc (14) and Stefani (14 a), who used methyl and ethyl radicals. The disproportionation rate constant k_d was measured relative to the combination rate constant k_c:

$$C_2H_5\cdot \; + \; C_2H_5\cdot \; \underset{k_d}{\overset{k_c}{\diagdown}} \; \begin{matrix} C_4H_{10} \\[4pt] C_2H_4 \; + \; C_2H_6 \end{matrix}$$

Since both reactions are second order, little difference in k_d/k_c is to be expected on changing the phase. In fluid alkane solvents, such as iso-octane or n-pentane, k_d/k_c is virtually identical with the gas-phase value. In less fluid solvents k_d/k_c was generally found to be greater, though not dramatically so, than the gas-phase value (see Table 6-1).

In the gas phase k_d/k_c is independent of temperature for the great majority of radicals, but a small temperature variation has been observed in solution for ethyl radicals (14 a) and for methyl with ethyl radicals (14). The temperature coefficient of the solution-phase process is attributed to the change in internal volume of the solvent with temperature (14 a), and not to an activation-energy barrier in the combination reaction. Similarly, the increase in k_d/k_c on going to less fluid solvents is explicable simply in terms of the internal volume factor. Stefani (14 a) has obtained a linear correlation of log k_d/k_c against the square root of the internal pressure of the solvent for the ethyl–ethyl disproportionation/combination data of Table 6-1 (cf. equation 3).

TABLE 6-1   *Disproportionation/combination rates in various solvents*

$$R \cdot + R' \cdot \xrightarrow{k_c} RR'$$

$$R \cdot + R' \cdot \xrightarrow{k_d} RH + R'(-H)$$

Radical	Solvent	Temp. °C	k_d/k_c	Reference
$CH_3 \cdot + C_2H_5 \cdot$	Gas phase	—	0.04	Table 8-3
	Isooctane	85°	0.054	14
	Glycol	85°	0.084	14
$C_2H_5 \cdot + C_2H_5 \cdot$	Gas phase	—	0.14	Table 8-3
	Isooctane	65°	0.144	14a
	Limonene	65°	0.158	14a
	Ethylbenzene	65°	0.156	14a
	m-Xylene	65°	0.165	14a
	Toluene	65°	0.167	14a
	Butan-2-ol	65°	0.168	14a
	Propan-1-ol	65°	0.181	14a
	Aniline	65°	0.195	14a
	Acetonitrile	65°	0.200	14a
	Glycol	65°	0.241	14a
$n\text{-}C_3H_7 \cdot + n\text{-}C_3H_7 \cdot$	Gas phase	—	0.15	Table 8-3
	n-Pentane	30°	0.15	11
	Decalin	30°	0.13	11
$n\text{-}C_4H_9 \cdot + n\text{-}C_4H_9 \cdot$	Gas phase	—	0.15	Table 8-3
	n-Pentane	30°	0.14	11
	Decalin	30°	0.13	11
$i\text{-}C_3H_7 \cdot + i\text{-}C_3H_7 \cdot$	Gas phase	—	0.69	Table 8-3
	Decalin	30°	1.2	11

6.3   SOLVATION EFFECTS

When specific interaction between the active participants in a free-radical reaction and the solvent occurs, then the solution is no longer 'regular'. The solvation effects which can occur as a consequence frequently outweigh all internal volume factors. The course of a free-radical reaction is represented by curve a in the potential energy diagram of Fig. 6-1, in a solvent where neither of the reactants nor the transition state is solvated, and the activation energy is E_a. In a different solvent where one or more of the reactants is solvated, their energy is lowered by an amount ΔH_s and the course of the reaction now follows curve b. The activation energy of the process is increased to $E_b = E_a + \Delta H_s$ and hence the rate decreases.

Curve a in Fig. 6-2 represents reaction in a solvent where neither reactants nor transition state are solvated. In a different solvent, where

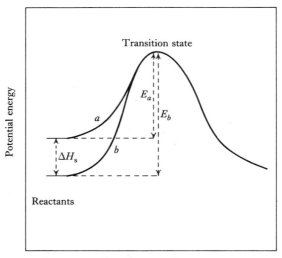

Fig. 6-1. Potential energy diagram for a reaction in which the reactants are initially unsolvated, *a*, but become solvated in another solvent, *b*.

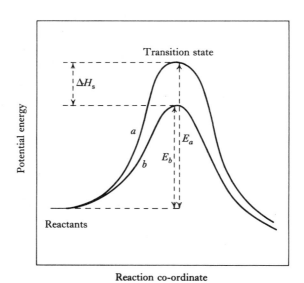

Fig. 6-2. Potential energy diagram for a reaction in which the transition state is unsolvated, *a*, but becomes solvated in another solvent, *b*.

the transition state is solvated, but not the reactants, then curve *b* represents the course of the reaction. The energy of the transition state is lowered by an amount ΔH_s, and hence the activation energy of the process is less and the rate increases. Provided other factors remain constant, solvation of the reactants leads to a retardation in rate, but solvation of the transition state leads to an increase in rate. If both reactants and transition state are solvated, the two effects will compensate each other and little change in the rate can be anticipated. Solvation of the products of the reaction has little direct influence on the rate.

Only relative rates are available for the majority of free-radical reactions. Thus in hydrogen abstraction the rate is expressed as a Relative Selectivity. The more reactive a radical, the less selective it is generally found to be, i.e. an increase in the rate of abstraction leads to a decrease in selectivity and vice versa. Solvation of the reactants would lead to an increase in selectivity therefore, provided other factors are constant, and solvation of the transition state would decrease the selectivity.

Up to this point we have considered the activation-energy change on solvation, i.e. the enthalpy change only. Solvation of a free radical will also have a big influence on the steric requirements of the reaction. This will also affect the entropy factor, i.e. the *A*-factor in the rate constant. Thus solvation of the reactant free radical leads to an increase in the activation energy, but if the solvated radical proceeds to reaction through an unsolvated transition state (Fig. 6-1) a positive increase in entropy of the system is expected and hence the *A*-factor will be greater. The increased activation energy may therefore be partly or wholly counterbalanced by the increased *A*-factor, and the rate be partly or wholly unaffected by solvation of the reactants. Such compensations between the entropy and enthalpy factors are very common in practice. In fact there is only one temperature for a particular system at which the entropy change exactly compensates the enthalpy change. This is known as the isokinetic temperature, β, and is defined by the relationship (17):

$$\Delta H^* = \Delta H_0^* + \beta \Delta S^*$$

Above this temperature, changes in the rate are largely due to changes in the entropy factor, whereas below this temperature, changes in the enthalpy factor are predominant. It is clear that measurements of the rate constant of a process at one temperature are of limited value, since if the chosen temperature happens to be near the isokinetic temperature the solvent effects will be masked.

Solvent effects

6.3.1 Hydrogen abstraction reactions

Hydrogen abstraction by chlorine atoms is the most thoroughly investigated process from the point of view of solvent effects [18]. In their pioneering work Hass, McBee and Weber showed that in many solvents chlorination of alkanes was less selective than in the gas phase [19]. More recently Russell [20] showed that chlorination of alkanes was much more selective in certain solvents:

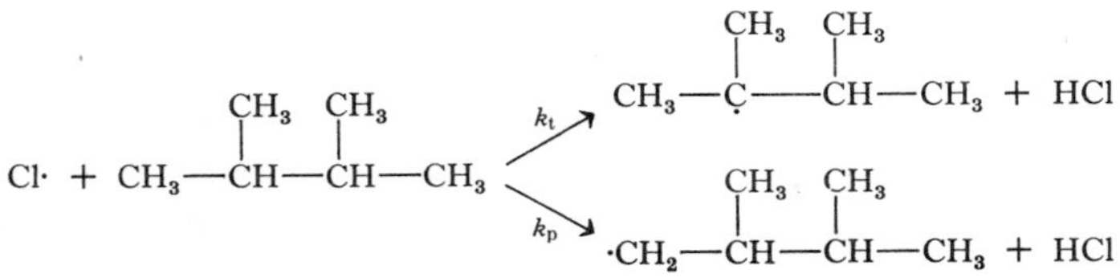

$$Cl\cdot + CH_3-\overset{\overset{\displaystyle CH_3}{|}}{CH}-\overset{\overset{\displaystyle CH_3}{|}}{CH}-CH_3 \underset{k_p}{\overset{k_t}{\lessgtr}} \begin{cases} CH_3-\overset{\overset{\displaystyle CH_3}{|}}{\underset{|}{C}}-\overset{\overset{\displaystyle CH_3}{|}}{CH}-CH_3 + HCl \\ \cdot CH_2-\overset{\overset{\displaystyle CH_3}{|}}{CH}-\overset{\overset{\displaystyle CH_3}{|}}{CH}-CH_3 + HCl \end{cases}$$

In the chlorination of 2,3-dimethylbutane the relative selectivity RS_p^t (i.e. the relative rate per hydrogen atom k_t/k_p) at a particular temperature was very much greater in aromatic solvents. The relative selectivities found by Russell for some solvents are given in Table 6-2.

TABLE 6-2   *Effect of solvents on the chlorination of 2,3-dimethylbutane (after Russell, 20)*

Solvent	RS_p^t (55 °C)
2,3-Dimethylbutane	3.7
Carbon tetrachloride	3.5
Cyclohexene	3.6
Nitrobenzene	4.9
Fluorobenzene	10.3
Benzophenone	8.8
Benzene	14.6
Toluene	15.4
Anisole	18.4

Aliphatic solvents had virtually no effect on the relative selectivity, but it dramatically increased in the aromatic solvents. Since the selectivity is increased, this effect is attributed to solvation of the reactants (Fig. 6-1), in this case the chlorine atom. It is noticeable in Table 6-2 that the most effective solvents in increasing the selectivity, are those with electron-repelling substituents which increase the electron density of the aromatic nucleus. The least effective solvents are those with electron-attracting substituents, which decrease the electron density of the aromatic ring. It seems likely, therefore, that the electrophilic chlorine atom is associated with the aromatic solvent as a 'π-complex' which is in equilibrium with solvent and uncomplexed chlorine atoms (6). The more electron-rich the solvent becomes, the further to the right the equilibrium goes, and the more selective the process becomes.

$$\text{Cl}\cdot + \text{C}_6\text{H}_5\text{X} \rightleftharpoons [\text{Cl}\cdots\cdots\text{C}_6\text{H}_5\text{X}] \tag{6}$$

$$[\text{Cl}\cdots\cdots\text{C}_6\text{H}_5\text{X}] + \text{RH} \longrightarrow \text{R}\cdot + \text{HCl} + \text{C}_6\text{H}_5\text{X}$$

A small solvation effect was also observed with oxygen-containing solvents such as dioxan and n-butyl ether. This might result from the complexing of the chlorine atom with the oxygen of the solvent. Sulphur-containing solvents such as carbon disulphide and diphenyl sulphide show a greater tendency to complex with chlorine, presumably by expansion of the valence shell of sulphur; and high relative selectivities were again observed:

$$\text{Cl}\cdot + \text{S}{=}\text{C}{=}\text{S} \rightleftharpoons \text{Cl}{-}\dot{\text{S}}{=}\text{C}{=}\text{S}$$

It is significant that olefinic solvents, such as cyclohexene, which might also be expected to form 'π-complexes' with chlorine atoms, showed no tendency to increase the selectivity in the chlorination of 2,3-dimethylbutane. The explanation of this is probably that the π-complex does form, but this proceeds to give the addition product (7):

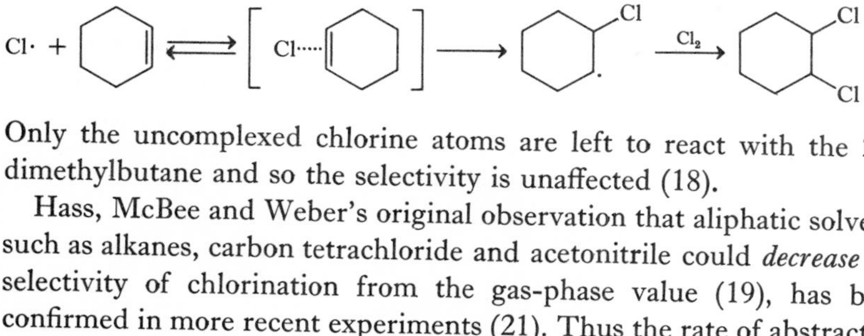

Only the uncomplexed chlorine atoms are left to react with the 2,3-dimethylbutane and so the selectivity is unaffected (18).

Hass, McBee and Weber's original observation that aliphatic solvents such as alkanes, carbon tetrachloride and acetonitrile could *decrease* the selectivity of chlorination from the gas-phase value (19), has been confirmed in more recent experiments (21). Thus the rate of abstraction from the secondary position in n-hexane relative to the rate of abstraction from the primary position, i.e. RS_p^s was found to be 3.1 in the gas phase at 40 °C but only 2.2 in CCl$_4$ solution at the same temperature. A study of the temperature dependence of the relative selectivity revealed, however, that the activation-energy difference for attack at the primary and secondary positions was greater in solution than in the gas phase.

Solvent effects

This would be expected if the chlorine atoms were solvated in the same way as with the strongly complexing aromatic solvents. The decrease in solution of RS_p^s is due to the lower ratio of the A-factors of the two processes. Even solvents such as CCl_4 and the alkanes are therefore weakly complexing, although the equilibrium (analogous to 6) must lie well to the left.

Solvation of the transition state also alters the nature of the directing influence of any substituents in the attacked alkane. Thus Singh and Tedder found that the inductive influence of the fluoroacyl group was not felt beyond the β-carbon atom of the chain

$$\text{X.CO}\overset{\alpha}{\text{—CH}_2}\overset{\beta}{\text{CH}_2}\text{(CH}_2)_n\text{CH}_3$$

in gas-phase chlorination, but that in solution the inductive effect of the analogous chloroacyl group was transmitted right down the chain even to positions seven carbon atoms distant (22). The results are shown in Table 7-13 and discussed in more detail in chapter 7, p. 183.

The effects of phase change on the abstraction and addition reactions of hydrocarbon radicals appear to be small, possibly because they are less polar and so less likely to be solvated than chlorine atoms or heteroradicals. Mayo (2) has compared the abstraction of hydrogen from alkanes by methyl radicals in the gas phase and in CCl_4 solution, and finds virtually no difference between the two sets of data. For the electrophilic trichloromethyl radicals, Martin and Gleicher have studied the addition to ω-phenylalkenes, $Ph(CH_2)_nCH{=}CR_2$ (23). The relative reactivities of compounds in which n was varied were measured. The results are in Table 6-3. The rate reached a maximum for both series of

TABLE 6-3   *Addition of trichloromethyl radicals to phenylalkenes in solution* (23)

$Ph(CH_2)_nCH{=}CH_2$		$Ph(CH_2)_nCH{=}CMe_2$	
	Rel. rate of		Rel. rate of
n	addn at 70 °C	n	addn at 94 °C
1	1.00	1	1.00
2	1.22	2	1.15
3	1.13	3	0.57
4	1.10	4	0.52
5	1.00	5	0.38
6	0.82	6	0.27

compounds when n was equal to two. Martin and Gleicher proposed that the trichloromethyl radical was complexed with both the phenyl group and the double bond of the olefin, i.e. that the radical was sandwiched between the two. When two or three methylene groups were in the chain, the conformation of the complex was most favourable for transfer of the CCl_3· from the aromatic ring wholly onto the double bond, so that addition would be fastest in that case.

Szwarc and co-workers investigated the addition of trifluoromethyl radicals to a series of mono-olefins in the gas and liquid phases (24). The rate of the addition step was measured relative to the rate of abstraction from 2,3-dimethylbutane (i.e. k_8/k_9):

$$CF_3· + Ol \longrightarrow CF_3Ol· \tag{8}$$

$$CF_3· + RH \longrightarrow CF_3H + R· \tag{9}$$

which acted as the solvent in the liquid phase, and which was added in great excess to the gas phase experiments. The results are in Table 6-4. The activation energies and A-factors are very similar in the two phases, which indicates that the hydrocarbon solvent affects the addition and abstraction processes to about the same extent. On the whole, the relative rate is slightly greater in the liquid phase, so that solvation of the transition state in the addition reaction (8) may be slightly more important.

TABLE 6-4   *Addition of trifluoromethyl radicals to olefins in the gas and liquid phases* (24)

$$CF_3· + Ol \xrightarrow{k_8} CF_3Ol· \quad CF_3· + RH \xrightarrow{k_9} CF_3H + R·$$

Olefin	Gas phase $-(E_8-E_9)$	A_8/A_9	Liquid phase $-(E_8-E_9)$	A_8/A_9	$\dfrac{(k_8/k_9)_L}{(k_8/k_9)_G}$ at 65 °C
Vinyl chloride	7.69	2.40	7.94	3.70	1.71
Ethylene	9.82	2×1.17	9.49	2×1.64	1.30
Propylene	11.54	1.52	11.62	2.19	1.44
Isobutene	14.71	1.46	15.34	1.58	1.36
Styrene	14.42	2.45	15.13	2.10	1.15
α-Methylstyrene	16.13	2.14	16.47	2.17	1.24
Isoprene	16.93	2×1.35	17.31	2×1.58	1.29

Activation energies in kJ mol^{-1}.

6.3.2 *Solvent effects on alkoxy radical reactions*

Walling and Jacknow studied the effect of solvent on the hydrogen abstraction reactions of the tertiary butoxy radical generated from tertiary butyl hypochlorite (25). The selectivity of the radical in abstracting primary and secondary hydrogens from n-butane and primary and tertiary hydrogens from 2,3-dimethylbutane was investigated in several solvents. The results are given in Table 6-5. There was a definite increase in the relative selectivity when solvents like CS_2 and aromatics were used, but nothing like the dramatic effect observed for chlorination in the same solvents.

TABLE 6-5   *Hydrogen abstraction from alkanes by*
$Bu^tO\cdot$ in various solvents at 40 °C (25)

Solvent	RS_p^s	RS_p^t
None	7.89	44.4
Carbon tetrachloride	—	42.8
Carbon disulphide	8.58	49.8
Benzene	9.43	55.3
t-Butylbenzene	9.96	61.2

RS_p^s: abstraction from n-butane; RS_p^t: abstraction from 2,3-dimethylbutane.

Walling and Wagner (26) later investigated the temperature dependence of the relative selectivity of $Bu^tO\cdot$ radicals. They found that although the rate in a particular solvent is little changed, there are big differences in the activation energies and A-factors which largely compensate each other. An increase in the activation-energy difference $E_p - E_t$ is counter-balanced by a decrease in the A-factor ratio A_p/A_t. This is clearly an example of the isokinetic relation (17). Walling and Wagner found, by plotting $E_p - E_t$ against $\log A_t/A_p$, that the isokinetic temperature β was 278 °K for these hydrogen abstraction reactions. There is, therefore, a pronounced solvent effect in the $Bu^tO\cdot$ radical abstractions, but it is not of the same kind as in chlorinations. The solvents which showed the greatest effect were polar compounds such as acetonitrile, acetone and chlorobenzene. The experimental observations can be explained if it is assumed that the $Bu^tO\cdot$ radical is solvated, and the transition state is solvated weakly in the case of abstraction of a primary hydrogen, and moderately in the case of tertiary hydrogen abstraction (18). Solvents which strongly complex the $Bu^tO\cdot$ radical should then show a greater $E_p - E_t$ than those which weakly complex $Bu^tO\cdot$. An increase in entropy is expected on going from the strongly solvated reactants to the weakly or moderately solvated transition state. Thus there is a greater increase in A_p than in A_t on going to strongly complexing solvents, and $\log A_t/A_p$ becomes negative.

TABLE 6-6   *Arrhenius parameters for hydrogen abstraction from 2,3-dimethylbutane by $Bu^tO\cdot$ radicals in various solvents* (26)

Solvent	RS_p^t (40 °C)	$E_p - E_t$	$\log A_t/A_p$
None	44	7.73	0.35
C_6H_6	55	8.32	0.35
$C_6H_5OCH_3$	65	9.36	0.22
C_6H_5Cl	54	10.78	-0.08
CH_3OCH_3	51	15.76	-0.92
CH_3CN	33	19.10	-1.67

Activation energies in kJ mol^{-1}.

Walling and Padwa (27) compared the fragmentation of the benzyl-dimethylmethoxy radical (*10*) with its hydrogen abstraction from cyclohexane (*11*):

$$PhCH_2C(Me_2)O\cdot \longrightarrow PhCH_2\cdot + MeCOMe \qquad (10)$$

$$PhCH_2C(Me_2)O\cdot + RH \longrightarrow PhCH_2C(Me_2)OH + R\cdot \qquad (11)$$

In cyclohexane the decomposition and abstraction reactions proceeded at about the same rates; whereas in certain olefin solvents such as cyclohexene, cyclopentene, pent-1-ene, hex-2-ene, hept-2-ene and oct-1-ene, the hydrogen abstraction process was almost completely suppressed and fragmentation only was observed. Aromatic solvents caused very little change in the rates of decomposition and abstraction. The suggestion was made that solvation of the alkoxy radical places the solvent in the vicinity of the oxygen atom. Thus there is a steric problem in the abstraction reaction because the solvent must first be released from the alkoxy radical. Hence the decomposition reaction, where there is no steric effect of the solvent in the transition state, is considerably favoured.

The complex clearly cannot be of the same type as in the chlorination reactions, since aromatic solvents had little or no effect. Walling

Solvent effects

suggested that a charge-transfer complex is formed between the alkoxy radical and the olefin:

$$
\text{PhCH}_2\text{C}(\text{Me}_2)\text{O}\cdot \;+\;
\begin{array}{c}\diagdown\;\diagup\\ \text{C}\\ \|\\ \text{C}\\ \diagup\;\diagdown\end{array}
\;\rightleftharpoons\;
\text{PhCH}_2\text{C}(\text{Me}_2)\text{O}^-
\begin{array}{c}\diagdown\;\diagup\\ ^+\text{C}\\ |\\ :\text{C}\\ \diagup\;\diagdown\end{array}
$$

The olefin double bond would be expected to be a better donor than the aromatic nucleus.

GENERAL REFERENCES

F. R. Mayo, *J. Amer. Chem. Soc.* 1967, **89**, 2654.

S. Glasstone, · K. J. Laidler and H. Eyring, *The Theory of Rate Processes*, McGraw-Hill, New York and London, 1941.

E. S. Huyser, *Adv. Free-Radical Chem.* 1965, **1**, 77.

SPECIFIC REFERENCES

1. J. H. Hildebrand and R. L. Scott, *Regular Solutions*, Prentice-Hall, Englewood Cliffs, N.J., 1962.
2. F. R. Mayo, *J. Amer. Chem. Soc.* 1967, **89**, 2654.
3. S. W. Benson, *The Foundations of Chemical Kinetics*, McGraw-Hill, New York and London, 1960, p. 504.
4. S. Glasstone, K. J. Laidler and H. Eyring, *The Theory of Rate Processes*, McGraw-Hill, New York and London, 1941, p. 413.
5. R. M. Noyes, *Progr. Reaction Kinetics*, 1961, **1**, 130.
6. R. M. Noyes, *J. Amer. Chem. Soc.* 1959, **81**, 566; G. A. Salmon and R. M. Noyes, *ibid.* 1962, **84**, 672.
7. D. J. Carlsson and K. U. Ingold, *J. Amer. Chem. Soc.* 1968, **90**, 7047.
8. E. Rabinowitch and W. C. Wood, *Trans. Faraday Soc.* 1936, **32**, 1381.
9. R. K. Lyon and D. H. Levy, *J. Amer. Chem. Soc.* 1961, **83**, 4290.
10. R. K. Lyon, *J. Amer. Chem. Soc.* 1964, **86**, 1907.
11. J. K. Kochi, *J. Amer. Chem. Soc.* 1970, **92**, 4395.
12. F. D. Greene, M. A. Berwick and J. C. Stowell, *J. Amer. Chem. Soc.* 1970, **92**, 867.
13. K. R. Kopecky and T. Gillan, *Canad. J. Chem.* 1969, **47**, 2371.
14. L. Herk, M. Feld and M. Szwarc, *J. Amer. Chem. Soc.* 1961, **83**, 2998; M. Matsuoka, P. S. Dixon and M. Szwarc, *ibid.* 1962, **84**, 304, and 1963, **85**, 2551; P. S. Dixon, A. P. Stefani and M. Szwarc, *ibid.* 1963, **85**, 3344.
14a. A. P. Stefani, *J. Amer. Chem. Soc.* 1968, **90**, 1694.
15. O. Dobis, J. M. Pearson and M. Szwarc, *J. Amer. Chem. Soc.* 1968, **90**, 278; K. Chakravorty, J. M. Pearson and M. Szwarc, *ibid.* 1968, **90**, 283.
16. H. Kiefer and T. G. Traylor, *J. Amer. Chem. Soc.* 1967, **89**, 6667.
17. J. H. Leffler, *J. Org. Chem.* 1955, **20**, 1202.
18. E. S. Huyser, *Adv. Free-Radical Chem.* 1965, **1**, 77.

19. H. B. Hass, E. T. McBee and P. Weber, *Ind. Eng. Chem.* 1935, **27**, 1190; 1936, **28**, 33.
20. G. A. Russell, *J. Amer. Chem. Soc.* 1957, **79**, 2977; 1958, **80**, 4987, 4997, 5002.
21. I. Galiba, J. M. Tedder and J. C. Walton, *J. Chem. Soc.* B, 1966, 604.
22. H. Singh and J. M. Tedder, *J. Chem. Soc.* B, 1966, 605.
23. M. M. Martin and G. J. Gleicher, *J. Amer. Chem. Soc.*, 1964, **86**, 233, 238, 242.
24. J. M. Pearson and M. Szwarc, *Trans. Faraday Soc.* 1964, **60**, 553; G. E. Owen, J. M. Pearson and M. Szwarc, *ibid.* 1964, **60**, 564.
25. C. Walling and B. B. Jacknow, *J. Amer. Chem. Soc.* 1960, **82**, 6108.
26. C. Walling and P. Wagner, *J. Amer. Chem. Soc.* 1964, **86**, 3368.
27. C. Walling and A. Padwa, *J. Amer. Chem. Soc.* 1962, **84**, 2845; 1963, **85**, 1593.
28. R. D. Burkhart, R. F. Boynton and J. C. Merrill, *J. Amer. Chem. Soc.* 1971, **93**, 5013.

7

Reactions of atoms

7.1 *INTRODUCTION*

In a general way, the chemistry of atoms with one unpaired electron, such as hydrogen or the halogens, is similar to that of more conventional alkyl radicals, but atoms also show some distinctive behaviour, peculiar to themselves, which sets them apart.

The experimental technique, now used for the quantitative study of radical behaviour, was largely developed for atom reactions. Many new methods of measuring atom concentrations and following atomic reactions have been developed in the last decade, and much benefit is expected when these methods are applied more widely in the field of radical chemistry. Comparison of atom with free-radical behaviour has always been most fruitful in providing a deeper understanding of radical chemistry. The field of atom reactivity is a favourite testing ground for theories of chemical kinetics, mainly because of the structural simplicity of atom systems, and this has given added impetus to the quantitative study of their behaviour.

Many atoms have low-lying electronic excited states, and it is found that an atom in an excited state may have quite different chemical behaviour to when it is in the ground state. Atoms possess fewer degrees of freedom than conventional radicals, so that they are particularly prone to 'hot', pressure-dependent behaviour. This, together with the availability of low-lying electronic states, may often lead to emission of radiation, and many chemiluminescent atom reactions are known. Similarly, the need for atoms to get rid of their excess energy, which cannot be shared amongst vibrational degrees of freedom, means that heterogeneous wall processes are also common.

We shall confine ourselves in this chapter to a consideration of the chemistry of doublet atoms, i.e. those such as H· or Cl· which have one unpaired electron. Singlet and triplet atoms are only introduced from time to time for purposes of comparison.

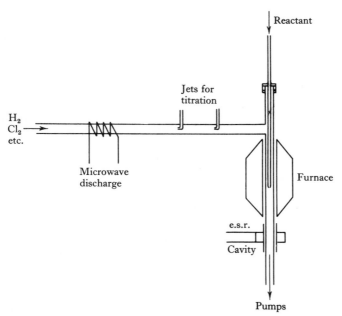

Fig. 7-1. Discharge-flow apparatus for the study of atoms at low pressures.

7.2 *PRODUCTION OF ATOMS*

Atoms are formed in many conventional thermal, photolytic or radiolytic processes. For kinetic studies, and for the elucidation of the mechanism of atom reactions, 'clean' atom sources are needed, i.e. sources which give the required atom as the only active species. Many quantitative results have been obtained from discharge-flow methods (1). In the discharge-flow system, atoms are generated by subjecting a fast flow of the diatomic gas, H_2, Cl_2, Br_2 etc., to an electric or microwave discharge. The microwave discharge has the advantage that contamination from electrodes is eliminated. The flow system is usually operated at pressures of a few torr, but can be used at higher pressures if an inert gas diluent, e.g. argon, is added to the flow. The discharged gas, containing the atom species, then flows down a tube which serves as the reactor, and any reactants can be added to the gas flow downstream from the discharge. The atom reactions can then be observed by physical and chemical methods.

Many hydrogen atom reactions have been studied by discharge-flow tube experiments with H_2. There are no low-lying excited states of the hydrogen atom so that all atoms are formed in the ground state (^2S). Other sources have, however, been extenisvely used, including mercury-photosensitized decomposition of H_2 and photolysis of HCHO, HBr and H_2S. Explosion-limit studies, flame experiments and shock tube methods have also been employed for observing hydrogen atoms under a wide variety of physical conditions.

Fluorine atoms ($^2P_{\frac{3}{2}}$, $^2P_{\frac{1}{2}}$) have been generated by thermolysis of F_2, but they can also be formed by flash photolysis of UF_6 and a wide variety of organic fluorides (2). The majority of chlorine atom studies, in the gas phase and in solution, have been carried out using photolysis of Cl_2 as the atom source (3). A microwave discharge in molecular chlorine also yields ground-state chlorine atoms ($^2P_{\frac{3}{2}}$), and discharge-flow methods are now being used (4). Bromine and iodine atoms ($^2P_{\frac{3}{2}}$) are most frequently formed by thermolysis or photolysis of Br_2 and I_2 respectively. But many other organic bromides and iodides can be useful atom sources, because of their weak carbon–bromine and carbon–iodine bonds. Both bromine (5) and iodine (6) atoms have been extensively studied by flash photolysis of the molecular halogen, which generally gives both ground- ($^2P_{\frac{3}{2}}$) and excited-state ($^2P_{\frac{1}{2}}$) halogen atoms (7).

7.3 ESTIMATION OF ATOM CONCENTRATIONS

Recent advances in technique have made it possible to measure and follow absolute atom concentrations in some cases. Direct kinetic information can thus be obtained, and much accurate rate data is available for atom systems. Relatively high atom concentrations are needed for this purpose, so that these methods are frequently used in discharge-flow systems where the atom species are pumped at high speed through the reaction zone and then measured. The products may be collected and analysed, or led directly into a mass spectrometer. High atom concentrations can also be produced in flash photolysis, pulse radiolysis and shock tube systems, where the atoms are usually measured by observing their electronic spectra. Many kinetic and mechanistic studies of atom reactions have also been carried out by the conventional photolysis or thermolysis methods, followed by product analysis.

7.3.1 Isothermal wire calorimeters

An isothermal wire calorimeter usually consists of a coil of silver-coated platinum wire mounted so as to be movable along the axis of the reaction zone in the reaction tube (8). The highly catalytic surface of the wire causes recombination of the atoms, and the energy released on combination heats the wire probe, the amount of heat released being proportional to the atom concentration. Some workers have used double probes, one following the other in the reaction zone (9), to ensure that all the atoms have reacted. The method is simple, accurate and reproducible, but has the disadvantage that it is not specific to one atom: any active species in the reaction zone may also be deactivated by the catalytic surface.

7.3.2 *Chemiluminescent reactions and gas-phase titration*

Many atom combination and addition reactions are accompanied by emission of visible or infrared radiation. The intensity of the emission is proportional to the concentrations of the species involved and can be used to determine relative or absolute atom concentrations. The atom to be studied is usually produced in a gas stream in a discharge-flow system, and the reactant, if any, is added through a jet downstream from the discharge. The light produced in the reaction can be visually observed tailing off downstream (10). The intensity of the emission is then measured with a photomultiplier fitted with an appropriate filter or monochromator.

Hydrogen atoms can be titrated with NO (11), NO_2 (12) or NOCl (13). The red emission, given with all three titrants, is due to HNO formed by:

$$H \cdot + NO \xrightarrow{\text{M}} HNO^* \longrightarrow HNO + h\nu \qquad (1)$$

In the NOCl procedure hydrogen atoms react rapidly and stoichiometrically to give NO:

$$H \cdot + NOCl \longrightarrow NO + HCl \qquad (2)$$

which is followed by reaction (1). The intensity of the emission is proportional to both $[H \cdot]$ and $[NO]$. The end point of the titration occurs when just sufficient NOCl is added to extinguish the red emission, and the added NOCl is then equal to the original hydrogen atom concentration.

Chlorine atom combination is also accompanied by red and infrared emission of radiation:

$$Cl \cdot + Cl \cdot \xrightarrow{\text{M}} Cl_2^* \longrightarrow Cl_2 + h\nu \qquad (3)$$

and this is the basis of the NOCl titration procedure (14). Chlorine atoms react rapidly with NOCl:

$$Cl \cdot + NOCl \longrightarrow Cl_2 + NO \qquad (4)$$

so that when a flow of NOCl, just equal to the original chlorine atom flow, is added to the gas stream, the chlorine combination afterglow (*3*) is extinguished. The emission intensity is proportional to $[Cl\cdot]^2$ at short wavelengths and proportional to $[Cl\cdot]$ at long wavelengths: because two different excited states of Cl_2 are formed (15). A more accurate procedure for finding the chlorine atom concentration is to plot the square root of the emission intensity at short wavelength against NOCl added, and extrapolate to $I^{\frac{1}{2}} = 0$. The concentration of NOCl is here equal to the original chlorine atom concentration.

Many variants of this gas-phase titration technique and procedure are now in use. In principle almost any chemiluminescent reaction, which can be suppressed by a fast, stoichiometric alternative reaction, can be put to this use. Atom combinations with several simple diatomic species, such as SO and CO, are chemiluminescent, and many other atom combinations, e.g. $F\cdot$ (16), $Br\cdot$ (15), are known to emit radiation. As yet, however, titration procedures for these species have not been extensively used.

7.3.3   *Electron spin resonance*

E.s.r. methods are also being successfully employed in the estimation of atom concentrations in gas discharge-flow systems, where the atom concentrations are relatively high. A full description of the e.s.r. method is given in chapter 3. E.s.r. spectra of many atomic and diatomic (17) species have been observed in the gas phase, and this appears to be a method which is particularly promising for adaptation to the study of the kinetics of larger radicals in the gas phase and in solution.

The method developed slowly owing to experimental difficulties, but is highly specific to a particular atom or radical. The signals from a particular atom are easily identifiable; thus H (^2S) gives a widely spaced doublet.

Kinetic measurements in the gas phase were first attempted by Krongelb and Strandberg (18) for atomic oxygen, and they also developed the method of comparing the atomic signal with that of molecular oxygen to obtain absolute atom concentrations. The method was developed, and its scope extended, by Westenberg and de Haas (19). Temperature variation studies of atomic reactions can also be carried out if the e.s.r. cavity is situated outside the furnace containing the reaction tube of the discharge flow system (20, 21). Many reactions of oxygen, hydrogen and other atoms have now been studied using this technique.

7.3.4   *Other methods for the measurement of atom concentrations*

The electronic absorption spectra of most atoms occur in the far ultra-violet, so that considerable experimental difficulties attend the use of these spectra for atom concentration measurements. It is more usual in a flash photolysis experiment with atoms to observe the growth of the absorption lines of the products formed from the atomic reactions, rather than the decay of the short-wavelength atomic lines themselves. Kinetic absorption spectroscopy in the vacuum ultraviolet has, however, been used to study iodine (22) and bromine (7) atom reactions, where the halogen atoms were generated by flash photolysis of organic halides. The method is highly specific to a particular atom, and has the added advantage that excited electronic states of one atom can be separately observed, and information about energy transfer between excited states obtained.

Atom concentrations can be determined with a mass spectrometer using a beam inlet system arranged so that atoms can pass directly into the ionizing region of the mass spectrometer without colliding with any surfaces (23, 24). Absolute atom concentrations can be obtained provided wall loss is negligible and the mass spectrometer is operated at low ionizing voltages, where the parent molecules (O_2, H_2, N_2, etc.) are not ionized and make no contribution to the atom mass peak. More accurate results can be obtained if a beam-chopping device is incorporated, which enables atoms in the beam to be distinguished from those which have encountered surfaces (25).

7.4   *ATOM COMBINATION REACTIONS*

When two atoms collide, the energy produced in forming a bond cannot be dissipated immediately, since the atoms lack vibrational and rotational degrees of freedom to absorb the energy. The atoms fall apart again before they can be collisionally stabilized. Atom combination can, however, occur through the intermediacy of a 'third body' or 'chaperon', usually given the symbol M. The energy produced appears principally as translational, vibrational and rotational energy of the third body. The overall combination is therefore third-order, second-order in atom concentration, and first-order in M. The reaction is usually expressed by the general equation:

$$A\cdot + A\cdot + M \longrightarrow A_2 + M \qquad (5)$$

and $d[A_2]/dt = k_5[A\cdot]^2[M]$ defines the rate constant k_5. Atom combination can also occur if the energy produced is lost by emission of radiation. Most third-order atom combination reactions are accompanied by emission of radiation, but simple bimolecular combination followed by emission can also occur. Generally emission occurs from a higher vibrational level of the diatomic molecule in the bimolecular process. Combination at surfaces can also occur, the reaction being accompanied by the emission of radiation in some cases.

Atom combination reactions have received intensive study, and the rates of some have been measured over very wide ranges of temperature and pressure. Shock-tube studies usually give values for the rate constant of dissociation of the diatomic molecule A_2 at high temperatures in the thousands of degrees range. This can be combined with the known equilibrium constant to give the combination rate constant. Discharge-flow and flash photolysis techniques give direct values of the combination constant at lower temperatures and pressures.

7.4.1 *Heterogeneous combination*

Atom combinations at surfaces are less well understood, and there is less systematic information than for homogeneous combination. There are three distinct ways whereby atom combination can occur:

$$A\cdot + A\cdot + M \longrightarrow A_2 + M \qquad (5)$$

$$A\cdot + Wall \longrightarrow \tfrac{1}{2}A_2 + Wall \qquad (6)$$

$$A\cdot + A\cdot + Wall \longrightarrow A_2 + Wall \qquad (7)$$

Most atoms appear to recombine in a wall process which is first-order in atom concentration. The rate of the surface reaction depends on the activity of the surface, the total pressure and the atom mean free path. The activity of the surface is generally expressed by the surface combination coefficient γ which is related to the rate constant of wall reaction (6): $\gamma = 2rk_6/\bar{c}$, where r is the diameter of the reaction tube, and $\bar{c}$ the mean atom velocity. Some atoms also combine in a heterogeneous process which is second-order in atom concentration. Nitrogen atoms, for instance, are known to do so (25), and hydrogen atoms also take part in a second-order wall process at high and low temperatures (26). Oxygen (1) and chlorine atoms (27) however, appear to take part in a first-order process only.

Measurement of γ can be carried out at low pressures by Smith's side-arm method (28) in which the atoms are generated in a discharge-flow tube with a closed side arm. A steady state is produced in the side arm since diffusion of atoms into it is balanced by surface and volume combination. The atom concentration in the side arm is then measured, usually with a calorimetric probe, thus enabling γ to be calculated. Alternatively, γ can be found from concentration variation experiments in conventional discharge-flow systems. Atoms are removed by reactions (5), (6) and (7) and the total rate of atom removal is given by the relation:

$$-d\ln[\text{A·}]/dt = k_6 + 2(k_7 + k_5[\text{M}])[\text{A·}] \qquad (8)$$

A plot of $d\ln[\text{A·}]/dt$ against $[\text{A·}]$ then gives k_6 as the intercept and γ may thus be obtained. Accurate values of k_6 and hence γ are obtained by working at very low pressures, when the surface recombination predominates. A plot of $2(k_7 + k_5[\text{M}])$, obtained as the gradient of the plot of equation (8), against total pressure $[\text{M}]$ then gives k_7, the second-order heterogeneous rate constant, as the intercept, and k_5, the homogeneous combination rate constant, as the gradient (27).

TABLE 7-1   *Surface recombination coefficient, γ, of various atoms at room temperature*

Atom	Surface	γ	Reference
O	Pyrex	3–4.5×10^{-5}	29
	H_3PO_4	0.3–5.0×10^{-5}	1
N	Pyrex	2×10^{-5}	30
	H_3PO_4	0.2×10^{-5}	31
H	Pyrex	2×10^{-5}	26
	H_3PO_4	2×10^{-5}	26
Cl	H_3PO_4	0.6–3.0×10^{-5}	27

Some typical values of γ for various atoms are shown in Table 7-1. Generally, unpoisoned pyrex surfaces give poor reproducibility, probably due to spots of high and low activity. Better reproducibility and lower surface activity are obtained from surfaces coated with phosphoric acid (32), or 'dri-film', which is a mixture of dimethyldichlorosilane and trimethylchlorosilane, or Teflon (26). Metal surfaces have very much higher activity, the value of γ approaching 1.0 for copper, silver and platinum surfaces. It seems probable that first-order combination on a glass surface occurs when atoms from the gas phase combine with loosely bound atoms at the surface. The loosely bound atoms are then

Atoms

rapidly replaced from the gas phase (29). For pyrex or quartz tubular reactors having $\gamma \sim 10^{-5}$ and diameters of a few centimetres, wall recombination becomes appreciable at total pressures of less than about 2 or 3 torr. At even lower pressures wall recombination is predominant. Most workers studying homogeneous combination (5) minimize wall effects by coating with one of the above materials, and measure γ *in situ*.

7.4.2 *Homogeneous combination*

Third-order homogeneous combination reactions have been studied for many atoms. The rate of combination is much less than for a simple alkyl radical such as methyl, because of the third-body restriction. The rate constants, which are generally in the region of $10^9 \ l^2 \ mol^{-2} \ s^{-1}$, are also less than for simple alkyl radicals, and depend on the nature of the third body present. The parent diatomic molecule A_2 is usually a very efficient third body. This is shown in Table 7-2 and is particularly apparent for iodine and bromine atoms.

TABLE 7-2   *Homogeneous combination rate constants*

$$A\cdot + A\cdot + M \longrightarrow A_2 + M \quad (5)$$

Atoms	M	Temp. (°K)	$k_5 \times 10^{-9}$ ($l^2 \ mol^{-1} \ s^{-1}$)	E_5 (kJ mol⁻¹)	Method	Reference
H+H	H_2	291	6.0	−4.6	Discharge-flow	9
H+H	Ar	291	4.6	−5.0	Discharge-flow	9
O+O	N_2	298	1.1	−5.9	Discharge-flow	25
O+O	Ar	298	0.6	−5.9	Discharge-flow	25
O+N	N_2	298	3.9	−1.2	Discharge-flow	25
O+N	Ar	298	3.0	—	Discharge-flow	25
N+N	N_2	298	3.5	−4.2	Discharge-flow	25
N+N	Ar	298	1.4	—	Discharge-flow	25
S+S	Ar	298	1.0	—	Discharge-flow	33
F+F	Ar	1000	0.03	—	Shock-tube	34
Cl+Cl	Cl_2	298	20	−6.7	Discharge-flow	27
Cl+Cl	Ar	298	4.4	−7.6	Discharge-flow	27
Br+Br	Br_2	298	50	—	Flash photolysis	5
Br+Br	Ar	298	2.6	—	Flash photolysis	5
I+I	I_2	300	1600	−18.5	Flash photolysis	35
I+I	Ar	300	3.0	−5.5	Flash photolysis	35

For all the atoms so far studied the combination rate constant has a small negative temperature coefficient. For iodine (35) and bromine atoms (36) the size of the temperature coefficient appears to correlate with the efficiency of the third bodies, a more efficient third body giving a larger negative temperature coefficient. For other atoms the differences in the temperature coefficients with various third bodies are very small, and it is not at present known if this correlation is generally true.

At first sight a negative temperature coefficient would appear to contravene the known laws of chemical kinetics, but it can be understood in terms of the 'atom–molecule' mechanism (35). According to this mechanism, the combination is not a single elementary reaction, but takes place via a series of steps. When two atoms meet, because they lack vibrational and rotational degrees of freedom, they fall apart again before the excited diatomic molecule can be collisionally stabilized. However, when an atom collides with a third-body molecule M, the resultant excited atom–molecule complex AM* can live long enough to be stabilized because of its extra degrees of freedom. In the presence of excess third body the mechanism will be:

$$A\cdot + M \underset{k_{10}}{\overset{k_9}{\rightleftharpoons}} AM^* \quad K_{10} \qquad\qquad (9, 10)$$

$$AM^* + M \underset{k_{12}}{\overset{k_{11}}{\rightleftharpoons}} AM + M \quad K_{12} \qquad\qquad (11, 12)$$

$$AM + A\cdot \overset{k_{13}}{\longrightarrow} A_2 + M \qquad\qquad (13)$$

where the atoms exist in equilibrium with the atom–molecule complex. The measured third-order rate constant is defined by k_5 where:

$$d[A_2]/dt = k_5[A\cdot]^2[M]$$

and hence:
$$k_5 = k_{13}K_{10}K_{12}$$

Clearly the third-order rate constant will depend on the nature of the third body, and the measured temperature coefficient E_5, which is given by a combination of terms, $E_5 = E_{13} + E_{11} + E_9 - (E_{10} + E_{12})$, may easily have a negative value. When the third body M is an inert gas such as argon, no specific chemical forces hold the atom–molecule complex together, just the weak Van der Waals interaction. Steps (9 to 13) are therefore simple collision processes occurring at the statistical rate. The difference in k_5 (in argon) for the various atoms is mainly due to their different collision diameters (see Table 7-2). Very efficient chaperons, such as I_2, probably interact chemically with the atoms in the intermediate atom–molecule complex.

When the species combining are polyatomic radicals R, the dimer R_2 has a longer lifetime than the diatomic molecule A_2 because of its vibrational and rotational degrees of freedom. The dimers formed from radicals such as methyl are normally sufficiently long-lived so that all of them are stabilized, and equilibrium between radicals and excited dimer is not set up:

$$R\cdot + R\cdot \longrightarrow R_2^* \qquad (14)$$

$$R_2^* + M \longrightarrow R_2 + M \qquad (15)$$

It is only at low total pressures, where the concentration of M is small, that reaction (15) becomes rate-determining, and radical combination is also third-order (cf. p. 199). Between these two extremes, intermediate cases exist, and many atom diatomic-radical combinations and atom polyatomic-radical combinations are known to be third-order at normal pressures. In general, for these intermediate cases both radical–molecule complexes RM* and excited dimer species R_2^* can contribute, the predominant combination route depending on the pressure, the natures of M and R·, etc.

The dissociation rates of many diatomic molecules have been measured at high temperatures in shock tube experiments. The combination rate constant can be calculated from the dissociation rate constant, since they are related through the equilibrium constant $K_{17} = k_{16}/k_{17}$.

$$A\cdot + A\cdot + M \underset{k_{17}}{\overset{k_{16}}{\rightleftharpoons}} A_2 + M \quad K_{17} \qquad (16, 17)$$

The equilibrium constants can be calculated from thermodynamic data, or measured experimentally. The temperature dependence of the combination rate constants of some atoms have been determined over extremely wide temperature ranges. In most of the cases studied the Arrhenius equation is adequate for representing the temperature variation of k_5. In some instances however, notably with bromine atoms, deviations from this simple law have been observed (36) and some workers have preferred to give their rate constants in terms of a power series such as:

$$\log k = a + b \log (T/300) + c \log^2 (T/300) + \cdots \qquad (18)$$

The rate of combination of fluorine atoms is only known at high temperatures from indirect experiments on the dissociation of F_2 (34). Several groups of workers have, however, studied the combination of ground-state chlorine atoms, and the accompanying emission of radiation:

$$Cl\cdot + Cl\cdot + M \longrightarrow Cl_2^* + M \qquad (19)$$

$$Cl_2^* \longrightarrow Cl_2 + h\nu$$

Hutton and Wright (37) have supported the suggestion that chlorine atom recombination occurs through the formation and removal of Cl_3:

$$Cl\cdot + Cl_2 \rightleftharpoons Cl_3 \qquad (20, 21)$$

$$Cl_3 + Cl\cdot \longrightarrow Cl_2 + Cl_2 \qquad (22)$$

The measured third-order rate constant, k_{19}, would then depend on the chlorine atom concentration, since this mechanism predicts:

$$1/k_{19} = [Cl\cdot]/k_{20} + k_{21}/k_{20}k_{22}$$

However, Clyne and Stedman (27) have shown that k_{19} is independent of chlorine atom concentration, and combination occurs by the normal third-order process. They further measured the temperature dependence of k_{19} and showed that extrapolation of k_{19}, with argon as third body, to high temperatures using the Arrhenius equation, leads to a value of k_{19} at about the mean of the values calculated from shock tube experiments. Extrapolation using a power series, however, gives a much lower value of k_{19} at high temperature than the shock tube data. Either there is uncertainty in the shock tube value, or else the form of the temperature dependence differs at high and low temperatures.

Ip and Burns (36) have studied the combination of bromine atoms by one direct technique, flash photolysis, over a very wide temperature range. They find that the combination rate constant k_{23}, measured directly at high temperatures, is only about 30% of the value calculated from the rate of dissociation of Br_2 (measured in shock tube experiments) using the equilibrium constant K_{24}:

$$Br\cdot + Br\cdot + M \rightleftharpoons Br_2 + M \quad K_{24} \qquad (23, 24)$$

They attribute this to the fact that the equilibrium constant is only equal to the ratio of the rate constants, when the rate constants are measured *at equilibrium*, i.e. $K_{24} = k_{23}$ (equil.)$/k_{24}$ (equil.). The dissociation rate constant k_{24} is measured far from equilibrium in shock tube experiments, so that it is not appropriate for using with K_{24} to calculate k_{23}, thereby accounting for the discrepancy. Some support for this view comes from work on the chlorine–hydrogen equilibrium (38):

$$H_2 + Cl\cdot \rightleftharpoons HCl + H\cdot$$

where a similar effect has been reported, but this is still a matter of controversy.

Iodine atom combination was one of the first reactions to be examined by the flash photolysis method (35). The combination rate constant was shown to have a negative temperature coefficient, whose value depended on the efficiency of the third body. Iodine molecules themselves were found to be extremely efficient as chaperons (35, 39), and in general a rough correlation exists between the efficiencies of the third bodies and their boiling points (40). This is good evidence for the non-specific, Van der Waals nature of the interaction between I· and M. Porter suggests that the great efficiency of molecular iodine as a chaperon is because a charge transfer complex is formed between it and atomic iodine. In support of this suggestion he has observed the charge transfer spectral band (35).

7.4.3 *Atom combination in solution*

Atom combination is second-order in solution because of the great excess of solvent available to act as third body. The rate of combination is therefore faster than in the gas phase, and similar in magnitude to that of simple alkyl radicals. The rates of combination of both bromine (41) and iodine (42) atoms have been measured in solution, and the rate constant is about 1×10^{10} l mol^{-1} s^{-1} for both atoms. Fast reactions occurring at about this rate are generally diffusion-controlled in solution, and there is good evidence that this is so for combination of both bromine and iodine atoms at room temperature. A more detailed discussion of this phenomenon is given in chapter 6 (p. 131).

7.5 *ATOM ADDITION REACTIONS*

The primary addition step in an atom addition process is exothermic, so that the adduct acquires excess energy. The excess energy can be lost by collision, the fate of the primary adduct depending on the number of collisions (and hence the total pressure), and the efficiency of each collision. If it is not collisionally stabilized the adduct may fragment by one or more pathways, the importance of fragmentation also being pressure-dependent. Some fragmentations and internal rearrangements occur too fast to be influenced by collisions, and these are pressure-independent processes. Pressure-dependent fragmentations are more important in atom addition reactions than in radical addition reactions, since in the latter case the primary adduct has more degrees of freedom to absorb the energy.

7.5.1 *Hydrogen atoms*

Most mechanistic studies of the reactions of hydrogen atoms with olefins have been carried out using photolytic sources of the atoms, such as mercury photosensitized decomposition of H_2, photolysis of H_2S or photolysis of HBr (43). Discharge-flow experiments, where the total pressure is much lower, have mainly yielded rate data (44).

When hydrogen atoms (^2S) add to olefins, excited alkyl radicals are produced. These excited radicals can dissociate back into atom and olefin, or dissociate into other products. Alternatively, the radicals may combine and disproportionate. At higher total pressures the main products are those of alkyl radical combination and disproportionation, whereas at low pressures fragmentation and rearrangement products predominate. Thus hydrogen atoms add to ethylene to give an excited ethyl radical (25), which may decompose (26) or be collisionally stabilized (27). The reaction products are then those of ethyl radical combination with hydrogen atoms (28) (which is probably a third-order process at normal pressures) and butane. Ethane and ethylene are formed by ethyl radicals combining and disproportionating with themselves (29 and 30):

$$H\cdot + C_2H_4 \longrightarrow C_2H_5{}^*\cdot \qquad (25)$$

$$C_2H_5{}^*\cdot \longrightarrow C_2H_4 + H\cdot \qquad (26)$$

$$C_2H_5{}^*\cdot + M \longrightarrow C_2H_5\cdot \qquad (27)$$

$$C_2H_5\cdot + H\cdot + M \longrightarrow C_2H_6 + M \qquad (28)$$

$$C_2H_5\cdot + C_2H_5\cdot \longrightarrow C_4H_{10} \qquad (29)$$

$$C_2H_5\cdot + C_2H_5\cdot \longrightarrow C_2H_6 + C_2H_4 \qquad (30)$$

Turner and Cvetanović studied the reaction of deuterium atoms with C_2H_4, and hydrogen atoms with C_2D_4. In the latter case, the only butane formed was $C_4D_8H_2$, showing that only $C_2D_4H\cdot$ radicals were present. In the reaction of deuterium atoms with C_2H_4, however, the butanes formed indicated that $C_2H_4D\cdot$ radicals, and also $C_2H_5\cdot$ and $C_2D_3H_2\cdot$ radicals, were participating. When deuterium atoms add to C_2H_4, the hot $C_2H_4D^*\cdot$ radicals possess enough energy to split off a hydrogen atom (32):

$$D\cdot + C_2H_4 \rightleftharpoons C_2H_4D^*\cdot \qquad (31)$$

$$C_2H_4D^*\cdot \longrightarrow C_2H_3D + H\cdot \qquad (32)$$

$$H\cdot + C_2H_4 \longrightarrow C_2H_5{}^*\cdot \qquad (33)$$

$$D\cdot + C_2H_3D \longrightarrow C_2H_3D_2{}^*\cdot \qquad (34)$$

Atoms

so that all three radicals can be formed (*33, 34*). When hydrogen atoms add to C$_2$D$_4$, however, there is not enough energy to split off a deuterium atom at room temperature and normal pressures. Step (*32*) clearly demonstrates the existence of 'hot' alkyl radical intermediates, and comparison of the two cases shows that a large kinetic isotope effect is in operation here.

Hydrogen atoms add predominantly at the terminal methylene in propylene. Addition to the terminal site produces isopropyl radicals and addition to the central site gives n-propyl radicals. The relative yields of the combination and disproportionation products of these two radicals enables the amount of each radical present to be determined (*43*). The proportion of non-terminal addition found in this way at room temperature for propene, but-1-ene and isobutene is shown below. Scott

	CH$_3$CH=CH$_2$	C$_2$H$_5$CH=CH$_2$	(CH$_3$)$_2$C=CH$_2$
% Non-terminal addition	5.7	5.7	0.5

and Jennings (49) have also shown that hydrogen atoms add predominantly at the least fluorinated end of vinyl fluoride, 1,1-difluoroethylene and trifluoroethylene.

TABLE 7-3 *Relative rate constants for the addition of hydrogen atoms to olefins at room temperature* (51, 52)

Olefin	k (rel.)	Olefin	k (rel.)
Ethylene	1.00	Trimethylethylene	1.57
Propylene	1.53	Tetramethylethylene	1.28
But-1-ene	1.58	1,3-Butadiene	7.5
Pent-1-ene	1.36	Vinyl fluoride	0.79
Isobutene	3.85	1,1-Difluoroethylene	1.45
cis-But-2-ene	0.72	cis-1,2-Difluoroethylene	0.70
trans-But-2-ene	0.90	trans-1,2-Difluoroethylene	1.16
cis-Pent-2-ene	0.60	Trifluoroethylene	1.65
trans Pent-2-ene	0.67	Tetrafluoroethylene	1.69

The relative rates of addition of hydrogen atoms to a series of hydrocarbon olefins (see Table 7-3) are now known fairly precisely from competitive experiments carried out mainly by Cvetanović and colleagues (50, 51). Measurements of the absolute rate of addition of hydrogen atoms to ethylene in discharge-flow systems (44) enable these relative rate constants to be put on an absolute scale. The general trend in the rates of hydrogen atom additions is very different from that of electronegative atoms (Tables 7-4 and 7-5) and there is no steady

TABLE 7-4   *Relative rate constants for the addition of electrophilic atoms to olefins at 25 °C*

Olefin	O (^3P) a	S (^3P) b	Se (^3P) c	Te (^3P) d	I.P. (eV)
Ethylene	1	1	1	1	10.62
Propylene	5.8	7.8	2.6	10	9.84
But-1-ene	5.8	11	7.1		9.75
cis-But-2-ene	24	18	23.9		9.34
trans-But-2-ene	28	23	56		9.27
Isobutene	25	56	44.7		9.26
Buta-1,3-diene	24	100	114		9.20
Pent-1-ene		11	5.0		9.67

a: Ref. 43; *b*: Ref. 45; *c*: Ref. 46; *d*: Ref. 47.

TABLE 7-5   *Absolute rates of addition of various atoms to ethylene in the gas phase*

Atom	$k(300 °C)$ (l mol^{-1} s^{-1})	A (l mol^{-1} s^{-1})	E (kJ mol^{-1})	Reference
O (^3P)	3–6×10^8	1.1×10^{10}	6.7	48
S (^3P)	7.2×10^8			48
Se (^3P)	1.0×10^8	1.1×10^{10}	11.8	46
Te (^3P)	2.0×10^7			47
H (^2S)	3×10^7			51
Br (^2P)	0.9×10^7			60

increase with the number of alkyl substituents in the olefin. The rate decreases from isobutene to trimethylethylene and from trimethyl-ethylene to tetramethylethylene. Also, in contrast to the electronegative atoms, the rate of addition of H· atoms increases with increasing fluorine content of the olefin (52). The trend in the rates with both alkylated and fluorinated olefins is similar to that shown by methyl radicals (see chapter 8), and hydrogen atoms behave as typical electroneutral radical species. This is particularly interesting in view of the fact that the unpaired electron on hydrogen is in an *s*-type orbital, and hydrogen atoms are the archetype σ radicals, while methyl radicals are typical π-species. The logarithms of the rate constants of hydrogen atom additions correlate with the olefin atom localization energies, in the same way as methyl radicals and other electroneutral radical species (43). A more detailed discussion of this topic is given in chapter 8 (p. 233).

7.5.2 *Fluorine atoms*

The reaction of fluorine with most organic compounds is very exothermic, and often leads to ignition or explosion. Fluorinations are usually moderated, for example, by dissolving the organic reactant in a relatively inert solvent, such as carbon tetrachloride, and bubbling through a mixture of nitrogen and fluorine; alternatively the reaction is carried out in the presence of a metal packing such as copper gauze or turnings (53). In the gas phase controlled fluorination can only be accomplished in the presence of a large excess of inert diluent. It seems likely that in many liquid-phase experiments, the main reaction occurs in the bubbles of gas, or at the gas/liquid interface, and the process is not wholly homogeneous. Pure molecular fluorine is also difficult to obtain and handle. These experimental difficulties have limited the number of studies carried out, and made interpretation of the results difficult, so that atomic reactions of fluorine are not as well understood as those of other halogens.

The mechanism of fluorine addition to hydrocarbon olefins has not been studied, but the reaction with halogeno-olefins has been examined in both gas and liquid phases (53). Addition occurs by a chain process, but the initiation step is still a matter of dispute. Fluorine atoms, generated thermally or photochemically, could be the initiating species (*35*). Alternatively the process could be initiated by molecular fluorine itself (*36*):

$$F_2 \xrightarrow[\text{or } h\nu]{\Delta} 2F\cdot$$

$$F\cdot + XCH{=}CH_2 \longrightarrow X\dot{C}HCH_2F \qquad (35)$$

$$F_2 + XCH{=}CH_2 \longrightarrow X\dot{C}HCH_2F + F\cdot \qquad (36)$$

Both reactions (*35*) and (*36*) are highly exothermic and appear plausible on thermodynamic grounds. The adduct difluoride is formed in the chain-propagating step (*37*):

$$X\dot{C}HCH_2F + F_2 \longrightarrow XCHFCH_2F + F\cdot \qquad (37)$$

With halo-olefins, appreciable quantities of halogen-substituted products are formed together with the normal adduct difluoride. For instance, in the liquid-phase fluorination of $CClF{=}CClF$, the adduct $CClF_2CClF_2$ is formed together with appreciable yields of $CClF_2CF_3$ and CCl_2CClF_2. Another interesting feature of these reactions is the formation, in good yield, of dimers:

$$2CClF_2CClF\cdot \longrightarrow CClF_2CClFCClFCClF_2$$

Olefins such as trichloro- and tetrachloroethylene, 1,2-dichloroethylene or 1,2-dichlorodifluoroethylene all gave appreciable yields of dimers, the yield being larger at lower reaction temperatures (54) The relatively high yield of dimers obtained at low temperatures has been interpreted as evidence of initiation by molecular fluorine (cf. chapter 2, p. 18). Atomic initiation would be expected to give very long chains (length $\sim 10^6$) and consequently undetectably small dimer yields. Unambiguous evidence on this point is still lacking.

7.5.3 *Chlorine atoms*

Hydrocarbon olefins react rapidly with molecular chlorine in non-polar solvents even in the dark at room temperature. These reactions were considered to go by a polar, carbonium ion pathway until recently. It has now been shown that chlorination of olefins in the dark is a radical process for linear olefins, but is polar for tri- and tetra-substituted alkenes (55). For linear olefins the radical pathway is replaced by an ionic reaction in the presence of oxygen, or when the solution is diluted with sufficient solvent. Dissociation of molecular chlorine at room temperature is insignificant, so these dark additions may involve initiation by molecular chlorine itself, as was suggested for molecular fluorine.

The initiation step appears to be second-order in olefin S:

$$S + Cl_2 \longrightarrow SCl_2 \xrightarrow{S} 2SCl\cdot \qquad (38)$$

The main product of the reaction is the adduct dichloride, formed in a chain process, but substitution products, formed by abstraction of all possible hydrogen atoms, accompany the adduct:

$$SCl\cdot + Cl_2 \longrightarrow ClSCl + Cl\cdot \qquad (39)$$

Thus in the radical chlorination of but-1-ene, the product distribution is shown below (55):

$$CH_3CH_2CHClCH_2Cl + CH_3CH=CHCH_2Cl + CH_3CHClCH=CH_2$$
$$85.5\% \qquad\qquad 7.5\% \qquad\qquad 2.5\%$$
$$+ CH_2ClCH_2CH=CH_2$$
$$4.5\%$$

In the chlorination of either *cis*- or *trans*-but-2-ene the same mixture of *meso*- and *racemic*-dichlorides is obtained. The radical intermediate is probably of the form (1), in which free rotation about the carbon–

carbon bond can occur, and not the cyclic (i.e. chlorine-bridged) intermediate (2), which would be expected to lead to products of stereo-specific addition.

$$
\begin{array}{cc}
\underset{\text{H Cl}}{\overset{\text{CH}_3 \quad \text{CH}_3}{\diagup \diagdown \diagup}}{\text{C}-\text{C}}\diagdown \text{H} & \underset{\text{H Cl H}}{\overset{\text{CH}_3 \quad \text{CH}_3}{\diagup \diagdown \diagup}}{\text{C}-\text{C}}\diagdown \\
(1) & (2)
\end{array}
$$

The proportion of hydrogen abstraction products increases as the temperature is raised, until in the vapour phase at temperatures greater than 400 °C it predominates. Allylic chlorination of propylene at 400 °C is an efficient industrial route to allyl chloride, and the derived glycerol and epichlorohydrin. Chlorine atom addition is faster than hydrogen abstraction, but at high temperatures the reverse of the addition step (-40) is also fast, whereas the substitution reaction is not appreciably reversible (84).

$$
\text{Cl·} + \text{CH}_3\text{CH}{=}\text{CH}_2
\underset{k_{42}}{\overset{k_{40},\,k_{-40}}{\rightleftarrows}}
\begin{cases}
\text{CH}_3\dot{\text{C}}\text{HCH}_2\text{Cl} \xrightarrow[k_{41}]{\text{Cl}_2} \text{CH}_3\text{CHClCH}_2\text{Cl} + \text{Cl·} \\[2mm]
\dot{\text{C}}\text{H}_2\text{CH}{=}\text{CH}_2 \xrightarrow[k_{43}]{\text{Cl}_2} \text{CH}_2\text{ClCH}{=}\text{CH}_2 + \text{Cl·} \\
+ \text{HCl}
\end{cases}
\quad (40\text{–}43)
$$

The ratio of the rates of substitution $R(\text{Subs.})$ to the rate of addition $R(\text{Add.})$ is given by:

$$
\frac{R(\text{Subs.})}{R(\text{Add.})} = \frac{k_{43}[\text{Cl}_2][\dot{\text{C}}\text{H}_2\text{CH}{=}\text{CH}_2]}{k_{41}[\text{Cl}_2][\text{CH}_3\dot{\text{C}}\text{HCH}_2\text{Cl}]} = \frac{k_{42}}{k_{40}}\left(1 + \frac{k_{-40}}{k_{41}[\text{Cl}_2]}\right) \quad (44)
$$

At low temperatures k_{40} will be larger than k_{42} or k_{-40} and addition predominates, but increasing the temperature increases k_{42} and k_{-40} far more than k_{40}, since the abstraction step (42) and radical decomposition (-40) have higher activation energies, i.e. $E_{-40} > E_{42} > E_{40}$. So that at high temperatures substitution is favoured. It is also clear from the equation (44) that allylic substitution is favoured by low molecular halogen concentrations.

The relative reactivities of various hydrocarbon olefins have been measured from competitive experiments. There is little change in olefin reactivity with increasing alkyl substitution, although a slight deactivating effect is observed as would be expected for electrophilic chlorine atoms.

The addition of chlorine atoms to halo-olefins has been studied in both gas (43) and liquid phases (55). In the gas phase the reaction occurs as a chain:

$$Cl\cdot + S \rightleftharpoons SCl\cdot \qquad (45)$$

$$SCl\cdot + Cl_2 \longrightarrow ClSCl + Cl\cdot \qquad (46)$$

$$2SCl\cdot \longrightarrow ClS_2Cl \qquad (47)$$

$$SCl\cdot + Cl\cdot \longrightarrow ClSCl \qquad (48)$$

$$Cl\cdot + Cl\cdot + M \longrightarrow Cl_2 + M \qquad (49)$$

In general, termination occurs mainly by reaction (47), making step (46), the abstraction of a chlorine atom from Cl_2 by the adduct radical, rate-determining. At low olefin concentrations however, termination occurs by steps (48) and (49) and the rate law is complex. Dainton and co-workers (43) studied the isomerization of pure *cis*- and of pure *trans*-1,2-dichloroethylene relative to the rate of chlorination proceeding at the same time. The rate of isomerization was expected to be extremely small, since the addition step (45) is not appreciably reversible at temperatures below about 150 °C. They found, however, that isomeriza-tion was fast even at room temperature, and that the rate of isomerization was practically temperature-independent. The explanation of these effects is that chlorine atoms add to give an initial adduct radical which is 'hot'. The original olefin configuration is lost by rotation about the carbon–carbon bond in this radical which, since it is hot, can decompose in a temperature-independent process. Hot radical intermediates are also formed in chlorine addition to other halogeno-olefins. Chlorine atoms are similar to hydrogen or oxygen atoms in this respect.

The rate-determining step is abstraction of chlorine from Cl_2 by the adduct radicals (46) at normal olefin concentrations, so that study of the atom addition step (45) is difficult in the gas phase. In solution however, excited radicals are very rapidly deactivated by collision with solvent molecules, and relative reactivities of chloro-olefins have been deter-mined in various solvents (see Table 7-6). It is usually assumed that the chlorine atom adds to the least substituted end of the olefin, but no definite information on this point has been obtained. The rate of addition decreases with the number of chlorine substituents on the olefin. It is often supposed that the rate of addition is controlled by the

TABLE 7-6   *Relative rates of addition of chlorine atoms to chloro-olefins at 25 °C (55)*

Olefin	k (rel.) Non-complexing solvent	k (rel.) CS$_2$
CCl$_2$=CH$_2$	1.2	7.0
cis-CHCl=CHCl[b]	0.6	0.9
trans-CHCl=CHCl[b]	0.5[a]	0.5[a]
CHCl=CCl$_2$	0.7	0.9
CCl$_2$=CCl$_2$[b]	0.2	0.03

[a] Other values relative to this.
[b] The rate of addition to olefins with two identical sites has been divided by two.

stability of the addend radical, which is determined by the extent of delocalization of the unpaired electron. Clearly, this is not the case here. With 1,1-dichloro-, trichloro- or tetrachloroethylene the unpaired electron in the adduct radical is probably sited on a –CCl$_2$ group, and the reactivities should be about equal on this basis. In 1,2-dichloro-ethylene the electron is sited on the CHCl group, alpha to only one electronegative chlorine atom, so this olefin would be expected to be less reactive. It seems more likely that the addition rate is controlled by steric and polar forces between the chlorine atom, and the site of attack in the olefin. Thus, with 1,1-dichloroethylene addition is at the CH$_2$ site and polar repulsion is minimal. In both 1,2-dichloroethylene and trichloroethylene addition is at the =CHCl group and polar repulsion is approximately the same, whilst with tetrachloroethylene addition is at the =CCl$_2$ group where repulsion would be greatest. The reactivity order is in much better accord with this picture, although it is probably an oversimplification.

Chlorine atoms also add to acetylenes in a fast radical process, which can become explosive. With acetylene itself, both *cis*- and *trans*-dichloroethylene with some tetrachloroethane are formed. Liquid-phase chlorination of but-1-yne at −9 °C is a radical process and addition to the triple bond is the major reaction giving mostly the *trans*-dichloro-butene. Other products include those of hydrogen abstraction from the primary and secondary carbon atoms, but these are less important than in the case of but-1-ene, because the triple bond deactivates the abstraction of adjacent hydrogen atoms.

Radical chlorination of butadiene gives the 1,2-addition product

3,4-dichlorobut-1-ene, and the 1,4-addition product 1,4-dichloro-*trans*-but-2-ene in a ratio of about 1 to 3.5. Only traces of the *cis*-1,4 product are formed. Butadiene exists predominantly in the *transoid* form (3) and addition to this produces the *trans*-radical which then reacts to give predominantly the *trans*-1,4 product, together with some of the 1,2-product.

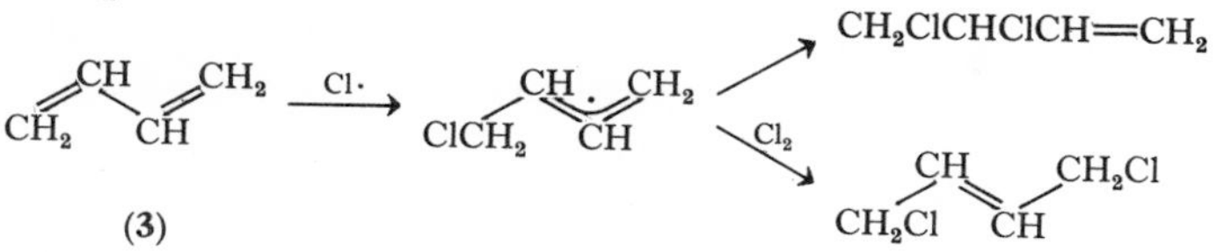

7.5.4 *Bromine atoms*

Both hydrogen bromide and bromine can add to olefins in a radical chain process whose first step is the addition of a bromine atom to the olefin double bond. In solution, non-radical processes may compete. The products of the ionic process with bromine are usually identical with the products of bromine radical addition, so that the results of HBr addition give a clearer picture of the mechanism. Both reagents have been extensively used in the synthesis of bromides and dibromides from olefins (56). The two chain-propagating steps in the radical addition of HBr to olefins (50) and (51) are exothermic:

$$Br\cdot + S \rightleftharpoons BrS\cdot \qquad (50)$$
$$BrS\cdot + HBr \longrightarrow BrSH + Br\cdot \qquad (51)$$

This probably explains why radical addition is facile with HBr, but virtually unknown with HF, HCl or HI. For these latter one of the propagating steps is endothermic.

Bromine atoms generally add to the least substituted end of an olefin, and this is true not only for alkyl substituents, but for fluorine substituents too. This orientation of the addition, leading to the 'anti-Markovnikov' product, was first observed by Kharasch and Mayo, and explained by them, and independently by Hey and Waters in terms of radical intermediates as distinct from ionic species.

The addition step is reversible at room temperature and even below, so that the rate of addition has an apparently negative temperature coefficient:

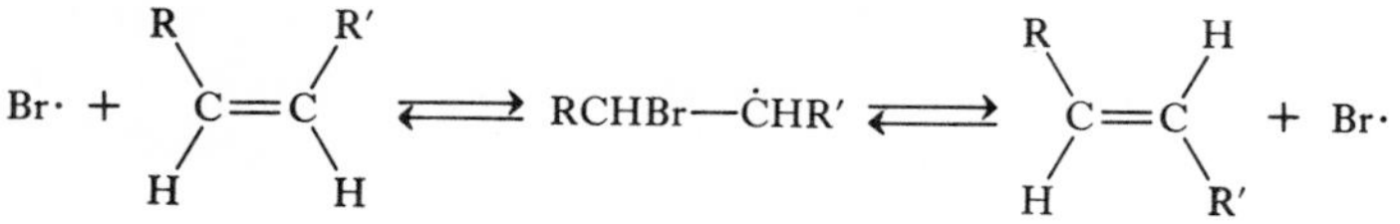

Reversibility of the addition also leads to isomerization of the olefin if a *cis*- or *trans*-1,2-disubstituted olefin is used (cf. chapter 12, p. 472). Study of the stereochemistry of the addition is thus complicated. By working at very low temperatures and high HBr concentrations, the reversibility of the addition is minimized, cf. equation (*44*). The addition of DBr to *cis*- and *trans*-but-2-ene at −70 °C gave almost completely stereospecific product formation (57). Thus more than 90% yield of the *threo*-3-deuterio-2-bromobutane was obtained from the *cis*-olefin and a similar yield of the *erythro*-isomer from the *trans*-olefin.

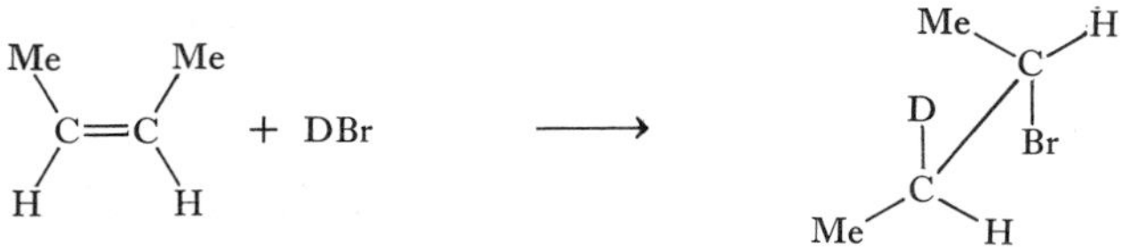

cis-but-2-ene   95% *threo*-3-deuterio-2-bromobutane

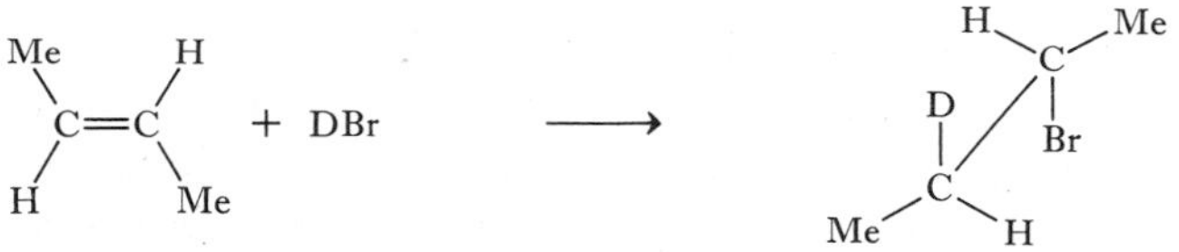

trans-but-2-ene   98% *erythro*-3-deuterio-2-bromobutane

Goering *et al.* have examined the stereochemistry of the addition to cycloalkenes, where olefin isomerization cannot occur (58). Radical addition of hydrogen bromide to 1-bromocyclohexene, or 1-chlorocyclohexene, gave exclusively the product of *trans*-addition, i.e. the *cis*-1,2-dihalocyclohexane. Other 1-substituted cycloalkenes, such as 1-methylcyclopentene, 1-methylcycloheptene and 1-bromocyclobutene, also gave exclusive, or predominant *trans*-addition of HBr to form the *cis*-product. These observations can be explained in two ways. The bromine atom approaches the double bond axially (**4**) to give an intermediate radical (**5**), which reacts rapidly with HBr before it can invert (Scheme 1). The basis of this argument is therefore that hydrogen abstraction from HBr is faster than ring inversion. Alternatively, the stereospecificity of the addition may be due to participation of a 'bridged' structure for the intermediate radical (**6**) which fixes it in the configuration of the original olefin until the abstraction step occurs. The exclusive *trans*-addition probably occurs because the approach of HBr to the intermediate radical (**5** or **6**) from an axial direction is least sterically hindered,

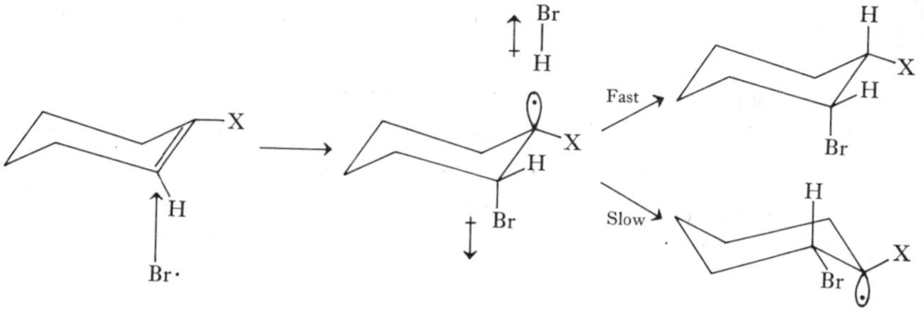

Scheme 1

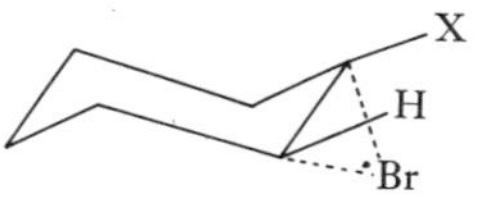

(6)

and the HBr dipole is then orientated away from the dipole of the C–Br bond so that polar interaction is also minimized.

The relative reactivities of a series of olefins towards HBr addition in the gas phase have been measured by a competitive technique (59). The addition step (50) is reversible, so the olefin reactivity is a function of the forward and reverse steps, and does not measure directly the rate of bromine atom addition. The rate of atom addition can only be obtained from series of experiments in which the reactant pressures are varied, or from atom concentration measurements. The absolute rate of bromine atom addition to ethylene has been measured in this way (60). The reactivity measurements indicate that bromine atoms behave as strongly electrophilic radicals, and the order of reactivity parallels that for other electrophilic species such as O (^3P) (cf. p. 161).

Radical addition of molecular bromine to olefins is broadly similar to hydrogen bromide addition. In solution both radical and ionic processes can occur, and both routes give the same product dibromide. It is not yet known if spontaneous radical initiation by molecular bromine can occur, as seems probable with chlorine and fluorine:

$$Br_2 + S \longrightarrow BrS\cdot + Br\cdot$$

cf. reactions (36) and (38). The addition step in the free-radical process is reversible, and the amount of reverse reaction increases with temperature. This leads to *cis–trans*-isomerization of the reactant olefins, so that studying the stereochemistry of the addition step is difficult, especially as there may be a concurrent ionic addition which is known to occur stereospecifically. The radical addition step, carried out under conditions where isomerization of the olefin does not take place, is apparently stereospecific. Thus predominant *trans*-addition was shown to occur in the light-initiated reaction of bromine with cyclohexene, *cis*- and *trans*-but-2-ene-1,4-diols and several other olefins (58).

7.5.5 *Iodine atoms*

Unlike the other halogens, molecular iodine does not add readily to olefins. This is because of the instability of the diiodides, and iodoalkyl radicals:

$$I\cdot + RCH{=}CH_2 \rightleftharpoons R\dot{C}HCH_2I$$
$$R\dot{C}HCH_2I + I_2 \rightleftharpoons RCHICH_2I + I\cdot$$

The equilibria are shifted to the left in both the addition and abstraction steps and no product is obtained. That the initial addition step does occur can be demonstrated, since geometrical isomerization of olefins can be induced by photolysis of I_2–olefin mixtures.

7.6 *ATOM ABSTRACTION REACTIONS*

7.6.1 *By hydrogen atoms*

Much effort has been put into the measurement of accurate rate constants for the reaction of hydrogen atoms with molecular hydrogen (61), because the abstraction step is one of the simplest chemical reactions, and the most amenable to theoretical treatment. The products of the reaction are identical with the reactants (52), so rates are measured with isotopically substituted species (53, 54), or else the rate of conversion of *ortho*- into *para*-hydrogen is measured (55).

$$H\cdot + H_2 \longrightarrow H_2 + H\cdot \tag{52}$$
$$H\cdot + D_2 \longrightarrow HD + D\cdot \tag{53}$$
$$D\cdot + H_2 \longrightarrow HD + H\cdot \tag{54}$$
$$H\cdot + o\text{-}H_2 \longrightarrow p\text{-}H_2 + H\cdot \tag{55}$$

The rates of reactions (53 and 54) obey the Arrhenius law at high temperatures, but curvature is shown in the low-temperature region (62). Such curvature is usually attributed to quantum-mechanical tunnelling (61), but Westenberg and de Haas show that theoretical treatment including tunnelling gives no better agreement with experiment than calculations neglecting tunnelling (62).

Hydrogen atoms also abstract hydrogen from alkanes (54, 26), the initial product being an alkyl radical:

$$H\cdot + RH \longrightarrow H_2 + R\cdot \qquad (56)$$

$$H\cdot + R\cdot \longrightarrow RH \qquad (57)$$

$$H\cdot + R\cdot \longrightarrow H_2 + \text{olefin} \qquad (58)$$

$$R\cdot + R\cdot \longrightarrow R_2 \qquad (59)$$

$$R\cdot + R\cdot \longrightarrow RH + \text{olefin} \qquad (60)$$

At low pressures, as in discharge-flow experiments, the alkyl radical R· may fragment, but normally the main products of the reaction are hydrogen, together with products of alkyl radical combination and disproportionation (59 and 60), and the products from the reaction of alkyl radicals with hydrogen atoms (57, 58). Well-established rate constants for a series of alkanes are still lacking, but it appears that the rate increases for alkanes containing secondary hydrogens and increases further when tertiary hydrogens are present.

Hydrogen atoms also react with haloalkanes when the main reaction is abstraction of hydrogen or halogen. The mechanism is similar to that with unsubstituted alkanes: for example with fluorotrichloromethane, both chlorine and fluorine are abstracted (63):

$$CCl_3F + H\cdot \longrightarrow HCl + CCl_2F\cdot \qquad (61)$$

$$CCl_3F + H\cdot \longrightarrow HF + CCl_3\cdot \qquad (62)$$

$$CCl_2F\cdot + H\cdot \longrightarrow CCl_2FH^* \qquad (63)$$

$$CCl_3\cdot + H\cdot \longrightarrow CCl_3H^* \qquad (64)$$

$$CCl_2FH^* \longrightarrow CClF: + HCl \qquad (65)$$

$$CCl_3H^* \longrightarrow CCl_2: + HCl \qquad (66)$$

$$CCl_2FH^* + M \longrightarrow CCl_2FH + M \qquad (67)$$

$$CCl_3H^* + M \longrightarrow CCl_3H + M \qquad (68)$$

One might expect the trihalomethyl radicals thus formed (61, 62) to abstract a hydrogen atom from the starting H_2. This step is slightly endothermic, however, and consequently the trihalomethyl radicals prefer to combine with hydrogen atoms (63, 64) since this step is exothermic, and hydrogen atoms are present in fairly high concentration in the discharge-flow apparatus. The excited halomethane thus formed can decompose to a carbene (*65, 66*), or be collisionally stabilized (*67, 68*). This is in fact analogous to the combination and disproportionation process (*57, 58*) observed with hydrocarbon alkanes. Clark and Tedder measured the relative ease of abstraction of various halogens from polyhalomethanes (*63*). They found the order of reactivity to be:

$$H > Br > Cl > F.$$

The rate of abstraction of a particular atom depended to a small extent on its near chemical environment, but was not simply proportional to the strength of the bond being broken.

7.6.2 By halogen atoms

The abstraction reactions of different halogen atoms show far-reaching similarities, and comparison of their behaviour is particularly interesting. With all the halogens the most widely used atom source is the molecular halogen itself, X_2, which dissociates on photolysis or thermolysis. Other halogenating reagents have found quite wide use in synthetic chlorination and bromination, and they are discussed later (p. 188).

The abstraction of a hydrogen atom from an alkane (*70*) is an exothermic process with fluorine atoms.

$$X_2 \xrightarrow[\text{or } \Delta]{h\nu} 2X\cdot \tag{69}$$

$$X\cdot + C_2H_6 \longrightarrow C_2H_5\cdot + HX \tag{70}$$

$$C_2H_5\cdot + X_2 \longrightarrow C_2H_5X + X\cdot \tag{71}$$

$$X\cdot + X\cdot + M \longrightarrow X_2 + M \tag{72}$$

$$\Delta H^0 \text{ (kJ mol}^{-1})$$

$X =$	F	Cl	Br	I
Reaction (*70*)	160	20	-46	-113
Reaction (*71*)	~ 285	97	92	71

It becomes approximately thermoneutral with chlorine atoms, and is increasingly endothermic from bromine to iodine. Thus, in bromination the abstraction step (*70*) may be reversible at high temperatures and low bromine concentrations. In iodination it is reversible under most initial conditions. The second propagation step in the chain halogenation is the abstraction from the molecular halogen by the alkyl radical (*71*). This step is exothermic with all the halogens, but becomes less so from fluorine to iodine; it may also be reversible under certain conditions in iodination.

In fluorination the reaction is explosive, and a large excess of inert solvent must be used in the liquid phase, or inactive diluent in the gas phase. Under these conditions, steps (69) to (72) represent the halogenation mechanism. The chain termination step is atom combination (72), and other radical–radical processes are negligible. The products of the reaction are the alkyl halide RX, and halogen acid HX. In bromination and iodination elimination of HX may also occur from the product halide to give an olefin A.

$$RX \longrightarrow A + HX \tag{73}$$

Only traces of the olefins are isolated in bromination, as it reacts quickly with HBr to give back the alkyl bromide, or with Br_2 to give the dibromide. No reaction occurs, however, between the olefin and iodine or hydrogen iodide (see section 7.5.5).

$$A + HBr \longrightarrow RBr$$
$$A + Br_2 \longrightarrow RBr_2$$

The experimental method, mechanism and products are particularly straight-forward in chlorination and bromination, and these processes are amongst the most intensively studied of all free-radical reactions. Experimental difficulties have limited the study of fluorinations, and kinetic results are few. It is now well established that iodination of alkanes occurs entirely by the atom mechanism (64), and not as a reaction with molecular iodine, and an increasing number of iodination studies are appearing in the literature.

7.6.3 *Halogenation of alkanes*

Halogenation of a straight or branched alkane gives a mixture of all possible halides formed by abstraction of the various hydrogen atoms present in the alkane. Rearrangements of the carbon chain do not occur. The rate of abstraction of hydrogen atoms in the molecule depends on the chemical environment of the hydrogen itself and the kind of halogen atom in use. Relative reactivities in halogenation are usually given as relative selectivities, a useful concept first introduced by Hass, McBee and Weber (65). The relative selectivity is the relative rate of abstraction per hydrogen atom. If there are x hydrogen atoms at position x in the alkane and y hydrogen atoms at position y, then the relative selectivity per hydrogen atom at y compared to that at x is RS_x^y.

$$RS_x^y = \frac{k_y}{k_x} \cdot \frac{x}{y}$$

TABLE 7-7   *Relative selectivities in hydrogen abstraction by
halogen atoms at 300 °K*

Halogen	—CH_3	=CH_2	≡CH	Reference
F	1	1.3	1.8	66
Cl	1	4.4	6.7	55
Br	1	80	1600	58
I	1	1850	210000	67

where k_x and k_y are the rate constants of hydrogen abstraction at x and y respectively.

Tertiary hydrogens are more easily abstracted than secondary, and secondary more easily than primary by all the halogen atoms. Some relative selectivities for primary, secondary and tertiary hydrogen atoms in simple alkanes are shown in Table 7-7. Fluorine atoms are the least selective and show little discrimination between the various types of hydrogen, whereas iodine atoms react almost exclusively with tertiary hydrogens. The tertiary:secondary:primary ratios decrease as temperature is increased. For fluorine and chlorine atoms the activation energy differences $E_p - E_s$ and $E_s - E_t$ are very small, generally less than about one kJ mol^{-1}. Thus the more reactive the halogen atom, the less selective it is. There is little difference in atom selectivity from the gas to liquid phase as long as the halogenation is carried out in neat alkane or other non-complexing solvent. In general, the reaction is slightly more selective in the gas phase, but the effect is small. In complexing solvents, such as benzene, toluene or carbon disulphide, the halogenation may become much more selective (see p. 138).

In chlorination of n-alkanes it has been demonstrated that the reactivity of secondary hydrogen atoms differs little with their position in the chain. Thus the RS_p^s for each $-CH_2-$ group at C_2, C_3, C_4, etc. in n-hexane, n-heptane, n-octane or n-decane varies only slightly along the chain (55). The actual value of RS_p^s can be greatly increased by carrying out the chlorination in a complexing solvent such as benzene or CS_2, but here again the same increased RS_p^s is maintained down the chain. In the chlorination of branched alkanes some hydrogen atoms are apparently deactivated, and these usually are in sterically inaccessible positions. Conversely, however, some hydrogens in inaccessible positions are abstracted at quite normal rates. In addition to steric effects, the hydrogen atoms of large branched alkanes are susceptible to intramolecular hydrogen transfer. The most common type is 1,5-hydrogen migration to the radical site, in which the transition state attains a favourable geometry (see p. 520). Abnormal reactivity can be expected wherever this situation can occur (see also p. 182).

TABLE 7-8   *Arrhenius parameters for hydrogen abstraction from alkanes by halogen atoms*

			RH+X· ⟶ R·+HX					
	F[a]		Cl[a]		Br[b]		I[c]	
RH	Log A	E	Log A	E	Log A	E	Log A	E
H₂	10.7	7.1	10.9	23.1	11.5	82.7	11.4	143
CH₃—H	10.5	5.0	10.6	16.0	11.0	78.1	11.5	141
C₂H₅—H	10.0*	1.3*	11.0	4.2	11.6	58.8	11.6	115
C₃H₇—H	9.7	0.0	11.0	4.2	—	—	11.6	121
(CH₃)₂CH—H	9.8	0.0	10.9	2.9	11.0	43.7	11.3	102
(CH₃)₃C—H	9.8	0.0	10.3	0.4	10.6	32.8	11.0	92

Log A in l mol⁻¹ s⁻¹, E in kJ mol⁻¹.
[a] Ref. 3, [b] Ref. 68 and 69, [c] Ref. 67.
* Assumed values: all other fluorine atom results relative to these.

In Table 7-8 some absolute values of the Arrhenius parameters for hydrogen abstraction are shown. The high value of the A-factors for all the reactions indicates a loose transition state, with virtual free rotation of the attacking halogen about the alkane, nearly every encounter leading to product formation. For each alkane the activation energy increases from fluorine to iodine, as would be expected since the reaction becomes successively more endothermic. For the iodine and bromine atom reactions, which are endothermic, the reaction is controlled by the strength of the C–H bond being broken. Both sets of reactions give good agreement with the Polanyi equation, which predicts a linear relationship between activation energy and enthalpy of a reaction:

$$E = \alpha\Delta H^0 + C$$

For bromine atoms α is 0.86 (3) and for iodine atoms α is 0.91 (67). These high values of α clearly indicate the controlling influence of the strength of the bond being broken, and suggest that the transition state may resemble the products more than the reactants, i.e. the C–H bond is almost fully broken and the H–X bond is formed, as in (7). In the exothermic fluorine and chlorine atom reactions, correlation of the activation energy with ΔH is less good and the values of α are small, about 0.1 for chlorination (55). Here the strength of the bond being broken is not the controlling influence and the transition state resembles the reactants rather than the products as in (**8**).

$$C \ldots\ldots\ldots H \ldots I(Br) \qquad\qquad C \ldots H \ldots\ldots\ldots F(Cl)$$

$$(7) \qquad\qquad\qquad\qquad (8)$$

Chlorination of cycloalkanes in several different solvents has been studied by Russell (70). Unrearranged chlorocycloalkane products are obtained from the C_4 to C_8 compounds. The reactivity of the ring hydrogens is comparable to that of secondary hydrogens in open-chain compounds. The relative reactivities of the cycloalkanes, per hydrogen atom, are shown in the text-table for chlorination in CCl_4 solution and in the gas phase. Whittle (72) has studied bromination of cyclo-

Relative rates of abstraction of hydrogen from cycloalkanes

	C_4	C_5	C_6	C_7
Cl· in CCl₄ soln	—	1.03	1.00	1.12
Cl· in gas phase at 350 °K	—	1.11	1.00	—
Br· in gas phase at 350 °K	0.091	3.21	1.00	21.4

alkanes where again only the expected cycloalkyl bromides are formed from the C_4 to C_7 cycloalkanes. The relative rates of abstraction, per hydrogen atom, are shown in the text-table. In both chlorination and bromination a minimum in rate is observed for the C_6 compound. The reactivity order in bromination undoubtedly follows the order of the $D(C–H)$ in the cyclanes, which is $C_7 < C_5 < C_6 < C_4$. In chlorination the bond dissociation energy plays a less dominant role, because the extent of bond breaking in the transition state is less, and the minimum at C_6 is much shallower than in bromination.

Chlorination and bromination of cyclopropane both give the straight-chain 1,3-dihalopropane, in addition to the expected cyclopropyl halide (Scheme 2).

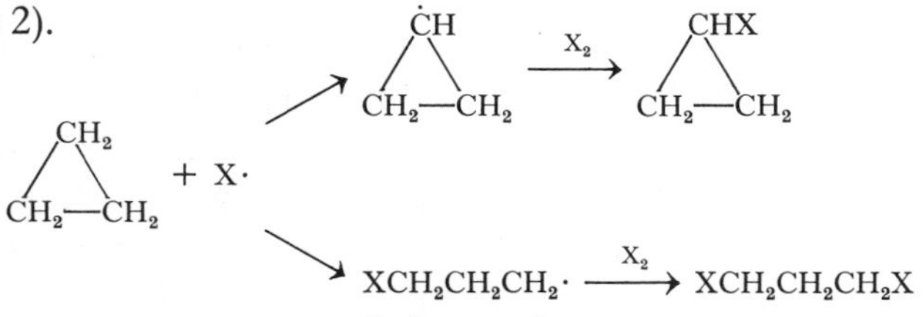

Scheme 2

The ring-opening process can be regarded as a halogen atom displacement on carbon, though it is perhaps more plausible to consider it as an *addition* of the halogen atom to the 'pseudo-π-system' of the cyclopropane ring. Alkyl-substituted cyclopropanes and spiroalkanes such as spiropentane (9), and spirohexane (10), also give ring-opened products together with normal hydrogen abstraction products from the alkyl side chain, or from the larger ring in a spiroalkane (55).

(9)   (10)

Generally, the hydrogens of a two-carbon bridge in a bicycloalkane are the most reactive to halogenation. Norbornane (11) gives a 3-to-1 mixture of the *exo*-2-halide and *endo*-2-halide in chlorination or brom-

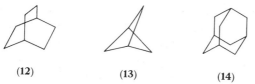

(11)

ination. There is little attack at the tertiary bridgehead hydrogen, or at the secondary hydrogens of the one-carbon bridge. Formation of the *exo*-product is favoured because approach of the halogen from the *exo*-side is least hindered sterically. Chlorination of other bicycloalkanes such as bicyclo[2,2,2]octane (12) or bicyclo[1,1,1]pentane (13) has shown

(12)   (13)   (14)

that attack at the bridgehead can occur, and that the bridgehead radical, unlike the analogous bridgehead carbonium ion, *can* be formed fairly readily. Chlorination of the strain-free adamantane (14) gives both possible monochlorides such that $RS_s^t = 1.9$; which is almost the same as the value found with open-chain compounds. Bromination of bi- and tricyclo compounds leads to formation of dibromo-products in addition to the monobromides, and the overall mechanism and relative reactivities are unclear.

7.6.4 *Halogenation of aralkanes*

Halogenation of aralkanes can lead to either substitution of the aromatic nucleus or of the alkyl side chain, depending on the reaction conditions. Radical conditions favour the latter process. Chlorination and bromination of series of *meta-* and *para*-substituted toluenes have been investigated (55, 58). The relative reactivities of the compounds in hydrogen abstraction from the side chain, are not simply explicable in terms of the strength of the bond being broken, but imply that 'polar effects' are of major importance. The polar effect may be the lowering of the energy of the transition state by an inductive effect of the substituent, or else by participation of resonance structures in which an electron is transferred to the halogen. Halogen atoms are electrophilic and so show increased reactivity for positions in a molecule to which electrons are supplied, and decreased reactivity when electron-withdrawing substituents are present.

The importance of the polar effect in atomic halogenations is demonstrated by the observation of Hammett sigma–rho correlations. In both chlorination and bromination of substituted toluenes correlations of the rate constant ratios with σ^+ are observed. In dilute carbon tetrachloride solution, where complexing with the aromatic reactants is minimized, chlorination gives a ρ value of -0.66. The value of ρ observed in bromination depends somewhat on reaction conditions, such as temperature and concentration of HBr formed (HBr can react with the benzyl radicals: $R\cdot + HBr \longrightarrow RH + Br\cdot$). Values of about -1.76 are obtained at low HBr concentrations and at room temperature.

The negative ρ values show clearly that the reactivity of the methyl hydrogens increases when the substituent in the aromatic ring is electron-releasing, and decreases due to the unfavourable polar effect when the substituent is electron-withdrawing. The larger absolute value of ρ observed in bromination shows that the polar effect has an even greater influence here, and is a further indication of the greater extent of bond breaking in the bromine transition state, than in the chlorine atom case (see (7) and (8)).

7.6.5 *Halogenation of substituted alkanes*

Hydrogen abstraction from halogenated methanes and ethanes has been studied for fluorination, chlorination and bromination. The reactivity of the methane increases with the number of halogen substituents

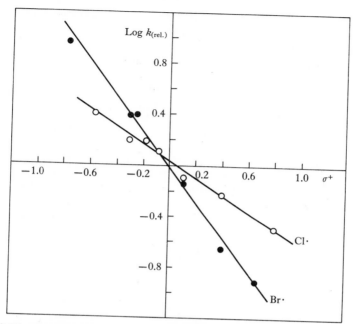

Fig. 7-2. Hammett–Brown correlations for chlorination and bromination of substituted toluenes. Open circles: chlorination data of Russell and Williams at 40 °C in CCl₄ solution (*J. Amer. Chem. Soc.* 1964, **86**, 2357). Filled circles: bromination data of Pearson and Martion at 80 °C (*J. Amer. Chem. Soc.* 1964, **85**, 3142).

present, but the rate of increase becomes less as more halogen substituents are added. Arrhenius parameters for the methanes and ethanes are shown in Table 7-9. There is a distinct minimum in the activation energies for all three types of halogenation on going from methane itself to a trihalomethane. The C–H bond dissociation energy also passes through a minimum in the series CH_4 to CHF_3, but the whole of the experimental effect cannot be attributed to bond dissociation energy changes, since there is a steady decrease in $D(C–H)$ from CH_4 to $CHCl_3$, yet the minimum in the activation energies is still apparent. The most likely explanation seems to be that there are two opposing effects; the decrease in $D(C–H)$ is responsible for the decreasing activation energies down the series CH_4 to CH_2X_2, but this is opposed by a repulsive polar effect, which offsets it. In the final member of each series, CHX_3, the opposing polar effect is predominant and the activation energy rises. The repulsive polar effect is most probably electrostatic repulsion between

TABLE 7-9   *Arrhenius parameters for hydrogen abstraction from haloalkanes*

RH	F[a] Log A	E	Cl[b] Log A	E	Br[c] Log A	E
CH_4	10.5	5.1	10.6	16.1	11.0	77.7
CH_3F	—	4.5	—	—	10.7	67.3
CH_2F_2	—	10	—	—	10.4	69.4
CHF_3	—	—	—	35.1	10.1	93.2
CH_3Cl	10.0	4.5	10.0	13.8	10.8	61.4
CH_2Cl_2	9.3	0.7	10.1	12.5	—	—
$CHCl_3$	9.5	2.8	9.8	14.0	9.4	38.9
C_2H_6	10.0†	1.3†	11.0	4.2	11.1	57.3
$CH_3\overset{*}{C}H_2F$	—	—	—	—	10.1	48.5
$CH_3\overset{*}{C}HF_2$	—	—	—	—	10.3	61.0
CH_3CF_3	—	—	—	—	11.0	98.2
CF_3CH_2F	—	—	—	—	10.6	81.5
CF_3CHF_2	—	—	—	—	10.2	75.2
CHF_2CHF_2	—	—	—	—	10.6	81.1

Log A in $l\,mol^{-1}\,s^{-1}$, E in $kJ\,mol^{-1}$.
[a] Ref. 66, [b] Ref. 3, [c] Ref. 68.
* Indicates H abstracted.
† Assumed Arrhenius parameters, other fluorine results relative to these.

the incipient products of the reaction, charge–charge and dipole–dipole repulsions as in (15) and (16):

$$\overset{\delta-\ \ \delta+}{X_3C\cdots\cdot H}\cdot\cdot\overset{\delta+\ \ \delta-}{X}$$
$$(15)$$

$$\overset{\longleftarrow\ +}{X_3C\cdots\cdot H}\cdot\cdot\overset{+\ \longrightarrow}{X}$$
$$(16)$$

Whittle's results on the bromination of fluoroalkanes shed some further light on the operation of the polar effect (72). The introduction of one fluorine atom into CH_4, C_2H_6 or CF_3CH_3 causes a reduction in the activation energy (see Table 7-9), i.e. hydrogens α to the fluorine substituent are activated. Hydrogens β to the substituent are strongly deactivated, however, and the activation energy increases greatly from CH_3CH_3 to CH_3CF_3. The reactivity of CHF_2CHF_2 is intermediate between that of CH_3CH_3 and CH_3CF_3 because each pair of fluorine atoms deactivates the hydrogen atom on the adjacent carbon atom. Similarly, only the product of hydrogen abstraction from the α-C atom is observed in bromination of CH_3CH_2F and CH_3CHF_2, again because the β-hydrogens are strongly deactivated. The deactivation of the

β-hydrogens is probably largely due to polar dipole–dipole repulsion between the radical $CF_3CH_2\cdot$ and HBr, although there may also be an increase in the bond dissociation energy of the β-C–H bond.

TABLE 7-10   *Relative selectivities RS_p^x for the fluorination of substituted butanes at 20 °C in the gas phase* (73)

X	CH$_2$X——CH$_2$——CH$_2$——CH$_3$			
H	1.0	1.3	1.3	1.0
F	<0.3	0.8	1.0	1.0
Cl	—	1.7		1.0
	CH$_3$——CHX——CH$_2$——CH$_3$			
F	0.54	—	1.1	1.0
Cl	0.12	—	0.57	1.0

TABLE 7-11   *Relative selectivities RS_p^x for the chlorination of substituted butanes at 50 °C in the gas phase* (74)

X———CH$_2$——CH$_2$——CH$_2$——CH$_3$			
H—	1	3.6	3.6	1
F—	0.9	1.7	3.7	1
Cl—	0.8	2.1	3.7	1
Br—	0.4	—	3.6	1
F$_3$C—	0.04	1.2	4.3	1
F.CO—	0.08	1.6	4.2	1
Cl.CO—	0.2	2.1	3.9	1
CH$_3$O.CO—	0.4	2.4	3.6	1
HCO.O—	—	1.5	4.1	1
CH$_3$.CO.O—	0.1	2.5	4.0	1
CF$_3$.CO.O—	0.2	1.4	4.0	1
NC—	0.2	1.7	3.9	1
CH$_3$O—	3.5	0.7	4.4	1
Me$_3$C—	2.9	3.7	5.3	1
Ph[a]—	6.5	1.0	—	1
CH$_2$=CH—	4.4	1.1	3.6	1

[a] From chlorination of 1-phenylpropane.

Tedder and co-workers have examined the gas-phase fluorination, chlorination and bromination of an extensive series of functionally substituted alkanes (73, 74). The few fluorination results (Table 7-10) show the same general pattern as those of chlorination. The relative selectivities are less in fluorination due to the greater reactivity of the fluorine atom, but substituent effects are similar in character to the chlorination case. In the chlorination of n-butyl derivatives all possible monochloro-isomers are formed, and the relative reactivities of the various hydrogens are given in Table 7-11. Hydrogen atoms on both C_α and C_β are deactivated by electron-withdrawing substituents in both chlorination and fluorination. In the exothermic chlorination and fluorination reactions, bond dissociation energies play little part, because the extent of C–H bond breaking in the transition state is small. The polar inductive effect (which causes dipole–dipole repulsions between the incipient substituted radical and the forming hydrogen halide) leads to deactivation at both C_α and C_β. The decrease in the C–H bond dissociation energy, due to the substituent, has only a minor effect on the reactivity of position α. Electron-repelling substituents such as $-CMe_3$ apparently cause activation of both C_α and C_β due to the favourable inductive effect. The methoxy- substituent is some-what anomalous, and with the n-butyl methyl ether attack at the α-position leads to formation of n-butanal by decomposition of the radical (reactions *74* and *75*):

$$CH_3CH_2CH_2CH_2OCH_3 + Cl\cdot \longrightarrow CH_3CH_2CH_2\dot{C}HOCH_3 + HCl \quad (74)$$

$$CH_3CH_2CH_2\dot{C}HOCH_3 \longrightarrow CH_3CH_2CH_2CHO + CH_3\cdot \quad (75)$$

The relative selectivity at this position was calculated from the amount of n-butanal formed. Decomposition of the radical formed by attack at C_β probably accounts for the low selectivity found at this site.

In these gas-phase experiments the inductive effect has no influence beyond C_β and the relative selectivities at C_γ and C_δ are the same as those in n-butane for virtually all the substituents. The only exception to this is with the $-CMe_3$ substituent where the RS% is anomalously high. It seems certain however, that hydrogens at position γ in 2,2-dimethylhexane are favourably placed for 1,5-intramolecular hydrogen abstraction (**17**) from the three methyl groups of the $-CMe_3$ substituent (cf. chapter 13, p. 521).

(**17**)

In the endothermic bromination reaction, the degree of bond breaking in the transition state is much greater, and here the decrease in the C_α–H bond dissociation energy outweighs the unfavourable inductive effect at C_α and this position is slightly activated (see Table 7-12). The decrease in $D(C_\alpha$–H$)$ is due to the mesomeric effect of the substituent which can only affect the α-position.

TABLE 7-12   *Relative selectivities RS_p^x for the bromination of substituted butanes at 160 °C in the gas phase (74)*

X—	—CH$_2$—	—CH$_2$—	—CH$_2$—	—CH$_3$
H—	1	80	80	1
F—	9	7	90	1
Cl—	34	32	80	1
CF$_3$—	1	7	80	1
F.CO—	30	25	80	1
Cl.CO—	30	30	80	1
CH$_3$O.CO—	40	25	80	1
CH$_3$CO.O—	20	30	70	1
CF$_3$CO.O—	2	7	70	1
NC—	20	8	80	1

$$R\dot{C}H\text{---}\ddot{\ddot{X}} \longleftrightarrow R\ddot{\ddot{C}}H\text{---}\overset{+}{\ddot{X}}$$

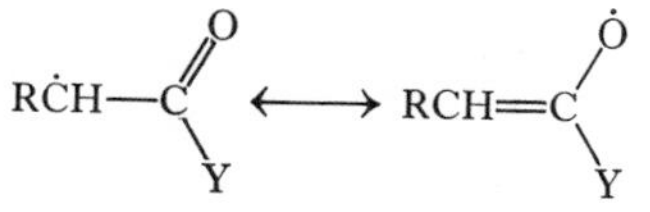

At C_β the major factor is again the unfavourable inductive effect from the electron-withdrawing substituents and C_β is deactivated. The behaviour of n-butyl derivatives at C_α and C_β is very similar to that of halogen-substituted methanes and ethanes, which also showed activation at C_α and deactivation at C_β on bromination. The inductive effect of the substituent is also not transmitted beyond C_β in bromination, and the relative selectivities at C_γ and C_δ are virtually the same as in n-butane for all the substituents.

Singh and Tedder have studied the chlorination of n-pentanoyl, n-hexanoyl and n-heptanoyl halides in the vapour and liquid phases (74). In the vapour phase the relative selectivities at the terminal methyl groups in the three acid fluorides are virtually identical, and the relative selectivities at the methylene groups beyond the β-carbon atom are also constant (see Table 7-13). This clearly demonstrates that in the vapour

TABLE 7-13 *Extent of the polar inductive effect in vapour- and liquid-phase chlorination of acid halides* (74)

Conditions	X	XCO——CH$_2$	——CH$_2$	——CH$_2$	——CH$_3$
Vapour (60 °C)	F	0.02	0.37	1.0	0.24
CH$_3$CN (52 °C)	Cl	0.06	0.40	1.0	0.44

Conditions	X	XCO——CH$_2$	——CH$_2$	——CH$_2$	——CH$_2$	——CH$_3$
Vapour (60 °C)	F	0.04	0.37	1.0	1.1	0.25
CH$_3$CN (52 °C)	Cl	0.09	0.32	1.0	1.4	0.62

Conditions	X	XCO—CH$_2$	—CH$_2$	—CH$_2$	——CH$_2$	——CH$_2$	——CH$_3$
Vapour (60 °C)	F	0.05	0.4	1.0	1.0	1.1	0.25
CH$_3$CN (52 °C)	Cl	0.06	0.5	1.0	1.3	1.5	1.1

phase the inductive polar effect of the fluoroacyl group does not extend beyond the β-carbon atom. In acetonitrile solution, the corresponding acid chlorides have been examined. The increase in the relative selectivity at the terminal methyl groups with increasing chain length, and at the methylene groups beyond the β-carbon atom, shows that in solution inductive effects are transmitted far down the hydrocarbon chain. The increased range of the inductive effect in solution is probably due to the more highly polar nature of the transition state which is caused by solvation of the hydrogen chloride formed:

$$\overset{\delta +}{R \cdots H} \cdots\cdots\cdots \overset{\delta -}{Cl}$$
$$\text{solv.}$$

In solution the reaction may also proceed by an ionic mechanism, and to minimize competition from this pathway non-polar solvents are chosen. If the compound to be halogenated can coexist with the molecular halogen in the absence of radical initiators, this is good evidence that ionic reactions are absent. Alkyl halides, carboxylic acids, acid halides, esters and nitriles all react by the free-radical pathway, but with alcohols, aldehydes and ketones competition from the polar route is serious.

Major differences between solution and vapour-phase halogenation usually occur only when complexing solvents are used. The relative reactivities of hydrogen atoms in n-pentanoyl chloride are shown for various solvents in Table 7-14. The relative selectivities in the neat liquid or in carbon tetrachloride solution differ from the vapour-phase

TABLE 7-14 *Solvent effects on the chlorination of n-pentanoyl chloride: relative selectivities* RS_p^x

Conditions	ClCO—	—CH$_2$—	—CH$_2$—	—CH$_2$—	—CH$_3$	Reference
Vapour (60 °C)	0.16	2.08	3.96	1.0		74
Neat (20 °C)	—	0.80	2.51	1.0		75
CCl$_4$ soln (52 °C)	0.15	0.80	2.11	1.0		74
CH$_3$CN soln (50 °C)	0.13	0.90	2.26	1.0		74
C$_6$H$_6$ soln (20 °C)	—	1.50	7.71	1.0		75

results mainly in the extent of the inductive deactivation by the acyl chloride substituent. Changing the solvent to acetonitrile, which has a much higher dielectric constant, also has little effect on the reactivities. In benzene, a complexing solvent, the chlorine atoms become much more selective and preferentially abstract at the γ-position.

Bruylants and co-workers (82) have studied the chlorination, in solution, of series of aliphatic acids, esters and nitriles. They adopt a 'linear free energy' approach to the interpretation of their results. They use the Hammett–Taft equation in the form:

$$\text{Log } RS_b^a = \rho^*(\Sigma\sigma_a{}^* - \Sigma\sigma_b{}^*) + h(N_a - N_b)$$

where $\Sigma\sigma_a{}^*$ and $\Sigma\sigma_b{}^*$ represent the sum of the σ inductive constants of the substituents at the reaction centres a and b respectively. N_a and N_b are the numbers of hydrogen atoms hyperconjugated to centres a and b respectively, and h measures the magnitude of the hyperconjugative effect for a particular reaction series. They obtained good straight-line correlations using this equation with negative values of ρ^* which indicate the electrophilic nature of the chlorine atom.

7.6.6 Halogenation of substituted cycloalkanes

Russell and co-workers have studied the photochlorination of halogen-substituted cyclopentanes and cyclohexanes in CCl$_4$ and CS$_2$ solutions (70). They observed that attack occurred predominantly *trans*, so that the *trans*-dihalo-products were formed preferentially for both the C$_5$ and C$_6$ rings. The *trans/cis* ratio was largest for attack at the 2-position, and less for the 3- and 4-positions. Ashton and Tedder (71) chlorinated a series of substituted cycloalkanes in the vapour phase. They also found preferential *trans* attack. Some of their results are shown in Table 7-15. In the gas phase the *trans* effect is large only at the 2-position, it is even

TABLE 7-15   *Chlorination of mono-substituted cycloalkanes at 100 °C in the gas phase:* trans/cis *ratios* (71)

Compound	$\dfrac{trans\text{-}1,2}{cis\text{-}1,2}$	$\dfrac{trans\text{-}1,3}{cis\text{-}1,3}$	$\dfrac{trans\text{-}1,4}{cis\text{-}1,4}$
Fluorocyclohexane	40.5	1.04	1.79
Chlorocyclopentane	16.1	1.76	[a]
Chlorocyclohexane	10.5	2.01	0.77
Chlorocycloheptane	5.1	1.10	1.09
Methylcyclopentane	1.3	—	[a]

[a] Not applicable.

reversed at the 4-position in chlorocyclohexane, and the size of the effect decreases as the ring size increases.

Two major factors appear to control the stereochemistry of the chlorination (71). Firstly, abstraction of a hydrogen atom from the 1-position produces initially the σ-halocycloalkyl radical (**18**), but this converts quickly to the π-radical (**19**). Overlap between the half-filled

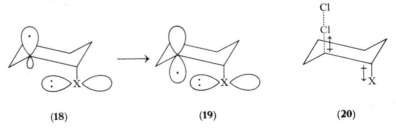

(**18**)   (**19**)   (**20**)

carbon 2p-orbital and the filled non-bonding p-orbital on the substituent fixes the ring conformation. Reaction of the π-radical with molecular chlorine is thus restricted to the side of the molecule *trans* to the substituent. Overlap of the carbon 2p-orbital at the 2-position would be more effective with 2p-orbitals of a fluorine substituent, and less so with the larger 3p-orbitals of a chlorine substituent. For a methyl substituent overlap would be virtually non-existent. The decrease in the *trans/cis* ratios from fluorocyclohexane to methylcyclopentane (Table 7-15) can thus be explained. Secondly, Little (76) and Price (77) have proposed that *trans* attack is also favoured by the inductive interaction between the substituent and the incipient carbon–chlorine bond (**20**). This polar factor may also explain the slight preference for *trans* attack in the 3- and 4- positions (77).

TABLE 7-16   *Relative selectivities[a] from the gas-phase chlorination of substituted cycloalkanes at 100 °C (71)*

Compound	1—	2—	3—	4—
Chlorocyclopentane	1.0	6.0	4.7	—
Chlorocyclohexane	1.4	4.7	4.4	6.2
Chlorocycloheptane	1.5	3.5	4.9	7.2
Fluorocyclohexane	1.4	3.4	2.4	3.2
1-Chloropentane	0.8	2.1	3.7	3.7

[a] From competitive experiments with 1-chlorobutane; reactivities relative to the primary position in 1-chlorobutane, i.e. $RS^{1; x}_{1; 4}$.

A major difference between the straight-chain and cyclic compounds is the reactivity at the β-position. In the open-chain compound deactivation occurs at C_β due to the substituent inductive effect. In the cyclic compounds no such effect is visible, although the reactivity at the β-position is clearly a function of ring conformation. Table 7-16 shows that the relative selectivity at the β-position decreases from the cyclopentane case, where the ring conformation is more rigidly fixed, to the cycloheptane case where freer movement is possible. Greater movement is also possible in fluorocyclohexane because of the small substituent size, and the reactivity at C_β begins to approach the open-chain value.

Alkanes, both open-chain and cyclic, having bromine substituents, give high yields of the 1,2-dihalo-product on halogenation in solution. This is particularly pronounced in bromination. With bromocyclohexane for example, the product of bromination is predominantly the *trans*-1,2-dibromocyclohexane. Skell and others (58) attribute this to the participation of bromine-bridged intermediates (21). In the gas-

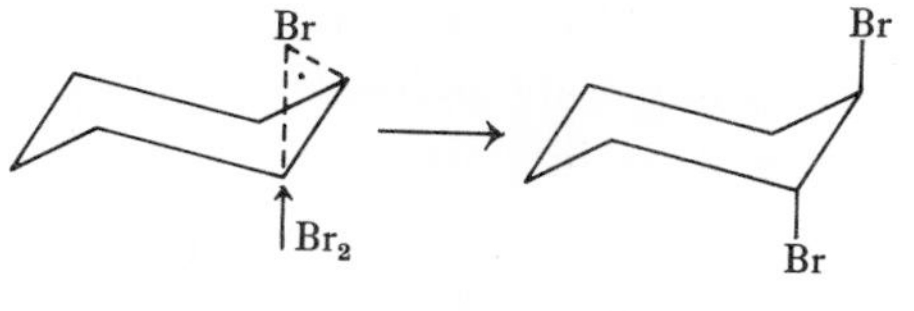

(21)   *trans*-dibromide

phase halogenation the high yields of the *trans*-1,2-products are due to formation of the olefin by reaction of the halocycloalkane with HBr at a surface. In the above example cyclohexene is formed which then adds Br_2 stereospecifically. The initial bromoalkanes are stable to thermolysis and photolysis, but in the presence of HBr or another halogen acid catalytic decomposition to the olefin occurs. Thus, in the gas phase photobromination of chlorocyclohexane, cyclohexene together with *trans*-1,2-dibromocyclohexane, 3-bromocyclohexene and bromocyclohexane can be isolated in addition to the normal bromochloro-abstraction products. Attack by Br_2 and HBr on the cyclohexene produces the *trans*-1,2-dibromocyclohexane, 3-bromocyclohexene and bromocyclohexane. Similar 'olefin-type' products can be isolated from the gas-phase photobromination of other cycloalkylhalides (80). However, olefin formation cannot be the whole explanation since solution-phase bromination of 1-bromobutane gives very high yields of 1,2-dibromobutane and if the original substituent bromine is labelled no exchange is observed (81).

Skell's work with optically active alkyl halides is more convincing evidence for the existence of bridged-bromine radicals. Thus (+)-1-bromo-2-methylbutane yielded (−)-1,2-dibromo-2-methylbutane on bromination and (−)-1-chloro-2-methylbutane also gave an optically active product (78) (cf. chapter 4, p. 81). Chlorination of the same optically active molecules gave racemic products. In chlorination, inversion of the intermediate radical is faster than the abstraction of a chlorine atom from molecular chlorine, whereas in bromination

(22)

abstraction of bromine occurs before inversion. This may be due to fixing of the configuration by interaction of the halogen substituent with the free-radical centre, possibly as in the bridged radical (22). The bromination of the 1-bromo-2-methylbutane has recently been repeated (81), and the retention of configuration confirmed.

7.6.7 *Other halogenation reagents*

Free-radical chlorination of hydrocarbon chains gives a mixture of all possible monochlorides formed by abstraction of the various hydrogen atoms. This lack of specificity in chlorination limits its usefulness as a synthetic method, and this has led to development of other, more specific reagents than molecular chlorine itself. Bromination with molecular bromine is much more selective and less effort has been devoted to the development of substitutes. With most of these halogenating reagents, the reaction proceeds through the intermediacy of another radical which is less reactive and hence more selective than the halogen. Chapter 9 on heteroradicals deals in greater detail with the chemistry of these species. Fluorination and iodination have been much less studied, although the recent discovery of chemical laser action in the reaction of fluorine atoms with alkanes (2) has prompted study of other fluorination reagents, notably UF_6.

Sulphuryl chloride, trichloromethanesulphonyl chloride and trichloromethanesulphenyl chloride have all been used as chlorinating reagents (55). Sulphuryl chloride is useful synthetically under conditions which favour polar reaction with molecular chlorine. The mechanism of chlorination probably involves a mixed chain, both chlorine atoms and chlorosulphonyl radicals acting as chain carriers (see Scheme 3). The

$$SO_2Cl_2 \longrightarrow \cdot SO_2Cl + Cl\cdot$$

$$Cl\cdot + RCH{=}CH_2 \longrightarrow R\dot{C}HCH_2Cl \xrightarrow{SO_2Cl_2} RCHClCH_2Cl$$

$$\cdot SO_2Cl + RCH{=}CH_2 \longrightarrow R\dot{C}HCH_2SO_2Cl \longrightarrow RCHClCH_2SO_2\cdot$$

$$\qquad\qquad\qquad\qquad\qquad\qquad\qquad\qquad\Big\downarrow RCH{=}CH_2$$

$$RCHClCH_2SO_2CH_2\dot{C}HClR \xleftarrow{SO_2Cl_2} RCHClCH_2SO_2CH_2\dot{C}HR$$

Scheme 3

selectivity is about the same as molecular chlorine itself in complexing solvents. Trichloromethanesulphonyl chloride is considerably more selective than molecular chlorine, and is useful with hydrocarbons which give reactive radicals. The mechanism of halogenation probably involves both trichloromethyl radicals and trichloromethanesulphonyl radicals

$$R\cdot + CCl_3SO_2Cl \longrightarrow RCl + CCl_3SO_2\cdot$$

$$CCl_3SO_2\cdot \longrightarrow CCl_3\cdot + SO_2$$

$$CCl_3\cdot + RH \longrightarrow CCl_3H + R\cdot$$

$$CCl_3SO_2\cdot + RH \longrightarrow CCl_3SO_2H + R\cdot$$

$$\Big\downarrow$$

$$CCl_3H + SO_2$$

Scheme 4

(see Scheme 4). Trichloromethanesulphenyl chloride is also more selective than molecular chlorine, probably because the chain-carrying radical is the trichloromethanethiyl radical rather than the chlorine atom (see Scheme 5).

$$CCl_3S\cdot + RH \longrightarrow R\cdot + CCl_3SH$$

$$R\cdot + CCl_3SCl \longrightarrow RCl + CCl_3S\cdot$$

$$CCl_3SH + CCl_3SCl \longrightarrow HCl + CCl_3SSCCl_3$$

Scheme 5

N-Chlorosuccinimide (**23**), *N*-chlorosulphonamides (**24**), phosphorus pentachloride and iodobenzene dichloride have also found limited use in the chlorination of certain hydrocarbons (55) as have copper(II) halides (83).

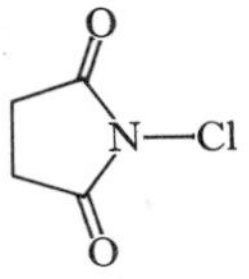

| (23) | (24) |

Allylic chlorination is best achieved by using t-butyl hypochlorite and this reagent has been investigated by Walling and co-workers, see ref. (58). The reaction probably proceeds via t-butoxy radicals as chain carriers:

$$Bu^tO\cdot + RH \longrightarrow Bu^tOH + R\cdot \qquad (76)$$

$$R\cdot + Bu^tOCl \longrightarrow RCl + Bu^tO\cdot \qquad (77)$$

There is little decomposition of t-butoxy radicals provided the temperature is low and CCl_4 is used as the solvent:

$$Me_3CO\cdot \longrightarrow Me\cdot + MeCOMe$$

Allylic bromination is best accomplished with *N*-bromosuccinimide. There has been considerable controversy about the mechanism of halogenation with this reagent (58). According to the original 'Bloomfield mechanism' the specificity of this reagent in abstracting the allylic hydrogens, rather than adding to the double bond, was due to the special nature of the succinimydyl radical which acted as the chain carrier (Scheme 6). In the 'Goldfinger mechanism', however, no special effects need to be invoked, and the bromination with *N*-bromosuccinimide is just a particular instance of a very general type of reaction. The function of the *N*-bromosuccinimide is to ensure a constant, but very low concentration, of molecular bromine (78) (see Scheme 7). Atomic bromine then acts as the chain-carrying radical and it can abstract allylic hydrogen (*80*) or add to the double bond (*82*), but only the addition step is reversible (see section 5.4). At low halogen concentrations allylic substitution is favoured, see equation (*44*). In support of the Goldfinger

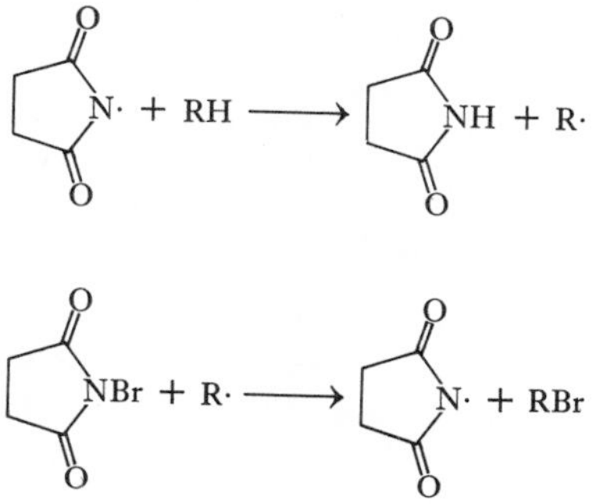

Scheme 6

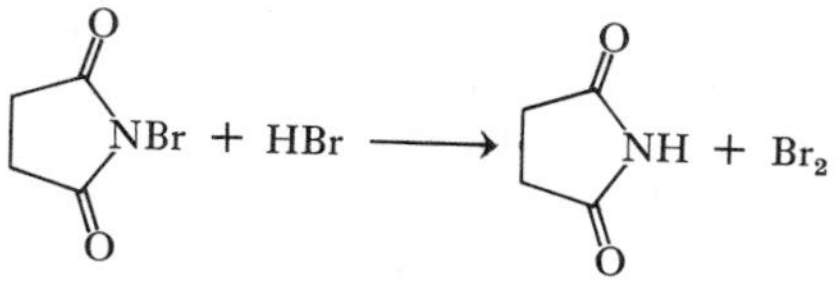

$$\text{NBr} + \text{HBr} \longrightarrow \text{NH} + \text{Br}_2 \tag{78}$$

$$\text{Br}_2 \longrightarrow 2\text{Br}\cdot \tag{79}$$

$$\text{RCH}_2\text{CH}{=}\text{CH}_2 + \text{Br}\cdot \longrightarrow \text{R}\dot{\text{C}}\text{HCH}{=}\text{CH}_2 + \text{HBr} \tag{80}$$

$$\text{R}\dot{\text{C}}\text{HCH}{=}\text{CH}_2 + \text{Br}_2 \longrightarrow \text{RCHBrCH}{=}\text{CH}_2 + \text{Br}\cdot \tag{81}$$

$$\text{RCH}_2\text{CH}{=}\text{CH}_2 + \text{Br}\cdot \rightleftharpoons \text{RCH}_2\dot{\text{C}}\text{HCH}_2\text{Br} \tag{82}$$

$$\text{RCH}_2\dot{\text{C}}\text{HCH}_2\text{Br} + \text{Br}_2 \longrightarrow \text{RCH}_2\text{CHBrCH}_2\text{Br} + \text{Br}\cdot \tag{83}$$

Scheme 7

mechanism it is found that pure *N*-bromosuccinimide does not react, and an initiator is essential. Since the chain-carrying radical is a bromine atom, the selectivities of bromination by molecular bromine, and *N*-bromosuccinimide should be identical, and several groups of workers have confirmed this (58). The Goldfinger mechanism also implies that allylic bromination by low concentrations of molecular bromine itself should be possible, and this has also been demonstrated, see ref. (58).

Atoms

GENERAL REFERENCES

R. J. Cvetanović, *Adv. in Photochem.* 1963, **1**, 115.
M. L. Poutsma, *Methods in Free-Radical Chemistry*, 1968, **1**, 79.
W. A. Thaler, *Methods in Free-Radical Chemistry*, 1969, **2**, 121.

SPECIFIC REFERENCES

1. F. Kaufman, *Progr. Reaction Kinetics*, 1961, **1**, 1.
2. K. L. Kompa, J. H. Parker and G. C. Pimentel, *J. Chem. Phys.* 1968, **49**, 4257.
3. G. C. Fettis and J. H. Knox, *Progr. Reaction Kinetics*, 1964, **2**, 1.
4. M. A. A. Clyne and D. H. Stedman, *Trans. Faraday Soc.* 1968, **64**, 1816.
5. J. K. K. Ip and G. Burns, *Discuss. Faraday Soc.* 1967, **44**, 241.
6. G. Porter in *Technique of Organic Chemistry*, ed. A. Weissberger, Inter-Science, New York, 1960, vol. VIII, pt II, p. 1055.
7. R. J. Donovan and D. Husain, *Trans. Faraday Soc.* 1966, **62**, 2023, 2643.
8. F. L. Tollefson and D. J. LeRoy, *J. Chem. Phys.* 1948, **16**, 1057; L. Elias, E. A. Ogryzlo and H. J. Schiff, *Canad. J. Chem.* 1959, **37**, 1690.
9. F. S. Larkin, *Canad. J. Chem.* 1968, **46**, 1005.
10. B. A. Thrush, *Science*, 1967, **156**, 470.
11. M. A. A. Clyne and B. A. Thrush, *Trans. Faraday Soc.* 1961, **57**, 1305.
12. M. A. A. Clyne and B. A. Thrush, *Trans. Faraday Soc.* 1961, **57**, 2176.
13. M. A. A. Clyne and D. H. Stedman, *Trans. Faraday Soc.* 1966, **62**, 2164.
14. F. Hutton and M. Wright, *Trans. Faraday Soc.* 1965, **61**, 78.
15. M. A. A. Clyne and D. H. Stedman, *Trans. Faraday Soc.* 1968, **64**, 1.
16. D. Rapp and H. S. Johnson, *J. Chem. Phys.* 1960, **33**, 695.
17. A. Carrington, *Chem. in Britain*, 1970, **6**, 71.
18. S. Krongelb and H. W. P. Strandberg, *J. Chem. Phys.* 1959, **31**, 1196.
19. A. A. Westenberg and N. de Haas, *J. Chem. Phys.* 1964, **40**, 3087.
20. A. A. Westenberg and N. de Haas, *J. Chem. Phys.* 1967, **46**, 490.
21. J. M. Brown and B. A. Thrush, *Trans. Faraday Soc.* 1967, **63**, 630.
22. R. S. Donovan and D. Husain, *Trans. Faraday* Soc. 1966, **62**, 11.
23. S. N. Foner and R. F. Hudson, *J. Chem. Phys.* 1953, **21**, 1374, 1608.
24. J. T. Herron, *J. Phys. Chem.* 1966, **70**, 2803; J. T. Herron and R. E. Huie, *ibid.* 1968, **72**, 2538.
25. S. N. Foner and R. F. Hudson, *J. Chem. Phys.* 1962, **37**, 2676.
25a. I. M. Campbell and B. A. Thrush, *Proc. Roy. Soc.* A, 1967, **296**, 201; M. A. A. Clyne and D. H. Stedman, *J. Phys. Chem.* 1967, **71**, 3071.
26. B. A. Thrush, *Progr. Reaction Kinetics*, 1965, **3**, 65.
27. M. A. A. Clyne and D. H. Stedman, *Trans. Faraday Soc.* 1968, **64**, 2698.
28. W. V. Smith, *J. Chem. Phys.* 1943, **11**, 110.
29. J. C. Greaves and J. W. Linnett, *Trans. Faraday Soc.* 1959, **55**, 1346.
30. R. A. Young, *J. Chem. Phys.* 1961, **34**, 1295.
31. B. Brocklehurst and K. R. Jennings, *Progr. Reaction Kinetics*, 1967, **4**, 1.
32. E. A. Ogryzlo, *Canad. J. Chem.* 1961, **39**, 2556.
33. R. W. Fair and B. A. Thrush, *Trans. Faraday Soc.* 1969, **65**, 1208.

34. C. D. Johnson and D. Britton, *J. Phys. Chem.* 1964, **68**, 3032; D. J. Seery, *J. Phys. Chem.* 1966, **70**, 1684, 4074.
35. G. Porter, *Discuss. Faraday Soc.* 1962, **33**, 198.
36. J. K. K. Ip and G. Burns, *J. Chem. Phys.* 1969, **51**, 3414.
37. E. Hutton and M. Wright, *Trans. Faraday Soc.* 1965, **61**, 78.
38. A. A. Westenberg and N. de Haas, *J. Chem. Phys.* 1968, **48**, 4405
39. D. L. Bunker and N. Davidson, *J. Amer. Chem. Soc.* 1958, **80**, 5085, 5090.
40. K. E. Russell and J. Simons, *Proc. Roy. Soc.* A, 1953, **217**, 271.
41. R. L. Strong, *J. Amer. Chem. Soc.* 1965, **87**, 3563.
42. G. A. Salmon and R. M. Noyes, *J. Amer. Chem. Soc.* 1962, **84**, 672.
43. R. J. Cvetanović, *Adv. Photochem.* 1963, **1**, 115; W. E. Falconer and W. A. Sunder, *Internat. J. Chem. Kinetics*, 1971, **3**, 395.
44. A. A. Westenberg and N. de Haas, *J. Chem. Phys.* 1969, **50**, 707.
45. H. E. Gunning and O. P. Strausz, *Adv. Photochem.* 1966, **4**, 143.
46. A. B. Callear and W. J. R. Tyerman, *Trans. Faraday Soc.* 1966, **62**, 371 and 2760.
47. J. Connor, G. Greig and O. P. Strausz, *J. Amer. Chem. Soc.* 1969, **91**, 5695.
48. R. J. Donovan, D. Husain, R. W. Fair, O. P. Strausz and H. E. Gunning, *Trans. Faraday Soc.* 1970, **66**, 1635.
49. P. M. Scott and K. R. Jennings, *J. Phys. Chem.* 1969, **73**, 1521.
50. G. R. Wooley and R. J. Cvetanović, *J. Chem. Phys.* 1969, **50**, 4697.
51. R. J. Cvetanović and L. C. Doyle, *J. Chem. Phys.* 1969, **50**, 4705.
52. R. D. Penzhorn and H. L. Sandoval, *J. Phys. Chem.* 1970, **74**, 2065.
53. J. M. Tedder, *Adv. Fluorine Chem.* 1961, **2**, 104.
54. W. T. Miller, R. L. Ehrenfeld, J. M. Phelan, M. Prober and S. K. Reed, *Ind. Eng. Chem.* 1947, **39**, 401.
55. M. L. Poutsma, *Methods in Free-Radical Chem.* 1969, **1**, 79.
56. F. W. Stacey and J. F. Harris, Jr, *Org. Reactions*, 1963, **13**, 154.
57. P. S. Skell and R. G. Allen, *J. Amer. Chem. Soc.* 1959, **81**, 5383.
58. W. A. Thaler, *Methods in Free-Radical Chem.* 1969, **2**, 189.
59. P. I. Abell, *Trans. Faraday Soc.* 1964, **60**, 2214.
60. K. T. Wong and D. A. Armstrong, *Canad. J. Chem.* 1970, **48**, 2426.
61. A. F. Trotman-Dickenson, *Adv. Free-Radical Chem.* 1965, **1**, 1.
62. A. A. Westenberg and N. de Haas, *J. Chem. Phys.* 1967, **47**, 1393.
63. D. T. Clark and J. M. Tedder, *Trans. Faraday Soc.* 1966, **62**, 393, 399, 405.
64. S. W. Benson and P. S. Nangia, *J. Amer. Chem. Soc.* 1964, **86**, 2770, 2773.
65. H. B. Haas, E. T. McBee and P. Weber, *Ind. Eng. Chem.* 1935, **27**, 1190; 1936, **28**, 33.
66. R. Foon and N. A. McAskill, *Trans. Faraday Soc.* 1969, **65**, 3005.
67. J. H. Knox and R. G. Musgrave, *Trans. Faraday Soc.* 1967, **63**, 2201.
68. J. C. Amphlett and E. Whittle, *Trans. Faraday Soc.* 1968, **64**, 2130.
69. K. D. King, D. M. Golden and S. W. Benson, *Trans. Faraday Soc.* 1970, **66**, 2794.
70. G. A. Russell and A. Ito, *J. Amer. Chem. Soc.* 1963, **85**, 2983; G. A. Russell, *ibid.* 1958, **80**, 4997; G. A. Russell, A. Ito and R. Konaka, *ibid.* 1963, **85**, 2988.
71. D. S. Ashton and J. M. Tedder, *J. Chem. Soc.* B, 1970, 1031; 1971, 1719, 1723.

72. J. W. Coomber and E. Whittle, *Trans. Faraday Soc.* 1966, **62**, 1553; K. C. Ferguson and E. Whittle, *ibid.* 1971, **67**, 2618.
73. P. S. Fredricks and J. M. Tedder, *J. Chem. Soc.* 1960, 144; 1961, 3520.
74. H. Singh and J. M. Tedder, *J. Chem. Soc.* C, 1966, 605, 608, 612; V. R. Desai, A. Nechvatal and J. M. Tedder, *J. Chem. Soc.* B, 1969, 30; 1970, 386.
75 P Smit and J. H. den Hertog, *Rec. Trav. chim.* 1964, **83**, 891.
76. J. C. Little, Yu-Lan Chang Tong and J. P. Heeschen, *J. Amer. Chem. Soc.* 1969, **91**, 7090.
77. C. C. Price, C. D. Beard and K. Akune, *J. Amer. Chem. Soc.* 1970, **92**, 5916.
78. P. S. Skell, D. L. Tuleen and P. D. Readio, *J. Amer. Chem. Soc.* 1963, **85**, 2849.
79. T. L. Gilchrist and C. W. Rees, *Carbenes, Nitrenes and Arynes*, Nelson, London, 1969.
80. D. D. Tanner, M. W. Mosher, N. C. Das and E. V. Blackburn, *J. Amer. Chem. Soc.* 1971, **93**, 5846.
81. D. D. Tanner, H. Yabuuchi and E. V. Blackburn, *J. Amer. Chem. Soc.* 1971, **93**, 4802; C. Ronneau, J. Ph. Soumillion, P. DeJaifvre and A. Bruylants, *Tetrahedron Letters*, 1972, 317.
82. J. Wautier and A. Bruylants, *Bull. Soc. chim. belges*, 1963, **72**, 222; R. Rouchard and A. Bruylants, *ibid.* 1967, **76**, 50; J. Ph. Soumillion and A. Bruylants, *ibid.* 1969, **78**, 169, 425, 435; J. Ph. Soumillion, *Ind. chim. belge*, 1970, **35**, 1065.
83. D. C. Nonhebel, *Essays in Free-Radical Chemistry*, *Chem. Soc. Special Publ.* no. 24, 1970, p. 409.
84. A. Nechvatal, *Adv. Free-Radical Chem.* 1972, **4**, 175.

8

Reactions of alkyl and substituted-alkyl radicals

In this chapter the term 'alkyl radical' will be taken to mean all radicals in which the unpaired electron is wholly or principally localized on a carbon atom, e.g. simple alkyl radicals such as $C_nH_{2n+1}\cdot$ and radicals in which the alkyl chain is substituted by halogen-, oxygen-, nitrogen-, etc., functional groups.

8.1 RADICAL–RADICAL REACTIONS

In every chemical system where free radicals are formed, the radicals take part in radical–radical reactions which remove them, as stable molecules, from the reaction sequences. There are two types of radical–radical reaction: combination in which the radicals combine to give a molecule which is the dimer of the radical (1); and disproportionation

$$A\cdot + A\cdot \longrightarrow A_2 \tag{1}$$

$$A\cdot + A\cdot \longrightarrow AH + A(-H) \tag{2}$$

in which a hydrogen atom is transferred from one radical to the other giving a saturated molecule AH and an olefin $A(-H)$ (2). With substituted radicals the atom transferred is not always hydrogen, but may be a halogen. When the radicals are taking part in a chain reaction, combination and disproportionation are chain-terminating processes. If the chains are long, the radical–radical products form a very small part of the total products and may be undetectable. The nature of the chain-termination step is, however, important in determining the overall order of the reaction.

The dimerization reaction (1) is a very useful reference reaction, and has been used in many competitive studies. The same radical is allowed to react with a series of substrates, and the rate of reaction with the

Alkyl radicals

substrate can be measured relative to the combination which occurs at the same time. These reactions are also useful because they provide a measure of the stationary radical concentration. The rates of dimer formation $R(A_2)$ and disproportionation product formation $R(AH)$ and $RA(-H)$ are given by:

$$R(A_2) = k_1[A\cdot]^2$$

$$R(AH) = R(A(-H)) = k_2[A\cdot]^2$$

So that once k_1 or k_2 is known, the stationary radical concentration can be calculated from the rate of formation of the dimer or the disproportionation product.

8.1.1 *Combination in the gas phase*

To determine the combination rate constant k_1, a knowledge of the radical concentration $[A\cdot]$ is needed. In a stationary-state photolysis or pyrolysis, the radical concentration is normally too small for measurement by the conventional physical techniques. Electronic spectra of alkyl radicals are usually weak and diffuse, so that the flash photolysis method, which was so successful with atoms, has only been applied in a very few cases (1). Methyl radical combination has been studied in a flow system using mass-spectrometric detection (2), but as yet neither mass spectrometry nor e.s.r. spectroscopy has received much application in gas-phase combination studies. The great majority of combination rates have been determined by the intermittent illumination method (3).

This indirect method is applicable to any free-radical system where the rate of reaction depends on the light intensity to some power other than unity. This condition is satisfied as long as the termination step is bimolecular, in which case the reaction rate is proportional to the light intensity to the half power. As an example we consider the addition of an alkyl halide to an olefin S:

$$RX + h\nu \longrightarrow R\cdot + X\cdot \qquad \text{Rate} = \phi I_a$$

$$R\cdot + S \xrightarrow{k_a} RS\cdot$$

$$RS\cdot + RX \xrightarrow{k_t} RSX + R\cdot$$

$$R\cdot + R\cdot \xrightarrow{k_c} R_2$$

When the illumination is switched on the radical concentration builds up to its steady-state value:

$$\frac{d[R\cdot]}{dt} = \phi I_a - 2k_c[R\cdot]^2$$

When the illumination is switched off the radical concentration dies away to zero:

$$\frac{d[R\cdot]}{dt} = -2k_c[R\cdot]^2$$

The average radical concentration over a series of light and dark periods is therefore a function of the frequency of the intermittent light source.

The average radical concentration in intermittent light $[R\cdot]_i$ relative to the average radical concentration in steady light $[R\cdot]_s$ can be measured experimentally by determining the rates of formation of the adduct RSX in intermittent and steady light:

$$\frac{\text{Rate (RSX)}_i}{\text{Rate (RSX)}_s} = \frac{k_a[R\cdot]_i[S]}{k_a[R\cdot]_s[S]} = \frac{[R\cdot]_i}{[R\cdot]_s}$$

By integrating the differential rate equations given above, and averaging over a large number of dark and light periods, an expression for the relative average radical concentration in terms of k_c, ϕI_a and the length of the light period λ can be obtained, i.e.

$$\frac{[R\cdot]_i}{[R\cdot]_s} = f(k_c, \phi I_a, \lambda)$$

The rate of initiation ϕI_a is obtained from the rate of dimer formation:

$$\phi I_a = 2\,\text{Rate}\,(R_2)$$

The best value of k_c can then be found by comparing the experimentally measured relative radical concentration at various flash times λ, with the values computed from $f(k_c, \phi I_a, \lambda)$.

The method is not capable of giving very precise values of k_c because the relative radical concentration does not vary by very much with flash time. The method gives initially the square root of k_c so that errors are doubled in obtaining the rate constant itself. In spite of this it has been very widely used, for want of a better method, in both the gas and solution phase for obtaining combination rate constants for atoms, alkyl radicals and heteroradicals. Polymerization termination rates are also determined by modifications for this method.

The most thoroughly investigated combination reaction is that of methyl radicals. This has been studied by flash photolysis (1), mass spectrometry (2), shock tube experiments (5), and several variants of the intermittent illumination method (6, 7). The rate constant is very large and independent of temperature. The intermittent illumination method

Alkyl radicals

is not sufficiently accurate to reveal small temperature dependencies of combination rate constants. Experimental results do indicate, however, that the activation energy of the combination step is less than about 8 kJ mol^{-1} for all the alkyl and substituted-alkyl radicals so far studied, and that it is probably close to zero in the gas phase. The absolute values of the combination rate constants for a series of alkyl radicals are shown in Table 8-1. The combination rate constants for alkyl radicals in general are very large, and the process is faster than the third-order combination of atoms.

TABLE 8-1 *Combination rate constants for alkyl radicals in the gas phase*

Radical	k_1 (l mol^{-1} s^{-1})	Z (l mol^{-1} s^{-1})	Reference
$CH_3 \cdot$	2.0×10^{10}	1.0×10^{11}	1, 4, 7
$CF_3 \cdot$	2.3×10^{10}	1.5×10^{11}	8
$CF_2Cl \cdot$	1.2×10^{10}	—	9
$CCl_3 \cdot$	$0.4–8.0 \times 10^{10}$	2.0×10^{11}	10, 11
$C_2H_5 \cdot$	$0.04–2.0 \times 10^{10}$	1.7×10^{11}	12, 155
$C_2H_3Cl_2 \cdot$[a]	0.5×10^{10}	—	13
$C_2H_2Cl_3 \cdot$[a]	0.3×10^{10}	—	13, 14
$C_2HCl_4 \cdot$[a]	0.3×10^{10}	—	13
$C_2Cl_5 \cdot$[a]	0.05×10^{10}	—	14
$(CH_3)_2CH \cdot$[b]	6×10^{10}	1.3×10^{11}	15
$(CH_3)_3C \cdot$[b]	0.3×10^{10}	1.4×10^{11}	16
c-$C_6F_{10}Cl \cdot$	0.03×10^{10}	—	14

Z is the collision frequency calculated at 373 °K.

[a] Rate constant here is the total termination rate including combination and disproportionation.

[b] See also text.

The experimental rate constants are not very different from the calculated collision frequencies, also shown in Table 8-1, and since three out of every four collisions lead to formation of a repulsive triplet state, the collision frequency Z should be multiplied by the electronic multiplicity factor $\frac{1}{4}$ before comparing with the experimental rate constants. It is clear that reaction occurs on virtually every 'potentially productive' collision, at least for the smallest alkyl radicals.

Benson has recently suggested a new method for determining alkyl radical combination rates by a competitive technique in which the unknown radical rate is compared with that of methyl (154). Methyl

radicals are generated in the presence of an alkyl iodide which thus sets
up a rapid equilibrium:

$$Me\cdot + RI \xrightleftharpoons{K} MeI + R\cdot$$

$$Me\cdot + Me\cdot \xrightarrow{k_M} C_2H_6$$

$$Me\cdot + R\cdot \xrightarrow{k_{MR}} MeR$$

$$R\cdot + R\cdot \xrightarrow{k_R} RR$$

The relative yields of the three alkanes are governed by the equilibrium
constant K and the rate constants of the combination steps. This
mechanism leads to the following expression (154):

$$\frac{k_R}{k_M} = \frac{4[RR]^2[MeI]^2}{[RMe]^2[RI]^2 K^2}$$

The unknown combination rate constant k_R can thus be found from the
yields of the alkanes and a knowledge of the methyl combination rate
constant and the equilibrium constant. This latter can be calculated
from the known thermodynamic properties of the two alkyl radicals and
alkyl iodides. The experiments carried out with ethyl, isopropyl and
t-butyl radicals indicated a somewhat lower rate constant for these
radicals than for methyl. They suggested the following order of reactivity
in the combination reaction:

$$Me\cdot > Et\cdot > Pr^i\cdot > Bu^t\cdot$$

A reinterpretation of the available kinetic data for ethyl radicals has also
suggested a lower value for the combination rate constant (155), i.e.
$k(Et\cdot) = 4 \times 10^8 \text{ l mol}^{-1}\text{ s}^{-1}$. Radical combination rate constants in
general appear to decrease as the size of the radical increases, and a
rough correlation between radical molecular weight and log k_1 has been
observed (14).

The similarity between the combination rate constant and the
collision number, means that during an encounter, the radicals must be
free to rotate and attain an orientation favourable to bond formation.
The transition state of the reaction must therefore be rather 'loose',
resembling the reactant radicals more than the product dimer.

The combination rate of methyl radicals becomes pressure-dependent
at total pressures less than about 20 torr, and the process becomes third-
order:

$$CH_3\cdot + CH_3\cdot + M \longrightarrow C_2H_6 + M \tag{3}$$

Alkyl radicals

It might be expected, by analogy with atom combinations, that this third-order reaction would have a negative temperature coefficient. A large negative activation energy has been reported (17), but there is little agreement in the literature as to the rate of reaction (3), or even the form of the pressure dependence. At higher temperatures the onset of third-order behaviour moves to higher pressures, and third-order behaviour has also been observed in the combination of $CF_3\cdot$ radicals at 773 °K at pressures up to 250 torr. For larger radicals third-order combination can only be expected at very low pressures.

In a system containing two unlike radicals A· and B· there are three possible combination reactions:

$$A\cdot + A\cdot \xrightarrow{k_{AA}} A_2$$

$$A\cdot + B\cdot \xrightarrow{k_{AB}} AB$$

$$B\cdot + B\cdot \xrightarrow{k_{BB}} B_2$$

The kinetic analysis of cross-combination reactions is usually stated in terms of the cross-combination rate constant ratio $k_{AB}/(k_{AA}k_{BB})^{\frac{1}{2}}$, often given the symbol Φ. This quantity is much easier to measure than the absolute rate of cross-combination and is derived simply from the rates of formation for the combination products:

$$R(AB)/[R(A_2)R(B_2)]^{\frac{1}{2}} = k_{AB}/(k_{AA}k_{BB})^{\frac{1}{2}} = \Phi$$

The cross-combination rate constant ratios of a number of radicals are shown in Table 8-2. The values of Φ are found to be independent of temperature, as would be expected for reactions with zero activation energy. Simple collision theory predicts that Φ should be about 2, and the experimental results in Table 8-2 are close to this value for a wide variety of radicals. This is further evidence that combination in the gas phase occurs at the maximum rate, i.e. the collision rate, and that the transition state is very loose.

8.1.2 *Disproportionation in the gas phase*

Disproportionation of two radicals in the gas phase normally gives an alkane and an olefin. Two ethyl radicals disproportionate to give ethane and ethylene:

$$CH_3CH_2\cdot + CH_3CH_2\cdot \begin{cases} \xrightarrow{k_c} CH_3CH_2CH_2CH_3 \\ \xrightarrow{k_d} CH_3CH_3 + CH_2{=}CH_2 \end{cases}$$

TABLE 8-2   *Cross-combination rate constant ratios for
alkyl radicals in the gas phase*

Radicals	Φ	Reference
$CH_3\cdot + CF_3\cdot$	2.5	19
$CH_2F\cdot + CHF_2\cdot$	2.2	20
$CHF_2\cdot + CF_3\cdot$	2.4	19
$CHF_2\cdot + CF_2Cl\cdot$	2.4	21
$CHCl_2\cdot + CCl_3\cdot$	2.2	22
$CF_3\cdot + CCl_3\cdot$	2.0	23
$CF_2Cl\cdot + CFCl_2\cdot$	~ 2	24
$CH_3\cdot + C_2H_5\cdot$	1.9	25
$CH_3\cdot + n\text{-}C_3H_7\cdot$	~ 2	26
$CF_3\cdot + C_2F_5\cdot$	2.0	27
$CF_3\cdot + n\text{-}C_3F_7\cdot$	1.8	28
$CF_2Cl\cdot + CF_2ClCF_2\cdot$	2.2	29
$CCl_3\cdot + C_2H_5\cdot$	2.0	30
$CCl_3\cdot + C_2H_4Cl\cdot$	2.5	30
$C_2H_5\cdot + n\text{-}C_3H_7\cdot$	1.9	26
$C_2H_5\cdot + i\text{-}C_3H_7\cdot$	2.0	26
$C_2H_5\cdot + i\text{-}C_4H_9\cdot$	1.9	31
$C_2H_5\cdot + t\text{-}C_4H_9\cdot$	1.9	31
$n\text{-}C_3H_7\cdot + n\text{-}C_4H_9\cdot$	2.0	26
$i\text{-}C_3H_7\cdot + n\text{-}C_4H_9\cdot$	2.2	26

If both of the radicals are smaller than ethyl, then the unsaturated
product may be a carbene, e.g. with $CF_3\cdot$ and $CF_2H\cdot$ radicals:

$$CF_3\cdot + CF_2H\cdot \longrightarrow CHF_3 + CF_2:$$

or alternatively, an olefin may still be formed, e.g.

$$CF_3\cdot + CH_3\cdot \longrightarrow CF_2{=}CH_2 + HF$$

When two unlike radicals disproportionate there are two possible sets
of products. Thus with ethyl and isopropyl radicals, either propylene
or ethylene may be formed.

$$CH_3CH_2\cdot + CH_3\dot{C}HCH_3 \longrightarrow C_2H_6 + CH_3CH{=}CH_2$$

$$CH_3\dot{C}HCH_3 + CH_3CH_2\cdot \longrightarrow CH_3CH_2CH_3 + CH_2{=}CH_2$$

Disproportionation of unlike radicals is usually called cross-dispropor-
tionation. Experimentally disproportionation rates are measured relative
to the combination rate, and the kinetic analysis is given in terms of the
disproportionation/combination ratio, i.e. k_d/k_c

$$k_d/k_c = R(A(-H))/R(A_2) = R(AH)/R(A_2)$$

Alkyl radicals

The rate constant ratio can be found either from the rate of olefin $A(-H)$ formation, or from the rate of alkane formation (AH). The rate of olefin formation usually gives the most reliable value since the alkane is more likely to be formed in competing reactions. Some typical values of k_d/k_c are given in Table 8-3. The disproportionation rates of the majority of the radicals are very fast: comparable in magnitude to the combination rates. Most of the k_d/k_c ratios in Table 8-3 are independent of temperature, which suggests that disproportionation, like combination, occurs with zero activation energy.

TABLE 8-3 *Disproportionation of alkyl radicals in the gas phase*

Radicals[a]	k_d/k_c	No.[b]	Reference
$CH_3 \cdot + C_2H_5 \cdot$	0.04	1	31
$CH_3 \cdot + i\text{-}C_3H_7 \cdot$	0.16	2	31
$CH_3 \cdot + n\text{-}C_3H_7 \cdot$	0.06	3	31
$C_2H_5 \cdot + C_2H_5 \cdot$	0.14	4	31
$C_2H_5 \cdot + i\text{-}C_3H_7 \cdot$	0.18	5	31
$C_2H_5 \cdot + n\text{-}C_3H_7 \cdot$	0.07	6	31
$C_2H_5 \cdot + i\text{-}C_4H_9 \cdot$	0.04	—	31
$C_2H_5 \cdot + t\text{-}C_4H_9 \cdot$	0.50	7	31
$i\text{-}C_3H_7 \cdot + C_2H_5 \cdot$	0.12	8	31
$i\text{-}C_3H_7 \cdot + i\text{-}C_3H_7 \cdot$	0.69	9	31
$i\text{-}C_3H_7 \cdot + n\text{-}C_3H_7 \cdot$	0.41	10	31
$n\text{-}C_3H_7 \cdot + C_2H_5 \cdot$	0.06	11	31
$n\text{-}C_3H_7 \cdot + n\text{-}C_3H_7 \cdot$	0.15	12	31
$n\text{-}C_4H_9 \cdot + n\text{-}C_4H_9 \cdot$	0.14	—	31
$i\text{-}C_4H_9 \cdot + i\text{-}C_4H_9 \cdot$	0.08	—	31
$t\text{-}C_4H_9 \cdot + t\text{-}C_4H_9 \cdot$	2.32	13	31
$CH_3 \cdot + CF_3 \cdot$	0.72	—	19
$CH_2F \cdot + CF_2H \cdot$	0.06	—	20
$CF_2H \cdot + CF_2H \cdot$	0.20	—	20, 21
$CF_3 \cdot + CF_2H \cdot$	0.09	—	32
$CF_2Cl \cdot + CF_2Cl \cdot$	0.20	—	9
$CCl_3 \cdot + C_2H_5 \cdot$	0.22	—	30
$CCl_3 \cdot + C_2H_4Cl \cdot$	0.12	—	30
$C_2F_5 \cdot + C_2H_5 \cdot$	0.56	—	34
$CH_3CHF \cdot + CH_3CHF \cdot$	0.21	—	33
$CH_3CF_2 \cdot + CH_3CF_2 \cdot$	0.55	—	33
$CH_2{=}CHCH_2 \cdot + CH_2{=}CHCH_2 \cdot$	0.008	14	35

[a] The radical stated first is the one to which a hydrogen or halogen is transferred.

[b] Number of the process in Fig. 8-1.

There has been considerable controversy over the question of the nature of the transition state in disproportionation reactions. The reaction is clearly not a straightforward case of hydrogen abstraction from one radical by the other, because the rates of hydrogen abstraction reactions are known to be at least an order of magnitude slower than the disproportionation process. Similarly hydrogen abstraction reactions have activation energies of, typically, about 30 kJ mol⁻¹ whereas the disproportionation step is temperature-independent.

Mass-spectroscopic investigation of the ethylene fraction arising from the disproportionation of $CH_3CD_2\cdot$ radicals showed it to be exclusively $CH_2{=}CD_2$ (36). Thus the transferred hydrogen atom comes mainly or wholly from the β-carbon atom of the radical. The k_d/k_c ratios for all the alkyl radicals in Table 8-3 also correlate well with the number of hydrogen atoms in the β-position of the radical available for transfer. For example, k_d/k_c increases from $CH_3\cdot + C_2H_5\cdot$ to $CH_3\cdot + i\text{-}C_3H_7\cdot$. Disproportionation combination ratios involving the $t\text{-}C_4H_9\cdot$ radical, which has nine β-hydrogens, are also all large. It appeared, therefore that the transition state in disproportionation was similar to that of combination. The two radicals formed an activated dimer (1) on collision, which could be collisionally stabilized (4) or dissociate into the disproportionation fragments (5). Bradley (37) proposed that transfer of the β-hydrogen

$$C_2H_5\cdot + C_2H_5\cdot \longrightarrow C_4H_{10}^* \begin{array}{l} \xrightarrow{\ M\ } C_4H_{10} \qquad (4) \\[1em] \searrow \\ \quad C_2H_4 + C_2H_6 \qquad (5) \end{array}$$

$$(1)$$

atom occurred as a rearrangement of the activated dimer involving a four-centre transition state such as (2).

$$\begin{array}{c} \overset{\displaystyle H}{} \\ \underset{CH_3}{\overset{\displaystyle CH_2}{\diagdown}} \underset{\underset{\alpha}{CH_2}}{\overset{CH_2\ \beta}{\diagup}} \end{array}$$

$$(2)$$

Bradley further suggested that the relative probability of disproportionation to combination could be decided by the relative stabilities of the products. Since the dimer is activated in the transition state, the relative stabilities of the products could best be measured by the difference in entropy of the disproportionation and combination products, i.e. $\Sigma S_d^0 - S_c^0$ which is the high-temperature limit of the free-energy difference of the products. Bradley's original correlation (37) has since been extended and improved as more experimental results

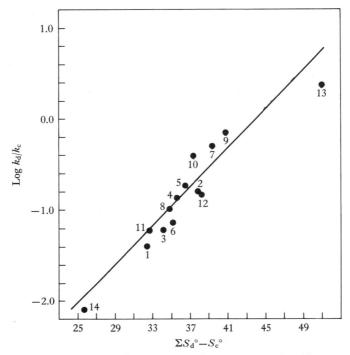

Fig. 8-1. Correlation of disproportionation/combination ratio with entropy difference of products. Key to the numbers is given in Table 8-3. Full line from Terry and Futrell (31). Log $k_d/k_c = 0.11 \ (\Sigma S_d^\circ - S_c^\circ) - 4.88$.

have become available (38, 39, 34) and a plot of the present results is shown in Fig. 8-1. The correlation is quite good for unsubstituted alkyl radicals, but it is found that the correlation does not hold for polar radicals, such as haloalkyl radicals (34) or for large branched alkyl radicals (32a).

In spite of its initial success the four-centre transition state has been strongly criticized, especially by Benson (40). The basic dilemma in both combination and disproportionation reactions is how to reconcile the experimental results, which indicate reaction at virtually every collision, and hence negligible steric difficulties, with the model of the transition state, which suggests that a very specific orientation of the reactants must be attained. Benson (40) has pointed out that in the four-centre transition state (2), the radicals must satisfy particularly stringent steric conditions before reaction. The excited dimer first formed may decompose (5) or it may be collisionally stabilized (4), hence the disproportionation/combination ratio should be pressure-dependent. However, in many cases the k_d/k_c ratio appears to be independent of pressure. The values of k_d/k_c measured in solution, where there is a great excess of solvent to act as the third body, are also little different from the gas-phase values.

Benson concluded that the transition state of the combination reaction is different from that for disproportionation. He proposed that bonding in the transition state must be partly ionic in character. Ionic bonds are non-directional and so impose no steric constraints, unlike the covalent bonds in the excited dimer which are highly directional. Structures such as **(3)** and **(4)** may be written to represent the transition state in the case

(3)   (4)

of two ethyl radicals. Both the 'excited dimer' and 'partly ionic' transition states receive support from time to time in the literature. The problem may be solved by the experimental studies of more diverse radicals which are now being published, although neither theory can satisfactorily rationalize all the data from this source at present (34).

8.1.3 *Combination and disproportionation in solution*

There are two distinct ways in which combination or disproportionation can occur in solution. When the radicals are generated singly, in the solvent medium, they must diffuse together before radical–radical reactions can occur. The rate of diffusion of the radical species is then the maximum rate, analogous to the collision rate of gas-phase combinations. If, however, the radicals are generated in pairs, as for instance in the thermal or photochemical decomposition of azoalkanes, then the two radicals may react before they diffuse apart. The solvent cage, within which they are formed, constrains them to undergo numerous collisions before they can separate. Thus there will be a 'cage' combination or disproportionation rate in addition to the rate of reaction of those radicals which escape from the solvent cage.

The measurement of absolute rate constants of combination in solution has mostly beeen carried out using the intermittent illumination method, and variants thereof. In particular Fessenden has pioneered the

TABLE 8-4 *Termination rate constants for alkyl radicals in solution at 300 °K*

Radical	Solvent	$k \times 10^{-9}$ (l mol^{-1} s^{-1})	$k(\text{Diff.}) \times 10^{-9}$ (l mol^{-1} s^{-1})	Reference
CH$_3$·	Cyclohexane	4.4	—	42
CCl$_3$·	Cyclohexane	0.1–0.6	—	43, 44
C$_2$H$_5$·[a]	Ethane	3.2	—	41
n-C$_3$H$_7$·	Cyclohexane	1.7	1.8[b]	46a
(CH$_3$)$_2$CH·	Benzene	9.6	—	45
(CH$_3$)$_3$C·	Cyclohexane	1.1	~8	42
(CH$_3$)$_2$ĊCN	Benzene	0.9	~7	45
n-C$_5$H$_{11}$·	Cyclohexane	0.2	0.3[b]	46a
n-C$_6$H$_{13}$·	Cyclohexane	1.1	~8	42
c-C$_6$H$_{11}$·	Cyclohexane	1.3	~8	42
CH$_2$(CH$_2$)$_4$ĊCN	Benzene	1.8	—	45
PhCH$_2$·	Benzene	4.1	~7	46
Ph$_2$CH·	Benzene	2.4	—	45
C$_6$H$_5$Ċ(CH$_3$)$_2$	Benzene	~8	—	45
p-CH$_3$C$_6$H$_4$Ċ(CH$_3$)$_2$	Benzene	~4	—	45
p-(CH$_3$)$_2$CHC$_6$H$_4$Ċ(CH$_3$)$_2$	Benzene	1.2	—	45
p-BrC$_6$H$_4$Ċ(CH$_3$)$_2$	Benzene	0.4	—	45
C$_6$H$_5$Ċ(CH$_3$)CH(CH$_3$)$_2$	Benzene	0.2	—	45

[a] Extrapolated from results at about 77 °K in liquid ethane.

[b] $k(\text{Diff.})$ calculated by use of an experimental diffusion coefficient for the radical (46a).

use of e.s.r. spectroscopy, coupled with a pulsed source of illumination for measuring radical concentrations (41). Absolute combination rates measured by these methods are given in Table 8-4. The experimental rates make no distinction between cage combination and combination in the bulk of the medium. In most cases no allowance has been made for disproportionation, and the rate constant represents the total termination rate. The termination rate constant differs little from 2×10^9 l mol^{-1} s^{-1}, for all the alkyl radicals examined, although it is well known that for larger radicals, studied in polymerization systems, the termination rate decreases as the size of the radicals increases (47, 48). Temperature coefficients of the rate constants are very small, usually too small to be measured by the intermittent illumination method. For ethyl radicals, the Arrhenius activation energy has been found to be 3.5 kJ mol^{-1} (41) and this low value is probably fairly typical of small alkyl radicals.

The experimental termination rates can be compared with the

diffusion rates of the radicals, calculated from equation (6). The derivation of this equation is given in chapter 6.

$$k(\text{Diff.}) = 4\pi\sigma D_{AB}(N/1000)/[1 + 4\pi\sigma D_{AB}N/1000k] \qquad (6)$$

Some values of the diffusion rate constant $k(\text{Diff.})$ are also shown in Table 8-4. It is evident that the experimental termination rates are smaller, by a factor of about four, than the diffusion rates. There are at least three possible reasons for this: (i) Approximations used in calculating $k(\text{Diff.})$. The relative diffusion coefficient D_{AB} is usually estimated from the Stokes–Einstein equation, $D = kT/6\pi\eta r$, where η is the solvent viscosity and r the diffusion radius. The diffusion radius is usually equated with the Van der Waals radius of the radicals, i.e. σ, and this assumption may introduce error. Experiments in which D_{AB} has been measured by the space intermittency method, for a number of alkyl radicals, show that the true values are less than these estimates. Good agreement has been obtained between the calculated and observed termination rates when the true D_{AB} values are used (46a) (see Table 8-4). (ii) Only one collision in four of the radicals leads to formation of an attractive singlet electron state. In solution, once the radicals diffuse together, they will be constrained to undergo numerous collisions by the caging effect of the surrounding solvent, so the spin correction factor of $\frac{1}{4}$ is normally not applied. It may be, however, that total neglect of this factor is not justified. (iii) Combination or disproportionation of the radicals can only be expected to occur at every encounter if they are free to rotate in the transition state, and attain the orientation favourable for reaction. Solvent molecules surrounding the two radicals may impede this free rotation. The diffusion rate, as calculated by equation (6), is the translational diffusion rate only, and takes no account of impeded rotation.

In spite of the poor agreement between the experimental and calculated rate constants, there is little doubt that the termination rates are diffusion-controlled. Incorporating the Stokes–Einstein equation in the diffusion rate equation (6) and making the approximation $4\pi\sigma D_{AB}N/1000k \ll 1$ gives:

$$k(\text{Diff.}) \approx 8RT/3000\eta \qquad \text{l mol}^{-1}\,\text{s}^{-1}$$

It is well known that alkyl radical termination rates do decrease with increasing solvent viscosity as this equation implies (49). Carlsson and Ingold were able to show, by measurements in several solvents, that the absolute termination rate of t-butyl radicals was proportional to $\eta^{-0.8}$, in reasonable agreement with the theory (42).

Weiner and Hammond have attempted to find the underlying factors controlling the termination rates (45). They point out that the relative rates for the substituted cumyl radicals (last five entries in Table 8-4) rule out pure electronic effects as the major factor. Thus although *p*-substituents on cumyl radicals have a marked effect on k_t, the difference in rate for *p*-CH$_3$ and *p*-(CH$_3$)$_2$CH substituents is too great to be accounted for solely by electronic effects. Nor can the rate data be explained in terms of radical stabilities. It might be expected that the termination rates of the more stable radicals would be slower. However, this is not the case, and k_t for cumyl radicals is greater than for t-butyl radicals although cumyl radicals would be expected to be more stable. Cumyl radical termination is also faster than benzyl, although again cumyl radicals are the more stable. Other reversals in the expected order, using this criterion, can be seen in Table 8-4. Weiner and Hammond also attempted to correlate the rate of termination with the proportion of termination occurring immediately within the solvent cage, but again no positive trend was found. Their results indicated that the termination rate constants do show variations with the nature of the solvent, and they tentatively suggested that k_t may be controlled by solute–solvent interactions.

Cage combination reactions have been studied by several different approaches. The radical source must be capable of generating two radicals simultaneously. If the source is chosen so that two *different* radicals are generated, e.g. an unsymmetrical azo compound:

$$R—N{=}N—R' \longrightarrow R\cdot + N_2 + R'\cdot$$

then cage combination produces the unsymmetrical dimer RR', whereas the symmetrical dimers R$_2$ and R'$_2$ are only formed in the bulk of the medium, when two like radicals diffuse together. In the same way two symmetrical radical sources may be decomposed at the same time, e.g.:

$$(RCO_2)_2 \longrightarrow R\cdot + 2CO_2 + R\cdot$$

$$(R'CO_2)_2 \longrightarrow R'\cdot + 2CO_2 + R'\cdot$$

Symmetrical dimers, R$_2$ and R'$_2$, are formed by cage combination and the unsymmetrical compound RR' is produced in the bulk of the medium after diffusion. The amount of cage combination depends on the solvent, and generally cage reaction forms a greater percentage of the total combination in more viscous solvents. In very viscous solvents such as decalin, cage combination may account for as much as 80 % of the total reaction, as compared to only 15 % in the less viscous pentane (50).

Sheldon and Kochi (50), by photolysing diacyl peroxides RCO_2O_2CR' and peresters $RCO_2OC(CH_3)_2R'$ in various solvents, have shown that the disproportionation/combination ratio k_d/k_c is essentially the same in decalin as in pentane, although the amount of cage reaction in decalin is much greater. It appears, therefore, that the ratio k_d/k_c is the same for radicals free in the solution, as for caged radicals. The measured k_d/k_c ratios are also strikingly similar to the gas-phase values for the same radicals.

Szwarc and co-workers have taken a different approach in the study of cage combinations (51). They carried out experiments on the photolysis of hexafluoroazomethane in the presence of radical scavengers, or in a suitably reactive solvent, so that all the $CF_3\cdot$ radicals diffusing out of the original solvent cage were removed. The C_2F_6 formed in the system then comes entirely from cage combination, and the ratio $[C_2F_6]/[N_2] = P_c$, represents the probability of cage combination.

$$CF_3\text{—}N\text{=}N\text{—}CF_3 \longrightarrow CF_3\cdot + N_2 + CF_3\cdot$$

They showed that for a wide variety of solvents and over a range of temperatures, the probability of cage combination obeyed the relationship $1/P_c \propto T^{0.5}/\eta$ as is predicted from theory (52). Small deviations from this linear correlation were attributed to hindrance of radical rotation by solvent molecules (factor (iii) above, p. 207). Further evidence for the importance of factor (iii) came from similar studies with methyl radicals. The probability of cage combination for $CH_3\cdot$ radicals obeyed a similar relationship. It was found that under similar conditions $CH_3\cdot$ radical combination was about five times as probable as $CF_3\cdot$ radical combination. This is most likely due to the faster rate of rotation of $CH_3\cdot$ radicals.

Another interesting question about cage combination has been posed by Kopecky and Gillan (53), and by Greene *et al.* (54). In a radical combination occurring rapidly within the solvent cage between two radicals generated close to each other, would the reaction be stereospecific? This question was investigated by thermally decomposing an unsymmetrical azoalkane having one optically active centre (53), viz. (S)—(−)-1,1′-diphenyl-1-methylazomethane (5). If inversion of configuration were slow in comparison with cage combination, then the product (6) would be formed with retention of configuration, and

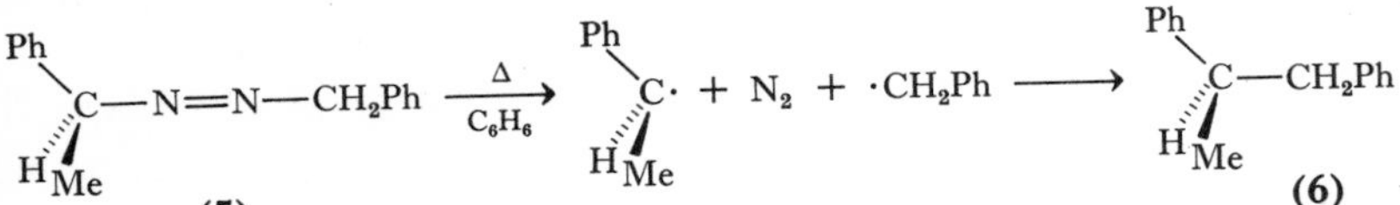

(5)   (6)

consequent optical activity. The amount of combination occurring, after diffusion out of the solvent cage, was corrected for by analysing the symmetrical products PhCH(Me)CH(Me)Ph and PhCH$_2$CH$_2$Ph formed. The authors observed about 15 % retention of configuration, the exact amount depending on the solvent, and they estimated that rotation of the radicals was about sixteen times as fast as combination. A great deal of randomization of radical orientation prior to combination was also observed in the thermal decomposition of (S,S)-(−)-azobis-1-phenyl-ethane (7), and the analysis again indicated that radical rotation was about fifteen times as fast as combination (54).

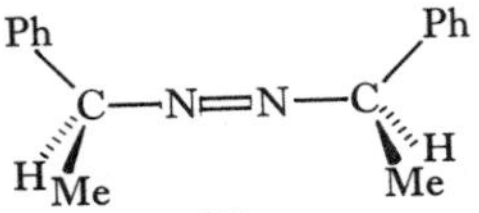

(7)

8.2 ALKYL RADICAL ADDITION REACTIONS

8.2.1 General features

Alkyl radicals will undergo addition reactions with many unsaturated compounds such as olefins, acetylenes, aromatic compounds, azo compounds and carbonyl compounds. The addition reaction with olefins is by far the most extensively investigated, and most of the kinetic studies have been carried out with simple alkyl radicals such as methyl, trifluoromethyl, trichloromethyl and ethyl.

Alkyl radical addition very commonly occurs as part of a chain sequence. Thus bromotrichloromethane will add across the double bond of an olefin in a chain process (7, 8), eventually giving the adduct

$$CCl_3Br + h\nu \longrightarrow CCl_3{\cdot} + Br{\cdot}$$

$$CCl_3{\cdot} + RCH{=}CH_2 \longrightarrow R\dot{C}HCH_2CCl_3 \qquad (7)$$

$$R\dot{C}HCH_2CCl_3 + CCl_3Br \longrightarrow RCHBrCH_2CCl_3 + CCl_3{\cdot} \qquad (8)$$

RCHBrCH$_2$CCl$_3$, when the initial addend radical abstracts bromine from the starting material (*8*). This process is typical of a whole class of reactions in which the radical source, CCl$_3$Br in the above case, adds to the olefin giving a new carbon–carbon bond and yielding a substituted alkane. These radical-chain addition reactions are commonly known as Kharasch additions, after M. S. Kharasch who discovered them and established the radical-chain nature of the mechanism. The addend radical does not necessarily abstract from the starting material. It may decompose back to olefin and alkyl radical, by the reverse of (*7*), i.e.

$$\dot{R}CHCH_2CCl_3 \longrightarrow CCl_3\cdot + RCH{=}CH_2 \qquad (-7)$$

or it may decompose into other products. If the concentration of olefin is high in comparison with the CCl$_3$Br, or if the olefin is very reactive, then the addend radical may add to another olefin unit:

$$\dot{R}CHCH_2CCl_3 + RCH{=}CH_2 \longrightarrow \dot{R}CHCH_2CH(R)CH_2CCl_3 \quad (9)$$

The two-to-one telomer is formed when the new radical abstracts bromine from the CCl$_3$Br (*10*):

$$\dot{R}CHCH_2CH(R)CH_2CCl_3$$
$$+ CCl_3Br \longrightarrow RCHBrCH_2CH(R)CH_2CCl_3 + CCl_3\cdot \quad (10)$$

Three-to-one, four-to-one, etc., telomers of the general form CCl$_3$(CH$_2$CHR)$_n$Br can also be formed by similar processes.

The chains are terminated by radical–radical reactions. In the example given above, three principal termination steps are possible:

$$CCl_3\cdot + CCl_3\cdot \longrightarrow C_2Cl_6 \qquad (11)$$

$$CCl_3\cdot + \dot{R}CHCH_2CCl_3 \longrightarrow CCl_3CH(R)CH_2CCl_3 \qquad (12)$$

$$\dot{R}CHCH_2CCl_3 + \dot{R}CHCH_2CCl_3 \longrightarrow CCl_3CH_2CH(R)CH(R)CH_2CCl_3$$
$$(13)$$

The relative importance of these combination steps is determined by the relative concentrations of the radicals (this depends on the concentration of starting materials), and the combination rate constants of the radicals involved.

There is little difference between the combination rate constants of most small radicals (see Tables 8-1 and 8-4), but the rate constants do decrease for large radicals. In most cases, therefore, the major termination reaction follows from the concentrations of starting materials. In the above example, if excess CCl_3Br is used, then the concentration of $CCl_3\cdot$ radicals will be high and (*11*) is likely to be the main termination reaction. If excess olefin is used, the $CCl_3\cdot$ radicals are more likely to add as in step (*7*); the concentration of $R\dot{C}HCH_2CCl_3$ radicals will be high and step (*13*) becomes the major mode of termination. In excess olefin the telomerization reactions (*8*) etc. are also favoured, so that the concentration of telomer radicals, $R\dot{C}H(CH_2CHR)_nCH_2CCl_3$, also increases and they may also take part in termination steps. The rate constants of large radicals like these are smaller, so that very high olefin concentrations have to be used before such terminations become important.

Chain lengths in alkyl radical additions are not usually as long as in the corresponding chlorinations and brominations, but values of 10^2 to 10^3 are common. In the addition step itself, a double bond is broken, but a single carbon–carbon bond is formed:

$$R_3C\cdot + CH_2{=}CHR \longrightarrow R_3C{-}CH_2{-}\dot{C}HR \qquad (14)$$

The addition step is therefore exothermic for simple alkyl radicals reacting with mono-olefins. It can become endothermic if the radical contains highly electron-withdrawing substituents, so that the new carbon–carbon bond is weak, or in additions to conjugated olefins, aromatic compounds, etc. when resonance energy is also lost in forming the addend radical.

There are good reasons for believing that the addition of alkyl radicals to mono-olefins is not appreciably reversible at temperatures below about 200 °C. The rate of adduct formation *increases* with increasing temperature so that the activation energy of the addition step is positive. In a reversible reaction the measured activation energy is the difference between that of the forward and that of the backward reaction, i.e. $E_{14} - E_{-14}$:

$$R_3C{-}CH_2{-}\dot{C}HR \longrightarrow R_3C\cdot + CH_2{=}CHR \qquad (-14)$$

The activation energies of unimolecular decompositions such as (-14) are generally greater than those of additions, i.e. $E_{-14} > E_{14}$ so that commonly the measured activation energy of a reversible process is negative, and the rate of adduct formation decreases with increasing temperature.

Szwarc and co-workers have used two techniques for establishing the irreversibility of $CF_3\cdot$ additions with olefins (55). The $CF_3\cdot$ radicals, generated by photolysis of hexafluoroazomethane, could then abstract hydrogen from the hydrocarbon solvent RH, or add to an olefin S which was added to the solution:

$$CF_3N{=}NCF_3 \xrightarrow{h\nu} 2CF_3{\cdot} + N_2$$

$$CF_3{\cdot} + S \longrightarrow CF_3S{\cdot} \qquad (15)$$

$$CF_3{\cdot} + RH \longrightarrow CF_3H + R{\cdot}$$

The quantity of nitrogen formed gives a measure of the number of $CF_3{\cdot}$ radicals produced. The addend radicals $CF_3S{\cdot}$ are removed from the system in radical–radical reactions:

$$CF_3S{\cdot} + \text{Radical} \longrightarrow \text{Stable product}$$

$$CF_3S{\cdot} \longrightarrow CF_3{\cdot} + S \qquad (-15)$$

If the addition were reversible, more $CF_3{\cdot}$ radicals would be available to form CF_3H. Decreasing the concentration of azo compound would diminish the number of radicals formed and also the amount of nitrogen; so that if (15) were reversible, more $CF_3S{\cdot}$ radicals would decompose (-15) rather than combine with other radicals, hence the ratio $[CF_3H]/[N_2]$ would increase. Szwarc showed from experiments in both the gas and liquid phases that, in fact, there is no change in the $[CF_3H]/[N_2]$ ratio for a wide variation of $CF_3N_2CF_3$ concentration.

The second technique adopted by Szwarc *et al.* was to carry out the reaction with a *trans*-1,2-disubstituted olefin. In the addend radical free rotation about the carbon–carbon bond, which was the double bond of the olefin, can occur; so that if decomposition to olefin plus $CF_3{\cdot}$ radicals were important, the unreacted olefin remaining at the end of the reaction would be a mixture of *cis*- and *trans*-isomers (cf. chapter 12). The reaction was carried out in solution with *trans*-dichloroethylene, and no isomerization of the unreacted olefin was observed. In the gas phase it was not possible to obtain positive evidence from this process because a side reaction formed chlorine atoms which themselves isomerized the olefin.

Tedder and Walton showed that $CCl_3{\cdot}$ radical addition is not reversible by a different method (56). The addition of $CCl_3{\cdot}$ radicals to vinyl chloride yields the radical $CCl_3CH_2\dot{C}HCl$ (8). This same radical was generated by chlorination of 1,1,1,3-tetrachloropropane:

$$CCl_3CH_2CH_2Cl + Cl{\cdot} \longrightarrow CCl_3CH_2\dot{C}HCl \longleftarrow CCl_3{\cdot} + CH_2{=}CHCl$$
$$\text{(8)}$$
$$\updownarrow$$
$$CCl_3{\cdot} + CH_2{=}CHCl$$

Alkyl radicals

Chlorine atoms attack predominantly at carbon atom 3, thus giving (**8**). Decomposition of this radical, by the reverse of the radical addition step, would give vinyl chloride and CCl_3· radicals, which could be detected in the chlorination reaction, but not in the CCl_3· addition step, since they are starting materials. No products incorporating CCl_3· or $CH_2{=}CHCl$ were found even when the chlorination was carried out at temperatures as high as 250 °C.

It seems reasonable to assume, therefore, that the addition of most alkyl and substituted-alkyl radicals to mono-olefins is irreversible. This conclusion does not apply to *any* unsaturated substrate, and evidence exists that the addition to aromatic molecules is reversible. With benzene, for instance, the reaction with CF_3· radicals in the gas phase becomes appreciably reversible at 150 °C (55). Robb and co-workers have also concluded from work with CF_2Cl· radicals and aromatic substrates, that the addition is reversible at higher temperatures (57).

8.2.2 Orientation of alkyl radical addition

Kharasch observed that HBr addition to olefins occurred in two ways. In the normal ionic addition, the bromine atom appeared attached to the most substituted carbon atom of the adduct. This is usually called normal, or Markovnikov addition:

$$RCH{=}CH_2 + H^+ \longrightarrow R\overset{+}{C}HCH_3$$

$$R\overset{+}{C}HCH_3 + Br^- \longrightarrow RCHBrCH_3$$

In the presence of peroxides or u.v. light the orientation was reversed. Kharasch explained this reversal in terms of a radical-chain mechanism. The active species was now the bromine atom, which also attacked the least substituted carbon atom:

$$RCH{=}CH_2 + Br· \longrightarrow R\overset{·}{C}HCH_2Br$$

$$R\overset{·}{C}HCH_2Br + HBr \longrightarrow RCH_2CH_2Br + Br·$$

Other reagents, such as polyhalomethanes, which reacted by the free-radical pathway, were also found to give 'Anti-Markovnikov' products, where the radical itself had added to the least substituted carbon atom (58). Modern work, using sophisticated analytical methods such as gas chromatography, has, in the main, confirmed the original findings. Free-radical addition is seldom completely specific to one carbon atom of the double bond, but the predominant direction of attack can nearly always be predicted by the simple empirical rule that it will occur at the 'least substituted' carbon atom. For all radicals so far studied, this rule appears to hold with few exceptions, irrespective of the nature of the substituents on the olefin.

Haszeldine and co-workers (59) have carried out a comprehensive investigation of the orientation of trifluoromethyl radical addition to unsaturated systems. The olefin, together with excess CF_3I, was sealed into a quartz or glass tube. The reactions were initiated either photochemically or thermally, and allowed to proceed until virtually all the olefin was consumed. The orientation of the addition was then determined by analysis of the addition products. Adduct isomers were separated and characterized by physical and chemical methods, and the relative amounts of the two isomeric one-to-one adducts gave a measure of the preferred orientation. With trifluoroethylene, for example, two one-to-one adducts were isolated:

$$CF_3I \xrightarrow{h\nu} CF_3 \cdot + I \cdot$$

$$CF_3 \cdot + CHF = CF_2 \longrightarrow CF_3CHFCF_2 \cdot$$

$$CF_3 \cdot + CF_2 = CHF \longrightarrow CF_3CF_2CHF \cdot$$

$$CF_3CHFCF_2 \cdot + CF_3I \longrightarrow CF_3CHFCF_2I + CF_3 \cdot$$

$$CF_3CF_2CHF \cdot + CF_3I \longrightarrow CF_3CF_2CHFI + CF_3 \cdot$$

– the ratio of the adduct formed by addition to $=CF_2$ to that of addition to $=CHF$, i.e. $CF_3CF_2CHFI : CF_3CHFCF_2I$, being $1:2.1$. Further results of this study are given in Table 8-5.

Tedder and co-workers (60) have studied the orientation of trichloromethyl radical addition. Excess CCl_3Br together with the olefin was photolysed, wholly in the gas phase, and the products identified and analysed. The reaction was allowed to proceed to only a few per cent consumption of starting materials, so that kinetic measurements were possible once the mechanism was established. For example, with vinyl fluoride:

$$CCl_3Br \xrightarrow{h\nu} CCl_3 \cdot + Br \cdot$$

$$CCl_3 \cdot + CH_2 = CHF \longrightarrow CCl_3CH_2CHF \cdot$$

$$CCl_3 \cdot + CHF = CH_2 \longrightarrow CCl_3CHFCH_2 \cdot$$

$$CCl_3CH_2CHF \cdot + CCl_3Br \longrightarrow CCl_3CH_2CHFBr + CCl_3 \cdot$$

$$CCl_3CHFCH_2 \cdot + CCl_3Br \longrightarrow CCl_3CHFCH_2Br + CCl_3 \cdot$$

Alkyl radicals

In a similar way, the orientation of addition of heptafluoropropyl radicals, generated in the gas phase by photolysis of C_3F_7I, was investigated. The results of these experiments are also given in Table 8-5.

TABLE 8-5   *Orientation of electrophilic-radical addition to olefins*

Olefin	CF$_3$·[a]	CCl$_3$·[b]	C$_3$F$_7$·[c]
CH$_2$=CHF	1:0.12[d]	1:0.077	1:0.050
CH$_2$=CF$_2$	1:0.0	1:0.012	1:0.009
CH$_2$=CHCl	1:0.0	1:0.0	—
CH$_2$=CHCH$_3$	1:0.12[d]	1:0.071	—
CH$_2$=CFCH$_3$	—	1:0.007	—
CHF=CF$_2$	1:0.48	1:0.29	1:0.25
CHF=CHCF$_3$	1:0.33	—	—
CHCl=CF$_2$	1:11.5	1:25	—
CHCl=CCl$_2$	—	1:0.033	—
CH$_3$CH=CF$_2$	1:0.0	—	—
CF$_3$CH=CF$_2$	1:0.67	—	—
CH$_3$CH=CHCF$_3$	1:0.25	—	—
CF$_2$=CFCl	1:0.25	1:0.0	—
CF$_2$=CFBr	—	1:0.03	—
CF$_2$=CFCF$_3$	1:0.25	1:0.0	—
CF$_2$=CCl$_2$	—	1:0.2	—

In cases where none of an adduct isomer could be detected, i.e. the addition was apparently exclusive to the other carbon atom of the double bond, the value zero is given in the table.

[a] CF$_3$I addition in sealed tubes at 200 °C (59).
[b] CCl$_3$Br addition in the gas phase at 150 °C (60).
[c] C$_3$F$_7$I addition in the gas phase at 150 °C (61).
[d] Values supported from CF$_3$CN addition (62).

The ratio of the rates of attack at each end of an olefin varies with temperature, and as the temperature is raised, the ratio approaches closer to 1:1. Values at a single temperature only are given in the Table. The CF$_3$· radical results were obtained from experiments where most of the reaction occurred in the liquid phase, whereas the CCl$_3$· and C$_3$F$_7$· radical results are from gas-phase work. There are very few results which give a direct comparison of orientation in the liquid and gas phases. It would be expected, however, by analogy with hydrogen abstraction reactions, that differences between the two phases should be minor. The selectivity in chlorination or bromination changes little from the gas to the liquid phase, provided non-complexing solvents are used (see p. 184). The orientation of CF$_2$Br· radicals in addition to CHF=CF$_2$ has been investigated in both phases (63), and it is found that the ratio of the rates of attack at the two ends is the same in both phases, i.e. 1:0.42 at 100 °C, which provides some support for this view.

Inspection of Table 8-5 shows that in all cases where comparison is possible, $CF_3\cdot$, $CCl_3\cdot$ and $C_3F_7\cdot$ radicals show the same orientation in addition to a given olefin. With trifluoroethylene all three radicals prefer to add at the $=CHF$ end. The ratio of the rates of addition at the two ends decreases from $CF_3\cdot$ to $C_3F_7\cdot$, i.e. $CCl_3\cdot$ and $C_3F_7\cdot$, are more selective than $CF_3\cdot$ radicals. The only olefin for which the simple rule: 'addition occurs preferentially at the least substituted carbon atom' breaks down, is that of 1,1-difluorochloroethylene where both $CF_3\cdot$ and $CCl_3\cdot$ radicals react faster at the $=CF_2$ end.

The orientation of radical addition has been interpreted in terms of four main factors. (i) The strength of the bond being formed, i.e. the radical attaches itself preferentially at the end of the olefin which leads to the strongest bond. (ii) Steric hindrance, i.e. the radical avoids sterically crowded sites, either because approach of the radical is hindered and/or because the bond being formed is weakened by steric interactions in the addend radical. (iii) Polar effects, e.g. radicals containing halogen atoms may be polarized as shown in (9), i.e. they will

$$
\begin{array}{c}
X_3 \quad \delta- \\
| \\
\overset{\cdot}{C} \quad \delta+ \\
R \qquad \qquad H \\
\diagdown \qquad \diagup \\
C = C \\
\diagup \quad \delta+ \quad \diagdown \\
X \qquad \qquad H \\
\delta-
\end{array}
$$

(9)

be electrophilic. When the olefin contains an electron-withdrawing substituent, such as halogen, electrostatic repulsions, charge–charge and/or dipole–dipole, may be expected. This polar effect is of the same kind as was invoked to explain some features of hydrogen abstraction reactions by halogens (p. 178). (iv) Stability of the addend radicals, i.e. the radical attaches itself at the end of the olefin which leads to the more stable product radical. The stability of the addend radical is usually measured by the extent to which the unpaired electron is delocalized (64).

By itself, none of the above effects can rationalize all the orientation data in Table 8-5, and combinations of them must be used. Two such combinations will be considered here. Firstly, the orientation can be explained in terms of bond strengths and steric effects. It is well known that α-halogen substituents weaken a bond, i.e. X_3C—CFH would be weaker than X_3C—CH_2.

The radical would always be expected to add at the site bearing the least number of halogen atoms, if the strength of the bond were the only consideration. The predominant addition at $=CH_2$ in the first three olefins of Table 8-5 is thus readily understandable. With propene and 2-fluoropropene steric repulsion between the attacking radical and the CH_3 substituent would induce the radical to add to the other end. With 2-fluoropropene this would be further helped since the bond to $=CH_2$ would be stronger than that to $=CFCH_3$. The orientation in the next group of seven olefins can be understood in a similar way. The most interesting case of this group is 1,1-difluoro-2-chloroethylene. The strength of the bond formed at $=CF_2$ might well be less than that at $=CHCl$, but here the steric effect of the chlorine atom apparently overrides, and both CF_3· and CCl_3· prefer to add at the $=CF_2$ end. In the final group of four olefins, both carbon atoms are fully substituted by highly electronegative substituents in each olefin. Differences in the strengths of the bonds are therefore minor, and the radical adds in each case to the least sterically crowded position, i.e. $=CF_2$.

Alternatively, the results can be rationalized in terms of polar and steric effects. When an electrophilic radical attacks an olefin with a halogen substituent, the polar force is repulsive, as already explained. Hence the three radicals CF_3·, CCl_3· and C_3F_7· would be expected to add at the end of the olefin carrying the least number of halogen substituents, if polar forces predominate. The preferred orientation of addition to $=CH_2$ of the first three olefins is thus easy to understand. In the case of propene and 2-fluoropropene the polar effect due to the methyl substituent would be slightly attractive, since alkyl groups are weakly electron-repelling (**10**). However, this slight attraction is out-

$$X_3 \quad \delta-$$
$$|$$
$$C \quad \delta+$$
$$H$$
$$\backslash$$
$$C\cdots CH_2$$
$$/ \quad \delta-$$
$$Me$$
$$\delta+$$

(10)

weighed by the steric repulsion, so that addition still occurs at $=CH_2$. With 2-fluoropropene there is also a polar repulsion, due to the fluorine substituent at the $=CFCH_3$ end, so that addition at $=CH_2$ is favoured even more (ratio is 1:0.007 for $CCl_3\cdot$) than with propene (ratio is 1:0.071). Similar arguments can be applied in the next group of seven olefins. With 1,1-difluoro-2-chloroethylene, the polar repulsion would be greatest at $=CF_2$, but again this is outweighed by the steric effect of the chlorine, and the majority of addition occurs at $=CF_2$. The steric effects of the methyl and trifluoromethyl groups would be similar, so that in 1,1,1-trifluorobut-2-ene, the final olefin in this group, the direction of attack should be determined by polar forces alone. As expected, the majority of addition occurs at $CH_3CH=$, rather than $CF_3CH=$. In the final group of olefins, all four are fully substituted by strongly electron-withdrawing substituents, so that polar effects will be similar at the two ends of the olefin. The direction of attack will there-fore be decided by steric factors, and the predominant site of addition is $=CF_2$ in each case, as expected.

The results of other studies of the orientation of radical addition can also be interpreted by either of the above arguments. Cadogan *et al.* (65) have determined the orientation of addition of $CCl_3\cdot$ radicals to a series of *meta-* and *para-*substituted stilbenes (**11**). The substituent R is

$$RC_6H_4CH=CHC_6H_5$$
$$(\mathbf{11})$$

remote from the reaction site, so that virtually no change in the strength of the bond being formed, or the polar force at the attacked carbon, is to be expected for different substituents. Similarly, the steric crowding at the two ends of the olefin is about the same. It would be predicted, therefore, that the ratio of the two adducts should be about 1:1 and independent of the nature of R. Cadogan has shown that for R = 4-NO$_2$, 4-Br, 4-MeO, 3-NO$_2$, 3-Me, 3-Br and 3-MeO both adducts are formed, and that the ratio of the two differs very little from 1:1 in every case.

The orientation of addition to a given olefin also depends on the nature of the attacking radical. Some differences in the adduct ratio obtained from $CF_3\cdot$, $CCl_3\cdot$ and $C_3F_7\cdot$ radical attack on particular olefins are visible in Table 8-5. The adduct ratios for addition of a series of radicals to trifluoroethylene, the only olefin for which such data are available, are shown in Table 8-6. If the relative orientation were governed solely by the stabilities of the addend radicals (**12** and **13**), then

TABLE 8-6   *Orientation of radical addition to trifluoroethylene*

Radical	Temp. (°C)	CHF=CF$_2$	Reference
CF$_3$·	200	1:0.48	59
CF$_2$Cl·	150	1:0.4	69
CF$_2$Br·	150	1:0.46	63
CFBr$_2$·	150	1:0.35	69
CCl$_3$·	150	1:0.30	60
C$_3$F$_7$·	150	1:0.25	61

virtually no change in the orientation should be observed when a different radical is used, since the unpaired electron is still situated on

$$CX_3CHFCF_2· \qquad\qquad CX_3CF_2CHF·$$
$$(12) \qquad\qquad\qquad (13)$$

—CF$_2$ or —CHF irrespective of the nature of the attacking radical. In fact, however, there is quite an appreciable change in the adduct ratio going down the Table from CF$_3$· to C$_3$F$_7$· radicals. It would be very difficult to predict what changes to expect from one radical to another, if the orientation were judged in terms of the strength of the bond formed, or in terms of polar forces. The difference in the polar forces cannot be accurately judged for the fully halogenated radicals of Table 8-6. There does seem to be a decrease in the adduct ratio, as the size of the attacking radical increases, so that for these particular electrophilic radicals, the decisive factor appears to be steric although this is obviously an oversimplification. It is not clear whether the steric effect acts to hinder approach of the larger radicals, or whether the strength of the bond formed is decreased by steric interaction of the large radical with substituents present in the olefin.

An interesting consequence of the above methods of rationalizing orientation can be expected for nucleophilic radicals. In the addition of a nucleophilic radical, steric effects are of the same kind as for electrophilic species, but the polar effect is reversed. In the addition to CH$_2$=CHF, where steric interaction will be minimal at either end, the polar force between an approaching nucleophilic radical and the =CHF reaction site is attractive. The predominant position of attack may therefore be =CHF, and not =CH$_2$ as was the case with the electrophilic radicals of Table 8-5. It is unfortunate that experimental orientation studies of nucleophilic radicals are practically non-existent.

Unsubstituted alkyl radicals are slightly nucleophilic. Getty, Kerr and Trotman-Dickenson have studied the orientation of alkyl radical addition to propyne (66). The polar force at $CH_3C\equiv$ would be repulsive, and steric effects also favour addition at the other end, i.e. $\equiv CH$. They found that at least 90 % of methyl radical addition does occur at $\equiv CH$, and that isopropyl and t-butyl radicals also add almost exclusively at that end. Methyl and trifluoromethyl radicals both add exclusively to the terminal $=CH_2$ in allene (67) as do other alkyl radicals (68). However, polar effects in this case would be very small, and the orientation may simply be controlled by the more favourable steric situation at the terminal positions. No case is yet known where a nucleophilic alkyl radical adds predominantly at the other end from an electrophilic radical, but this may simply be due to lack of experimental evidence. Nucleophilic hetero-radicals are known to add at the other end from electrophilic radicals (see p. 276).

8.2.3   *Structure and reactivity in alkyl radical addition reactions*

The problem of relating olefin and radical structure to their reactivity in the addition reaction is one of continuing interest to organic chemists. Experimentally the main technique has been to react one radical with a series of olefins exhibiting various structural features, and then attempt to correlate the relative olefin reactivities with the radical and olefin properties. Competitive techniques have been employed enabling the addition reaction to be compared with an abstraction reaction of known rate. Alternatively the addition step is measured relative to the radical combination rate, e.g. in the addition of bromodifluoromethyl radicals to ethylene. The radicals are generated by photolysis of dibromodifluoromethane.

$$CF_2Br_2 \xrightarrow{h\nu} CF_2Br\cdot + Br\cdot$$

$$CF_2Br\cdot + CH_2{=}CH_2 \longrightarrow CF_2BrCH_2CH_2\cdot \qquad (16)$$

$$CF_2BrCH_2CH_2\cdot + CF_2Br_2 \longrightarrow CF_2BrCH_2CH_2Br + CF_2Br\cdot \quad (17)$$

$$CF_2Br\cdot + CF_2Br\cdot \longrightarrow CF_2BrCF_2Br \qquad (18)$$

$$Br\cdot + Br\cdot + M \longrightarrow Br_2 + M \qquad (19)$$

$$Br_2 + CH_2{=}CH_2 \longrightarrow CH_2BrCH_2Br \qquad (20)$$

The rate of adduct formation is given by:

$$R_{Add.} = k_{15}[CF_2Br\cdot][CH_2{=}CH_2]$$

Alkyl radicals

and the rate of dimer formation by:

$$R_c = k_{17}[CF_2Br\cdot]^2$$

so that $k_{15}/k_{17}^{\frac{1}{2}} = R_{Add.}/[CH_2{=}CH_2]R_c^{\frac{1}{2}}$, and the rate of addition, relative to that of combination, can then be found from the rates of formation of the adduct and dimer.

By the application of this kind of method the *relative* rates of addition of several alkyl radicals to series of olefins have been determined. Absolute values of the rates and Arrhenius parameters of addition can be deduced only if the absolute rate of the reference reaction, i.e. combination, or hydrogen abstraction, is known. There is thus quite a high degree of uncertainty in the absolute values quoted in tables below, but their relative rates are fairly precise.

It is much more difficult to measure the reactivity of a series of radicals with an olefin, chiefly because details of the reaction mechanism change from one system employing one radical source to another. Adjusting the conditions to minimize side reactions is usually a tedious and time-consuming process, which has to be carried out for each new radical. Nevertheless the rates of addition of a number of radicals to ethylene, which is a convenient basis or reference olefin for gas-phase work, have been determined and these are shown in Table 8-7. The rate of addition varies very little from radical to radical, which is to be expected since the strength of the bond formed, polar effects, steric effects and the stabilities of the addend radicals, will be similar in all cases. There is particularly little variation in the rate of addition of unsubstituted alkyl radicals, which is again just what we expect, since bond strength

TABLE 8-7 *Arrhenius parameters for the addition of alkyl radicals to ethylene*

$$R_3C\cdot + CH_2{=}CH_2 \longrightarrow R_3CCH_2CH_2\cdot$$

Radical	Log k (164 °C) (l mol^{-1} s^{-1})	Log A (l mol^{-1} s^{-1})	E (kJ mol^{-1})	Reference
$CF_3\cdot$	7.2	8.4	10.1	70
$CCl_3\cdot$	5.3	8.3	26.4	60
$CH_2F\cdot$	5.4	7.6	18.1	71
$CH_3\cdot$	4.6	8.1	29.4	72
$C_2H_5\cdot$	4.4	8.2	31.9	73
$n\text{-}C_3H_7\cdot$	4.8	7.4	21.4	72
$i\text{-}C_3H_7\cdot$	4.7	8.2	29.0	73
$n\text{-}C_4H_9\cdot$	4.0	7.3	28.1	73
3-Methylbutyl	4.3	7 5	26.9	73

222

differences, polar effects, etc., will be minimal in this case. There is, however, a general tendency for the activation energy to decrease, and hence the rate to increase, as the radical becomes more electrophilic, e.g. $CF_3 \cdot$ radical addition is much faster than $CH_3 \cdot$ radical addition. It also seems that the A-factors of the larger radicals like $CCl_3 \cdot$ or $n\text{-}C_4H_9 \cdot$ tend to be lower, which is a good indication of their increased steric interaction with the olefin.

The activation energy of the addition step is quite low, less than 40 kJ mol^{-1}, for all the radicals in the Table. This provides good evidence that the process is exothermic for these simple radicals. This idea is further strengthened since the A-factors are high, only about two orders of magnitude less than the collision frequency. We can immediately expect therefore, by application of the Hammond postulate, that the transition state will resemble the reactants rather than the products. The new carbon–carbon bond between the approaching radical and the olefin will hardly have started to form in the transition state.

8.2.4 *Secondary kinetic isotope effects in addition reactions*

When the olefin is substituted with deuterium atoms at the reaction centre, then these isotopic atoms do not undergo chemical transformation in a radical addition reaction, and the effect on the reaction rate is termed a secondary kinetic isotope effect. Szwarc and co-workers (75)

have determined secondary deuterium isotope effects for methyl and trifluoromethyl radical addition reactions by comparing the rate of reaction for a terminal olefin with the rate for the deuterium-substituted olefin. Stefani and co-workers (76) have measured the isotope effect for cyclopropyl radical addition by a similar technique. The results of these studies are summarized in Table 8-8.

Streitwieser suggested that the major contribution to this isotope effect would come from the change in the soft out-of-plane bending vibration of the olefin C—H into a harder bending vibration when the carbon atom was rehybridized from sp^2 to sp^3 by addition of a radical (74). On this basis the calculated value of the kinetic isotope effect is

TABLE 8-8   *Kinetic isotope effect in alkyl radical addition reactions*

$$=CH_2 + \cdot CR_3 \xrightarrow{k_H} -CH_2CR_3$$
$$=CD_2 + \cdot CR_3 \xrightarrow{k_D} -CD_2CR_3$$

Olefin	$CH_3 \cdot$[a] k_D/k_H	$CF_3 \cdot$[a] k_D/k_H	$c\text{-}C_3H_5 \cdot$[b] k_D/k_H
$CD_2{=}CD_2$	1.05	1.07	1.11
$CH_3CH{=}CD_2$	1.12	1.07	1.09
$CD_3CD{=}CD_2$	1.16	1.09	1.08
$PhCD{=}CD_2$	1.11	1.10	—
$CD_2{=}CDCD{=}CD_2$	1.20	1.09	—

Reactions carried out at 65 °C in iso-octane solution.
[a] Ref. 75.   [b] Ref. 76.

$k_D/k_H = 1.82$, assuming that rehybridization to the tetrahedral state is complete in the transition state (75). If this value is compared with the very low experimental values given in Table 8-8 the conclusion would be that very little rehybridization had occurred in the transition state, and that the transition state resembled the reactants.

Strauss *et al.* (153) have carried out detailed transition-state calculations to determine the secondary kinetic isotope effect for the addition of sulphur atoms to deuteriated ethylenes. Their conclusions, which are quite general for radical addition reactions, suggest that the simple Streitwieser picture is not valid. Strauss *et al.* find that the most important contribution to the isotope effect is the net gain in the isotope-sensitive normal vibrational modes during passage from reactants to the transition state. For the addition of a polyatomic radical to an ethylene the gain in normal modes is six, of which one will coincide with the reaction co-ordinate. From the remaining five, at least one, the CH_2 twist, will always be isotope-sensitive and will generate a substantial isotope effect. For the ethylene molecule there is a substantial barrier

to the twist motion, which is considerably lowered in the transition state. Applying the lower energy for the twist mode leads to a decrease in the calculated isotope effect. Calculations which neglect the change in the CH_2 twist will always overestimate the isotope effect. Strauss *et al.* also found substantial contributions to the isotope effect arising from the changes in the mass and moment of inertia of the reactants. The position of the transition state along the reaction co-ordinate can only be determined by a detailed analysis of each reaction.

8.2.5　*The transition state in radical addition reactions*

If the transition state of a radical addition reaction resembles the reactants rather than the products, the reactant properties will have the greatest influence on the orientation and rate of the reaction. Polar and steric interactions between the radical and olefin would be decisive, whilst factors like the stability of the product radical should be minor.

TABLE 8-9　*Relative rates of addition of alkyl radicals to olefins*

Olefin	$CH_3 \cdot$[a] 180 °C	$c\text{-}C_3H_5 \cdot$[c] 65 °C	$CCl_3 \cdot$[d] 100 °C	$CF_3 \cdot$[e] 65 °C	$C_3F_7 \cdot$[g] 164 °C	Ionization potential (eV)
$CH_2{=}CH_2$	1.00	1.00	1.00	1.00	1.00	10.6
$MeCH{=}CH_2$	0.67	0.44	1.47	1.25[f]	—	9.8
$EtCH{=}CH_2$	0.42	0.41	—	1.52[f]	—	9.8
cis-$MeCH{=}CHMe$	0.18	0.26	—	3.54[f]	—	9.3
trans-$MeCH{=}CHMe$	0.28	0.24	—	2.24[f]	—	9.3
$Me_2C{=}CH_2$	1.03	0.28	—	3.80[f]	—	9.3
$Me_2C{=}CHMe$	0.24	0.15	—	3.47[f]	—	8.8
$Me_2C{=}CMe_2$	0.08	0.07	—	7.08[f]	—	8.3
$ClCH{=}CH_2$	—	1.74	0.94	0.49	—	—
$MeCF{=}CH_2$	—	—	0.66	0.42	—	—
$CHF{=}CH_2$	—	—	0.69	0.09	0.54	—
$CF_2{=}CH_2$	—	—	0.17	—	0.20	—
$CF_2{=}CHF$	—	—	0.13	—	0.08	—
$CF_2{=}CF_2$	11.4[b]	—	0.06	0.12	0.20	—
$CF_3CF{=}CF_2$	15.5[b]	—	0.02	—	—	—

[a] Methyl radicals in the gas phase (77).
[b] From ref. 81.
[c] Cyclopropyl radicals in iso-octane solution (76).
[d] Trichloromethyl radicals in the gas phase (78).
[e] Trifluoromethyl radicals in the gas phase (79).
[f] Solution-phase results (80).
[g] Heptafluoropropyl radicals in the gas phase (61).

In Table 8-9 the relative rates of addition of some alkyl radicals to hydrocarbon and halo-substituted olefins are given. The relative rates are the combined rates for addition at both ends of the olefin. The radicals fall naturally into two groups: firstly with $CCl_3 \cdot$, $CF_3 \cdot$ and $C_3F_7 \cdot$ the relative rate tends to increase as the number of alkyl substituents in the olefin increases, but decreases as the number of halogen substituents in the olefin increases. These radicals are clearly electrophilic in character, and show a marked similarity in their behaviour compared to the electrophilic atoms O, S, Se, etc., compare with Table 7-5. The nucleophilic methyl and cyclopropyl radicals, however, show the reverse trends, i.e. the relative rate decreases with increasing alkyl substitution but increases with halogen substitution.

The classification of alkyl radicals as electrophilic or nucleophilic involves the tacit assumption that polar forces are of importance. Several features of the results in Table 8-9 support this assumption. Plots of the logarithm of the rate constant against the number of electron-repelling alkyl substituents in the olefin, are quite good straight lines for methyl and cyclopropyl radicals (76). A plot of log k(rel.) against the number of alkyl substituents in the olefin is also a good straight line for trifluoromethyl radicals, but the slope of the plot is opposite to that of CH_3· or c-C_3H_5· radicals. The conclusion clearly must be that the polar effect of the alkyl substituent outweighs its steric effect in determining the overall rate of addition. The fact that the slopes of the plots differ in sign supports the idea that the unsubstituted alkyl radicals are nucleophilic relative to the electrophilic CF_3· radicals.

The electron-releasing power of an olefin is related to its ionization potential. The more electron-repelling alkyl substituents there are on the olefin, the lower is its ionization potential. For the addition of CH_3·, CF_3· and c-C_3H_5· radicals to alkyl-substituted olefins, a good correlation of the logarithm of the rate constant with olefin ionization potential can be obtained (see Fig. 8-2). Again, the slope of the CF_3· plot is opposite to that of CH_3· or c-C_3H_5·, emphasizing their opposite response to the increasing electronegativity of the olefin double bonds. It is not valid to extend this correlation to the halogen-substituted olefins, however, since here the first ionization potential refers to ionization of one of the lone-pair electrons on the halogen itself, and does not give a true measure of olefin double bond electronegativity.

The rate of addition apparently responds simply to the number of electron-releasing or -attracting groups in the olefin. This is shown most clearly for the reactions of the nucleophilic CH_3· and c-C_3H_5· radicals with di-, tri- and tetra-methyl substituted ethylenes. For the electrophilic CCl_3· radicals the rate decreases from trifluoroethylene to tetrafluoroethylene to hexafluoropropene. This evidence, together with the ionization potential correlations, has led some workers to suggest that it

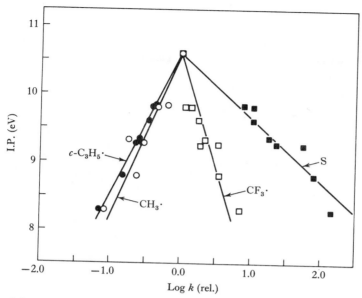

Fig. 8-2. Correlation of the relative rates of addition of cyclopropyl radicals (filled circles), methyl radicals (empty circles), trifluoromethyl radicals (empty squares) and S (^3P) atoms (filled squares) with olefin ionization potentials.

is the properties of the olefinic double bond which influence the rate of addition, i.e. that the initial addition of the radical occurs to the double bond, to form a π-complex, and that localization of the radical at one or other of the carbon atoms of the bond occurs only at a later stage in the reaction.

It is also clear from an examination of the data in Table 8-9 that steric effects are by no means negligible. The most dramatic demonstration of this appears with isobutene, which has two alkyl substituents and a $=CH_2$ site where radicals can add unhindered. With both methyl and cyclopropyl radicals, the relative rate of addition is faster than to other di-substituted olefins. A steep increase in rate is also observed for the addition of the electrophilic CF_3· radicals. The increase in the rate of addition observed with CF_3· radicals as the number of alkyl substituents increases, is also far less pronounced than the decrease in rate observed with CH_3· or c-C_3H_5· radicals. The explanation of this is almost certainly that the addition of CF_3· radicals to alkyl-substituted olefins is retarded by the unfavourable steric situation: but the decreasing rate of addition of $CH_3\cdot$ and $c\text{-}C_3H_5\cdot$ radicals is accelerated by this effect. Further support for this explanation comes from the addition of electrophilic triplet atoms, O, S, Se, etc. In the addition of an atom steric factors are much smaller, and as expected the rate of addition of the electrophilic oxygen or sulphur increases much more steeply with increasing alkyl substitution of the olefin than for $CF_3\cdot$ radicals. This is shown in Table 7-5 and illustrated for sulphur atoms in Fig. 8-2. The trends in reactivity shown in Table 8-9 are thus well accounted for, in a qualitative way, in terms of polar and steric effects.

It is much less easy to construct an explanation in terms of the stabilities of the addend radicals. On these grounds we might expect the relative reactivities of isobutene, trimethylethylene and tetramethyl-ethylene to be about the same, since in each case the unpaired electron in the addend radical is situated at a $=CMe_2$ site. The decrease in rate for $CH_3\cdot$ or $c\text{-}C_3H_5\cdot$ radicals and the increase in rate for $CF_3\cdot$ radicals are thus unaccounted for. Similarly the rate of addition of $CCl_3\cdot$ radicals to 1,1-difluoroethylene, trifluoroethylene and tetrafluoroethylene de-creases strongly, although the unpaired electron in the addend radical is situated on $=CF_2$ in each case.

Other studies with a wider range of olefins have in general supported the idea that polar and steric effects are important. James and co-workers have studied the addition reactions of the ethyl radical (82) with a series of polar olefins. Ethyl radicals were generated in the gas phase by photolysis of diethyl ketone, and the rate of their addition was measured by adding the appropriate olefin and quantitatively analysing the products. The results are given in Table 8-10. The rate of addition is enhanced by polar substituents, in the order: $NC— > C_6H_5—$ $\sim CH_2{=}CH— > CH_3CO.O— > C_4H_9O— >$ alkyl. This is about the order in which the substituents would be expected to influence the rate of addition of a nucleophile.

Cadogan and Sadler (83) have measured the relative reactivity of series of substituted stilbenes ($RC_6H_4CH{=}CHC_6H_5$) and substituted styrenes ($XC_6H_4CY{=}CHZ$) towards the $CCl_3\cdot$ radical in solution. The reactivity trend was in accord with the electrophilic character of $CCl_3\cdot$ radicals. They showed that the relative reactivities of the olefins obeyed the Hammett–Brown equation, $\log k/k_0 = \rho\sigma^+$, a straight-line plot being obtained. The value of ρ was negative ($\rho = -0.7$) showing up the polar, electrophilic nature of the effects involved. The magnitude of ρ is quite small in comparison with that observed from ionic reactions,

TABLE 8-10 *Addition of the ethyl radical to polar olefins* (82)

$$C_2H_5\cdot + CH_2{=}CHR \longrightarrow C_2H_5CH_2CHR\cdot$$

Olefin	k(rel.) (100 °C)	Log A (1 mol^{-1} s^{-1})	E (kJ mol^{-1})
Oct-1-ene	1.0	8.0	31.9
2-Methylpent-1-ene	1.6	8.0	30.7
n-Butyl vinyl ether	1.5	7.3	25.6
Vinyl acetate	1.8	7.8	29.0
Cyclohexa-1,3-diene	14.1	7.7	21.8
2,3-Dimethylbuta-1,3-diene	45.9	7.8	18.9
Styrene	34.1	7.5	17.2
cis-CH$_3$CH=CHCN	4.9	7.1	21.0
trans-CH$_3$CH=CHCN	6.3	7.4	21.8
Acrylonitrile	140	7.7	14.3

but is comparable with values obtained from other radical reactions, e.g. chlorination and bromination (see chapter 7, section 6).

Studies of the addition reactions of trichloromethyl radicals with other series of substituted alkenes, such as that of Martin and Gleicher (84) with 3-phenyl-1-propenes and 4-phenyl-1-butenes, or that of Sakurai *et al.* with alkenylsilanes (85), have provided similar evidence for the importance of polar effects.

Szwarc and his collaborators (79) have obtained evidence for the importance of steric effects in the addition of CF$_3$· radicals to substituted olefins. The Arrhenius A-factors were determined for symmetrically substituted olefins, and these were then compared with the A-factor for addition to ethylene itself; some of the results are given in Table 8-11. Substituents at the reaction centre restrict the rotation of CF$_3$· in the transition state. The entropy factor A is therefore decreased. It is clear from the experimental results that replacing one of the hydrogens at the reaction centre by methyl or chlorine, reduces the A-factor about five-fold; as expected the reduction caused by a fluorine substituent is rather less.

Arrhenius parameters have also been measured for the addition of several radicals to alkyl-substituted olefins, and to halogen-substituted olefins. A difficulty encountered with olefins containing alkyl substituents is that the radical may abstract hydrogen from the alkyl side chain as well as adding to the double bond:

$$R_3C\cdot + XCH_2CH{=}CH_2 \longrightarrow R_3CH + X\dot{C}HCH{=}CH_2$$
$$R_3C\cdot + XCH_2CH{=}CH_2 \longrightarrow XCH_2\dot{C}H{-}CH_2CR_3$$

TABLE 8-11 *Arrhenius parameters for the addition of trifluoromethyl radicals to symmetrically substituted olefins in the gas phase* (79)

$$CF_3\cdot + RR'C{=}CRR' \longrightarrow CF_3RR'C\dot{C}RR' \qquad (14)$$
$$CF_3\cdot + Me_2CHCHMe_2 \longrightarrow CF_3H + Me\dot{C}CHMe_2 \qquad (15)$$

Olefin	A_{14}/A_{15}	$\dfrac{(A_{14})\text{Ethylene}}{(A_{14})\text{Olefin}}$
$CF_2{=}CF_2$	0.25	4.4
cis-CHCl=CHCl	0.32	3.4
trans-CHCl=CHCl	0.22	5.0
$CCl_2{=}CCl_2$	0.20	6.0
cis-MeCH=CHMe	0.20	5.5
trans-MeCH=CHMe	0.24	4.6
$Me_2C{=}CMe_2$	0.15	8.0

Most competitive techniques measure the overall rate of reaction of the olefin with the radical. Deducing the rate of addition can be difficult, especially with heavily alkylated olefins like tetramethylethylene, where hydrogen abstraction is of major importance.

For many of the reactions, e.g. those of $CH_3\cdot$ and $CF_3\cdot$ radicals in Table 8-12, the final products were not isolated so that the position of addition is not known. Equating the overall rate of addition with the rate of addition to the 'least substituted' carbon atom of the olefin is probably justified in most cases however.

TABLE 8-12 *Arrhenius parameters for the addition of substituted-alkyl radicals to hydrocarbon-olefins in the gas phase*

Olefin	No.	$E(CH_3\cdot)$[a] (kJ mol^{-1})	$E(CF_3\cdot)$[b] (kJ mol^{-1})	Log A $(CH_3\cdot)$[a] (l mol^{-1} s^{-1})	Log A $(CF_3\cdot)$[b] (l mol^{-1} s^{-1})
$CH_2{=}CH_2$	1	33.2	10.1	8.93	8.39
$MeCH{=}CH_2$	2	31.1	8.4	8.52	8.20
$EtCH{=}CH_2$	3	30.2	—	8.30	—
cis-MeCH=CHMe	4	31.5	5.5	7.95	7.62
trans-MeCH=CHMe	5	34.0	5.4	8.46	7.62
$Me_2C{=}CH_2$	6	29.0	5.2	8.46	8.20
$Me_2C{=}CHMe$	7	(25.6)	—	(7.48)	—
$Me_2C{=}CMe_2$	8	(28.6)	3.6	(7.30)	7.50
$CH_2{=}CH{-}CH{=}CH_2$	9	17.2	4.1	8.20	8.47
$PhCH{=}CH_2$	10	—	5.5	—	8.41

[a] Values of Cvetanović and Irwin (77), results in parenthesis are regarded by the authors as doubtful.
[b] Values of Szwarc and collaborators (79), put on absolute scale by use of Thynne's values of the rate constant of $CF_3\cdot$ addition to $CH_2{=}CH_2$ (70).

TABLE 8-13   *Arrhenius parameters for addition to specific sites in polar olefins*

Addition to =CH$_2$	No.	E(CCl$_3$·)[a] (kJ mol^{-1})	E(C$_3$F$_7$·)[b] (kJ mol^{-1})	Log A (CCl$_3$·)[a] (l mol^{-1} s^{-1})	Log A (C$_3$F$_7$·)[b] (l mol^{-1} s^{-1})
CH$_2$=CH$_2$	1	26.4	13.0[c]	8.3	8.6[c]
CH$_2$=CHF	11	26.8	16.3	8.1	8.4
CH$_2$=CF$_2$	12	32.3	24.3	8.4	9.2
CH$_2$=CHMe	2	27.2	—	9.6	—
CH$_2$=CHCl	13	27.2	—	9.1	—
CH$_2$=CFMe	14	26.4	—	8.7	—
Addition to =CHF					
CHF=CH$_2$	15	35.2	30.6	8.1	8.6
CHF=CF$_2$	16	38.5	25.9	9.1	8.7
Addition to =CF$_2$					
CF$_2$=CH$_2$	17	47.8	49.0	8.4	9.6
CF$_2$=CHF	18	42.7	30.2	9.1	8.7
CF$_2$=CF$_2$	19	38.5	32.7	9.7	10.1
CF$_2$=CFCF$_3$	20	39.0	—	8.2	—

[a] Gas-phase values of Tedder and Walton (60), revised in ref. 60*a*.
[b] Gas-phase values of Tedder *et al.* (61).
[c] Assumed values, others relative to this.

Tedder, Walton and collaborators have measured Arrhenius parameters for addition of CCl$_3$· and C$_3$F$_7$· radicals to specific sites in fluoro-olefins (60, 61). The rate of addition to a particular end of an olefin was measured by quantitative analysis of the product of addition at that end. An added advantage of working with fluoro-olefins was that abstraction from the olefin was negligible, due to the very strong C–F bond. The results of this work are in Table 8-13. For the first six olefins in the table, addition occurs at =CH$_2$, so that in the addend radical CCl$_3$CH$_2$–ĊX$_2$, the unpaired electron is sited on –ĊH$_2$, –ĊHF, –ĊF$_2$, –ĊHMe, –ĊHCl and –ĊFMe groups, yet the activation energy hardly changes. The unpaired electron would obviously be stabilized to very different extents by such widely differing substituents, hence the site of the unpaired electron in the addend radical can have very little influence on the rate of the addition step. This conclusion is also supported by the data for addition to =CH$_2$ in ethylene (No. 1), =CHF in vinyl fluoride (No. 15) and =CF$_2$ in 1,1-difluoroethylene (No. 17). In each of these reactions the unpaired electron is sited on –ĊH$_2$, so that the extent of delocalization is virtually the same in each case:

$$CCl_3\cdot + X_2C{=}CH_2 \longrightarrow CCl_3CX_2\dot{C}H_2$$

Alkyl radicals

However, large changes in the activation energy, from 26.4 to 35.2 to 47.8, were observed.

A second general conclusion can be drawn for $CCl_3\cdot$ and $C_3F_7\cdot$ radicals. In the addition to unsymmetrical olefins, the activation energies are different for the two ends of the olefin:

	$CH_2{=}CHF$		$CHF{=}CF_2$		$CH_2{=}CF_2$		
$CCl_3\cdot$	26.8	35.2	38.5	42.7	32.3	47.8	E (kJ mol^{-1})
$C_3F_7\cdot$	16.3	30.6	25.9	30.2	24.3	49.0	E (kJ mol^{-1})

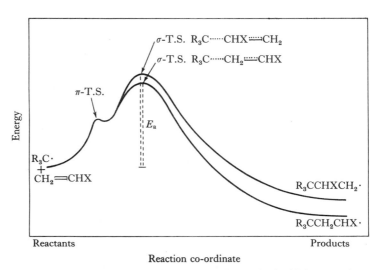

Fig. 8-3. Potential energy diagram for alkyl radical addition reactions.

The radical must therefore add, in the rate-determining step, to a specific atom of the olefin, and not simply to the double bond followed later by migration of the radical to one or other end of the olefin. The free-energy profile of the reaction may be envisaged as shown in Fig. 8-3.

In the first place, for the reasons already given (exothermic reaction, very small kinetic isotope effect), the transition state probably occurs towards the reactant end of the 'reaction coordinate', i.e. it resembles the reactants rather than the products. Secondly, for the reasons already detailed (dependence of the rate on total *number* of substituents, correlation of rate with ionization potential), it is clear that association of the radical with the π-bond occurs early on in the addition process, and that formation of this 'π-complex' influences the height of the potential barrier. There is of course only one π-complex, although the reaction ultimately leads to two different products.

Association of the radical with one specific carbon atom of the double bond is often called the σ-transition state. This transition state also resembles the reactants, and the incipient carbon-carbon bond is long. Separate σ-transition states exist for attack at each end of the olefin. The activation energy of the addition is the difference in energy between the reactants and the highest point of the potential energy curve. Since different activation energies are obtained experimentally for $CCl_3\cdot$ and $C_3F_7\cdot$ radicals, at the two ends of the olefin, the σ-transition states must be higher in energy than the π-transition state, for these radicals.

8.2.6 *Molecular orbital treatment of reactivity*

Numerous authors have discussed the types of correlations to be expected from radical and ionic reagents in addition reactions (86). The most successful reactivity index is undoubtedly the Localization Energy. In the localization process for radical addition, an electron is taken from the olefin double bond, and localized in a $2p$-orbital on one carbon atom of the olefin.

$$R-C_vH-C_\mu H_2 \quad \longleftarrow \quad R-C_vH-C_\mu H_2 \quad \longrightarrow \quad R-C_vH-C_\mu H_2$$

$$E_\pi'' \qquad\qquad E_\pi \qquad\qquad E_\pi'$$

The π-system of the olefin may be extended by interaction with π-orbitals on the substituent R, or by hyperconjugation. The remaining electron may therefore still form part of a π-system. The difference in energy between the original olefin π-system, E_π, and the energy of the localized system E_π' or E_π'' is known as the atom localization energy L_μ or L_ν.

$$L_\mu = E_\pi - E_\pi' \qquad L_\nu = E_\pi - E_\pi''$$

Different localization energies will be obtained, depending on which of the atoms ν or μ in the unsymmetrical olefin the electron is localized on.

Localization energies can be calculated from Hückel and other approximate molecular orbital methods (86, 87). Ideally the reaction rates should be correlated with properties of the transition state itself, i.e. the σ-transition state which represents the highest point of the potential energy profile (Fig. 8-3). In a real situation, the localization process will obviously be complete well after the highest point of the energy profile has been passed and further towards the product.

Alkyl radicals

Nevertheless it is reasonable to expect that the localization energy can reflect the height of the potential barrier: the higher the activation energy, the higher the localization energy, and a correlation of L_μ with activation energy can thus be expected. We can also expect that the correlation will be best for radical reactions which are less exothermic, i.e. those for which the σ-transition state is shifted more towards the products, and for which the localized state may be a better representation of the transition state.

Experimentally, relative rate constants can be determined more easily and with greater accuracy than activation energies. It is tempting, therefore, to correlate the logarithm of the relative rate constant, which should be proportional to the activation energy, with L_μ:

$$\log k(\text{rel.}) = \log A - E/2.3RT$$

This procedure is only justified, however, if steric effects are small and constant so that the $\log A$ term remains the same from one reaction to another, or varies by a constant increment. Correlation of $\log k(\text{rel.})$ with L_μ is therefore justified for series of reactions in which the radical adds to the same site, $=CH_2$ for instance, or $=C\diagup^{H}$ as in an aromatic compound. Log $k(\text{rel.})$ is also suitable if the attacking radical is so small as to be sterically unaffected by substituents at the reaction site, for example hydrogen atoms. The correlation of the activation energy with L_μ may also be upset by strong polar forces which come into play when there are substituents at the reaction site itself.

Coulson first demonstrated the existence of a linear correlation of $\log k_{CH_3}$. against L_μ for the addition of methyl radicals to polycyclic aromatic compounds (88). Szwarc and his collaborators extended the correlation for methyl radicals and showed an equally good straight line for trifluoromethyl radical addition to aromatics (89, 90). Correlations have often been less impressive with olefin addition reactions. This is probably because the reaction is more exothermic than with aromatics, and the localized state is a poorer representation of the transition state. For the addition of hydrogen atoms to alkyl-substituted olefins, steric and polar effects are small: good correlation with L_μ can therefore be anticipated. Cvetanović (91) measured the rates of addition and showed that a good correlation of $\log k(\text{rel.})$ with L_μ exists (Fig. 8-4). Yang (92) measured activation energies for the hydrogen-atom additions showing that these also correlate well with L_μ. The gas-phase data of Szwarc

234

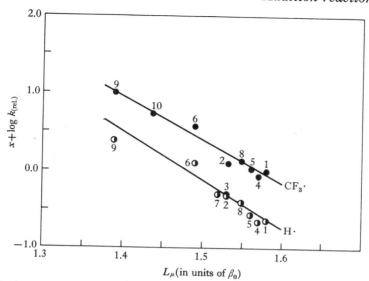

Fig. 8-4. Correlation of log k(rel.) with localization energy for hydrogen atoms (open circles, x = −0·5) and trifluoromethyl radicals (half-filled circles, x = 0) adding to alkyl-olefins. Key to the point numbers is in Table 8-12. Localization energies, which include overlap, are from Sato and Cvetanović (93).

(Table 8-12) for $CF_3\cdot$ radical addition to hydrocarbon olefins also correlate reasonably well with L_μ. The line for log k(rel.) is shown in Fig. 8-4, and the activation energy correlation in Fig. 8-5. The correlation of Cvetanović's data for methyl radicals (Table 8-12) is also shown in Fig. 8-5. It is interesting to note that whilst the methyl radical activation energies correlate quite well with L_μ, log k(rel.) for the same reactions shows little more than a general trend, suggesting that steric effects are quite important. For $CF_3\cdot$ radicals, however, both E and log k(rel.) correlate well with L_μ. This is perfectly consistent with the previous conclusion that in $CH_3\cdot$ radical addition, bond formation has proceeded further than in $CF_3\cdot$ radical addition, so that rotation of the methyl group in the transition state is impeded to a greater extent than that of the trifluoromethyl group.

In free-radical addition to olefins containing polar substituents, a large contribution to the activation energy from the polar inductive forces can be anticipated. In spite of this, James and co-workers have found a fair correlation of both log k(rel.) and activation energy with L_μ, for ethyl radical addition to a wide variety of polar olefins (Table 8-10). Similarly, Tedder and Walton have found an approximate

235

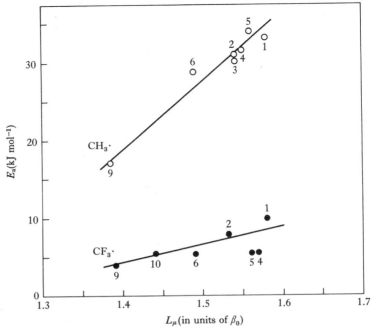

Fig. 8-5. Correlation of activation energy with localization energy for methyl and tri-fluoromethyl radicals. In the methyl radical plot the data for trimethylethylene and tetramethylethylene, which the authors regarded as doubtful (77), have been omitted.

correlation of activation energy with L_μ for trichloromethyl radical addition to halogen-substituted olefins (Table 8-13). In this case log k(rel.) does not correlate with L_μ, presumably because steric effects are more important with this larger radical.

Localization energies used in the figures are expressed in units of β_0, the resonance integral for a carbon–carbon bond in benzene of length 1.397 Å. The gradient of the activation energy–localization energy plot gives a value of the effective resonance integral β_{eff}. The difference between this β_{eff} and the true β_0 gives an indication of the departure of the real transition state from the localized state. The true β_0 value is known from spectroscopic and polarographic correlations (87) to be not less than 230 kJ mol^{-1}. Some values of β_{eff} derived from the correlation of gas-phase data are shown in Table 8-14. All the β_{eff} values are considerably less than β_0, showing that in radical addition to olefins localization has not proceeded far by the time the transition state is reached. If the activation energy for the addition of the radical to ethyl-

TABLE 8-14   *Values of the effective resonance integral* $\beta_{eff.}$
              from localization correlations

Radical	Olefin series	$\beta_{eff.}$ (kJ mol^{-1})	E^a (kJ mol^{-1})
CH$_3$·	Alkyl-olefins	−92	29
C$_2$H$_5$·	Polar-olefins	−38	32
CCl$_3$·	Fluoro-olefins	−42	26
H·	Alkyl-olefins	−34	11
CF$_3$·	Alkyl-olefins	−29	10

[a] Activation energy for the addition of the radical to ethylene.

ene is taken as a rough guide to the exothermicity of each series of reactions, it appears that $\beta_{eff.}$ approaches towards β_0 as the reactions become less exothermic (with the exception of ethyl radicals), and that the localized state becomes a better approximation to the true transition state.

When the olefin contains polar substituents, the reaction centre may carry a charge (9), and a significant polar contribution to the activation energy can be expected. The largest polar term will be the electrostatic repulsion of the charge on the olefin carbon atom q_S and the charge on the radical q_R. The energy will be proportional to $q_R q_S/r$, where r is the separation of the charges in the transition state. For the same radical adding to a series of similar sites, q_R/r should be approximately constant, so that the activation energy will depend on the charge at the olefin carbon atom, and a correlation of the activation energy with $L_\mu + Aq_S$ can be anticipated, where A is a constant.

The net atom charges δQ_μ on the olefin carbon atoms can also be calculated from Hückel molecular orbital theory (86). The experimental activation energies can then be correlated with $L_\mu + A\delta Q_\mu$ by adjustment of the constant A to give the best fit with experiment. The value of the constant A gives a measure of the importance of polar forces in the given series of reactions. For alternant hydrocarbons, such as polycyclic aromatics, δQ_μ is always zero, and the excellent correlations of log k(rel.) with localization energy alone, obtained for CH$_3$·, CF$_3$· and CCl$_3$· radicals, are easy to understand. In the addition of CCl$_3$· and C$_3$F$_7$· radicals to fluoro-olefins, polar forces are of major importance. In Fig. 8-6 the experimental activation energies are shown plotted against

237

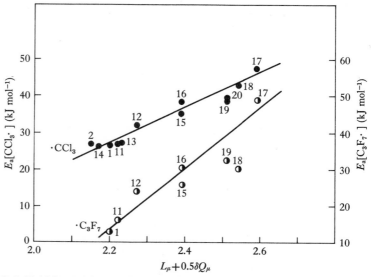

Fig. 8-6. Trichloromethyl radical addition to fluoro-olefins, correlation of activation energy with $L_\mu + 0.5\,\delta Q_\mu$. Key to the numbers is in Table 8-13. Localization energies are calculated neglecting overlap.

$L_\mu + A\delta Q_\mu$ and the correlation achieved is a considerable improvement over the simple L_μ plot. The large positive values of the constant A found to be necessary, are a good indication of the strong electrophilic character of $CCl_3\cdot$ and $C_3F_7\cdot$ radicals.

Cyclopropyl radicals provide an example of contrasting behaviour. With these radicals, Stefani (76) has shown that the reactivity with alkyl-olefins does not correlate with atom localization energies. Both $\log k$(rel.) and activation energy do, however, correlate with bond properties such as bond localization energy (93) and ionization potential. In this respect they are reminiscent of $O(^3P)$ and $S(^3P)$ atoms, although of course the reactivity trends are opposite in direction, c-$C_3H_5\cdot$ being nucleophilic and triplet O and S being electrophilic. Stefani has interpreted this behaviour as due to the fact that formation of the π-complex is the rate-determining step for these radicals, i.e. that the π-transition state of Fig. 8-3 is of higher energy than the σ-transition state. In view of this, it would be most interesting to have information about the orientation of cyclopropyl radical addition, and in particular to know whether the two possible adducts from an unsymmetric olefin are formed with the same activation energy.

8.3 *ALKYL RADICAL ABSTRACTION REACTIONS*

The most common abstraction reaction is hydrogen transfer:

$$R_3C\cdot + R'H \longrightarrow R'\cdot + R_3CH \qquad (21)$$

This has been extensively studied for alkyl radicals. Other atoms can also be abstracted, e.g. chlorine, bromine and iodine, but fluorine abstractions are seldom encountered at normal temperatures because of the exceptionally strong C–F bond. Normally single atoms are transferred, when the radical is attacking at a carbon centre, although transfer of groups of atoms can be postulated in certain cases, e.g. bromination or chlorination of cyclopropane (see p. 176). Transfer of groups can also occur when the radical attacks compounds such as disulphides, RSSR', or peroxides, ROOR', and this type of abstraction is discussed in chapter 15.

Experimentally, atom transfer reactions are studied by methods similar to those used for addition reactions. The range of radicals available for examination is again limited by the number of clean radical sources, and by far the most studies have been made with $CH_3\cdot$ and $CF_3\cdot$ radicals. The rate of the abstraction reaction is usually measured relative to some standard reaction, commonly the radical combination, e.g.

$$CF_3COCF_3 \xrightarrow{h\nu} 2CF_3\cdot + CO$$

$$CF_3\cdot + RH \longrightarrow CF_3H + R\cdot \qquad (22)$$

$$2CF_3\cdot \longrightarrow C_2F_6 \qquad (23)$$

From the rates of formation of the fluoroform and hexafluoroethane, the rate of abstraction relative to that of combination can be measured:

$$\frac{R(CF_3H)}{R^{\frac{1}{2}}(C_2F_6)} = \frac{k_{22}[RH]}{k_{23}^{\frac{1}{2}}}$$

8.3.1 *Hydrogen abstraction from alkanes*

The rate constants for hydrogen abstraction from ethane in the gas phase by a variety of radicals are shown in Table 8-15. The A-factors are all about normal for bimolecular reactions of this degree of complexity. The nature of the radicals changes quite markedly from the nucleophilic methyl to the electrophilic haloalkyl radicals, but in the reaction with a hydrocarbon such as ethane, no dramatic polar effects

Alkyl radicals

TABLE 8-15 *Arrhenius parameters for hydrogen abstraction from ethane by various alkyl radicals*

$$X_3C\cdot + CH_3CH_3 \longrightarrow X_3CH + CH_3CH_2\cdot$$

Radical	E	Log A	Log k (164 °C)	ΔH_{25}^0	Reference
$CH_3\cdot$	50.0	8.9	3.0	-25	94
$CHF_2\cdot$	52.5	9.6	3.3	-8	95
$CF_3\cdot$	35.3	7.7	3.5	-34	96
$CCl_3\cdot$[a]	60.1	9.7	2.6	$+8$	97, 60a
$C_2F_5\cdot$	38.2	8.3	3.8	-21	98
$C_3F_7\cdot$	38.6	7.4	2.8	-21	99

E and ΔH_{25}^0 in kJ mol^{-1}, A and k in l mol^{-1} s^{-1}.
[a] Results for abstraction of primary hydrogen from n-butane.

can be expected, and in fact the rate at 164 °C does change comparatively little from radical to radical.

The greatest unifying influence in the interpretation of radical abstraction reactions is still exercised by the Evans–Polanyi equation:

$$E_a = \alpha\Delta H^0 + C$$

which relates the activation energy of a reaction to the total enthalpy change ΔH^0 (100). The relationship is deduced by considering the potential energy curves of a series of closely related reactions. Provided the potential energy curve from one reactant to another does not change its basic shape, a geometric argument leads to the above equation in which α and C are constants (101). It can further be shown that α is less than one and that α should be larger for series of reactions with greater activation energies.

The results in Table 8-15 fit only very approximately the Evans–Polanyi equation. It is not surprising that the assumption of similar potential energy curve shapes is invalid for a series of reactions in which the nature of the attacking radical varies so widely.

Table 8-16 gives the Arrhenius parameters for attack of methyl and trifluoromethyl radicals on a series of alkanes. Many of the reactions have been checked by several groups of workers, and a reasonable degree of confidence can now be placed in them.

The A-factors remain fairly constant from reaction to reaction and the rates are controlled by the activation energy. Both $CH_3\cdot$ and $CF_3\cdot$ radicals abstract tertiary hydrogen atoms most easily, followed by

TABLE 8-16 *Arrhenius parameters for hydrogen abstraction from alkanes by alkyl radicals*

Alkane	No.	$CH_3\cdot$[a]		$CF_3\cdot$[b]	
		E	Log A	E	Log A
CH_4	1	60.9	8.8	47.0	7.9
CH_3CH_3	2	50.0	8.9	35.3	7.7
$\overset{*}{C}H_3CH_2CH_3$	3	48.7	9.1	—	—
$CH_3\overset{*}{C}H_2CH_3$	4	43.3	8.9	26.9	7.6
$\overset{*}{C}H_3CH_2CH_2CH_3$	5	49.1	9.1	—	—
$CH_3\overset{*}{C}H_2CH_2CH_3$	6	40.7	8.9	23.9	8.3
$(CH_3)_3\overset{*}{C}H$	7	34.4	8.4	19.7	8.1
$(CH_3)_4C$[c]	8	50.4	8.3	35.2	8.0
c-C_3H_6	9	55.9	8.8	36.5	8.5
c-C_4H_8	10	44.1	8.8	27.3	8.7
c-C_5H_{10}	11	39.9	8.9	26.0	9.3
c-C_6H_{12}	12	41.6	9.1	26.5	9.1
c-C_7H_{14}	13	39.1	9.1	24.8	9.4

* Indicates the hydrogen abstracted. E in kJ mol^{-1}, A in l mol^{-1} s^{-1}.
[a] Gas-phase results of Jackson, McNesby and Darwent (94), cycloalkane results from Gordon et al. (102).
[b] Gas-phase results of various authors, see ref. 96.
[c] $CH_3\cdot$ result (108), $CF_3\cdot$ result (109).

secondary and then primary. Trichloromethyl radicals are also known to follow a similar pattern (103). Alkyl radicals are similar to the halogen atoms in this respect (see p. 173) and in fact this order of reactivity in hydrogen abstraction appears to be the same for all radical species. The greater reactivity at the tertiary position is undoubtedly a result of the weaker tertiary C–H bond, and the order of reactivity as a whole follows the trend in the strength of the bond being broken. In the reaction of a radical with a series of alkanes, the assumptions underlying the Evans–Polanyi equation may be valid, as polar forces especially will be minimal. For this series of reactions the enthalpy, ΔH^0, is given by the difference in the strength of the bond being broken and the bond being formed, e.g. for methyl radicals:

$$CH_3\cdot + R\text{—}H \longrightarrow CH_4 + R\cdot \qquad \Delta H^0 = D(R\text{—}H) - D(CH_3\text{—}H)$$

Since the bond formed is the same in each case, i.e. CH_3—H, the Evans–Polanyi equation can be simplified to:

$$E_a = \alpha D(R\text{—}H) - C_2$$

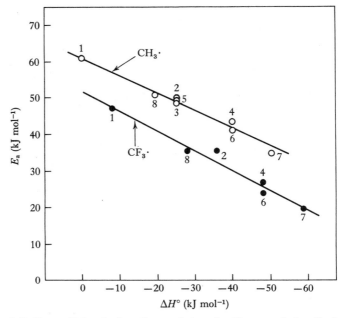

Fig. 8-7. Evans–Polanyi plots for methyl and trifluoromethyl radicals. Hydrogen abstraction from alkanes. Open circles: methyl radicals, data mainly of Jackson, McNesby and Darwent (94); closed circles: trifluoromethyl radicals, see Jones and Whittle (96). Key to the numbering is in Table 8-16.

Trotman-Dickenson showed that methyl radicals obey this equation (104). Jones and Whittle (96) have demonstrated a similar correlation for trifluoromethyl radicals. Kerr (105) has given values of the bond dissociation energies and shown similar correlations for a number of radicals. The $CH_3\cdot$ and $CF_3\cdot$ correlations with ΔH^0 are shown in Fig. 8-7. The best-fit equations of the straight lines are:

$$E_a(CH_3\cdot) = 0.49\Delta H^0 + 62 \text{ kJ mol}^{-1}$$
$$E_a(CF_3\cdot) = 0.53\Delta H^0 + 52 \text{ kJ mol}^{-1}$$

The points from the two radicals do not all lie on a single straight line as is implied in the original relationship, but rather special relationships apply for a given series of reactions having values of α and C characteristic of that series. The values of α are similar for the two radicals as would be expected for reactions having such similar activation energies. The α values are also considerably less than those of iodine or bromine atom reactions, $\alpha = 0.97$ and 0.86 respectively (105), in agreement with the original idea (101) that α would be greater for reactions of high activation energy.

Arrhenius constants for hydrogen abstraction from cycloalkanes are also given in Table 8-16. Trotman-Dickenson used the methyl radical activation energies to calculate carbon-hydrogen bond dissociation energies in the cycloalkanes, which are not known independently, by assuming they obeyed the same relationship given above (104). Whittle has used his results for hydrogen abstraction by $CF_3\cdot$ radicals and $Br\cdot$ atoms, to calculate two further sets of $D(C—H)$ for the cycloalkanes (106), also by assuming that the correlations for alkanes could be extended to the cycloalkanes. All three methods give essentially the same set of $D(C—H)$ values, which strongly supports the contention that the Evans–Polanyi equation is also obeyed by the cycloalkanes.

8.3.2 *Hydrogen abstraction from haloalkanes*

Alkyl radicals also abstract hydrogen atoms from substituted alkanes. Whittle has obtained quantitative results for a series of related halogen-substituted alkanes in the reaction with $CF_3\cdot$ radicals. Both hydrogen and halogen atoms are abstracted from the chloromethanes, methyl bromide and methyl iodide. Chlorine is abstracted less readily than hydrogen, but bromine and iodine more readily. The halogen abstraction reactions are discussed more fully in a later section (p. 247). The hydrogen abstraction results, with some data for methyl radicals, are given in Table 8-17.

The C–H bond dissociation energy probably decreases from CH_4 to CH_3X to CH_2X_2, as the number of halogen substituents increases. The decrease in the activation energy for abstraction by $CF_3\cdot$ radicals from the chloromethanes, and by $CH_3\cdot$ radicals from the fluoromethanes, is thus readily explained. The constancy of the E-values for hydrogen abstraction by $CF_3\cdot$ radicals from CH_4, CH_3F and CH_2F_2 is surprising, however, and it probably occurs because the decrease in $D(C—H)$ is opposed by a repulsive polar effect which increases as the number of halogen substituents increases. The two effects fortuitously cancel for this series of molecules. The polar effect would be expected to decrease as the halogen changes from F to Cl, to Br, etc., and indeed the activation energy decreases from CH_3F to CH_3Cl to CH_3Br to CH_3I. In the case of reactions between $CF_3\cdot$ radicals and chloromethanes, the polar effect is outweighed by the decrease in $D(C—H)$ and the activation energies decrease from CH_4 to $CHCl_3$. The results for abstraction by

Alkyl radicals

TABLE 8-17 *Hydrogen abstraction from halogen-substituted alkanes*

Alkane	$CF_3 \cdot$ [a]		$CH_3 \cdot$ [b]	
	E	Log A	E	Log A
CH_4	47.0	7.9	60.9	8.8
CH_3F	47.0	8.6	47.9	~8
CH_2F_2	47.0	8.3	42.8	~8
CHF_3	—	—	47.9	~8
CH_3Cl	44.5	7.8	—	—
CH_2Cl_2	31.9	6.8	—	—
$CHCl_3$	27.7	6.7	—	—
CH_3Br	43.7	7.7	—	—
CH_3I	31.1	7.1	—	—
CH_3CH_3	35.3	7.7	—	—
CF_3CH_3	56.7	9.4	—	—
CHF_2CHF_2	50.0	8.4	—	—
CF_3CHF_2	42.4	7.7	—	—

E in kJ mol^{-1}, A in l mol^{-1} s^{-1}.
[a] Gas-phase results from ref. 19.
[b] Gas-phase results from ref. 107.

$CF_3 \cdot$ radicals from fluoroethanes are very similar to those obtained in the bromination of the same molecules (p. 178). They can be interpreted in the same way, viz. fluorine substituents β to the attacked H-atom deactivate it strongly by a polar inductive effect, fluorine substituents α to the attached H-atom have a slight activating effect.

8.3.3 *Hydrogen abstraction from silanes*

There has been considerable interest recently in hydrogen abstraction reactions from silanes and substituted silanes, analogous to the alkanes already discussed. Arrhenius parameters for the reactions of $CH_3 \cdot$ and $CF_3 \cdot$ radicals are given in Table 8-18. A-factors are all in the range considered 'normal' for reactions of this complexity with the exception of those for $CH_3 \cdot$ with chlorosilanes. These results have recently been questioned by the authors themselves (114) and appear doubtful.

The first four silanes in the table contain only carbon and hydrogen in addition to silicon. For these molecules there is little reason to doubt that the activation energy is controlled by the Si–H bond dissociation energy in SiH_4 and Si_2H_6, and the C–H bond dissociation energy in Me_3SiH and Me_4Si. The value of $D(Si-H)$ is probably about 38 kJ mol^{-1} less than $D(C-H)$, so that the activation energies for abstraction by both $CH_3 \cdot$ and $CF_3 \cdot$ radicals follow the order expected from the Evans–

Table 8-18   *Hydrogen abstraction from silanes and group IV metal alkyls*

Substrate	$CH_3\cdot$			$CF_3\cdot$		
	E	Log A	Reference	E	Log A	Reference
SiH_4	29.0	8.8	110	21.0	8.9	111
Si_2H_6	23.6	9.0	110	—	—	—
Me_3SiH	33.2	8.4	111	23.5	9.3	111
Me_4Si	43.0	8.5	111	30.6	8.9	109
Me_4Ge	40.3	8.8	112	31.0	8.7	109
Me_4Sn	36.1	8.1	112	30.5	8.7	109
Me_4Pb	31.0	7.2	112	—	—	—
Me_3SiF	—	—	—	39.9	9.4	109
Me_2SiF_2	—	—	—	44.1	9.3	109
$MeSiF_3$	—	—	—	49.1	9.0	109
Me_3SiCl	(48.3	10.4)	113	38.2	9.3	109
Me_2SiCl_2	(48.7	10.2)	113	38.6	8.8	109
$MeSiCl_3$	(48.3	9.9)	113	39.5	8.3	109

E in kJ mol^{-1}, A in l mol^{-1} s^{-1}. Values in parenthesis are considered doubtful.

Polanyi equation. The Si–H bond dissociation energies are not known accurately, and these results have been used for estimating them. The methylfluorosilanes show evidence of a strong polar repulsion and the activation energy increases as the number of fluorine atoms on silicon increases. In the methylchlorosilanes there is no decrease in activation energy for abstraction by $CF_3\cdot$ or $CH_3\cdot$ radicals, although this would be expected at least for $CF_3\cdot$ radicals, because of polar repulsion. The explanation may be that back-coordination between the chlorine p_π-orbitals and the silicon d_π-orbitals is occurring. (This effect would be much smaller with fluorine p_π due to the smaller orbital size.) This explanation is supported to some extent by the fact that no chlorine abstraction occurs from the chlorosilanes, suggesting a strong Si–Cl bond, in contrast to the chloroalkanes where chlorine abstraction is facile.

Cheng and Szwarc showed that the rate constants for hydrogen abstraction from methylchlorosilanes correlate with the proton chemical shifts of the molecules (115). Similar correlations have also been found for $CH_3\cdot$ (112) and $CF_3\cdot$ (109) radicals abstracting from the group IV tetramethyls, and for $CF_3\cdot$ radicals abstracting from the methyl fluorosilanes (109). The proton chemical shift gives a measure of the electron density on the hydrogen atom being abstracted. Hence the rate decreases as the electron density decreases, and the correlation gives a good representation of the polar electron-withdrawing effect of halogen substituents.

Alkyl radicals

8.3.4 Hydrogen abstraction from nitrogen- and oxygen-containing compounds

Alkyl radicals also readily abstract hydrogen atoms from amines, hydrazines, amides, alcohols, esters, etc., and some systematic studies of these classes of compounds are now being published. It is possible, by means of isotopic substitution, to determine the rates of abstraction from different sites within the same substrate. For example, in the case of methylamine, the overall rate of hydrogen abstraction by $CF_3\cdot$ radicals is determined in the usual way by measurement of the amount of fluoroform formed:

$$CF_3\cdot + CH_3NH_2 \longrightarrow CF_3H + (CH_4N)\cdot$$

If CH_3ND_2 is now examined, any CF_3H formed (as opposed to CF_3D) must come by hydrogen abstraction from the methyl group, and so the rate of attack at this site can be determined:

$$CF_3\cdot + CH_3ND_2 \longrightarrow CF_3D + CH_3ND\cdot$$
$$CF_3\cdot + CH_3ND_2 \longrightarrow CF_3H + \cdot CH_2ND_2$$

The rate of abstraction from the —NH_2 site can then be found by subtraction from the overall rate. By this and similar methods, the selectivity of alkyl radicals in abstraction reactions has been found for a number of compounds (Table 8-19).

TABLE 8-19 *Selectivity in the abstraction of hydrogen from nitrogen- and oxygen-containing compounds by alkyl radicals*

Substrate	$CH_3\cdot$	$C_2H_5\cdot$[a]	$CF_3\cdot$	Reference
$\overset{*}{C}H_3NH_2$	1:2.0	1:1.7	1:0.2	116, 120
$(\overset{*}{C}H_3)_2NH$	1:21	1:70	—	116
$NH_2\overset{*}{C}H_2CH_2NH_2$	1:0.3	—	—	117
$\overset{*}{C}H_3CONH_2$	1:1.4	—	—	121
$(\overset{*}{C}H_3)_2NNH_2$	1:40	—	—	119
$\overset{*}{C}H_3NHNHCH_3$	1:12	—	—	119
$\overset{*}{C}H_3OH$	1:0.6	—	1:1.5	120
$\overset{*}{C}H_3COOCH_3$	1:0.1	—	1:0.09	118
$CH_3\overset{*}{C}H_2OH$	1:0.53	—	—	117

Gas-phase results at 164 °C. Starred position is the standard in each molecule. Figures give the relative rates of abstraction *per H atom*.
[a] Ethyl radical results at 150 °C from ref. 122.

246

The rate of hydrogen abstraction from nitrogen by methyl radicals is greater than from carbon, and the rate increases from primary to secondary nitrogen. The relative rates of abstraction from NH_3, $MeNH_2$ and Me_2NH are 1 to 11 to 168 at 150 °C (116) and in this respect the amines apparently follow the alkanes, the strength of the N–H bond determining to a large extent the rate of radical attack. Thynne *et al.* have pointed out however (116) that the A-factor for abstraction from $MeNH_2$ is low, so that steric effects also play a part. In ethylenediamine, abstraction from the methylene groups is about three times as fast as from $-NH_2$, and this also is accounted for in terms of steric hindrance at the terminal sites (117).

In methanol and ethanol abstraction from O—H is not the predominant process; appreciable attack also occurs at the C–H bonds. In methyl acetate both $CH_3\cdot$ and $CF_3\cdot$ radicals abstract predominantly from the acetyl, rather than the methoxy, group. This is attributed to the greater extent of resonance in the product radical from the former process (118).

8.3.5   *Halogen abstraction reactions*

Halogen abstraction reactions are well known in a qualitative way from the work of Kharasch (58), Haszeldine (59) and others on the radical chain addition of polyhalomethanes, CCl_3Br and CF_3I, to olefins, where halogen abstraction is part of the chain transfer process:

$$R_3C\cdot + CX_4 \longrightarrow R_3CX + CX_3\cdot$$

There is, however, relatively little quantitative kinetic information about the reaction. Such studies as there are indicate that the ease of halogen abstraction is in the order:

$$I > Br \sim H > Cl > F$$

The order of abstraction follows the strengths of the C–X bonds being broken with the exception of hydrogen abstraction.

Szwarc (123) has studied the abstraction of halogen atoms from a series of halides by methyl radicals, in toluene and iso-octane solutions. He showed that the difference in activation energy for bromine and iodine abstraction, $\Delta E_{Br,I}$, was approximately equal to half the difference in the C–Br and C–I bond dissociation energies, similarly $\Delta E_{Cl,Br}$ was equal to half the difference in the C–Cl and C–Br bond dissociation energies.

Alkyl radicals

Hydrogen abstraction occupied an anomolous position. This relationship implies that an approximate Evans–Polanyi equation applies to halogen abstraction, but that hydrogen abstraction does not fit the same correlation. Szwarc explained these results in terms of polar repulsion between the approaching $CH_3 \cdot$ radical and the filled orbitals of the halogen atom. This repulsion would be practically zero when a $CH_3 \cdot$ radical approaches hydrogen and hence the rate of hydrogen abstraction is greater than would be expected from the bond dissociation energy alone. The polar repulsion should decrease if the halide RX contains electron-withdrawing substituents, and indeed Szwarc found that $CH_3 \cdot$ radicals abstract iodine from CF_3I about 500 times faster than from CH_3I.

Whittle (124) has studied chlorine atom abstraction from a series of chloromethanes by $CF_3 \cdot$ radicals in the gas phase. The results of this study are given in Table 8-20. Trifluoromethyl radicals behave in a similar way to methyl radicals and the activation energies generally decrease as the carbon–halogen bond dissociation energy decreases. The exception again is hydrogen abstraction which has a lower activation energy than would be expected from $D(C—H)$ alone. This anomaly can again be explained in terms of polar repulsion between the attacking $CF_3 \cdot$ radical and the filled orbitals of the halogen. The polar repulsion should be less with $CF_3 \cdot$ since the radical contains electron-withdrawing substituents. Pritchard *et al.* found the activation energy for chlorine abstraction from CCl_4 by $CH_3 \cdot$ radicals to be 56.3 kJ mol⁻¹ (125) whilst that for abstraction by $CF_3 \cdot$ radicals was indeed smaller at 43.7 (124).

TABLE 8-20   *Halogen abstraction by CF₃· radicals in the gas phase* (124)

$$CF_3 \cdot + CH_3X \longrightarrow CF_3X + CH_3 \cdot$$

Halomethane	E	Log A	D(C—X)
CH_4[a]	47.0	7.9	437
CH_3Cl	71.4	6.3	351
CH_2Cl_2	49.6	7.0	335
$CHCl_3$	50.4	7.7	327
CCl_4	43.7	8.0	308
CH_3Br	34.9	6.3	292
CH_3I	13.9	6.1	236

[a] Arrhenius parameters for *hydrogen* abstraction from methane.
E in kJ mol⁻¹, A in l mol⁻¹ s⁻¹.

The results for the abstraction of chlorine (Table 8-20) show that the activation energy falls steeply from CH_3Cl to CH_2Cl_2 but then remains approximately constant, in spite of the fact that $D(C—Cl)$ almost certainly falls progressively from CH_2Cl_2 to CCl_4. The behaviour of the $CF_3\cdot$ radical in chlorine abstraction thus parallels its behaviour in hydrogen abstraction from fluoromethanes (Table 8-17) and that of the chlorine atom in abstracting hydrogen from chloromethanes (Table 7-9). The polar effect appears to be important in opposing the increased reactivity to be expected from the decrease in $D(C—Cl)$.

Other studies also lend support to the idea that polar effects are important. Danen and Winter (126) have studied the abstraction of iodine from a series of aliphatic iodides in solution. The rate of iodine abstraction by phenyl radicals was measured relative to the rate of abstraction of bromine from CCl_3Br, i.e. k_I/k_{Br}:

$$Ph\cdot + RCH_2I \xrightarrow{k_I} PhI + RCH_2\cdot$$

$$Ph\cdot + CCl_3Br \xrightarrow{k_{Br}} PhBr + CCl_3\cdot$$

The logarithms of relative rate constants correlate with the Taft polar substituent constants σ^*, in a Hammett-type plot, i.e. $\log k_I/k_{Br} = \rho\sigma^*$. It is noteworthy that the value of ρ ($+0.184$) is opposite in sign to that usually found in hydrogen abstraction reactions, which indicates that for this system at least, the polarization of the transition state is in the reverse direction to that of hydrogen abstraction.

8.3.6   *Primary kinetic isotope effects in hydrogen abstraction reactions*

When deuterium is substituted for hydrogen in an organic compound, then a primary kinetic isotope effect is observable in the hydrogen abstraction step:

$$X—H + R\cdot \xrightarrow{k_H} X\cdot + RH$$

$$X—D + R\cdot \xrightarrow{k_D} X\cdot + RD$$

Secondary deuterium isotope effects are usually too small to observe in radical abstraction reactions and are commonly assumed to be zero. Data on the primary isotope effect are usually given in the form of the rate constant ratio $k_H/k_D = A_H/A_D \exp(E_D - E_H)/RT$. This ratio and its temperature dependence, have now been determined for quite a wide variety of substrates, and a selection of values is given in Table 8-21.

TABLE 8-21   *Primary kinetic isotope effects in hydrogen abstraction reactions in the gas phase at 150 °C*

$$XH + R\cdot \xrightarrow{k_H} RH + X\cdot$$
$$XD + R\cdot \xrightarrow{k_D} RD + X\cdot$$

Reactants	$CH_3\cdot$ k_H/k_D	$CF_3\cdot$ k_H/k_D	Theory[a] $(k_H/k_D)_{calc.}$	Reference
H_2/D_2	—	4.2	9.3, 3.8[b]	127
HCl/DCl	—	2.2	3.8	128
H_2S/D_2S	2.9	1.9	3.8	129, 128
CH_4/CD_4	—	6.7	4.2	130
SiH_4/SiD_4	3.5	3.0	CH_3: 2.4[b], CF_3: 3.2[c]	110
C_2H_6/C_2D_6	5.9	—	4.2	110
NH_3/ND_3	2.6	4.3	4.7	119
N_2H_4/N_2D_4	7.0	—	4.7	119
$EtOH/EtOD$	3.2	—	5.9	117

[a] (k_H/k_D) calculated from zero-point energy difference.
[b] Value calculated from simple '3-atom' transition state model by the 'BEBO' method without inclusion of tunnelling correction.
[c] Value calculated by 'BEBO' method with inclusion of tunnelling correction.

The simplest interpretation of kinetic isotope effects is based on the assumption that, on passing from reactants to the transition state, the zero-point energy associated with the stretching vibration of the X–H bond is lost (131). If ΔE_0 is the difference in zero-point energy of the X–H and X–D stretching vibrations, then:

$$(k_H/k_D) = \exp \Delta E_0/RT$$

ΔE_0 can be calculated from the difference in the infrared stretching frequencies of the X–H and X–D bonds. If only ν_H, the hydrogen stretching frequency, is known then ν_D can be estimated from the approximate relation $\nu_H/\nu_D = (M_D/M_H)^{\frac{1}{2}}$, i.e. $\nu_D = \nu_H/\sqrt{2}$. Values of (k_H/k_D) calculated in this way are also given in Table 8-21. It is clear that the simple theory correctly predicts the order of magnitude of the effect. Gray *et al.* have determined k_H/k_D for a series of N–H bonds. The average value of 5.0 at 150 °C for all the N–H bonds, compares very favourably with the ratio of 4.7 calculated from zero-point energy differences at 150 °C. For individual compounds, however, values of k_H/k_D both higher and lower than the zero-point energy value are observed. The higher values can be explained in terms of quantum-mechanical tunnelling (132, 94), and the loss of bending vibrations in the transition state (133). Rate-constant ratios k_H/k_D have been calculated for several of the pairs of reactants by Johnston's bond-order bond-energy (BEBO) transition state method (134), which includes the bending vibrations. Quite good agreement with experiment can be achieved with or without tunnelling corrections (127, 110). The origin of low isotope effects is usually attributed to the motion of the central atom in the symmetrical vibration of the activated complex (135):

$$X \overset{\longleftarrow}{\cdots\cdots} H \overset{\longrightarrow}{\cdots\cdots} R$$

When the transition state is symmetrical the vibration is independent of the isotopic mass, since it involves no motion of the central atom. The more asymmetrical the transition state becomes, the greater is the mass-dependence of this vibration, and the smaller is k_H/k_D. Low values of k_H/k_D are therefore usually taken as an indication of asymmetry in the transition state. The low values for 1,2-dimethylhydrazine, ethylamine and 1,1-dimethylhydrazine of 2.1, 3.1 and 3.2 respectively, found for methyl radical attack (119), are therefore consistent with highly asymmetrical transition states for these reactions:

$$R_2N \cdots\cdots H \cdots CH_3$$

The low value for ammonia, $k_H/k_D = 2.6$, appears anomalous, since the transition state should be symmetrical. Morris and Thynne (136) have found complications due to secondary reactions when $CF_3 \cdot$ radicals react with ammonia, and it may be that the $CH_3 \cdot$ radical reaction needs reinvestigation.

8.3.7 *Structure–reactivity relationships in alkyl radical abstraction reactions*

The order of reactivity of primary, secondary and tertiary C–H bonds remains the same in hydrogen abstraction reactions irrespective of the nature of the attacking radical. It is clear that the strength of the bond being broken plays an important part in determining the rate of abstraction. The influence of bond strength is summed up in the Evans–Polanyi equation:

$$E = \alpha \Delta H^0 + C$$

The applicability of this equation is much more limited than its general form implies. For a given radical abstracting from alkanes or cycloalkanes a special relationship exists having values of α and C characteristic of the reaction series. Such special relationships have been found for iodine and bromine atoms, and for $CH_3 \cdot$ and $CF_3 \cdot$ radicals abstracting from hydrocarbons. There seems little reason to doubt that all or most alkyl radicals also obey special relationships for alkanes. Apart from the alkanes however, few reaction series obey the equation even in its limited form. It seems probable that the activation energies for hydrogen abstraction from silanes by $CF_3\cdot$ and $CH_3\cdot$ radicals also fit an Evans–Polanyi equation. Similarly a special relationship has also been found to cover abstraction from H_2S and hydrogen halides HX by $CF_3\cdot$ radicals (137), and an approximate equation fits the behaviour of $CF_3\cdot$ radicals with chloromethanes. Substituents in the molecules can influence the strength of the C–H bond dissociation energy, but they also exert a polar effect on the approach of the radical which can raise or lower the activation energy of abstraction. If the polar effect is constant, or varies by a constant amount from one molecule to another, then the basic shape of the potential energy curve does not change, and the Evans–Polanyi equation is obeyed. In the $CF_3\cdot/HX$ and $CF_3\cdot/CH_nCl_{4-n}$ series, polar effects probably change by a constant amount from one reactant to another.

A quantitative assessment of the polar effect can be achieved by the use of the Hammett equation, although here again use of the equation is restricted to abstraction from series of closely related compounds in which the site of abstraction is the same in each instance. Huyser has found a Hammett correlation for the abstraction of hydrogen from substituted toluenes by $CCl_3\cdot$ radicals in solution with a ρ value -1.46 (138). Bridger and Russell (139) found a correlation ($\rho = -0.1$) for phenyl radicals abstracting from toluenes. Gleicher and co-workers (140) studied abstraction of bridgehead hydrogen from 1-substituted adamantanes (14) and correlated the relative rates with Taft polar

(14)

substituent constants σ^* (141); the ρ value was -0.40. Negative ρ values indicate that the abstraction rate is increased by electron-repelling substituents and decreased by electron-withdrawing substituents. Positive ρ values have been obtained by Danen and Winter (126) for the abstraction of iodine atoms by phenyl radicals from aliphatic and aromatic iodides. In iodine abstraction the direction of the polar effect is evidently reversed.

The quantum-mechanical reactivity theory aims to predict reaction rates, or where this is not possible to interrelate the reactivities of a series of molecules. For most organic molecules it is still not practicable to carry out complete calculations of the activation energy by absolute rate theory. The quantum-mechanical theory therefore employs several drastic simplifications in order to deduce a 'reactivity index' which can be employed as a scale of the magnitude of the activation energy, and as such can be compared with the experimental values. Localization energy L_μ is such a reactivity index, which gives the difference in π-energy of an 'isolated' unsaturated molecule and the π-energy of the molecule in the hypothetical 'localized' state. It has been used with fair success for correlating the activation energies of radical addition reactions.

By analogy with localization theory of olefins, we might expect for a hydrogen abstraction reaction, that a correlation would exist between the activation energy, and the energy required to localize an electron on the hydrogen atom being abstracted. Localizing an electron on the hydrogen atom effectively means breaking the C–H bond of the alkane RH which is being attacked:

$$RCH{=}CH_2 \longrightarrow R\dot{C}H{-}\dot{C}H_2 \qquad\qquad R{-}H \longrightarrow R{\cdot}H{\cdot}$$
$$E_\pi \quad \text{'Localized state'}\ E_\pi' \qquad\qquad E_\sigma \quad \text{'Localized state'}\ E_\sigma'$$

The energy required to do this is the total σ-energy of the alkane E_σ minus the σ-energy of the radical formed, E_σ', i.e. $E_\sigma - E_\sigma' = \Delta F_\sigma$. This energy difference also equals the C–H bond dissociation energy, which we already know correlates with the activation energy.

Brown (142) has calculated total σ-energies for the alkanes and corresponding radicals by molecular orbital theory, using bond-orbitals as the basic set. The energy differences ΔE_σ are found to belong to four classes: methane-type, primary, secondary and tertiary, and hence the theory predicts, independently of any parameters, a uniform variation in reactivity of the four classes of alkanes, just as is observed experimentally for all radicals. The variation of ΔE_σ from one molecule to another was calculated by Brown, in terms of the integral parameter Φ. A plot of experimental activation energy for hydrogen abstraction by $CH_3{\cdot}$ and $CF_3{\cdot}$ radicals, and halogen atoms against ΔE_σ in units of Φ, is shown in Fig. 8-8. A clear correlation, though non-linear, can be seen for all five types of reaction.

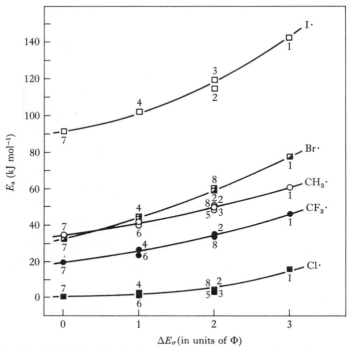

Fig. 8-8. Correlation of activation energy for abstraction from alkanes with Brown's ΔE_σ values. See ref. 142 for definitions of Φ and the constant F. Experimental results and references in Tables 7-8 and 8-16.

An alternative quantum-mechanical approach to the problem of the reactivity of alkanes and substituted alkanes in the hydrogen abstraction process has been developed by Fukui and his associates (143). In the transition state, at the approach of the reactants, the electrons will be rearranged so as to bring about a new electron distribution, different from that of the isolated reactants. The whole reacting system is treated as essentially one many-electron system like an ordinary molecule. Electron delocalization takes place to cause stabilization of the whole system, and hence the amount of such stabilization determines the activation energy of the reaction. Using the perturbation method, and employing several simplifying approximations such as treating only the valence orbitals of the radical and reactant, Fukui derived 'delocalizabilities' D_r as an index of the amount of stabilization. The correlation of the activation energies of hydrogen abstraction from the alkanes by $CH_3\cdot$ and $CF_3\cdot$ radicals, and by halogen atoms with the delocalizability

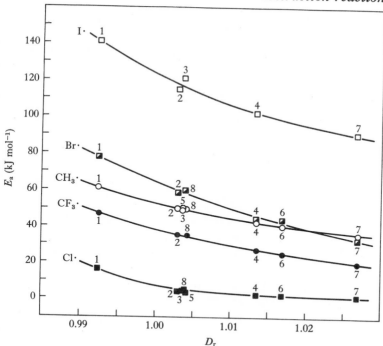

Fig. 8-9. Correlation of activation energy for abstraction from alkanes with Fukui's delocalizabilities, D_r. Experimental results and key to the numbers in Table 8-16; halogen atom results from Table 7-8.

D_r, is shown in Fig. 8-9. Good correlation, though non-linear, is again obtained.

Fukui has also calculated D_r values for hydrogen abstraction from haloalkanes, and for halogen abstraction from halomethanes (144). Recent experimental results do not correlate with D_r for these molecules. For radical addition reactions it was necessary to include a term in the reactivity index which allowed for polar interaction between the radical and olefin. A similar term, such as the net atom charge δQ_μ, could be included with the delocalizability (56) for polar substrates in hydrogen abstraction, although this suggestion has not yet been tested.

8.4 *ALKYL RADICALS IN ORGANIC SYNTHESIS*

A very useful reaction in organic synthesis is the addition of reagents across the double bond of olefins in a radical chain process (145), i.e.

Alkyl radicals

Kharasch addition. The direction of addition is almost invariably the opposite to that encountered in ionic Markovnikov additions, so that syntheses by the radical route are complementary to those of ionic addition.

$$A\text{---}B + \text{Initiator} \longrightarrow A\cdot + B\cdot \tag{24}$$

$$A\cdot + RCH{=}CH_2 \longrightarrow R\dot{C}HCH_2A \tag{25}$$

$$R\dot{C}HCH_2A + AB \longrightarrow RCHBCH_2A + A\cdot \tag{26}$$

$$A\cdot + A\cdot \longrightarrow A_2 \tag{27}$$

$$A\cdot + R\dot{C}HCH_2A \longrightarrow RCHACH_2A \tag{28}$$

$$R\dot{C}HCH_2A + R\dot{C}HCH_2A \longrightarrow ACH_2CHRCHRCH_2A \tag{29}$$

Initiation (24) can be accomplished photochemically if the addend, AB, absorbs light in the visible or ultraviolet region of the spectrum. This method of initiation is very useful for the addition of polyhalo-methanes, organic bromides and iodides. With these molecules light absorption leads to homolytic breaking of the weakest carbon–halogen bond. For other molecules such as aldehydes or ketones photolysis can lead to side reactions, and consequently chemical initiation has to be employed. There are two main classes of chemical initiators; peroxides and azobisnitriles. Di-t-butyl peroxide is a readily available dialkyl-peroxide which decomposes in the temperature range 110–150 °C to give t-butoxy and subsequently methyl radicals, both of which can act as chain initiators:

$$Me_3COOCMe_3 \longrightarrow 2Me_3CO\cdot$$

$$Me_3CO\cdot \longrightarrow Me\cdot + MeCOMe$$

Diacyl peroxides, like dibenzoyl peroxide and diacetyl peroxide, are also decomposed thermally to give phenyl and methyl radicals respectively:

$$(RCO_2)_2 \longrightarrow 2RCO_2\cdot \longrightarrow R\cdot + CO_2$$

The benzoyloxy radical is also sufficiently stable under certain conditions to participate. Some substrates such as ethers, and primary and secondary alcohols, induce decomposition of diacyl peroxides causing serious interference with the chain addition (cf. chapter 2, p. 15, and also chapter 15, p. 552). Azobisnitriles decompose to give cyanoalkyl radicals, e.g.:

$$\underset{\displaystyle Me_2\overset{\textstyle |}{C}\text{---}N{=}N\text{---}\overset{\textstyle |}{C}Me_2}{\overset{\displaystyle CN \qquad\qquad CN \qquad\qquad CN}{}} \longrightarrow 2Me_2\dot{C}\cdot + N_2$$

Azobisisobutyronitrile itself is useful in the temperature range 40–80 °C, but other azonitriles can be used for higher or lower temperatures. Other compounds such as peroxy carbonates have recently been used as low-temperature initiators (cf. chapter 11, p. 446).

The desired product RCHBCH$_2$A is formed by the chain-propagation steps *(25)* and *(26)*. The chain length must therefore be as long as possible so that termination products form a small part of the total products. Both the addition *(25)* and abstraction *(26)* must be very fast, since the terminations *(27, 28* and *29)* are very fast in solution. The structure of the olefin and adding species AB, will obviously be very important in deciding whether a synthesis is feasible. The effects of structure on reactivity have been discussed in the preceding two sections for addition and abstraction and the accompanying tables can serve as a guide to the conditions necessary. The chain length can be increased by working with low rates of initiation. For photochemical initiation low light intensities may help, and with chemical initiators small amounts and slow rates of decomposition produce the same effect. Working with a low initiation rate also has the advantage of reducing side reactions involving the initiator itself, but it increases the time required for complete reaction of the starting materials.

An important process which interferes with the formation of the one-to-one adduct is polymerization. The adduct-radical RĊHCH$_2$A may add to another olefin molecule instead of abstracting from the starting material:

$$R\dot{C}HCH_2A + RCH{=}CH_2 \longrightarrow R\dot{C}HCH_2CH(R)CH_2A \longrightarrow \text{etc.} \qquad (30)$$

A series of products containing 2,3, etc., olefin units, A(CH$_2$RCH)$_n$B, is eventually formed. These telomers are undesirable by-products in a synthesis of the one-to-one adduct, but their formation can usually be reduced or suppressed by using a high ratio of AB to olefin, or by working at higher temperatures. When the olefin contains allylic hydrogen atoms, allylic attack can also be a major complication.

$$A\cdot\ +\ RCH_2CH{=}CH_2 \Big\langle \begin{array}{l} RCH_2\dot{C}H{-}CH_2A \\[2mm] R\dot{C}HCH{=}CH_2\ +\ AH \end{array} \qquad (31)$$

It is usually important for cyclic and non-terminal olefins, but does not occur to a great extent with terminal olefins. The amount of allylic attack can frequently be reduced by working at lower temperatures.

Alkyl radicals

Polyhaloalkanes readily take part in radical chain additions. Normally the weakest C–X bond is broken so that abstraction occurs in the order I > Br > Cl, e.g.:

$$CFCl_2Br + RCH{=}CH_2 \longrightarrow RCHBrCH_2CFCl_2$$

The best-behaved haloalkanes are those with at least three halogen atoms. If there are less, the abstraction step (26) tends to be slow, and hydrogen abstraction from the haloalkane may occur, e.g. from CH_2Cl_2. The radical produced from a haloalkane is generally electrophilic so that addition occurs fast to electron-rich olefins such as hydrocarbons and vinyl ethers; the chains are long and good product yields can be obtained. If the haloalkane contains two or more bromine and/or iodine atoms decomposition to the carbene may occur (146) with formation of the cyclopropane, e.g.

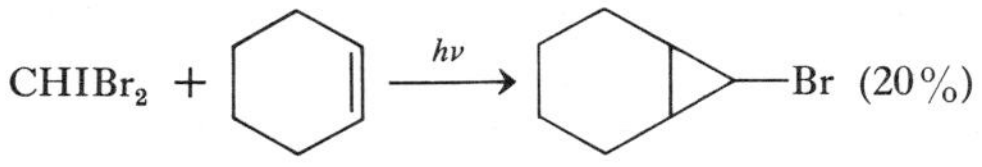

7-Bromonorcarane

Walling and Huyser have prepared a very useful survey of syntheses carried out by radical chain additions (145). Some examples are given below:

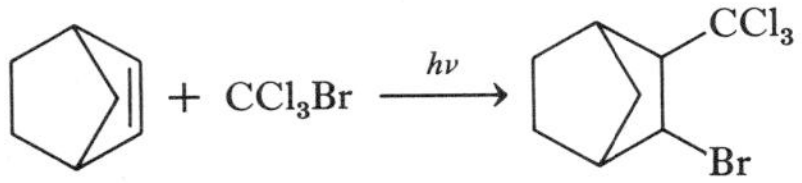

ref. 147

Norbornene

$$CF_3CF{=}CF_2 + CF_3I \xrightarrow{h\nu} CF_3CFICF_2CF_3 \quad (94\%) \quad \text{ref. 148}$$

$$CF_2{=}CFI + CH_2{=}CHF \xrightarrow{h\nu} CF_2{=}CFCH_2CHFI \quad (50\%) \quad \text{ref. 149}$$

$$CF_3C{\equiv}CH + CCl_3I \xrightarrow{h\nu} CCl_3CH{=}CICF_3 \quad (74\%) \quad \text{ref. 148}$$

Aldehydes add to olefins giving ketones by the radical chain mechanism. The overall reaction is:

$$R'CHO + CH_2{=}CHR \longrightarrow R'COCH_2CH_2R$$

The yield of ketone is reduced when the acyl radical tends to undergo decarbonylation:
$$R'\dot{C}O \longrightarrow R'{\cdot} + CO$$

This is particularly pronounced for acyl radicals which break down to tertiary or secondary alkyl radicals. The decarbonylation process can be minimized by working at lower temperatures ($< 100\ ^\circ$C) and by using olefins which are electron-accepting, e.g. perfluoro-olefins.

Primary and secondary alcohols will also add to olefins:

$$R'CH_2OH + CH_2=CHR \longrightarrow R'CH(OH)CH_2CH_2R$$
Primary alcohol   Secondary alcohol

$$\begin{array}{c} R^1 \\ \diagdown \\ CHOH + CH_2=CHR \longrightarrow \\ \diagup \\ R^2 \end{array} \qquad \begin{array}{c} R^1 \\ \diagdown \\ C(OH)CH_2CH_2R \\ \diagup \\ R^2 \end{array}$$

Secondary alcohol   Tertiary alcohol

Best yields are again obtained with electron-accepting olefins. Examples of aldehyde and alcohol additions are:

$$n\text{-}C_3H_7CHO + CH_2=CH(CH_2)_2COCH_3 \longrightarrow$$
$$n\text{-}C_3H_7CO(CH_2)_4COCH_3 \quad (70\%) \quad \text{ref. 150}$$

$$C_2H_5OH + n\text{-}C_3F_7CF=CF_2 \xrightarrow{(PhCO_2)_2} n\text{-}C_3F_7CHFCF_2CH(OH)CH_3$$
$$(70\%) \quad \text{ref. 151}$$

Amines also add to olefins by a radical chain process (152). The radical is formed mainly by abstraction of a hydrogen from the carbon atom adjacent to nitrogen, e.g.

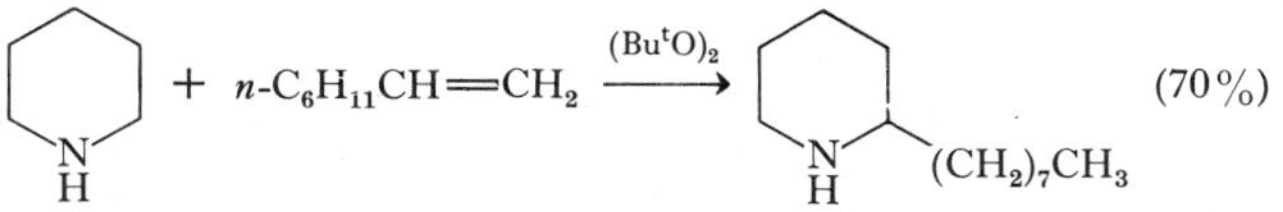

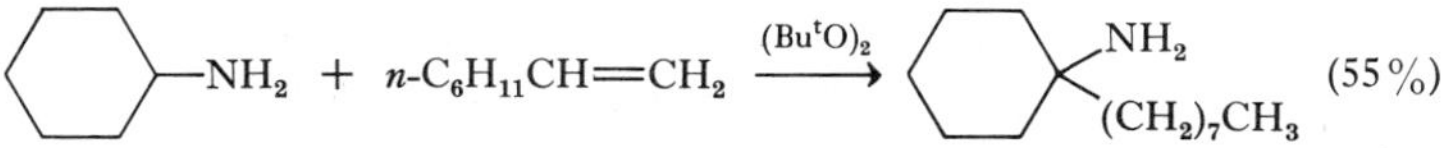

GENERAL REFERENCES

A. F. Trotman-Dickenson, *Gas Kinetics*, Butterworths, London, 1955.
C. Walling and E. S. Huyser, *Organic Reactions*, 1963, **13**, 91.

SPECIFIC REFERENCES

1. H. E. Van den Berg, A. B. Callear and R. J. Norstrom, *Chem. Phys. Letters*, 1969, **4**, 101; N. Basco, D. G. L. James and R. D. Suart, *Internat. J. Chem. Kinetics*, 1970, **2**, 215.
2. F. P. Lossing and A. W. Tickner, *J. Chem. Phys.* 1952, **20**, 907.

3. G. M. Burnett and H. W. Melville, *Technique of Organic Chemistry*, ed. S. L. Friess, E. S. Lewis and A. Weissberger, 2nd edn, 1963, vol. VIII, pt II, p. 1107.
4. A. Shepp, *J. Chem. Phys.* 1956, **24**, 944.
5. T. C. Clark, T. P. J. Izod, M. A. Di-Valentin and J. E. Dove, *J. Chem. Phys.* 1970, **53**, 2982.
6. G. B. Kistiakowsky and E. K. Roberts, *J. Chem. Phys.* 1953, **21**, 1637.
7. R. E. March and J. C. Polanyi, *Proc. Roy. Soc.* A, 1963, **273**, 360.
8. P. B. Ayscough, *J. Chem. Phys.* 1956, **24**, 944.
9. J. R. Majer, C. Olavesen and J. C. Robb, *Trans. Faraday Soc.* 1969, **65**, 2988.
10. G. R. DeMaré and G. H. Huybrechts, *Chem. Phys. Letters*, 1967, **1**, 64.
11. J. M. Tedder and J. C. Walton, *Trans. Faraday Soc.* 1967, **63**, 2464.
12. A. Shepp and K. O. Kutschke, *J. Chem. Phys.* 1957, **26**, 1020.
13. P. B. Ayscough, F. S. Dainton and B. E. Fleischfresser, *Trans. Faraday Soc.* 1966, **62**, 1838.
14. L. Bertrand, G. R. DeMaré, G. Huybrechts, J. Olbregts and M. Toth, *Chem. Phys. Letters*, 1970, **5**, 183.
15. E. L. Metcalfe and A. F. Trotman-Dickenson, *J. Chem. Soc.* 1962, 4620.
16. E. L. Metcalfe, *J. Chem. Soc.* 1963, 3560.
17. K. J. Hole and M. F. R. Mulcahy, *J. Phys. Chem.* 1969, **73**, 177.
18. N. L. Arthur and T. N. Bell, *Chem. Comm.* 1965, 166.
19. R. D. Giles and E. Whittle, *Trans. Faraday Soc.* 1965, **61**, 1425; R. D. Giles, L. M. Quick and E. Whittle, *ibid.* 1967, **63**, 662.
20. G. O. Pritchard and J. T. Bryant, *J. Phys. Chem.* 1968, **72**, 1603.
21. M. G. Bellas, O. P. Strausz and H. E. Gunning, *Canad. J. Chem.* 1965, **43**, 1022.
22. I. Unger and G. P. Semeluk, *Canad. J. Chem.* 1966, **44**, 1427.
23. W. G. Alcock and E. Whittle, *Trans. Faraday Soc.* 1966, **62**, 134.
24. R. Bowles, J. R. Majer and J. C. Robb, *Trans. Faraday Soc.* 1962, **58**, 2394.
25. J. Grotewold, E. A. Lissi and M. G. Neumann, *J. Chem. Soc.* A, 1968, 375.
26. J. A. Kerr and A. F. Trotman-Dickenson, *Progr. Reaction Kinetics*, 1961, **1**, 1.
27. E. R. Morris and J. C. J. Thynne, *Trans. Faraday Soc.* 1968, **64**, 3027.
28. G. O. Pritchard, G. H. Miller and J. R. Dacey, *Canad. J. Chem.* 1961, **39**, 1968.
29. R. Bowles, J. R. Majer and J. C. Robb, *Trans. Faraday Soc.* 1962, **58**, 1541.
30. B. C. Roquitte and M. H. J. Wijnen, *J. Amer. Chem. Soc.* 1963, **85**, 2053.
31. J. O. Terry and J. H. Futrell, *Canad. J. Chem.* 1967, **45**, 2327; *ibid.* 1968, **46**, 664.
32. G. O. Pritchard and M. J. Perona, *Internat. J. Chem. Kinetics*, 1969, **1**, 413.
32a. J. H. Georgakakos, B. S. Rabinovitch and C. W. Larson, *Internat. J. Chem. Kinetics*, 1971, **3**, 535.
33. P. M. Scott and K. R. Jennings, *J. Phys. Chem.* 1969, **73**, 1513, 1521.
34. P. Cadman, V. Inel and A. F. Trotman-Dickenson, *J. Chem. Soc.* A, 1970, 1207.

35. D. G. L. James and S. M. Kambanis, *Trans. Faraday Soc.* 1969, **65**, 1350.
36. J. R. McNesby, C. M. Drew and A. S. Gordon, *J. Phys. Chem.* 1960, **59**, 988.
37. J. N. Bradley, *J. Chem. Phys.* 1961, **35**, 748; J. N. Bradley and B. S. Rabinovitch, *ibid.* 1962, **36**, 3498.
38. R. A. Holroyd and G. W. Klein, *J. Phys. Chem.* 1963, **67**, 2273.
39. A. F. Trotman-Dickenson, *Proc. Chem. Soc.* 1964, 249.
40. S. W. Benson, *Adv. Photochem.* 1964, **2**, 1.
41. R. W. Fessenden, *J. Phys. Chem.* 1964, **68**, 1508.
42. D. J. Carlsson and K. U. Ingold, *J. Amer. Chem. Soc.* 1968, **90**, 7047.
43. H. W. Melville, J. C. Robb and R. C. Tutton, *Discuss. Faraday Soc.* 1953, **14**, 158.
44. W. I. Bengough and R. A. M. Thomson, *Trans. Faraday Soc.* 1961, **57**, 1928.
45. S. Weiner and G. S. Hammond, *J. Amer. Chem. Soc.* 1968, **90**, 1659; *ibid.* 1969, **91**, 986.
46. R. D. Burkhart, *J. Amer. Chem. Soc.* 1968, **90**, 1659.
46a. R. D. Burkhart, R. F. Boynton and J. C. Merrill, *J. Amer. Chem. Soc.* 1971, **93**, 5013.
47. A. M. North and G. A. Reed, *J. Polymer Sci.* 1963, **14**, 1311.
48. J. Hughes and A. M. North, *Trans. Faraday Soc.* 1964, **60**, 960.
49. A. M. North, *Quart. Rev.* 1966, **20**, 421.
50. R. A. Sheldon and J. K. Kochi, *J. Amer. Chem. Soc.* 1970, **92**, 4395, 5175.
51. O. Dobis, J. M. Pearson and M. Szwarc, *J. Amer. Chem. Soc.* 1968, **90**, 278; K. Chakravorty, J. M. Pearson and M. Szwarc, *ibid.* 1968, **90**, 283.
52. R. M. Noyes, *Progr. Reaction Kinetics*, 1961, **1**, 131.
53. K. R. Kopecky and T. Gillan, *Canad. J. Chem.* 1969, **47**, 2371.
54. F. D. Greene, M. A. Berwick and J. C. Stowell, *J. Amer. Chem. Soc.* 1970, **92**, 867.
55. H. Komazawa, A. P. Stefani and M. Szwarc, *J. Amer. Chem. Soc.* 1963, **85**, 2043.
56. J. M. Tedder and J. C. Walton, *Progr. Reaction Kinetics*, 1967, **2**, 39.
57. J. R. Majer, D. Phillips and J. C. Robb, *Trans. Faraday Soc.* 1965, **61**, 110, 122.
58. M. S. Kharasch, E. V. Jensen and W. H. Urry, *Science*, 1945, **102**, 128; M. S. Kharasch, O. Reinmuth and W. H. Urry, *J. Amer. Chem. Soc.* 1947, **69**, 1105; M. S. Kharasch, E. V. Jensen and W. H. Urry, *ibid.* 1947, **69**, 1100.
59. R. Gregory, R. N. Haszeldine and A. E. Tipping, *J. Chem. Soc.* C, 1970, 1750, and previous papers in the series.
60. D. P. Johari, H. W. Sidebottom, J. M. Tedder and J. C. Walton, *J. Chem. Soc.* B, 1971, 95, and previous papers in the series.
60a. H. W. Sidebottom, J. M. Tedder and J. C. Walton, *Internat. J. Chem. Kinetics*, 1972, **4**, 249.
61. J. Gibb, M. J. Peters, J. M. Tedder, J. C. Walton and K. D. R. Winton, *Chem. Comm.* 1970, 978.
62. G. J. Janz, N. A. Gac, A. R. Monohan and W. J. Leahy, *J. Org. Chem.* 1965, **30**, 2075; G. J. Janz and J. B. Flannery, *J. Phys. Chem.* 1966, **70**, 2061.

63. J. M. Tedder and J. C. Walton, *Trans. Faraday Soc.* 1970, **66**, 1135.
64. F. R. Mayo and C. Walling, *Chem. Rev.* 1950, **46**, 191.
65. J. I. G. Cadogan and P. W. Inward, *J. Chem. Soc.* 1962, 4170; J. I. G. Cadogan and E. G. Duell, *ibid.* 1962, 4154; J. I. G. Cadogan, E. G. Duell and P. W. Inward, *ibid.* 1962, 4164.
66. R. R. Getty, J. A. Kerr and A. F. Trotman-Dickenson, *J. Chem. Soc.* A, 1967, 1360.
67. H. G. Mennier and P. I. Abell, *J. Phys. Chem.* 1967, **71**, 1430.
68. R. R. Getty, J. A. Kerr and A. F. Trotman-Dickenson, *J. Chem. Soc.* A, 1967, 979.
69. J. M. Tedder, unpublished work.
70. J. M. Sangster and J. C. J. Thynne, *J. Phys. Chem.* 1969, **73**, 2746.
71. J. M. Sangster and J. C. J. Thynne, *Trans. Faraday Soc.* 1969, **65**, 2110.
72. L. Endrenyi and D. J. LeRoy, *J. Phys. Chem.* 1967, **71**, 1334.
73. K. W. Watkins and L. A. O'Deen, *J. Phys. Chem.* 1969, **73**, 4094.
74. A. Streitwieser and R. C. Fahey, *Chem. and Ind.* 1959, 1417; A. Streitwieser, R. H. Jagow, R. C. Fahey and S. Suzuki, *J. Amer. Chem. Soc.* 1958, **80**, 2326.
75. M. Feld, A. P. Stefani and M. Szwarc, *J. Amer. Chem. Soc.* 1962, **84**, 4451.
76. A. P. Stefani, Lan-Yuh Yang Chung and H. E. Todd, *J. Amer. Chem. Soc.* 1970, **92**, 4168.
77. R. J. Cvetanović and R. S. Irwin, *J. Chem. Phys.* 1967, **46**, 1694.
78. J. M. Tedder and J. C. Walton, *Trans. Faraday Soc.* 1966, **62**, 1859; 1967, **63**, 2678.
79. P. S. Dixon and M. Szwarc, *Trans. Faraday Soc.* 1963, **59**, 112; J. M. Pearson and M. Szwarc, *ibid.* 1964, **60**, 564; G. E. Owen, J. M. Pearson and M. Szwarc, *ibid.* 1965, **61**, 1722.
80. A. P. Stefani, L. Herk and M. Szwarc, *J. Amer. Chem. Soc.* 1961, **83**, 4732.
81. J. M. Sangster and J. C. J. Thynne, *Internat. J. Chem. Kinetics*, 1969, **1**, 571; 1971, **3**, 155.
82. J. E. Bloor, A. C. R. Brown and D. G. L. James, *J. Phys. Chem.* 1966, **70**, 2191, and previous papers in the series.
83. J. I. G. Cadogan and I. H. Sadler, *J. Chem. Soc.* B, 1966, 1191.
84. M. M. Martin and G. J. Gleicher, *J. Amer. Chem. Soc.* 1964, **86**, 233, 238, 242.
85. H. Sakurai, A. Hosomi and M. Kumada, *J. Org. Chem.* 1969, **34**, 1764.
86. See for example: F. H. Burkitt, C. A. Coulson and H. C. Longuet-Higgins, *Trans. Faraday Soc.* 1951, **47**, 553; J. N. Murrell, S. F. A. Kettle and J. M. Tedder, *Valence Theory*, Wiley, London, 1965, p. 255; M. J. S. Dewar, *The Molecular Orbital Theory of Organic Chemistry*, McGraw-Hill, New York and London, 1969.
87. A. Streitwieser, *Molecular Orbital Theory for Organic Chemists*, Wiley, New York, 1961.
88. C. A. Coulson, *J. Chem. Soc.* 1955, 1435.
89. J. H. Binks and M. Szwarc, *J. Chem. Phys.* 1959, **30**, 1494.
90. A. P. Stefani and M. Szwarc, *J. Amer. Chem. Soc.* 1962, **84**, 3661.
91. R. J. Cvetanović, *Adv. Photochem.* 1963, **1**, 116.
92. K. Yang, *J. Amer. Chem. Soc.* 1962, **84**, 3795.

93. S. Sato and R. J. Cvetanović, *J. Amer. Chem. Soc.* 1959, **81**, 3223.
94. W. M. Jackson, J. R. McNesby and B. deB. Darwent, *J. Chem. Phys.* 1962, **37**, 1610.
95. G. O. Pritchard and M. J. Perona, *Internat. J. Chem. Kinetics*, 1969, **1**, 509.
96. S. H. Jones and E. Whittle, *Internat. J. Chem. Kinetics*, 1970, **2**, 479.
97. J. M. Tedder and R. A. Watson, *Trans. Faraday Soc.* 1966, **62**, 1215.
98. J. D. Clarke, C. Pearce and D. A. Whytock, *Trans. Faraday Soc.* 1971, **67**, 1049.
99. G. Giacometti and E. W. R. Steacie, *Canad. J. Chem.* 1958, **36**, 1493.
100. M. G. Evans and M. Polanyi, *Trans. Faraday Soc.* 1938, **34**, 11; E. T. Butler and M. Polanyi, *ibid.* 1943, **39**, 19.
101. A. F. Trotman-Dickenson, *Gas Kinetics*, Butterworths, London, 1955, p. 228.
102. A. S. Gordon, S. R. Smith and C. M. Drew, *J. Chem. Phys.* 1962, **36**, 824; A. S. Gordon, *Canad. J. Chem.* 1965, **43**, 70; A. S. Gordon and S. R. Smith, *J. Phys. Chem.* 1962, **66**, 521.
103. B. P. McGrath and J. M. Tedder, *Bull. Soc. chim. belges*, 1962, **71**, 772.
104. A. F. Trotman-Dickenson, *Chem. and Ind.* 1965, 379.
105. J. A. Kerr, *Chem. Rev.* 1966, **66**, 465.
106. S. H. Jones and E. Whittle, *Internat. J. Chem. Kinetics*, 1970, **2**, 479.
107. G. O. Pritchard, J. T. Bryant and R. L. Tommarson, *J. Phys. Chem.* 1965, **69**, 664.
108. J. A. Kerr and D. Timlin, *J. Chem. Soc.* A, 1969, 1241.
109. T. N. Bell and U. F. Zucker, *Canad. J. Chem.* 1970, **48**, 1209; T. N. Bell and A. E. Platt, *Internat. J. Chem. Kinetics*, 1970, **2**, 299.
110. O. P. Strausz, E. Jakubowski, H. S. Sandhu and H. E. Gunning, *J. Chem. Phys.* 1969, **51**, 552; E. Jakubowski, H. S. Sandhu, H. E. Gunning and O. P. Strausz, *ibid.* 1970, **52**, 4242.
111. E. R. Morris and J. C. J. Thynne, *Trans. Faraday Soc.* 1970, **66**, 183.
112. A. U. Chaudhry and B. G. Gowenlock, *J. Organometallic Chem.* 1969, **16**, 221.
113. J. A. Kerr, D. H. Slater and J. C. Young, *J. Chem. Soc.* A, 1967, 134.
114. J. A. Kerr, A. Stephens and J. C. Young, *Internat. J. Chem. Kinetics*, 1969, **1**, 371, 339.
115. W. J. Cheng and M. Szwarc, *J. Phys. Chem.* 1968, **72**, 494.
116. P. Gray, A. Jones and J. C. J. Thynne, *Trans. Faraday Soc.* 1965, **61**, 474.
117. P. Gray and A. A. Herod, *Trans. Faraday Soc.* 1967, **63**, 2489; 1968, **64**, 1568, 2723.
118. K. C. Ferguson and J. T. Pearson, *Trans. Faraday Soc.* 1970, **66**, 910.
119. P. Gray, A. A. Herod, A. Jones and J. C. J. Thynne, *Trans. Faraday Soc.* 1966, **62**, 2774.
120. E. R. Morris and J. C. J. Thynne, *Trans. Faraday Soc.* 1968, **64**, 414, 2124, 3021.
121. P. Gray and L. J. Leyshon, *Trans. Faraday Soc.* 1969, **65**, 780.
122. D. A. Edwards, J. A. Kerr, A. C. Lloyd and A. F. Trotman-Dickenson, *J. Chem. Soc.* A, 1966, 1500.

Alkyl radicals

123. F. W. Evans and M. Szwarc, *Trans. Faraday Soc.* 1961, **57**, 1905; R. J. Fox, F. W. Evans and M. Szwarc, *ibid.* 1961, **57**, 1915.
124. W. G. Alcock and E. Whittle, *Trans. Faraday Soc.* 1965, **61**, 244; 1966, **62**, 134, 664.
125. D. M. Tompkinson, J. P. Galvin and H. O. Pritchard, *J. Phys. Chem.* 1964, **68**, 541.
126. W. C. Danen and R. L. Winter, *J. Amer. Chem. Soc.* 1971, **93**, 716.
127. C. L. Kibby and R. E. Weston Jr, *J. Chem. Phys.* 1968, **49**, 4825.
128. N. L. Arthur and P. Gray, *Trans. Faraday Soc.* 1969, **65**, 434.
129. N. Imai, T. Dohmaru and O. Toyama, *Bull. Chem. Soc. Japan*, 1965, **38**, 639.
130. P. Gray, N. L. Arthur and A. C. Lloyd, *Trans. Faraday Soc.* 1969, **65**, 775.
131. L. Melander, *Isotope Effects on Reaction Rates*, Ronald Press, New York, 1960.
132. F. S. Dainton, G. A. Creak and K. J. Irvin, *Trans. Faraday Soc.* 1962, **58**, 326.
133. M. Salomon, *Canad. J. Chem.* 1964, **42**, 610.
134. H. S. Johnston and C. Parr, *J. Amer. Chem. Soc.* 1963, **84**, 2544; H. S. Johnston, *Gas Phase Reaction Rate Theory*, Ronald Press, New York, 1966, pp. 207, 339.
135. F. H. Westheimer, *Chem. Rev.* 1961, **61**, 265.
136. E. R. Morris and J. C. J. Thynne, *Internat. J. Chem. Kinetics*, 1970, **2**, 257.
137. J. C. Amphlett and E. Whittle, *Trans. Faraday Soc.* 1967, **63**, 2695.
138. E. S. Huyser, *J. Amer. Chem. Soc.* 1960, **82**, 394.
139. R. F. Bridger and G. A. Russell, *J. Amer. Chem. Soc.* 1963, **85**, 3754.
140. P. H. Owens, G. J. Gleicher and L. H. Smith, *J. Amer. Chem. Soc.* 1968, **90**, 4122.
141. J. E. Leffler and E. Grunwald, *Rates and Equilibria of Organic Reactions*, Wiley, New York, 1963, p. 226.
142. R. D. Brown, *J. Chem. Soc.* 1953, 2615.
143. K. Fukui in *Modern Quantum Chemistry*, ed. O. Sinanoğlu, Academic Press, New York and London, 1965, pt 1, p. 49.
144. K. Fukui, H. Kato and T. Yonezawa, *Bull. Chem. Soc. Japan*, 1961, **34**, 1111; 1962, **55**, 1475.
145. C. Walling and E. S. Huyser, *Org. Reactions*, 1963, **13**, 91.
146. T. Marolewski and N. C. Yang, *Chem. Comm.* 1967, 1225.
147. M. S. Kharasch and H. N. Friedlander, *J. Org. Chem.* 1949, **14**, 239.
148. R. N. Haszeldine, *J. Chem. Soc.* 1953, 3559, 922.
149. J. D. Park, R. J. Seffl and J. R. Lacher, *J. Amer. Chem. Soc.* 1956, **78**, 59.
150. G. I. Nikishin, V. D. Vorob'ev and A. D. Petrov, *Izv. Akad. Nauk SSSR Otd. Khim. Nauk.* 1961, 882; *Chem. Abs.* 1961, **55**, 22093.
151. J. D. LaZerte and R. J. Koshar, *J. Amer. Chem. Soc.* 1955, **77**, 910.
152. W. H. Urry and O. O. Juveland, *J. Amer. Chem. Soc.* 1958, **80**, 3323.
153. O. P. Strausz, I. Safarik, W. B. O'Callaghan and H. E. Gunning, *J. Amer. Chem. Soc.* 1972, **94**, 1828.
154. R. Hiatt and S. W. Benson, *J. Amer. Chem. Soc.* 1972, **94**, 25; *Internat. J. Chem. Kinetics*, 1972, **4**, 151.
155. R. M. Marshall and J. H. Purnell, *Chem. Comm.* 1972, 764.

9

Reactions of heteroradicals

9.1 *INTRODUCTION*

Radical intermediates having the unpaired electron localized on elements other than carbon, are frequently encountered in organic chemistry. Reaction mechanisms involving heteroradicals are well established, and the behaviour of radicals having unpaired electrons localized on oxygen, RO· and ROO·, sulphur, RS·, or nitrogen, R_2N·, is almost as well known as that of alkyl radicals. The existence of free radicals centred on some other elements such as some group IIA metals is, however, still a matter of speculation.

Heteroradicals generally parallel alkyl radicals in their behaviour, and their reactions fall into the same classification as for alkyl radicals. Heteroradical reactions have been very widely exploited in synthesis, particularly the addition reaction with olefins, but as yet there have been relatively few kinetic studies. This is because 'clean sources' are needed for the study of elementary reactions, free from competing side reactions, and few such sources are available.

In this chapter we shall concentrate mainly on the reactions of radicals with unpaired electrons centred on elements of groups IVB, VB and VIB. Radicals centred on elements of groups II and III are also known to exist (1). However, there is little systematic information, and much of the evidence is, at present, speculative in nature. These radicals are only treated where the results are of use, in the opinion of the authors, to the main body of organic chemistry. Radicals centred on oxygen are well covered in chapter 10 on oxidation and only a few are included here for purposes of comparison.

9.2 *RADICAL–RADICAL REACTIONS*

Heteroradicals react with each other by the processes of combination and disproportionation which are so characteristic of any free-radical system.

Heteroradicals

9.2.1 *Group IVB elements* (1)

The radical–radical behaviour of group IVB radicals is governed mainly by the reluctance of the metals to form p_π–p_π bonds (2): combination is therefore favoured over disproportionation. Silyl radicals generated by pyrolysis of silane react by combination:

$$\cdot SiH_3 + \cdot SiH_3 \longrightarrow Si_2H_6 \qquad (1)$$

and disilane is detected in the products (3). Photochemical decomposition by mercury photosensitization of alkylsilanes containing at least one hydrogen atom attached to silicon, provides a clean source of alkylsilyl radicals (4). These radicals also mainly undergo combination:

$$Me_3Si\cdot + Me_3Si\cdot \longrightarrow Me_3SiSiMe_3 \qquad (2)$$

The rate of combination of trimethylsilyl radicals has been measured in the gas phase by the conventional intermittent illumination method (46). Cadman *et al.* generated the radicals by mercury-photosensitized decomposition of trimethylsilane and monitored the radical concentration by means of the reaction with ethyl chloride:

$$Me_3Si\cdot + EtCl \longrightarrow Me_3SiCl + Et\cdot$$

The combination rate was found to be: $k_2 = 1.8 \times 10^{11} \, \text{l mol}^{-1} \, \text{s}^{-1}$, which was independent of temperature. This combination rate is very similar to that of alkyl radicals in the gas phase, and suggests that silyl radicals combine at the maximum collision rate.

Frangopol and Ingold have measured the combination rate constant of trimethylsilyl radicals in solution (47), by observation of the decay of the e.s.r. signal of the radical. They found the combination rate to be $2.2 \times 10^9 \, \text{l mol}^{-1} \, \text{s}^{-1}$ at 298 °K; again this is in line with alkyl radicals in solution, and suggests that silyl radicals also combine at the maximum diffusion-controlled rate.

The experimental results indicate that silyl radical combination is analogous in most respects to alkyl radical combination, and is controlled by similar electronic factors. Thynne (6) estimated the trimethylsilyl radical combination rate constant to be about $10^2 \, \text{l mol}^{-1} \, \text{s}^{-1}$, from the rate of decomposition of hexamethyldisilane, and the entropy change in the reaction. The complete lack of agreement between this estimate and the experimental value probably indicates that the rate of decomposition of hexamethyldisilane or its entropy is in error.

Germyl and alkylgermyl radicals are known (1); the main radical–radical reaction is presumably combination, but this has been studied for very few cases.

Trialkyltin radicals are better known and can be produced by photolysis of trialkyltin hydrides, or by the reduction of an alkyl halide with an organotin hydride (5):

$$X\cdot + R_3SnH \longrightarrow R_3Sn\cdot + XH \qquad (3)$$

Combination is probably the main radical–radical process, but other bimolecular processes such as (5) are also known:

$$R_3Sn\cdot + R_3Sn\cdot \longrightarrow R_3SnSnR_3 \qquad (4)$$

$$R_3Sn\cdot + R_3Sn\cdot \longrightarrow R_4Sn + R_2Sn \qquad (5)$$

Carlsson and Ingold (7) have studied the recombination rate of a series of tin radicals in solution. The radicals were generated from organotin hydrides by reduction with alkyl halides (6). It was found that when alkyl chlorides were used, but not bromides or iodides, the abstraction of chlorine by trialkyltin radicals was rate-controlling:

$$R_3Sn\cdot + R'Cl \longrightarrow R_3SnCl + R'\cdot \qquad (6)$$

Combination of the tin radical as in (4) represented the major termination step. The combination rate constants were measured by the intermittent illumination method and are given in Table 9-1. The combination rate constants are of the same order as those of alkyl radicals in solution. The combination rate is rather less than the calculated rate of diffusion in solution ($\sim 7 \times 10^9$ l mol^{-1} s^{-1}), but the reaction is probably diffusion-controlled, since the rate constants vary little with radical structure.

TABLE 9-1 *Rate constants for tin–radical combinations in cyclohexane solution at 25 °C* (7)

$$R_3Sn\cdot + R_3Sn\cdot \xrightarrow{k_4} R_3SnSnR_3$$

Radical	$k_4 \times 10^9$ (l mol^{-1} s^{-1})
Me$_3$Sn·	1.6
Bu$_3^n$Sn·	0.7
Bu$_2^n$HSn·	0.8
Bu$_2^n$ClSn·	1.8
Ph$_3$Sn·	1.4

Heteroradicals

Heteroradicals with the unpaired electron centred on lead have been proposed as intermediates in the decomposition of lead alkyls. Hexaalkyldilead compounds are very unstable, and Pb=C bonds do not form, so that radical–radical reactions are unknown. They appear to undergo unimolecular decomposition (7) much more readily:

$$Et_3Pb\cdot \longrightarrow Et_2Pb + Et\cdot \qquad (7)$$

9.2.2 Group VB elements

Nitrogen radicals can both dimerize and disproportionate. The amino radical itself has been produced from hydrazine and ammonia, and the main radical–radical reactions appear to be:

$$\cdot NH_2 + \cdot NH_2 \longrightarrow N_2H_4 \qquad (8)$$

$$\cdot NH_2 + \cdot NH_2 \longrightarrow NH_3 + NH: \qquad (9)$$

although others have been proposed, e.g.

$$\cdot NH_2 + \cdot NH_2 \longrightarrow N_2 + 2H_2 \qquad (10)$$

The rates of reactions (8) and (9) have been estimated by kinetic spectroscopy of the decomposition of ammonia (8). The values are given in Table 9-2. Alkylamino radicals have also been extensively studied (9). Diethylamino radicals, for instance, both disproportionate and dimerize (11 and 12) and will also react with alkyl radicals:

$$2Et_2N\cdot \longrightarrow Et_2NNEt_2 \qquad (11)$$

$$2Et_2N\cdot \longrightarrow Et_2NH + EtN{=}CHCH_3 \qquad (12)$$

$$Et_2N\cdot + CH_3\cdot \longrightarrow Et_2NCH_3 \qquad (13)$$

$$Et_2N\cdot + CH_3\cdot \longrightarrow EtN{=}CHCH_3 + CH_4 \qquad (14)$$

$$2CH_3\cdot \longrightarrow C_2H_6 \qquad (15)$$

The cross-combination rate constant ratio $(k_{13} + k_{14})/[k_{15}(k_{11} + k_{12})]^{\frac{1}{2}}$, corrected for disproportionation, has been measured by Gowenlock and co-workers (10). They found values between 1.1 and 2.2 in reasonable agreement with the value of about 2 found for alkyl radicals (cf. chapter 8, p. 200).

The difluoroamino radical has been the subject of extensive and interesting kinetic studies. This radical exists, in detectable quantities, in equilibrium with its dimer even at room temperature:

$$\cdot NF_2 + \cdot NF_2 \rightleftharpoons N_2F_4 \qquad (16)$$

TABLE 9-2   *Rate constants for heteroradical combination and disproportionation reactions in the gas phase*

Reaction	Temperature (°K)	k (1 mol^{-1} s^{-1})	Reference
$\cdot OH + \cdot OH \longrightarrow H_2O + O$	300	5×10^8	17
$NO \cdot + \cdot CN \longrightarrow NOCN$	300	2×10^9	14
$\cdot NH_2 + \cdot NH_2 \longrightarrow N_2H_4$	300	3×10^9	8
$\cdot NH_2 + \cdot NH_2 \longrightarrow NH + NH_3$	300	5×10^8	8
$\cdot NF_2 + \cdot NF_2 \longrightarrow N_2F_4$	400	3×10^7	11
$\cdot NF_2 + Et \cdot \longrightarrow C_2H_5NF_2$	300–450	6×10^9	12
$Me_3Si \cdot + Me_3Si \cdot \longrightarrow Me_3SiSiMe_3$	317–399	2×10^{11}	46
$MeO \cdot + MeO \cdot \longrightarrow MeOOMe$	300	$(2 \times 10^9)^a$	18
$MeS \cdot + MeS \cdot \longrightarrow MeSSMe$	300	5×10^{10}	21

a Value estimated.

The decomposition of tetrafluorohydrazine has been studied in a shock tube, and the combination rate constant has been estimated by combining the decomposition rate of N_2F_4 with the known equilibrium constant (11) (see Table 9-2). Rate constants for the combination and disproportionation of $\cdot NF_2$ radicals with alkyl radicals have also been measured (12). The disproportionation/combination rate constant ratios for $\cdot NF_2$ with isopropyl and t-butyl radicals are 0.064 and 0.123 respectively.

$$\cdot NF_2 + C_3H_7^{i \cdot} \longrightarrow NF_2CHMe_2 \qquad (17)$$

$$\cdot NF_2 + C_3H_7^{i \cdot} \longrightarrow NF_2H + CH_3CH{=}CH_2 \qquad (18)$$

The combination and disproportionation show some temperature dependence. The values for $\cdot NF_2$ with alkyls fit reasonably well on to the same correlation of $\log k_d/k_c$ against $\Sigma S_d{}^0 - S_c{}^0$, as was found for alkyl–alkyl disproportionations (12), see Fig. 8-1.

In general, nitrogen radicals combine at a considerably slower rate than alkyl radicals (see Table 9-2). Simons (13) has carried out molecular orbital calculations for $\cdot NF_2$ which suggest that the unpaired electron is localized in a π-orbital perpendicular to the molecular plane. Encounter between two $\cdot NF_2$ radicals would lead initially to formation of a weak π-bond, and dimer formation can only proceed when an electron is promoted from a more tightly bound orbital having σ-symmetry. This may account for the slow combination rate of $\cdot NF_2$ radicals.

Phosphinyl radicals $\cdot PH_2$, and alkyl-substituted phosphinyl radicals, appear to disproportionate rather than dimerize (15). The phosphene species (:PH) then react to give hydrogen and elemental phosphorus:

$$\cdot PH_2 + \cdot PH_2 \longrightarrow PH_3 + \;:PH \qquad (19)$$

$$:PH + \;:PH \longrightarrow P_2 + H_2 \qquad (20)$$

Dimerization is observed however, for substituted phosphinyl radicals, e.g.

$$Ph_3P + h\nu \longrightarrow Ph_2P\cdot + Ph\cdot \qquad (21)$$

$$2Ph_2P\cdot \longrightarrow Ph_2PPPh_2 \qquad (22)$$

Small amounts of tetraphenyldiphosphine were isolated from the photolysis of triphenylphosphine in benzene solution (16). Similarly, small amounts of P_2Cl_4 can be isolated from the photolysis of PCl_3 in the gas phase, suggesting the dimerization (*23*):

$$\cdot PCl_2 + \cdot PCl_2 \longrightarrow P_2Cl_4 \qquad (23)$$

Very little is known about radicals with an unpaired electron centred on arsenic, antimony or bismuth. Such studies as there are (15) indicate that the radicals undergo stepwise decomposition giving eventually the metal itself:

$$Me_3M \longrightarrow Me_2M\cdot + Me\cdot \qquad (24)$$

$$Me_2M\cdot + Me_2M\cdot \longrightarrow Me_3M + MeM: \qquad (25)$$

$$MeM: \longrightarrow Me\cdot + M \qquad (26)$$

$$nMeM: \longrightarrow Polymer \qquad (27)$$

Dimerization seems very unlikely because of the weakness of the metal–metal bonds.

9.2.3 *Group VIB elements*

There are two major classes of radicals in which the unpaired electron is localized on oxygen: the oxy radicals $RO\cdot$ and the peroxy radicals $ROO\cdot$. Peroxy radical reactions are dealt with more fully in chapter 10 (p. 394) and will only be touched on here.

Oxy radicals can both combine and disproportionate. Direct experimental measurements of combination rates in the gas phase have been made for hydroxy radicals (17), and Gray *et al.* (18) have estimated the combination rate constant for methoxy radicals:

$$MeO\cdot + MeO\cdot \longrightarrow MeOOMe \qquad (28)$$

$$MeOOMe \longrightarrow MeO\cdot + MeO\cdot \qquad (29)$$

The combination rate constant was calculated from the known rate of peroxide decomposition k_{29}, by use of the expression:

$$\log k_{28} = \log A_{29} + \Delta S/2.3R + 1.8$$

where ΔS, the entropy change, was estimated from atomic additivity rules (19), and the factor 1.8 corrects for the change in the number of molecules when the radicals combine. The estimated combination rate constant, given in Table 9-2, is considerably smaller than those of alkyl radicals of comparable size.

Combination/disproportionation rate constant ratios have been determined for several alkoxy-radical pairs. For methoxy- and for ethoxy-radicals k_d/k_c was found to be 9.3 and 12 respectively (20). These high values show that disproportionation is very much favoured for these polar radicals. They do not fit into the correlation of $\log k_d/k_c$ with $\Sigma S_d{}^0 - S_c{}^0$ found for alkyl radicals (18).

The simple thiyl radicals interact mainly by combination, but disproportionation can also occur. Methylthiyl radical combination:

$$CH_3S\cdot + CH_3S\cdot \longrightarrow CH_3SSCH_3 \qquad (30)$$

giving the disulphide, has been studied by Sivertz and co-workers (21) who photolysed methyl mercaptan in the presence of olefins. From intermittent illumination experiments in the gas phase, they found $k_{30} = 5 \times 10^{10}\ l\ mol^{-1}\ s^{-1}$, which is of similar magnitude to that of a simple alkyl radical. Disproportionation of methylthiyl radicals has also been observed (22), and the disproportionation/combination

$$CH_3S\cdot + CH_3S\cdot \longrightarrow CH_3SH + CH_2S \qquad (31)$$

rate constant ratio k_{31}/k_{30} was found to vary from 0 at atmospheric pressures to 0.3 at low pressures. Other primary and secondary alkylthiyl radicals can also disproportionate to give a mercaptan and a thioketone (or aldehyde).

$$2R_2CHS\cdot \longrightarrow R_2CHSH + R_2CS \qquad (32)$$

Free-radical intermediates with the unpaired electron localized on selenium and tellurium have been proposed for various reactions (15), and the HSe· and HTe· radicals have been observed by e.s.r. spectroscopy (23). Diselenides and ditellurides are known, so that in principle combination reactions are possible. It can be anticipated, however, that disproportionation is unlikely because of the reluctance of selenium and tellurium to form double bonds.

9.3 *ADDITION REACTIONS*

The addition of heteroradicals to carbon–carbon multiple bonds is a process of very wide applicability (24, 25). Heteroradicals add to alkenes to form carbon–heteroatom bonds, so that the reaction is very useful as a method of synthesizing organic compounds containing carbon–silicon, carbon–nitrogen, carbon–phosphorus, carbon–sulphur, etc., bonds. Additions to other unsaturated molecules such as alkynes, aromatic and heterocyclic compounds are also of very common occurrence.

9.3.1 *Mechanism*

The mechanism of addition can of course vary greatly, depending on factors such as the radical source, pressure, nature of the radical and unsaturated molecule, etc. However, a wide class of reactions, in both the gas and solution phases, is covered by the normal chain sequence as for alkyl radicals. The heteroradical source RAX reacts with the initiator radical In·, which can be derived from any of the usual radical initiators, to produce the heteroradical RA· (*33*):

$$RAX + In· \longrightarrow RA· + InX \quad (Rate = R_i) \qquad (33)$$

$$RA· + Ol \longrightarrow RAOl· \qquad (34)$$

$$RAOl· + RAX \longrightarrow RAOlX + RA· \qquad (35)$$

$$RA· + RA· \longrightarrow RAAR \qquad (36)$$

$$RA· + RAOl· \longrightarrow RAOlAR \qquad (37)$$

$$RAOl· + RAOl· \longrightarrow RAOlOlAR \qquad (38)$$

The heteroradical adds to the unsaturated molecule (*34*) producing a new radical, which can abstract from the addend to give the adduct RAOlX and another heteroradical (*35*). The chains are terminated by combination of the radicals involved (*36*, *37*, and *38*), or by disproportionations.

For many heteroradical additions the carbon–heteroatom bond formed is weak, and the adduct radical decomposes, i.e. step (*34*) is reversible (cf. chapter 12):

$$RAOl· \longrightarrow RA· + Ol \qquad (39)$$

From the synthetic point of view this may have little importance, since experimental conditions can usually be arranged so that reasonable yields of the adduct are obtained. It does have far-reaching kinetic

272

consequences however, since the decomposition of RAOl· makes it very difficult to study the addition step in isolation. If we consider a system in which the main chain-terminating step is a bimolecular reaction between two heteroradicals (36), then application of the steady-state approximation leads to the following expression for the rate of adduct formation:

$$R(\text{Adduct}) = k_{34}[\text{Ol}](R_i/2k_{36})^{\frac{1}{2}}[1/(1+k_{39}/k_{35}[\text{RAX}])] \qquad (40)$$

The rate of adduct formation depends not only on the rate of addition (34), but also on the concentration of reactant RAX, and on the ratio of the rates of decomposition (39) to abstraction (35) of the adduct-radical RAOl·.

If a second olefin is added to the reaction mixture in a competitive experiment, then three new reactions must be considered:

$$RA· + Ol' \longrightarrow RAOl'· \qquad (34')$$

$$RAOl'· + RAX \longrightarrow RAOl'X + RA· \qquad (35')$$

$$RAOl'· \longrightarrow RA· + Ol· \qquad (39')$$

The ratio of the rates of formation of the two adducts is then given by:

$$\frac{R(\text{Adduct})}{R(\text{Adduct})'} = \frac{k_{34}[\text{Ol}]}{k_{34}'[\text{Ol}']}\left[\frac{1+k_{39}'/k_{35}'[\text{RAX}]}{1+k_{39}/k_{35}[\text{RAX}]}\right] \qquad (41)$$

Thus the rate of adduct formation (or olefin consumption) is not a direct measure of the rate of addition, even in a competitive experiment.

If the experiment is carried out in a great excess of RAX, or if the abstraction step (35) is very efficient in comparison to the decomposition (39), then these rate expressions simplify to:

$$R(\text{Adduct}) = k_{34}[\text{Ol}]\left(\frac{R_i}{2k_{36}}\right)^{\frac{1}{2}} \qquad (42)$$

and

$$\frac{R(\text{Adduct})}{R(\text{Adduct})'} = \frac{k_{34}[\text{Ol}]}{k_{34}'[\text{Ol}']} \qquad (43)$$

In practice, increasing the RAX concentration diminishes the yield of adduct, so that this method is not often of much use for kinetic studies.

When the decomposition (39) is rapid compared to the abstraction (35), equation (40) simplifies to:

$$R(\text{Adduct}) = \frac{k_{34}k_{35}[\text{Ol}][\text{RAX}]}{k_{39}}\left(\frac{R_i}{2k_{36}}\right)^{\frac{1}{2}} \qquad (44)$$

Heteroradicals

Since the decomposition reaction normally has a higher activation energy than the addition, the overall activation energy $(E_{34}+E_{35}-E_{39})$ may be negative, and a reversible addition step can often be detected when the rate of adduct formation *decreases* with increasing temperature.

The rate of the addition step (*34*) for a reversible reaction can be obtained from experiments in which the pressure of RAX is varied. From equation (*40*) we obtain:

$$\frac{1}{R(\text{Adduct})} = \left(1+\frac{k_{39}}{k_{35}[\text{RAX}]}\right)\Big/ k_{34}[\text{Ol}]\left(\frac{R_i}{2k_{36}}\right)^{\frac{1}{2}} \tag{45}$$

A plot of $1/R(\text{Adduct})$ against $1/[\text{RAX}]$ gives for the intercept:

$$\text{Int.} = 1/k_{34}[\text{Ol}]\left(\frac{R_i}{2k_{36}}\right)^{\frac{1}{2}}$$

and for the gradient:

$$\text{Grad.} = \text{Int.}\,\frac{k_{39}}{k_{35}}$$

The values of k_{34} and k_{39}/k_{35} can thus be calculated. This is a tedious procedure, and accounts, in part, for the reason why so few kinetic studies have been carried out with heteroradicals.

9.3.2 *Reversibility of addition*

There are two main methods for detecting a reversible addition step. First, the kinetic method referred to above, in which the rate of adduct formation is measured at various temperatures. A decrease in the rate with increasing temperature indicates that the addition is reversible. Secondly, the addition of the radical under study to a 1,2-disubstituted olefin may be examined. If a *cis*-olefin is taken, then addition produces the adduct radical which can undergo rotation about the carbon–carbon bond (*46*):

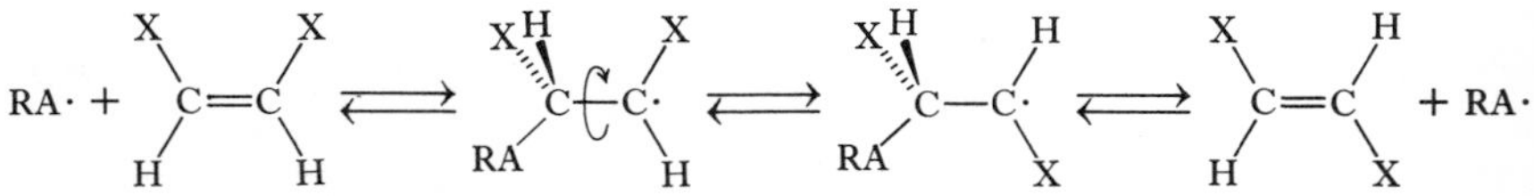

$$\tag{46}$$

Decomposition of the adduct radical can then lead to formation of either *cis*- or *trans*-olefin (cf. chapter 12). If therefore, the unconsumed olefin is found to be isomerized into a *cis/trans* mixture, this is good evidence that the addition is reversible.

274

Of the radicals formed by group IV elements, alkyl and silyl radicals do not give reversible addition under normal conditions in solution or in the gas phase. For germanium radicals this aspect of addition has yet to be investigated, but for tin radicals there is good evidence for reversibility. Kuivila and Sommer (26) have shown that the olefins *cis-* and *trans*-CHD=CH(CH$_2$)$_3$CH$_3$, *cis*-PhCH=CHD, and *cis-* and *trans*-but-2-ene are isomerized by tri-n-butyltin radicals derived from tri-n-butyltin hydride:

$$Bu_3^nSn\cdot + \quad \overset{R}{\underset{H}{\diagdown}}C=C\overset{D}{\underset{H}{\diagup}} \quad \rightleftharpoons \quad \overset{R\ H}{\underset{Bu_3^nSn}{}}C-\overset{D}{\underset{H}{C\cdot}} \quad \rightleftharpoons \quad \overset{R}{\underset{H}{\diagdown}}C=C\overset{H}{\underset{D}{\diagup}} \quad + \ Bu_3^nSn\cdot$$

For the radicals of group V elements, it is established that some nitrogen radicals add reversibly. For instance, the addition of $\cdot NF_2$ radicals to olefins is known to be reversible (27) from kinetic studies. There are also indications that $\cdot NO_2$ radicals derived from N_2O_4 or nitryl chloride, NO_2Cl, add reversibly (24). Numerous phosphorus radicals are known to undergo reversible addition, thus PhPH$\cdot$ and Bu$_2^n$P$\cdot$ (28) isomerize 1,2-disubstituted olefins, as do H$_2$P$\cdot$, Me$_2$P$\cdot$ and (CF$_3$)$_2$P$\cdot$ (29). For arsenic, antimony and bismuth radicals experimental evidence is lacking.

In group VI there is no evidence that oxygen radicals add reversibly. However, sulphur radicals are one of the most fully investigated systems and there can be little doubt that addition is reversible for virtually all sulphur radicals. Sivertz and co-workers (30) have shown that methyl-thiyl radicals isomerize *cis-* or *trans*-1,2-dideuterioethylene and but-2-ene, and that the rate of adduct formation decreases with temperature:

$$MeS\cdot + \quad \overset{D}{\underset{H}{\diagdown}}C=C\overset{D}{\underset{H}{\diagup}} \quad \rightleftharpoons \quad \overset{D\ H}{\underset{MeS}{}}C-\overset{D}{\underset{H}{C\cdot}} \quad \rightleftharpoons \quad \overset{D}{\underset{H}{\diagdown}}C=C\overset{H}{\underset{D}{\diagup}} \quad + \ MeS\cdot$$

Other sulphur radicals which are known to add reversibly include HS$\cdot$, AcS$\cdot$, PhS$\cdot$ and $\cdot SF_5$. For selenium, tellurium and polonium, experimental evidence is lacking.

It is to be expected that the strength of the carbon–heteroatom bond formed in the addition will be the main factor which influences the ease of addition. For a weak bond, the reverse process (*39*) will be favoured. In Table 9-3 some approximate values of carbon–heteroatom bond

TABLE 9-3 *Approximate carbon–heteroatom bond dissociation energies for the elements of groups IV, V and VI*

Bond	$D(\text{X–CH}_3)$ (kJ mol^{-1})	Comments
$\text{Me}_3\text{C–CH}_3$	327[a]	Not reversible
$\text{Me}_3\text{Si–CH}_3$	319[b]	Reversibility not observed
$\text{Me}_3\text{Ge–CH}_3$	(269)[c]	?
$\text{Me}_3\text{Sn–CH}_3$	(218)[c]	Reversible with $\text{Bu}_3{}^n\text{Sn·}$ etc.
$\text{Me}_3\text{Pb–CH}_3$	155[c]	?
$\text{Me}_2\text{N–CH}_3$	290[d]	Reversible with $\cdot\text{NF}_2$, $\cdot\text{NO}_2$
$\text{Me}_2\text{P–CH}_3$	(277)[e]	Reversible with $\cdot\text{PH}_2$, $\cdot\text{PMe}_2$, $\cdot\text{P(CF}_3)_2$, etc.
$\text{Me}_2\text{As–CH}_3$	(231)[e]	?
$\text{Me}_2\text{Sb–CH}_3$	(218)[e]	?
$\text{Me}_2\text{Bi–CH}_3$	(185)[e]	?
MeO–CH_3	319[a]	Not reversible
MeS–CH_3	302[a]	Reversible with $\cdot\text{SH}$, $\cdot\text{SMe}$, $\cdot\text{SF}_5$, $\cdot\text{SPh}$, etc.
$\text{MeSe–CH}_2\text{CH}_3$	(243)[e]	?

Values in brackets are average bond dissociation energies.
[a] Ref. 31, [b] Ref. 2, [c] Ref. 1, [d] Ref. 32, [e] Ref. 10.

dissociation energies for alkyl compounds of the elements of groups IV, V and VI are given. It is interesting to note that reversibility has been observed for radical addition of all the elements so far studied having $D(\text{X–CH}_3)$ less than about 300 kJ mol^{-1}. On this basis we would also expect germanium radical addition, for instance, to be reversible in solution at normal temperatures.

9.3.3 *Orientation of addition*

Addition of a heteroradical to an unsymmetrical olefin produces two isomeric one-to-one adduct radicals (*47, 48*) both of which may decompose (*49, 50*).

$$\text{RA·} + \text{CH}_2{=}\text{CHY} \rightleftharpoons \text{RACH}_2\text{CHY·} \qquad (47)\,(49)$$

$$\text{RA·} + \text{YCH}{=}\text{CH}_2 \rightleftharpoons \text{RACHYCH}_2\text{·} \qquad (48)\,(50)$$

$$\text{RACH}_2\text{CHY·} + \text{RAX} \longrightarrow \text{RACH}_2\text{CHYX} + \text{RA·} \qquad (51)$$

$$\text{RACHYCH}_2\text{·} + \text{RAX} \longrightarrow \text{RACHYCH}_2\text{X} + \text{RA·} \qquad (52)$$

Few experimental measurements of the ratio of the two one-to-one adducts have been made for heteroradicals.

Since the adduct radicals can decompose in reversible additions, the adduct ratio follows from equation (41), i.e.

$$\frac{[\text{RACH}_2\text{CHYX}]}{[\text{RACHYCH}_2\text{X}]} = \frac{k_{47}}{k_{48}}\left[\frac{1+k_{50}/k_{52}[\text{RAX}]}{1+k_{49}/k_{51}[\text{RAX}]}\right] \tag{53}$$

The adduct ratio is a function, not only of the rates of addition, but also of the starting material concentration, and the decomposition-to-abstraction ratio of the adduct radicals (k_{49}/k_{51} etc.). The ratio of the isomeric adducts will vary with temperature, starting material concentration and the type of radical source, so that direct comparison of one set of data with another must be carried out with caution.

It has been found that for terminal olefins, virtually all heteroradicals from groups IV, V and VI add predominantly or exclusively to the terminal $=\text{CH}_2$ site. In group IV exclusive addition of silanes and germanes to terminal sites has been reported (24):

$$\text{Cl}_3\text{Si}\cdot + \text{n-C}_5\text{H}_{11}\text{CH}=\text{CH}_2 \longrightarrow \text{n-C}_5\text{H}_{11}\dot{\text{C}}\text{HCH}_2\text{SiCl}_3$$

$$\text{Ph}_3\text{Si}\cdot + \text{n-C}_6\text{H}_{13}\text{CH}=\text{CH}_2 \longrightarrow \text{n-C}_6\text{H}_{13}\dot{\text{C}}\text{HCH}_2\text{SiPh}_3$$

$$\text{Et}_3\text{Ge}\cdot + \text{CH}_2=\text{CHCN} \longrightarrow \text{Et}_3\text{GeCH}_2\dot{\text{C}}\text{HCN}$$

Tin radicals also add exclusively to $=\text{CH}_2$ in terminal olefins. Kuivila, Rahman and Fish have studied the orientation of addition of trimethyltin radicals to allene and methyl allenes (33). With methyl allenes the proportion of central addition increased as the number of methyl groups increased:

	$\text{CH}_2=\text{C}=\text{CH}_2$	$\text{MeCH}=\text{C}=\text{CH}_2$	$\text{Me}_2\text{C}=\text{C}=\text{CH}_2$
Central addition	45%	87%	100%

	$\text{MeCH}=\text{C}=\text{CHMe}$	$\text{Me}_2\text{C}=\text{C}=\text{CHMe}$
Central addition	100%	100%

For the methyl-substituted allenes steric factors favour central addition, and the proportion of central addition increases as expected. If the stability of the adduct radical were important in determining the orientation of addition, a very high percentage of central addition would be expected with allene. Central addition would give the resonance-stabilized allylic radical (reaction 54) whereas no stabilization is obtained in the radical from terminal addition:

$$\text{Me}_3\text{Sn}\cdot + \text{CH}_2=\text{C}=\text{CH}_2 \longrightarrow \underset{\underset{\text{SnMe}_3}{|}}{\text{CH}_2=\text{C}-\dot{\text{C}}\text{H}_2} \longleftrightarrow \underset{\underset{\text{SnMe}_3}{|}}{\dot{\text{C}}\text{H}_2-\text{C}=\text{CH}_2} \tag{54}$$

In fact only 45% of the central adduct was observed (33).

Heteroradicals

Radicals centred on nitrogen also add preferentially at the terminal site in terminal olefins. Thus $F_2N\cdot$ was found to add mainly at $=CH_2$ in propene and isobutene (27). Similarly, $\cdot NO_2$ radicals derived from nitryl chloride add exclusively to the least substituted carbon atom (24). This is also the case for phosphorus radicals derived from phosphines, phosphorous acid, etc. (24). Fields, Haszeldine and co-workers have made a careful study of the addition of $Me_2P\cdot$, $H_2P\cdot$ and $(CF_3)_2P\cdot$ radicals to olefins (34). For most olefins the predominant adduct was that formed by addition of the radical to the least substituted site. For some fluoro-olefins both one-to-one adducts were isolated and the relative proportions are given in Table 9-4.

TABLE 9-4 *Isomer ratios from the addition of heteroradicals to unsymmetric olefins*

Olefin	Radical						
	$Me_2P\cdot$[a] $\alpha:\beta$	$H_2P\cdot$[a] $\alpha:\beta$	$(CF_3)_2P\cdot$[a] $\alpha:\beta$	$MeS\cdot$[b] $\alpha:\beta$	$CF_3CH_2S\cdot$[b] $\alpha:\beta$	$CF_3S\cdot$[b] $\alpha:\beta$	$F_5S\cdot$[c] $\alpha:\beta$
$\overset{\alpha}{CH_2}{=}\overset{\beta}{CHF}$	1:0.09	—	1:0.00	—	—	—	1:0.01
$CH_2{=}CF_2$	1:0.39	1:0.00	1:0.00	—	—	1:0.00	1:0.00
$CHF{=}CF_2$	1:1.08	1:0.19	1:0.02	1:0.33	—	1:0.02	1:0.10
$CF_2{=}CFCF_3$	—	1:0.50	—	1:0.10	1:0.43	1:1.24	—
$CF_2{=}CFCl$	—	1:0.00	—	1:0.00	—	1:0.00	—

Ratio of isomer formed by addition of the radical at carbon atom α to that formed by addition at β.
[a] Ref. 34. [b] Ref. 35. [c] Ref. 36.

The factors which influence orientation have already been discussed in chapter 8, p. 217. For the first three fluoro-olefins in Table 9-4 steric effects are probably minor, so that for trifluoroethylene for instance, the orientation is controlled by other factors, such as polar effects or resonance stabilization. The radical $Me_2P\cdot$ should be much more nucleophilic than $H_2P\cdot$ or $(CF_3)_2P\cdot$ radicals. The latter is probably strongly electrophilic, because of the presence of the powerful electron-attracting CF_3 groups. It is interesting to note that the nucleophilic $Me_2P\cdot$ radicals, in contrast to $(CF_3)_2P\cdot$ radicals, slightly prefer to add to $=CF_2$ in trifluoroethylene, and that the proportion of addition at $=CF_2$ increases as the nucleophilicity of the radical increases. Even for 1,1-difluoroethylene the $Me_2P\cdot$ radicals give an appreciable proportion of addition at $=CF_2$, whereas $H_2P\cdot$ and $(CF_3)_2P\cdot$ radicals add exclusively at the $=CH_2$ end. This strongly suggests that polar factors are one of the major influences in deciding the orientation of radical addition.

The orientation of addition of $AsCl_3$ to chlorotrifluoroethylene has been studied by Goldwhite (37). The major product was found to be $Cl_2AsCFClCF_2Cl$, in which the arsenic radical is attached to the $=CFCl$ end of the olefin. Virtually all other radicals prefer to add to the other end of this olefin (24). This led Goldwhite to suggest that the chain-carrying radical from $AsCl_3$ is, in fact, the chlorine atom, and that the product is formed as shown below:

$$Cl\cdot + CF_2{=}CFCl \longrightarrow CF_2Cl\dot{C}FCl \qquad (55)$$
$$CF_2Cl\dot{C}FCl + AsCl_3 \longrightarrow CF_2ClCFCl\dot{A}sCl_3 \qquad (56)$$
$$CF_2ClCFCl\dot{A}sCl_3 + CF_2{=}CFCl \longrightarrow CF_2ClCFClAsCl_2$$
$$+ CF_2Cl\dot{C}FCl \quad (57)$$

A similar mechanism has been suggested for the addition of PBr_3, which also shows abnormal orientation in addition to olefins such as propene (38):

$$PBr_3 + MeCH{=}CH_2 \longrightarrow CH_3CH(PBr_2)CH_2Br + CH_3CHBrCH_2PBr_2$$
$$\phantom{PBr_3 + MeCH{=}CH_2 \longrightarrow } 95\% 5\%$$

Of the radicals derived from elements in group VI, sulphur radical addition has been most extensively studied. Thiyl radicals, derived from thiols, thiol acids, thiol phosphoric acids, etc., all show preferential addition to the least substituted carbon atom of the olefin. Harris and Stacey (35) have studied the addition of $CH_3S\cdot$, $CF_3CH_2S\cdot$ and $CF_3S\cdot$ radicals, derived from the corresponding thiols, to fluoro-olefins and the adduct ratios they obtained are given in Table 9-4. Once again it is clear that for addition to trifluoroethylene, the proportion of addition at $=CF_2$ increases as the nucleophilicity of the radical increases. Similar behaviour is also shown by hexafluoropropene, where $CH_3S\cdot$ adds predominantly at $=CF_2$ while $CF_3S\cdot$ adds more readily to $=CFCF_3$. These results also imply that polar effects are important in determining orientation when steric limitations are absent.

Sulphenyl chlorides also add to olefins to give one-to-one adducts (24). The orientation is the reverse of that expected for a thiyl radical:

$$CCl_3SCl + PhCH{=}CH_2 \longrightarrow PhCH(SCCl_3)CH_2Cl$$
$$CF_3SCl + CHF{=}CF_2 \longrightarrow CF_3SCF_2CHFCl + CF_3SCHFCF_2Cl$$
$$\phantom{CF_3SCl + CHF{=}CF_2 \longrightarrow } 82\% 18\%$$

It is believed that the main adding species must be the chlorine atom.

Heteroradicals

9.3.4 Stereochemistry of addition

Study of the stereochemistry of heteroradical addition to olefins is complicated by the reversibility of the addition step, which leads to isomerization of the original olefin. It is usual to work at low temperatures and in the presence of a great excess of the adding agent to prevent this isomerization. With acyclic olefins, the addition is generally found to be non-stereospecific. For example, Skell (39) found that addition of methane thiol-D to *cis*- and *trans*-but-2-ene gave the same mixture of *threo*- and *erythro*-3-deuterio-2-methylthiobutanes. Stereospecific addition was achieved when deuterium bromide was added to the mixture at −78 °C. Thus *cis*-but-2-ene gave *threo*-3-deuterio-2-methylthiobutane, and *trans*-but-2-ene gave the *erythro*-isomer:

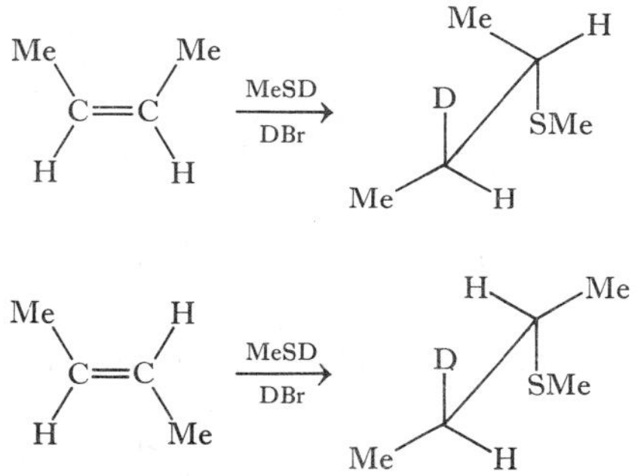

It appears that in this case, the abstraction step with DBr takes place so quickly that the intermediate radical has no time to equilibrate.

Detailed study of the stereochemistry of addition to acetylenes has not been undertaken. In many cases the major product of the addition is the *trans*-olefin (40). Dessau and Heiba report that addition of disulphides to acetylenes is less reversible than the corresponding addition to olefins. They obtained both *cis*- and *trans*-products from the addition to several alkynes, e.g.

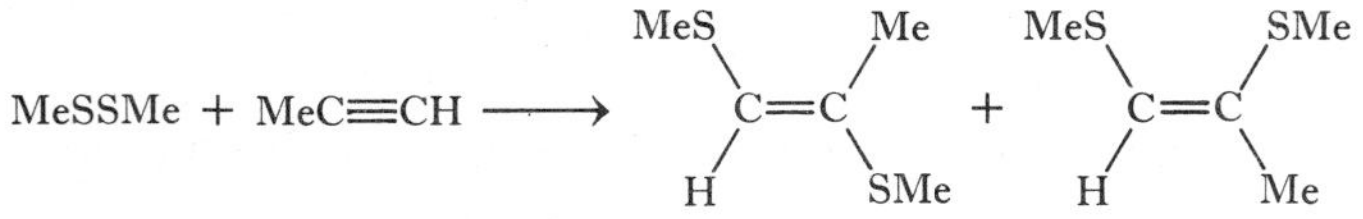

The predominant product was the *trans*-olefin in every case (40).

The addition of thiols to cyclic olefins is frequently highly stereoselective, but because of the reversibility of the addition step, the proportion of each adduct obtained depends on the concentrations of the starting materials and on the temperature. LeBel and DeBoer examined the addition of thiolacetic acid to 2-chloro-4-t-butylcyclo-hexene (41) and obtained all four possible stereoisomeric adducts (1 to 4):

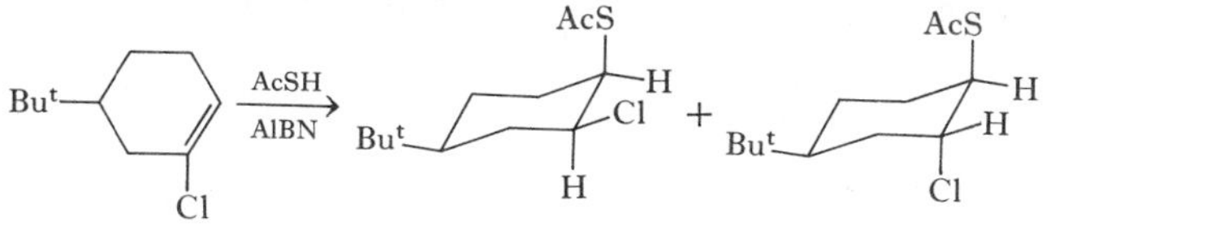

(1)   (2)

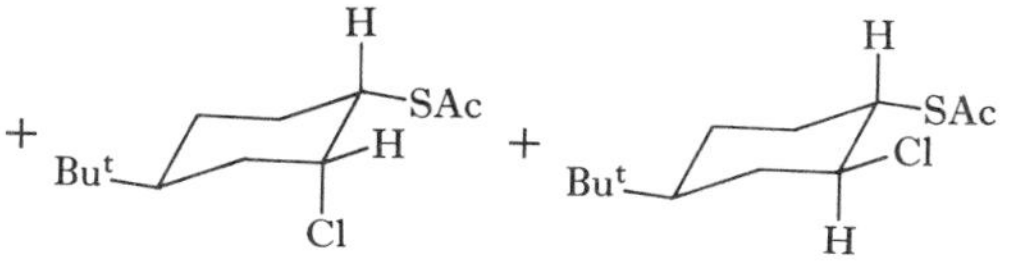

(3)   (4)

The exact proportions of the adducts depended on the reaction conditions, but overall it was found that *trans*-addition, giving *cis*-products, predominated, i.e. (1) > (2) and (3) > (4).

Bordwell *et al.* studied the addition of thiolacetic acid to 4-t-butyl-1-methylcyclohexene (42). The major product (6) was formed by *trans*-addition. The thiyl radical evidently approaches from above in an axial direction to give the intermediate radical:

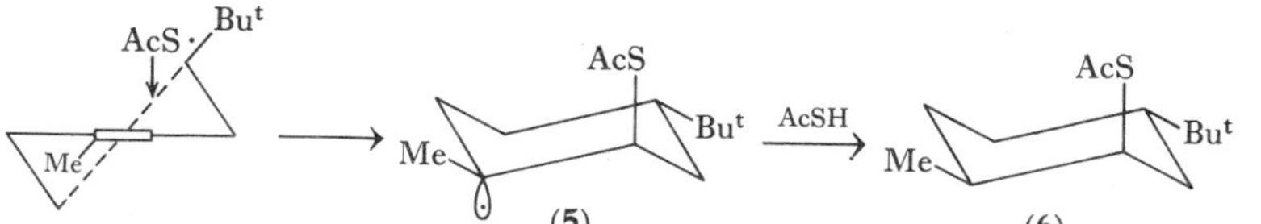

(5)   (6)

Approach of the radical from above is less hindered sterically than approach from below which leads to the minor product (7):

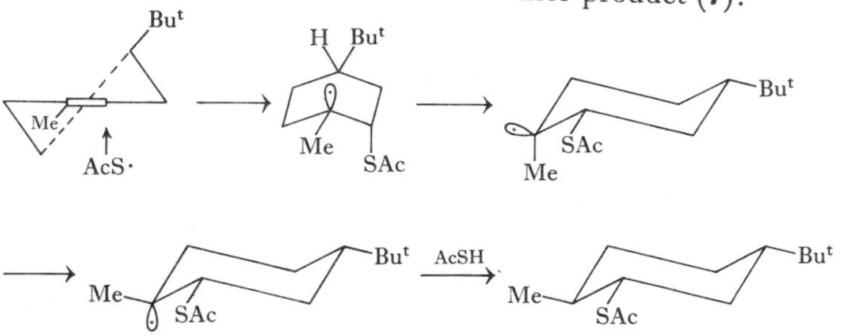

(7)

Heteroradicals

The abstraction step also appears to be axial, so that the major product is formed by a *trans*-diaxial process. The reasons for this are not clear as yet, since this route is sterically less favourable. It may be that polar interaction between the dipole at the saturated carbon atom and the developing dipole is most favourable.

A preference for *trans*-addition, leading to the *cis*-product, has been found in numerous other thiol additions, e.g. with thiolacetic acid and 1,4-dimethylcyclohexene (42), 1-methylcyclohexene and 1-methylcyclopentene (43).

Thiyl radical addition has been studied far more extensively than that of other radicals. However, there is evidence that *trans*-addition is also favoured by many other heteroradicals. Thus Selin and West (44) found that the radical addition of trichlorosilane to 1-methylcyclohexene, gave predominantly the *cis*-1-methyl-2-(trichlorosilyl)cyclohexane, corresponding to *trans*-addition (58):

$$\text{(cyclohexene with Me)} + \text{HSiCl}_3 \xrightarrow{h\nu} \text{(cyclohexane with Me and SiCl}_3) \tag{58}$$

Similarly, preferential *trans*-addition of radicals centred on nitrogen has been demonstrated for $\cdot NO_2$ derived from dinitrogen tetroxide. With 1-methylcyclohexene the product is exclusively *trans*-1-methyl-2-nitrocyclohexyl nitrite (45).

9.3.5 Structure–reactivity relationships

There have been a number of determinations of the 'relative reactivities' of series of olefins and aromatic substrates towards particular heteroradicals. When the olefin reactivity is determined at a fixed temperature by measurement of relative rate of product formation, or olefin consumption, the resulting 'relative reactivity' is a function, not only of the temperature, but also of the starting material concentration and the decomposition-to-abstraction ratio of the adduct radical, see equation (41). For a reversible addition such studies are of very limited value in aiding the understanding of the addition step, since the 'relative reactivity' is a function of so many elementary rate constants, all of which may change with olefin structure.

Sivertz and co-workers (21, 30) were the first to attempt to obtain rate data for the addition step itself, from the reversible addition of methanethiol to olefins. They obtained relative data for a limited range of olefins. Sidebottom *et al.* (36) studied the addition of sulphur chloride penta-fluoride to several olefins. The reaction involves the reversible addition of sulphur pentafluoride radicals to the olefins:

$$SF_5Cl + h\nu \longrightarrow SF_5\cdot + Cl\cdot \tag{59}$$

$$SF_5\cdot + Ol \rightleftharpoons SF_5Ol\cdot \tag{60)(61}$$

$$SF_5Ol\cdot + SF_5Cl \longrightarrow SF_5OlCl + SF_5\cdot \tag{62}$$

$$SF_5\cdot + SF_5\cdot \longrightarrow S_2F_{10} \tag{63}$$

$$Cl\cdot + Ol \longrightarrow ClOl\cdot \tag{64}$$

$$ClOl\cdot + SF_5Cl \longrightarrow ClOlCl + SF_5\cdot \tag{65}$$

The rate constant for the addition step was derived, by the use of equation (45), from a series of experiments for each olefin, in which the SF_5Cl concentration was varied. The addition rate constants, relative to the combination rate constant, are given in Table 9-5. The range of olefins studied is too limited to enable any generalizations about the behaviour of $SF_5\cdot$ radicals to be drawn. The low values of the addition activation energies suggest that $SF_5\cdot$ radicals are highly reactive, elec-trophilic species, which behave in a similar way to $CF_3\cdot$ or $CCl_3\cdot$ radicals on addition to olefins.

TABLE 9-5 *Arrhenius parameters for the addition of SF₅Cl to fluoro-olefins (36)*

$$SF_5\cdot + Ol \xrightarrow{k_{60}} SF_5Ol\cdot$$

$$SF_5\cdot + SF_5\cdot \xrightarrow{k_{63}} S_2F_{10}$$

Olefin	Adduct	$E_{60} - \frac{1}{2}E_{63}$ (kJ mol^{-1})	Log $(A_{60}/A_{63}^{\frac{1}{2}})$ (l$^{\frac{1}{2}}$ mol$^{-\frac{1}{2}}$ s$^{-\frac{1}{2}}$)
$CH_2{=}CH_2$	$SF_5CH_2CH_2Cl$	8.0	0.59
$CH_2{=}CHF$	SF_5CH_2CHFCl	8.8	0.62
$CHF{=}CF_2$	SF_5CHFCF_2Cl	14.3	1.40

Kerr, Trotman-Dickenson and associates have obtained absolute Arrhenius parameters for the addition of $NF_2\cdot$ radicals to a much more extensive series of olefins (27). The mechanism of the reaction was as shown below:

$$N_2F_4 \rightleftharpoons 2NF_2 \cdot \tag{66}$$

$$\cdot NF_2 + Ol \rightleftharpoons NF_2Ol^* \cdot \tag{67}$$

$$NF_2Ol^* \cdot + M \longrightarrow NF_2Ol \cdot + M \tag{68}$$

$$NF_2Ol \cdot \longrightarrow \cdot NF_2 + Ol \tag{69}$$

$$\cdot NF_2 + NF_2Ol \cdot \longrightarrow NF_2OlNF_2 \tag{70}$$

TABLE 9-6   *Arrhenius parameters for the addition of NF$_2$· radicals to olefins (27)*

$$\cdot NF_2 + Ol \xrightarrow{k_{67}} NF_2Ol.$$

No.	Olefin	k(rel.) (373 °K)	E_{67} (kJ mol^{-1})	Log A_{67} (l mol^{-1} s^{-1})
1.	Ethylene	1.0	65	7.6
2.	Propene	4.4	58	7.2
3.	But-1-ene	4.3	57	7.1
4.	*trans*-But-2-ene	10.8	50	6.5
5.	*cis*-But-2-ene	10.3	50	6.5
6.	Isobutene	19.6	50	6.8
7.	2-Methylbut-2-ene	33.6	42	6.0
8.	2,3-Dimethylbut-2-ene	87.5	35	5.3
	Cyclopentene	7.9	46	5.9
	Vinyl chloride	2.2	54	6.4
	Vinyl bromide	2.4	56	6.6

The rate data on the addition step were obtained from pressure variation experiments, and are given in Table 9-6. The rate of addition increases as the number of electron-repelling substituents in the olefin increases, so that ·NF$_2$ radicals are typical electrophilic species. The A-factor decreases steadily as the number of alkyl substituents increases, and the degree of crowding around the double bond increases. It is clear that the activation energy is a function of the number of alkyl substituents at the double bond, and is relatively insensitive to the exact structure; all three dimethylethylenes show the same activation energy. Kerr *et al.* derived the following relationship:

$$E_{67} = 64.3 - 5.2 \text{ (No. of alkyl substituents at double bond)} \tag{71}$$

In this respect ·NF$_2$ radicals resemble oxygen atoms, rather than alkyl radicals such as CH$_3$· or CF$_3$·. The rate of addition of ·NF$_2$ also corre-

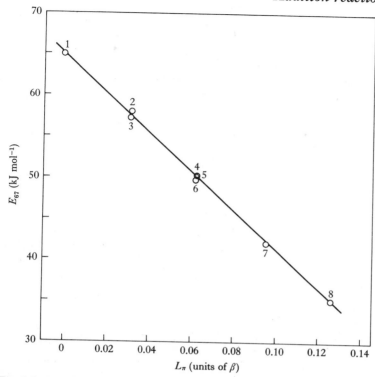

Fig. 9-1. Correlation of the activation energy of $\cdot NF_2$ addition to olefins E_{67} with bond localization energy L_π. Activation energies from Kerr *et al.* (27), bond localization energies from Sato and Cvetanović (49). For the key to the numbers see Table 9-6.

lates with other bond properties such as olefin ionization potential, suggesting that the rate-determining step is addition of the radical to the olefin π-system. The top of the potential energy pass in the addition reaction, is therefore the π-transition state for $\cdot NF_2$ radicals, oxygen atoms and cyclopropyl radicals. For $CH_3\cdot$, $CF_3\cdot$ and $CCl_3\cdot$ radicals, addition occurs at a specific carbon atom of the olefin double bond and the σ-transition state represents the top of the potential energy pass (see Fig. 8-3).

In conformity with this, Stefani *et al.* (48) have shown that the rate of $\cdot NF_2$ radical addition does not correlate with atom localization energy, but rather with bond localization energy. The correlation of the activation energy of $\cdot NF_2$ radical addition E_{67} with bond localization energy L_π is shown in Fig. 9-1. It would be of great interest to know why radicals apparently separate into two groups: those such as $\cdot NF_2$,

285

cyclopropyl radicals and $O(^3P)$ atoms which react by a π-transition state, and those such as $CH_3\cdot$, $CF_3\cdot$ and $CCl_3\cdot$ radicals which react by a σ-transition state. A possible explanation may be that radicals such as $\cdot NF_2$ and c-$C_3H_5\cdot$ containing the unpaired electron in an orbital having π- or pseudo-π-symmetry react via the π-transition state and other alkyl radicals react via the σ-transition state.

9.4 SYNTHETIC APPLICATIONS

9.4.1 *Additions to form carbon–silicon bonds*

Synthesis of organosilicon compounds can be accomplished by radical addition of silanes to olefins and acetylenes (24). Trichlorosilane was the first to be studied extensively (50 to 52), but the reaction is of wide applicability and any silane having one or more hydrogen atoms attached to silicon can be used. Initiation can be purely thermal at 400–600 °K, provided the products are stable at these temperatures. Peroxides, azo compounds, etc., when used as initiators, also give good yields of product. Ultraviolet irradiation gives very good yields in the addition to fluorinated olefins (53). With readily polymerizable olefins such as styrene and tetrafluoroethylene, telomer formation is a complication. With styrene for instance, only polymer is obtained, but for other olefins the yield of one-to-one adduct can be increased by increasing the amount of silane relative to olefin (54, 51):

$$HSiCl_3 + CF_2{=}CF_2 \longrightarrow Cl_3SiCF_2CF_2H + \text{higher telomers}$$
$$(50\%)$$

$$HSiMeCl_2 + C_2H_5CH{=}CHCH_3 \xrightarrow{\text{ROOR}} \text{n-}C_3H_7CH(CH_3)SiMeCl_2$$
$$(21\%)$$

Reaction of a silane with a chloroethylene frequently leads to the formation of a vinyl silane rather than the expected one-to-one adduct (55):

$$HSiMeCl_2 + CF_3CCl{=}CCl_2 \xrightarrow{\text{ROOR}} CF_3C(SiMeCl_2){=}CCl_2$$
$$(21\%)$$

The product is believed to arise from β-elimination of a chlorine atom from the intermediate radical, rather than from dehydrochlorination of the one-to-one adduct.

Addition of trichlorosilane to buta-1,3-diene yielded the 1,2- rather than the 1,4-adduct (52):

$$HSiCl_3 + CH_2{=}CHCH{=}CH_2 \xrightarrow{\Delta} CH_2{=}CH(CH_2)_2SiCl_3$$

Addition to hexafluorobuta-1,3-diene also gave mainly the 1,2-product (24).

Silanes also add to alkynes giving alkenylsilanes as products. The addition of trichlorosilane to 1-alkynes gives predominantly the *cis*-olefin by *trans*-addition (56):

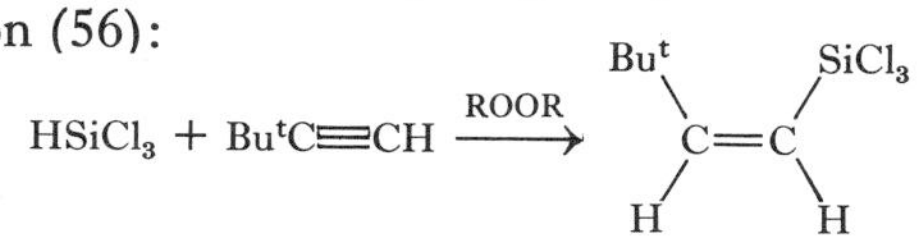

$$HSiCl_3 + Bu^tC{\equiv}CH \xrightarrow{ROOR}$$

$$\begin{matrix} Bu^t & & SiCl_3 \\ & C{=}C & \\ H & & H \end{matrix}$$

(39%)

9.4.2 *Additions to form carbon–germanium bonds*

Germanium hydrides add to olefins in a similar manner to silanes. The reaction has been studied with trichlorogermane, triphenylgermane and some trialkylgermanes (24, 25). The mechanism is less well established than with silanes, but appears to be free-radical in nature. Germanes are more reactive than the corresponding silanes, and sometimes add exothermically to olefins at room temperature in the absence of initiators (57, 58):

$$HGeCl_3 + CH_2{=}CHCN \longrightarrow NCCH_2CH_2GeCl_3$$
(53%)

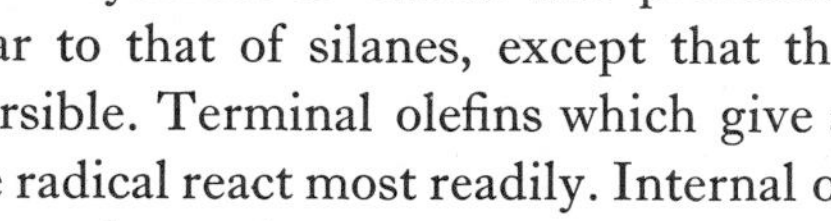

$$HGePh_3 + \text{(cyclohexene)} \xrightarrow{h\nu} \text{(cyclohexyl-GePh}_3)$$

(39%)

9.4.3 *Additions to form carbon–tin bonds*

The addition of tin hybrides to olefins also proceeds by a free-radical mechanism similar to that of silanes, except that the addition step is undoubtedly reversible. Terminal olefins which give a resonance-stabilized intermediate radical react most readily. Internal olefins do not react unless strained or conjugated:

$$HSnMe_3 + CH_2{=}C{=}CH_2 \longrightarrow$$
$$Me_3SnC(CH_3){=}CH_2 + Me_3SnCH_2CH{=}CH_2 \quad \text{Ref. (33)}$$
$$(30\%) \qquad\qquad (37\%)$$

Telomers and polymers are not formed in the reaction, which implies that the abstraction of hydrogen from the tin hydride by the intermediate radical is very fast.

With conjugated dienes both 1,2- and 1,4-addition is observed (59):

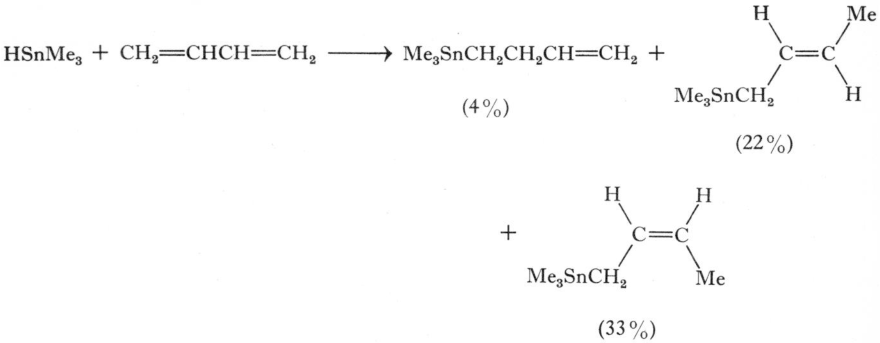

$$HSnMe_3 + CH_2{=}CHCH{=}CH_2 \longrightarrow Me_3SnCH_2CH_2CH{=}CH_2 +$$

$$
\begin{array}{c}
H \qquad Me \\
\diagdown \ \ \diagup \\
C{=}C \\
\diagup \qquad \diagdown \\
Me_3SnCH_2 \qquad H
\end{array}
$$

(4%)

(22%)

$$
+ \qquad
\begin{array}{c}
H \qquad H \\
\diagdown \ \ \diagup \\
C{=}C \\
\diagup \qquad \diagdown \\
Me_3SnCH_2 \qquad Me
\end{array}
$$

(33%)

Addition of trimethyl tin hydride to phenylacetylene gives mainly the *trans*-product (60):

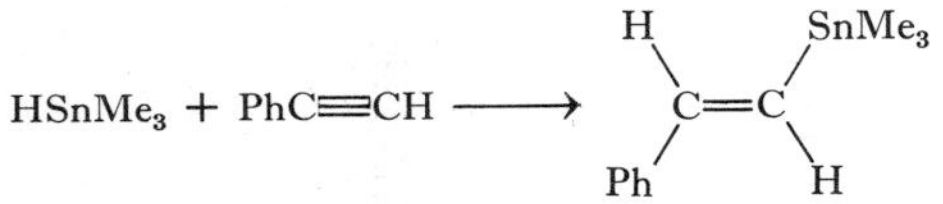

$$
HSnMe_3 + PhC{\equiv}CH \longrightarrow
\begin{array}{c}
H \qquad SnMe_3 \\
\diagdown \ \ \diagup \\
C{=}C \\
\diagup \qquad \diagdown \\
Ph \qquad H
\end{array}
$$

It is believed, however, that *trans*-addition occurs to give initially the *cis*-product, which then rearranges to the more stable *trans*-olefin.

9.4.4 *Additions to form carbon–lead bonds*

The carbon–lead bond is extremely weak, $D(\text{Pb–C}) \approx 155 \text{ kJ mol}^{-1}$, so that the addition step would be endothermic for many olefins, and addition reactions can only be expected when resonance-stabilized intermediate radicals are formed. Tributyllead hydride adds to several alkenes, including styrene and acrylonitrile (61), and also to phenylacetylene. Addition of trimethyllead hydride to ethylene, giving trimethyllead, will also occur under pressure (62).

9.4.5 *Additions to form carbon–nitrogen bonds*

Dinitrogen tetroxide adds to olefins by a free-radical process to give a vicinal dinitroalkane and a nitronitrite (63):

$$RCH{=}CH_2 + N_2O_4 \longrightarrow RCH(NO_2)CH_2NO_2 + RCH(ONO)CH_2NO_2$$

The nitronitrite is usually hydrolysed to the nitro-alcohol or oxidized to the nitronitrate. These are the products which are usually isolated, as well as the dinitroalkane. The reaction is of wide applicability and can occur with almost any olefin (64 to 66):

$$PhCH=CHPh + N_2O_4 \longrightarrow PhCH(NO_2)CH(NO_2)Ph$$
$$meso(21-25\%), (\pm)(32-39\%)$$
$$+ PhCH(OH)CH(NO_2)Ph$$
$$erythro(9-10\%), threo(13-18\%)$$

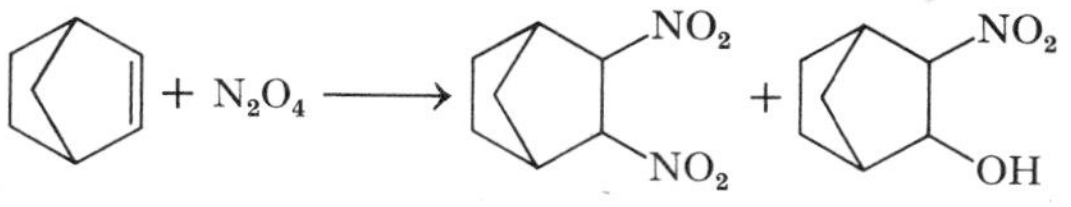

exo-cis 16–20%   trans 33–40%
trans 12–14%

Nitryl chloride will also add to olefins under free-radical conditions to give chloronitroalkanes and nitronitrites (isolated as alcohols) as the main products (67).

$$CH_2=CHCH_2Cl + NO_2Cl \longrightarrow O_2NCH_2CHClCH_2Cl$$
$$(61\%)$$

These products are usually rationalized in terms of a mechanism which involves initial attack of the $\cdot NO_2$ radical on the alkene, followed by abstraction of chlorine or ONO from the nitryl chloride by the intermediate radical. The chlorine atom can also add to the alkene, and in many cases minor amounts of the dichloroalkane have also been isolated (67).

$$Pr^iCH=CH_2 + NO_2Cl \longrightarrow Pr^iCHClCH_2NO_2 + Pr^iCHClCH_2Cl$$
$$(47\%) \qquad\qquad (26\%)$$

Other reagents which add to olefins giving carbon–nitrogen bonds include N_2O_4–halogen mixtures, nitrosyl chloride and tetrafluorohydrazine (24, 27).

9.4.6 *Additions to form carbon–phosphorus bonds*

A large number of phosphorus compounds containing P–H bonds will add to olefins by a free-radical route. Phosphines, derivatives of phosphorous, hypophosphorous and phosphinic acids, all behave in this way (29, 68, 69).

$$(CF_3)_2PH + C_2H_5CH=CH_2 \longrightarrow Bu^nP(CF_3)_2$$
$$(99\%)$$

$$Me_2PH + CFCl=CF_2 \longrightarrow CFHClCF_2PMe_2$$
$$(61\%)$$

$$HP(O)(OEt)_2 + CH_3CO_2CH=CH_2 \xrightarrow{\text{ROOR}} CH_3CO_2CH_2CH_2PO(OEt)_2$$
$$(14\%)$$

$$H_2P(O)ONa + CH_2=CHBu^n \xrightarrow{\text{ROOR}} \text{n-}C_6H_{13}PH(O)ONa$$
$$(100\%)$$

$$C_6H_5PH(O)OEt + \text{n-}C_6H_{13}CH=CH_2 \longrightarrow \text{n-}C_8H_{17}(C_6H_5)P(O)OEt$$
$$(37\%)$$

Dialkyl phosphites have been studied most extensively: they add to both terminal and internal olefins. At high olefin concentrations telomers are formed, especially with easily polymerizable olefins such as $CF_2=CF_2$.

Phosphorus trihalides also add to olefins in a free-radical process (70, 71):

$$PCl_3 + CH_2=CHC_6H_{13}\text{-n} \longrightarrow \text{n-}C_6H_{13}CHClCH_2PCl_2$$
$$(33\%)$$

$$PCl_3 + Me_2C=CH_2 \longrightarrow PCl_2CH_2CClMe_2 + CH_2ClC(Me_2)PCl_2$$
$$(36\%) \qquad\qquad\qquad (31\%)$$

Oxidized products are usually obtained unless oxygen is carefully excluded from the reaction (72):

$$PCl_3 + CH_3CH=CHCH_3 \xrightarrow{O_2} CH_3CHClCH(CH_3)POCl_2$$

A one-to-one adduct is also obtained in the addition of PBr_3 to olefins (38).

9.4.7 *Additions to form carbon–arsenic bonds*

Radical addition reactions to form C–As, C–Sb and C–Bi bonds have not been exploited to any great extent in organic synthesis. Cullen *et al.* (73) have reported the photolytic addition of several arsenic compounds, including $AsCl_3$, $MeAsCl_2$ and $Me_2AsAsMe_2$, to hexafluorobut-2-yne. Goldwhite (37) has also reported the addition of $AsCl_3$ to chlorotrifluoroethylene in what is believed to be a radical process:

$$AsCl_3 + CFCl=CF_2 \longrightarrow Cl_2AsCFClCF_2Cl$$

The one-to-one adduct was probably formed by initial addition of a chlorine atom. The process is clearly of potential synthetic utility.

9.4.8 *Additions to form carbon–oxygen bonds*

Relatively few oxygen-containing compounds add to olefins to give carbon–oxygen bonds. For alcohols and acids, initiation by the usual methods leads to formation of a radical centred on carbon. A C–H bond in the molecule is usually broken, presumably because of the greater strength of the O–H bond:

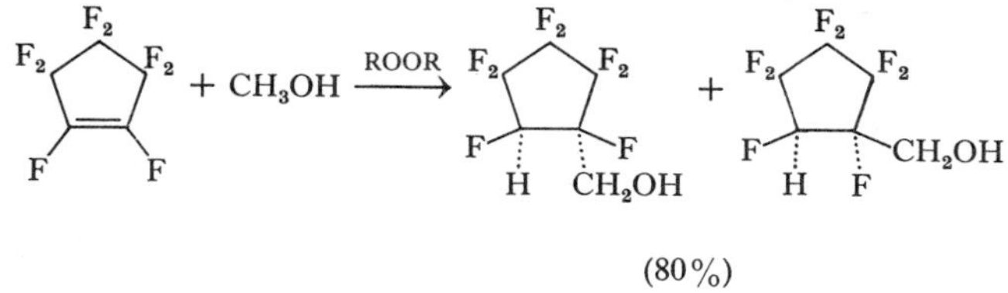

$$\text{(olefin)} + CH_3OH \xrightarrow{ROOR} \text{(product 1)} + \text{(product 2)} \qquad \text{Ref. (74)}$$

(80%)

A small yield of the one-to-one adduct can be obtained from the addition of t-butylhypochlorite to olefins, but the main products are usually those of allylic chlorination (75):

$$\underset{Me}{\overset{Me}{\diagdown}}C\!=\!CH_2 + Bu^tOCl \xrightarrow[40°C]{h\nu} CH_2ClCMe\!=\!CH_2 + Bu^tOCH_2CClMe_2$$

$$\qquad\qquad (83\%) \qquad\qquad (17\%)$$

9.4.9 *Additions to form carbon–sulphur bonds*

The most widely used thiyl radical sources are thiols, of which a great variety, including aliphatic, aromatic and heterocyclic, give good yields of the one-to-one adducts. The addition of the thiyl radical is reversible so that high temperatures are disadvantageous, and preparations are usually carried out below 400 °K. It is also usual to work with excess thiol over olefin, since this reduces the extent of telomerization and helps to reduce the amount of fragmentation of the adduct radical, see equation (40):

$$C_2H_5SH + CH_2\!=\!CHOBu^n \xrightarrow{RN_2R} C_2H_5SCH_2CH_2OBu^n \quad \text{Ref. (76)}$$

$$HSCH_2CO_2H + Me_3SiCH\!=\!CH_2 \longrightarrow Me_3SiCH_2CH_2SCH_2CO_2H$$

$$\qquad\qquad\qquad\qquad\qquad (25\%) \qquad \text{Ref. (77)}$$

Addition of *p*-thiocresol to norbornene gave only *exo*-norbornyl *p*-tolylsulphide, and none of the rearranged products characteristic of ionic additions to this olefin (78):

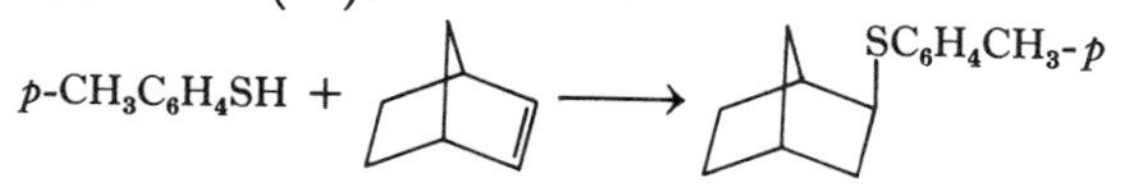

$$p\text{-}CH_3C_6H_4SH + \text{(norbornene)} \longrightarrow \text{(norbornyl } SC_6H_4CH_3\text{-}p)$$

Heteroradicals

Addition to norborna-2,5-diene gave two one-to-one adducts indicating the presence of two intermediate radicals (79):

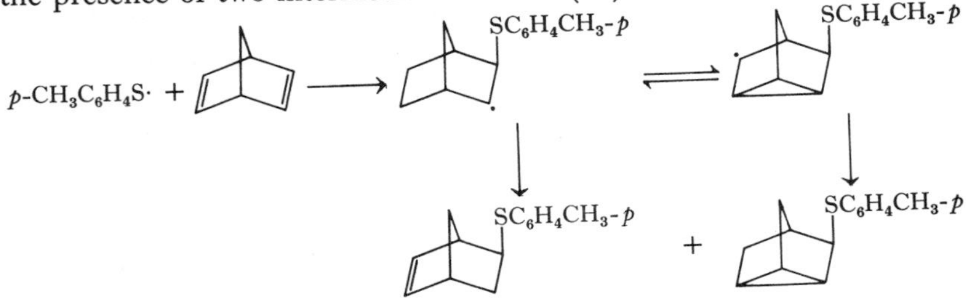

(36%)   (55%)

Addition of thiol acids has also been reported for a wide variety of olefins (80, 81):

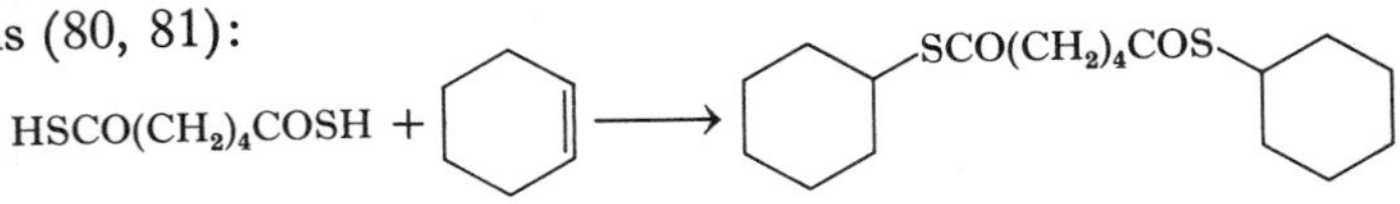

(56%)

$$(EtO)_2P(S)SH + CH_2{=}CHC_6H_{13}\text{-}n \xrightarrow{ROOR} (EtO)_2P(S)SC_8H_{17}\text{-}n$$
$$(75\%)$$

Many other sulphur compounds have been utilized in essentially the same way. Thus, bisulphite ion, sulphonyl and sulphuryl halides, disulphides and sulphur chloride pentafluoride are all known to undergo radical addition:

$$NaHSO_3 + n\text{-}C_5H_{11}CF{=}CF_2 \xrightarrow{ROOR} n\text{-}C_5H_{11}CHFCF_2SO_3Na$$
$$(73\%) \quad \text{Ref. (82)}$$

$$PhSO_2Br + CH_2{=}CHBr \longrightarrow PhSO_2CH_2CHBr_2 \quad \text{Ref. (24)}$$
$$(67\%)$$

$$SF_5Cl + CH_2{=}CHCH{=}CH_2 \longrightarrow SF_5CH_2CHClCH{=}CH_2 \quad \text{Ref. (83)}$$
$$(37\%)$$

Sulphur compounds also add readily to acetylenes, and in many cases the yield of one-to-one adduct is greater than for the corresponding olefin, as the addition step is less reversible for acetylenes (40).

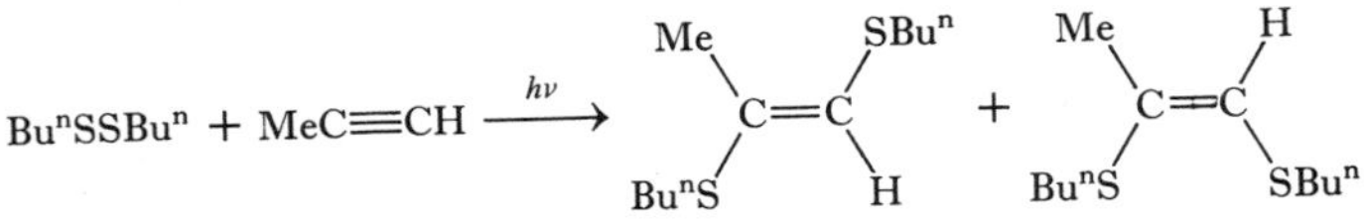

(64%)   (16%)

9.5 *ABSTRACTION REACTIONS*

9.5.1 *Hydrogen abstraction*

The abstraction of a hydrogen atom from an organic compound involves breaking a C–H bond and forming a hydrogen–heteroatom bond:

$$RA\cdot + R'H \longrightarrow RAH + R'\cdot \qquad (72)$$

Some approximate hydrogen–heteroradical bond dissociation energies are shown in Table 9-7. For radicals centred on the first members of

TABLE 9-7 *Approximate hydrogen–heteroatom bond dissociation energies for the elements of groups IV, V and VI*

Bond[a]	D(H–A) (kJ mol^{-1})	Bond[b]	D(H–A) (kJ mol^{-1})	Bond[b]	D(H–A) (kJ mol^{-1})
H_3C—H	437				
Me_3C—H	382	H_2N—H	437	HO—H	495
Me_3Si—H[d]	340	H_2P—H	357	HS—H	386
$Bu_3{}^nGe$—H[c]	340	H_2As—H	(281)	HSe—H	(306)
Me_3Sn—H	273	H_2Sb—H	—	HTe—H	(269)
Me_3Pb—H	(206)	H_2Bi—H	(197)		

Approximate halogen–heteroatom bond dissociation energies for the elements of group IV

Bond[a,d]	X=Cl	Br	I
H_3C—X	353	294	235
Me_3Si—X	382	340	302
$Bu_3{}^nGe$—X[c]	424	(277)	264
Me_3Sn—X	374	332	269

[a] ref. (1).   [b] ref. (1 and 15).   [c] ref (84).   [d] ref. (2).
Values in brackets are *mean* bond dissociation energies.

each group, i.e. carbon, nitrogen and oxygen radicals, the abstraction step is exothermic or thermoneutral, and consequently occurs with great facility. For the next member of each group, i.e. silicon, phosphorus and sulphur, the hydrogen abstraction step is endothermic and so does not occur easily, although it may occur on raising the reaction temperature. Abstraction from compounds having exceptionally weak C–H bonds, such as benzylic or tertiary hydrogen bonds, is also to be expected. Hydrogen abstraction is not, however, characteristic of these radicals. For heteroradicals of the elements further down each group, the hydrogen abstraction step is even more endothermic and hence this reaction is seldom encountered.

In group IV hydrogen abstraction is known to occur when silyl or germyl radicals are generated in organic solvents. Thus trimethylsilyl radicals react with toluene, which has specially weak α-carbon–hydrogen bonds, to produce benzyltrimethylsilane (1):

$$Me_3Si\cdot + PhCH_3 \longrightarrow Me_3SiH + PhCH_2\cdot \qquad (73)$$

$$Me_3Si\cdot + PhCH_2\cdot \longrightarrow PhCH_2SiMe_3 \qquad (74)$$

Dessy, Kitching and Chivers (85) produced triphenylgermyl radicals in glyme, as a solvent, by electrolysis of Ph_3GeCl and postulated hydrogen abstraction from the solvent to account for the production of Ph_3GeH. No systematic studies of these reactions have been undertaken.

In group V, nitrogen radicals readily abstract hydrogen, but the only radical for which comprehensive kinetic studies have been carried out is $\cdot NF_2$ (86). In the presence of an alkane the rate-determining reactions are:

$$N_2F_4 \rightleftharpoons 2NF_2\cdot \qquad (75)$$

$$\cdot NF_2 + RH \longrightarrow HNF_2 + R\cdot \qquad (76)$$

$$R\cdot + \cdot NF_2 \longrightarrow RNF_2 \qquad (77)$$

Trotman-Dickenson and co-workers have determined the rate of the abstraction step (76) by two methods: consumption of the alkane (86) and formation of the product difluoroamines (87). The latter method was believed to give the more accurate results and these are shown in Table 9-8. Difluoroamino radicals are clearly more selective and less reactive than alkyl radicals, and comparable in reactivity to bromine atoms. The first column in Table 9-8 shows the relative rates of abstraction of primary, secondary and tertiary hydrogen atoms, which are in the expected order primary < secondary < tertiary. Comparison of these results with the corresponding data for halogen atoms in Tables 7-7 and 7-8 shows that the $\cdot NF_2$ radical bears a close resemblance to Br$\cdot$ in reactivity.

The activation energies of the abstraction steps correlate well with the enthalpy of the reaction, as predicted by the Evans–Polanyi equation (86, 87):

$$E = \alpha\Delta H^0 + C$$

This implies that steric and polar effects are of negligible importance, and that the reaction rate is controlled by the strength of the bond being

294

TABLE 9-8   *Abstraction of hydrogen from alkanes by* $\cdot NF_2$ *radicals* (87)

$$\cdot NF_2 + RH \longrightarrow HNF_2 + R\cdot \qquad (76)$$

Alkane	k_{76} (100 °C) per H (relative)	E_{76} (kJ mol^{-1})	Log A_{76} (l mol^{-1} s^{-1})
CH₃CH₂CH₃*	1.0	109	8.8
(CH₃)₄C	5.0	112	10.2
CH₃CH₂CH₃*	7.6	94	7.4
CH₃CH₂CH₂CH₃*	19.1	103	9.3
c-C₅H₁₀	30.2	92	8.4
(CH₃)₃CH*	692	86	8.0

* Indicates the hydrogen abstracted.

broken. The value of α found by Cadman *et al.* was $\alpha(NF_2\cdot) = 0.90$ which can be compared with the values for halogen atoms $\alpha(I\cdot) = 0.91$, $\alpha(Br\cdot) = 0.86$ and those of alkyl radicals: $\alpha(CH_3\cdot) = 0.49$, $\alpha(CF_3\cdot) = 0.53$. The higher value of $\alpha(NF_2\cdot)$ is just what would be expected for the abstraction reactions which have higher activation energies than the corresponding reactions of $CH_3\cdot$ or $CF_3\cdot$ radicals. The high α-value also suggests that the transition state in $\cdot NF_2$ radical abstractions may resemble the products rather than the reactants.

The same research group has also measured the rates of hydrogen abstraction from a series of aldehydes by $\cdot NF_2$ radicals:

$$RCHO + \cdot NF_2 \longrightarrow R\dot{C}O + HNF_2 \qquad (78)$$

The rate of abstraction of the weakly bound aldehydic hydrogen was found to be considerably faster than the rate of abstraction from even a tertiary alkane (88). The activation energies and A-factors were found to be nearly constant from one aldehyde to another, which suggests that the strength of the RC(O)–H bond varies little from one aldehyde to another.

Hydrogen abstraction reactions by radicals centred on phosphorus, arsenic, antimony and bismuth, appear to be very rare (89, 15). It has been suggested that diphenylphosphine and biphenyl, which are the major products from the decomposition of triphenylphosphine in benzene, arise from hydrogen abstraction by diphenylphosphinyl radicals (90).

$$Ph_3P + h\nu \longrightarrow Ph_2P\cdot + Ph\cdot \qquad (79)$$

$$Ph_2P\cdot + C_6H_6 \longrightarrow Ph_2PH + Ph\cdot \qquad (80)$$

$$Ph\cdot + Ph\cdot \longrightarrow PhPh \qquad (81)$$

This appears most unlikely in view of the endothermicity of the reaction. There are, however, very few examples of hydrogen abstractions by radicals from this group, and this is not surprising in view of the weak P–H, As–H, etc., bonds. Conversely, hydrogen abstraction by carbon radicals from compounds containing P–H, As–H, etc., bonds is very facile, and this is one reason for the success of synthetic reactions which involve addition of R_2PH compounds to olefins, see p. 289.

Of the radicals centred on elements of group VI, oxygen radicals readily abstract hydrogen from alkyl groups in organic molecules. The evidence suggests that oxygen radicals bear many similarities to the halogen atoms in the abstraction reaction. For both HO· (91, 92) and ClO· (93) the ease of abstraction of hydrogen from C–H bonds is tertiary > secondary > primary. *Tert*-butoxy radicals have been studied extensively in solution, particularly by Walling and co-workers (94). It was found that the rate of abstraction from substituted toluenes correlates with the σ^+ constants of the substituents (94, 95), which again suggests the resemblance to the halogen atoms.

Sulphur, selenium and tellurium radicals show little tendency to abstract hydrogen, although examples of this kind of reaction by thiyl radicals are known for certain situations. The pyrolysis of thiols in the toluene-carrier technique leads to abstraction of the weakly bound benzylic hydrogens by thiyl radicals (96):

$$HS\cdot + CH_3Ph \longrightarrow H_2S + \cdot CH_2Ph \qquad (82)$$

Other types of reaction are favoured for most systems containing radicals from this group.

9.5.2 *Halogen abstraction*

The abstraction of a halogen atom from an organic halide by a hetero-radical, involves breaking a carbon–halogen bond and forming a halogen–heteroatom bond:

$$RA\cdot + R'X \longrightarrow RAX + R'\cdot \qquad (83)$$

For the elements in group IV the halogen–heteroatom bond is stronger than the halogen–carbon bond, so that the abstraction reaction (83) is exothermic, see Table 9-7. The reduction of an alkyl halide can, therefore, be brought about by a free-radical reaction of trialkylmetal hydrides of this group, both steps in the reaction being exothermic:

$$R_3M\cdot + R'X \longrightarrow R_3MX + R'\cdot \qquad (84)$$

$$R'\cdot + R_3MH \longrightarrow R_3M\cdot + R'H \qquad (85)$$

The halogen–heteroatom bond dissociation energies are not known for the elements of groups V and VI. It is probable they are also greater than the corresponding halogen–carbon bond dissociation energies, though halogen abstraction is not well known by heteroradicals from these groups. In this section we shall, therefore, be concerned mainly with group IV heteroradicals.

Haszeldine and Young (97) observed the formation of $CCl_3SiCF_2CH_2F$ in the reaction of trichlorosilane with CCl_3SiCF_2CHFCl, and attributed its formation to the chlorine abstraction reaction:

$$CCl_3Si\cdot + CCl_3SiCF_2CHFCl \longrightarrow Cl_4Si + CCl_3SiCF_2CHF\cdot \quad (86)$$

$$CCl_3SiCF_2CHF\cdot + Cl_3SiH \longrightarrow Cl_3Si\cdot + CCl_3SiCF_2CH_2F \quad (87)$$

Since this observation, numerous examples of this kind of reduction have been observed. Nagai and co-workers (98) found that the ease of abstraction of halogens by trialkyl- and triaryl-silanes was Br > Cl > F. They observed that the reduction occurred more easily for polyhalogenated groups, i.e. $CCl_3 > CHCl_2 > CH_2Cl$. Alkyl iodides are reduced only very slowly by silicon hydrides. Davidson (2) has suggested that this may be because the alkyl radicals formed in the halogen abstraction step (88) prefer to react with molecular iodine formed in the system, rather than continue the chain as in step (85).

$$R_3SiI + h\nu \longrightarrow R_3Si\cdot + I\cdot$$

$$R_3Si\cdot + R'I \longrightarrow R_3SiI + R'\cdot \quad (88)$$

$$I\cdot + I\cdot + M \longrightarrow I_2 + M \quad (89)$$

$$R'\cdot + I_2 \longrightarrow R'I + I\cdot \quad (90)$$

Kerr *et al.* (99) have studied the gas-phase reaction of trichlorosilane with a series of alkyl chlorides and bromides. Arrhenius parameters were obtained by a competitive technique relative to methyl chloride, see Table 9-9. The order of reactivity is clearly tertiary > secondary > primary for both the alkyl chlorides and bromides. The A-factors do not change much from one chloride to another, so that the rate is determined principally by the activation energy. It was found (99) that the activation energies approximately obey the Evans–Polanyi equation: $E = \alpha\Delta H^0 + C$. For the alkyl chlorides the value of α was found to be $\alpha(SiCl_3\cdot) = 0.65$ and for the bromides $\alpha(SiCl_3\cdot) = 0.42$. Similar approximate Evans–Polanyi relationships were found for the reactions of alkyl radicals with alkyl halides (p. 251). They imply that the controlling factor in the

TABLE 9-9 *Abstraction of halogen from alkyl halides by $Cl_3Si\cdot$ radicals* (99)

$$Cl_3Si\cdot + CH_3Cl \xrightarrow{k_0} Cl_4Si + CH_3\cdot$$

$$Cl_3Si\cdot + RX \xrightarrow{k} Cl_3SiX + R\cdot$$

Alkyl halide	k/k_0 275 °C	E_0-E (kJ mol^{-1})	A/A_0
MeCl	1.0	0.0	1.0
EtCl	2.6	4.7	0.94
Pr^{n}Cl	3.5	4.6	1.26
Bu^{n}Cl	4.2	5.3	1.32
Me$_2$CHCH$_2$Cl	3.8	5.5	1.14
Me$_3$CCH$_2$Cl	3.5	6.1	0.93
Pr^{i}Cl	9.2	7.9	1.65
Bu^{s}Cl	11.2	8.4	1.82
Bu^{t}Cl	30.6	13.6	1.60
MeBr	76.3	22.7	0.27
EtBr	117.5	26.8	0.34
Pr^{i}Br	420	29.6	0.53
Bu^{t}Br	925	33.3	0.64

abstraction reaction is the strength of the carbon–halogen bond being broken, and that steric, polar and resonance effects are of minor importance.

Trialkylgermanium hybrides also appear to reduce alkyl halides in essentially the same manner, although the reaction has received relatively little study. Jackson (1) attributes the formation of triethylbromogermane in the reaction of bis(triethylgermyl)mercury with bromobenzene to the halogen abstraction reaction:

$$Et_3Ge\cdot + PhBr \longrightarrow Et_3GeBr + Ph\cdot \qquad (91)$$

The reaction of trihalogermanes with alkyl iodides has also been noted (100). The trihalogermanes reacted in ether solution, by an ionic pathway:

$$HGeX_3 + RI \xrightarrow[-HI]{} RGeX_3 + RGeX_nI_{3-n}$$

Some of the expected radical products, i.e. CHCl$_3$ and GeX$_3$Br, were obtained with bromo- and iodotrichloromethane but the nature of the minor products made it clear that part, if not all, of this reaction also proceeded by an ionic route. Carlsson, Ingold and Bray (101) have carried out mechanistic studies of the reduction of CCl_4 and CH_3I with tri-n-butylgermanium hydride.

$$CCl_3 \cdot + Bu_3^{n}GeH \longrightarrow CHCl_3 + Bu_3^{n}Ge \cdot \qquad (92)$$

$$Bu_3^{n}Ge \cdot + CCl_4 \longrightarrow Bu_3^{n}GeCl + CCl_3 \cdot \qquad (93)$$

The abstraction of hydrogen from the germanium hydride by the trichloromethyl radical was found to be the rate-determining step, so that no kinetic data on the behaviour of germyl radicals was obtained.

The reduction of alkyl halides by organotin hydrides is a fast radical process of wide synthetic applicability (see chapter 10, p. 352). Kuivila and co-workers have examined the mechanism, scope and limitations of this reaction (102). Reduction can be successfully carried out with trialkyltin hydrides and triaryltin hydrides. Polyhalides are reduced stepwise since the reduction becomes progressively more difficult so that good yields of partly reduced products can be obtained, e.g.:

$$PhCCl_3 + Bu_3^{n}SnH \longrightarrow PhCHCl_2 + Bu_3^{n}SnCl$$
$$(88\%)$$

The order of reactivity of alkyl halides in these reactions is found to be: $RI > RBr > RCl > RF$ (102, 7). The iodides are more than a million times as reactive as the fluorides.

The abstraction of hydrogen from the organotin hydride by the intermediate alkyl radical is extremely fast, and increases in the order $Ph_3SnH > Bu_3^{n}SnH > Me_3SnH$. The rapidity of this step is illustrated by the reaction of triphenyltin hydride with 1,1,1-triphenylethyl chloride (103). The intermediate 2,2,2-triphenylethyl radical normally rearranges so rapidly to the 1,2,2-triphenylethyl radical, that only rearranged products are obtained (cf. chapter 13, p. 501).

$$Ph_3Sn \cdot + Ph_3CCH_2Cl \longrightarrow Ph_3SnCl + Ph_3CCH_2 \cdot \qquad (94)$$

$$Ph_3CCH_2 \cdot \longrightarrow Ph_2\dot{C}CH_2Ph \qquad (95)$$

$$Ph_3CCH_2 \cdot + Ph_3SnH \longrightarrow Ph_3CCH_3 + Ph_3Sn \cdot \qquad (96)$$

$$Ph_2\dot{C}CH_2Ph + Ph_3SnH \longrightarrow Ph_2CHCH_2Ph + Ph_3Sn \cdot \qquad (97)$$

However, in the triphenyltin hydride reduction, Kaplan obtained a mixture of unrearranged 1,1,1-triphenylethane, and the rearranged product 1,1,2-triphenylethane. The proportion of the unrearranged product increased at higher tin hydride concentrations. The rate of the abstraction reaction (96) is clearly comparable to that of the fast uni-molecular rearrangement (95).

Trialkyl- and triaryltin hydride reductions are of obvious synthetic utility (104–7), cf. reaction (98) and pp. 353–4.

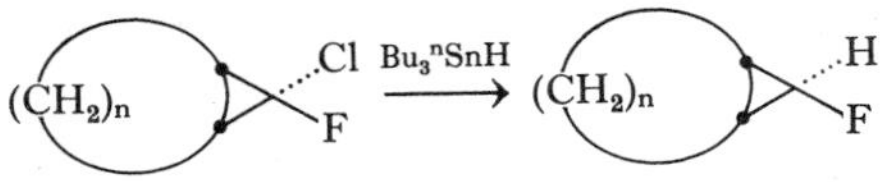

$$\tag{98}$$

Acid chlorides react with tin hydrides to give either the aldehyde or the ester depending on the nature of the acyl group (107). Benzoyl chloride gave a good yield of benzaldehyde:

$$PhCOCl + Bu_3{}^nSn\cdot \longrightarrow PhCO\cdot + Bu_3{}^nSnCl$$

$$PhCO\cdot + Bu_3{}^nSnH \longrightarrow PhCHO + Bu_3{}^nSn\cdot$$
$$(65\%)$$

Under comparable conditions acetyl chloride gives mostly the ester:

$$CH_3COCl + Bu_3{}^nSn\cdot \longrightarrow CH_3CO\cdot + Bu_3{}^nSnCl$$

$$CH_3CO\cdot + Bu_3{}^nSnH \longrightarrow CH_3CHO + Bu_3{}^nSn\cdot$$

$$CH_3CO\cdot + CH_3CHO \longrightarrow CH_3COO\dot{C}HCH_3$$

$$CH_3COO\dot{C}HCH_3 + Bu_3{}^nSnH \longrightarrow CH_3COOCH_2CH_3 + Bu_3{}^nSn\cdot$$
$$(95\%)$$

GENERAL REFERENCES

R. A. Jackson, *Adv. in Free-Radical Chem.* 1969, **3**, 231.
F. W. Stacey and J. F. Harris, *Org. Reactions*, 1963, **13**, 150.

SPECIFIC REFERENCES

1. R. A. Jackson, *Adv. Free-Radical Chem.* 1969, **3**, 231; N. J. Friswell and B. G. Gowenlock, *Adv. Free-Radical Chem.* 1965, **1**, 39.
2. I. M. T. Davidson, *Quart. Rev.* 1971, **25**, 111.
3. G. Fritz, *Z. Naturforsch.* 1952, **7b**, 207.
4. M. A. Nay, G. N. C. Woodall, O. P. Strausz and H. E. Gunning, *J. Amer. Chem. Soc.* 1965, **87**, 179.
5. C. R. Warner, R. J. Strunk and H. G. Kuivila, *J. Org. Chem.* 1966, **31**, 338.
6. J. C. J. Thynne, *J. Organometallic Chem.* 1969, **17**, 155.
7. D. J. Carlsson and K. U. Ingold, *J. Amer. Chem. Soc.* 1968, **90**, 7047.
8. J. D. Salzman and E. J. Bair, *J. Chem. Phys.* 1964, **41**, 3654.

9. B. G. Gowenlock and D. R. Snelling, *Free Radicals in Inorganic Chemistry*, 'Advances in Chemistry' Series no. 26, American Chemical Society 1962, p. 150.
10. B. G. Gowenlock, P. P. Jones and D. R. Snelling, *Canad. J. Chem.* 1963, **41**, 1911.
11. L. M. Brown and B. deB. Darwent, *J. Chem. Phys.* 1965, **42**, 2158.
12. P. Cadman, Y. Inel, A. F. Trotman-Dickenson and A. J. White, *J. Chem. Soc.* A, 1971, 1353, 2859.
13. J. P. Simons, *J. Chem. Soc.* A, 1965, 5406.
14. N. Basco and R. G. W. Norrish, *Proc. Roy. Soc.* A, 1965, **283**, 291.
15. N. J. Friswell and B. G. Gowenlock, *Adv. Free-Radical Chem.* 1967, **2**, 1.
16. M. L. Kaufman and C. E. Griffin, *Tetrahedron Letters*, 1965, 769.
17. J. E. Breen and G. P. Glass, *J. Chem. Phys.* 1970, **52**, 1082.
18. P. Gray, R. Shaw and J. C. J. Thynne, *Progr. Reaction Kinetics*, 1967, **4**, 63.
19. S. W. Benson and J. Bass, *J. Chem. Phys.* 1958, **29**, 546.
20. J. Heicklen and H. S. Johnston, *J. Amer. Chem. Soc.* 1962, **84**, 4030, 4394.
21. D. M. Graham, R. L. Mieville, R. H. Pallen and C. Siveitz, *Canad. J. Chem.* 1964, **42**, 2250.
22. A. Jones, S. Yamashita and F. P. Lossing, *Canad. J. Chem.* 1968, **46**, 833; R. P. Steer and A. R. Knight, *J. Phys. Chem.* 1968, **72**, 2145.
23. H. E. Radford, *J. Chem. Phys.* 1964, **40**, 2732.
24. F. W. Stacey and J. F. Harris, *Org. Reactions*, 1963, **13**, 150.
25. G. Sosnovsky, *Free Radical Reactions in Preparative Organic Chemistry*, Macmillan, New York, 1964.
26. H. G. Kuivila and R. Sommer, *J. Amer. Chem. Soc.* 1967, **89**, 5616.
27. A. J. Dijkstra, J. A. Kerr and A. F. Trotman-Dickenson, *J. Chem. Soc.* A, 1966, 582; *ibid.* 1967, 105, 864.
28. J. Pellon, *J. Amer. Chem. Soc.* 1961, **83**, 195.
29. R. Fields, R. N. Haszeldine and J. Kirman, *J. Chem. Soc.* C, 1970, 197.
30. C. Sivertz, *J. Phys. Chem.* 1959, **63**, 34; D. M. Graham, R. L. Mieville and C. Sivertz, *Canad. J. Chem.* 1964, **42**, 2239.
31. J. G. Calvert and J. N. Pitts, *Photochemistry*, Wiley, New York, 1967.
32. J. A. Kerr, *Chem. Rev.* 1966, **66**, 493.
33. H. G. Kuivila, W. Rahman and R. Fish, *J. Amer. Chem. Soc.* 1965, **87**, 2835.
34. R. Fields, H. Goldwhite, R. N. Haszeldine and J. Kirman, *J. Chem. Soc.* C, 1966, 2800; R. Fields, R. N. Haszeldine and J. Kirman, *ibid.* 1970, 197; R. Fields, R. N. Haszeldine and N. F. Wood, *ibid.* 1970, 774, 1370.
35. J. F. Harris and F. W. Stacey, *J. Amer. Chem. Soc.* 1961, **83**, 840.
36. H. W. Sidebottom, J. M. Tedder and J. C. Walton, *Trans. Faraday Soc.* 1969, **65**, 2103; *ibid.* 1970, **66**, 2038; *Chem. Comm.* 1970, 253.
37. H. Goldwhite, *Inorg. Nuclear Chem. Letters*, 1966, **2**, 5.
38. B. Fontal and H. Goldwhite, *J. Org. Chem.* 1966, **31**, 3804.
39. P. S. Skell and R. G. Allen, *J. Amer. Chem. Soc.* 1960, **82**, 1511.
40. R. M. Dessau and E. I. Heiba, *J. Org. Chem.* 1967, **32**, 3837.
41. N. A. LeBel and A. DeBoer, *J. Amer. Chem. Soc.* 1967, **89**, 2784.
42. F. G. Bordwell, P. S. Landis and G. S. Whitney, *J. Org. Chem.* 1965, **30**, 3764.

43. F. G. Bordwell and W. A. Hewett, *J. Amer. Chem. Soc.* 1957, **79**, 3493.
44. T. G. Selin and R. West, *J. Amer. Chem. Soc.* 1962, **84**, 1860.
45. J. C. D. Brand and I. D. R. Stevens, *J. Chem. Soc.* 1958, 629.
46. P. Cadman, G. M. Tilsey and A. F. Trotman-Dickenson, *Chem. Comm.* 1970, 1721.
47. P. T. Frangopol and K. U. Ingold, *J. Organometallic Chem.* 1970, **25**, C9.
48. A. P. Stefani, Lan-Yuh Yang Chuang and H. E. Todd, *J. Amer. Chem. Soc.* 1970, **92**, 4168.
49. S. Sato and R. J. Cvetanović, *J. Amer. Chem. Soc.* 1959, **81**, 3223.
50. L. H. Sommer, E. W. Pietrusa and F. C. Whitmore, *J. Amer. Chem. Soc.* 1947, **69**, 188.
51. C. A. Burkhard and R. H. Krieble, *J. Amer. Chem. Soc.* 1947, **69**, 2687.
52. A. J. Barry, L. DePree, J. W. Gilkey and D. E. Hook, *J. Amer. Chem. Soc.* 1947, **69**, 2916.
53. A. M. Geyer and R. N. Haszeldine, *J. Chem. Soc.* 1957, 1038 and 3925; A. M. Geyer, R. N. Haszeldine, K. Leedham and R. J. Marklow, *ibid.* 1957, 4472.
54. R. N. Haszeldine and R. J. Marklow, *J. Chem. Soc.* A, 1956, 962.
55. E. T. McBee, C. W. Roberts and G. W. R. Puerckhauer, *J. Amer. Chem. Soc.* 1957, **79**, 2329.
56. R. A. Benkeser, M. L. Burrous, L. E. Nelson and J. V. Swisher, *J. Amer. Chem. Soc.* 1961, **83**, 4385.
57. A. D. Petrov, V. F. Mironov and N. D. Dzhurinskaya, *Doklady Akad. Nauk. SSSR*, 1959, **128**, 302.
58. R. Fuchs and H. Gilman, *J. Org. Chem.* 1957, **22**, 1009.
59. R. H. Fish, H. G. Kuivila and J. J. Tyminski, *J. Amer. Chem. Soc.* 1967, **89**, 5861.
60. R. F. Fulton, *Diss. Abs.* 1962, **22**, 3397.
61. W. P. Newmann and K. Kuhlein, *Angew. Chem. Internat. Edn*, 1965, **4**, 784.
62. W. E. Becker and S. F. Cook, *J. Amer. Chem. Soc.* 1960, **82**, 6264.
63. N. Levy and C. W. Scaife, *J. Chem. Soc.* 1946, 1093, 1096, 1100; N. Levy, C. W. Scaife and A. E. Wilder-Smith, *J. Chem. Soc.* 1948, 52; N. Levy and J. D. Rose, *Quart. Rev.* 1947, **1**, 358.
64. J. J. Gardikes, A. H. Pagano and H. Shechter, *J. Amer. Chem. Soc.* 1959, **81**, 5420; *Chem. and Ind.* 1958, 632.
65. T. E. Stevens, *J. Amer. Chem. Soc.* 1959, **81**, 3593.
66. J. J. Gardikes, *Diss. Abs.* 1960, **21**, 757.
67. J. Ville and G. duPont, *Bull. Soc. chim. France*, 1956, 804.
68. R. L. McConnell and H. W. Coover, *J. Amer. Chem. Soc.* 1957, **79**, 1961.
69. A. N. Pudovik and I. V. Konovalova, *J. Gen. Chem. USSR (Engl. transl.)*, 1960, **30**, 2328.
70. M. S. Kharasch, E. V. Jensen and W. H. Urry, *J. Amer. Chem. Soc.* 1945, **67**, 1864.
71. J. R. Little and P. F. Hartmann, *J. Amer. Chem. Soc.* 1966, **88**, 96.
72. Yu. M. Zinov'ev and L. Z. Soborovskii, *J. Gen. Chem. USSR (Eng. Transl.)*, 1957, **26**, 611.
73. W. R. Cullen, D. S. Dawson and G. E. Styan, *J. Organometallic Chem.* 1965, **3**, 406.

74. N. F. Stockel and M. T. Beachem, *J. Org. Chem.* 1967, **32**, 1658.
75. C. Walling and W. Thaler, *J. Amer. Chem. Soc.* 1961, **83**, 3877.
76. E. N. Prilezhaeva and M. F. Shostakovskii, *Bull. Acad. Sci. USSR. Div. Chem. Sci. (Eng. Transl.)*, 1958, 1071.
77. C. A. Burkhard, *J. Amer. Chem. Soc.* 1950, **72**, 1078.
78. S. J. Cristol and G. D. Brindell, *J. Amer. Chem. Soc.* 1954, **76**, 5699.
79. S. J. Cristol, G. D. Brindell and J. A. Reeder, *J. Amer. Chem. Soc.* 1958, **80**, 635.
80. J. I. Cunneen, *Rubber Chem. Technol.* 1953, **26**, 370.
81. W. E. Bacon and W. M. LeSuer, *J. Amer. Chem. Soc.* 1954, **76**, 670.
82. R. J. Koshar, P. W. Trott and J. D. LaZerte, *J. Amer. Chem. Soc.* 1953, **75**, 4595.
83. J. R. Case, N. H. Ray and H. L. Roberts, *J. Chem. Soc.* 1961, 2066.
84. D. J. Carlsson, K. U. Ingold and L. C. Bray, *Internat. J. Chem. Kinetics*, 1969, **1**, 315.
85. R. E. Dessy, W. Kitching and T. Chivers, *J. Amer. Chem. Soc.* 1966, **88**, 453.
86. J. Grechowiak, J. A. Kerr and A. F. Trotman-Dickenson, *Chem. Comm.* 1965, 109.
87. P. Cadman, C. Dodwell, A. J. White and A. F. Trotman-Dickenson, *J. Chem. Soc. A*, 1971, 2967.
88. P. Cadman, C. Dodwell, A. F. Trotman-Dickenson and A. J. White, *J. Chem. Soc. A*, 1970, 2371 and 3189.
89. J. I. G. Cadogan, *Adv. Free-Radical Chem.* 1967, **2**, 203.
90. M. L. Kaufman and C. E. Griffin, *Tetrahedron Letters*, 1965, 769.
91. N. R. Greiner, *J. Chem. Phys.* 1967, **46**, 3389.
92. D. G. Horne and R. G. W. Norrish, *Nature*, 1967, **215**, 1373.
93. R. Shaw, *J. Chem. Soc. B*, 1968, 513.
94. C. Walling, IUPAC Symposium, *Free Radicals in Solution*, Michigan, 1966, Butterworths, London, p. 69.
95. H. Sakurai and A. Hosomi, *J. Amer. Chem. Soc.* 1967, **89**, 458; R. D. Gillon and J. R. Hawles, *Canad. J. Chem.* 1968, **46**, 2752.
96. A. H. Sehon and B. de B. Darwent, *J. Amer. Chem. Soc.* 1954, **76**, 4806.
97. R. N. Haszeldine and J. C. Young, *J. Chem. Soc.* 1960, 4503.
98. Y. Nagai, K. Yamazaki, N. Kobori and M. Kosugi, *Nippon Kag-aka Zasshi*, 1967, **88**, 793; Y. Nagai, K. Yamazahi and I. Shiojima, *J. Organometallic Chem.* 1967, **9**, 25.
99. J. A. Kerr, B. J. Smith, A. F. Trotman-Dickenson and J. C. Young, *J. Chem. Soc. A*, 1968, 510; P. Cadman, G. M. Tilsley and A. F. Trotman-Dickenson, *ibid.* 1969, 1370.
100. T. K. Gar, F. M. Berliner and V. F. Mironar, *J. Gen. Chem. USSR (Engl. Transl.)* 1971, **41**, 343.
101. D. J. Carlsson, K. U. Ingold and L. C. Bray, *Internat. J. Chem. Kinetics*, 1969, **1**, 315.
102. H. G. Kuivila, L. W. Menapace and C. R. Warner, *J. Amer. Chem. Soc.* 1962, **84**, 3584; H. G. Kuivila and L. W. Menapace, *J. Org. Chem.* 1963, **28**, 2165; L. N. Menapace and H. G. Kuivila, *J. Amer. Chem. Soc.* 1964, **86**, 3047.

Heteroradicals

103. L. Kaplan, *J. Amer. Chem. Soc.* 1966, **88**, 4531.
104. D. B. Denny, R. M. Hoyte and P. T. MacGregor, *Chem. Comm.* 1967, 1241.
105. S. J. Cristol and R. V. Barbour, *J. Amer. Chem. Soc.* 1968, **90**, 2832.
106. T. Ando, F. Namigata, H. Yamanaka and W. Fanasaka, *J. Amer. Chem. Soc.* 1967, **89**, 5719.
107. H. G. Kuivila, *Adv. Organometallic Chem.* 1964, **1**, 47; H. G. Kuivila and E. J. Walsh, *J. Amer. Chem. Soc.* 1966, **88**, 571, 576.

10

Radical oxidations and reductions

10.1 *INTRODUCTION*

This chapter will be devoted to the consideration of oxidations and reductions which involve the intervention of radicals as intermediates, including reactions in which the radicals have little discrete existence. Discussion will cover autoxidation as well as metal-catalysed oxidations and reductions and the somewhat related electrochemical reactions.

The distinctive feature about homolytic oxidation and reduction is that the oxidation or reduction step involves the transfer of a single electron or its equivalent, e.g. a chlorine atom (reactions *1* and *2*) rather than two electrons (reaction *3*):

$$ArO^- + Ce^{4+} \longrightarrow ArO\cdot + Ce^{3+} \qquad (1)$$

$$Ar\cdot + CuCl_2 \longrightarrow ArCl + CuCl \qquad (2)$$

$$R_2\overset{\overset{\displaystyle H}{|}}{C}\text{---}O\text{---}CrO_3H + B: \longrightarrow R_2CO + HCrO_3^- + BH^+ \qquad (3)$$

Two distinct steps can be discerned for metal-catalysed oxidations and reductions and analogous electrochemical reactions:

(1) oxidation or reduction of the organic compound to give a radical,
(2) subsequent reaction of the radical, which may either involve its further oxidation or reduction, or alternatively radical–radical reactions may occur.

These steps can be illustrated by reference to the oxidation of phenols:

$$ArO^- + Ce^{4+} \longrightarrow ArO\cdot + Ce^{3+} \qquad (4)$$

$$ArO\cdot + Ce^{4+} \longrightarrow ArO^+ + Ce^{3+} \qquad (5)$$

$$ArO\cdot + ArO\cdot \longrightarrow Dimers \qquad (6)$$

In this chapter we shall be concerned with the mechanism of the oxidation (or reduction) steps as well as with a number of important types of oxidation and reduction including phenol oxidation.

10.2 MECHANISM OF OXIDATION OF ORGANIC COMPOUNDS TO RADICALS

There are two distinct mechanisms for oxidations involving metal ions depending on whether or not the electron transfer is accompanied by the transfer of a ligand. The mechanism of the oxidation is said to be outer-sphere if direct electron transfer occurs (reaction 7), and inner-sphere if the electron transfer involves the transfer of a ligand (reaction 8):

$$Fe(o\text{-phen})_3{}^{3+} + Fe(o\text{-phen})_3{}^{2+} \rightleftharpoons Fe(o\text{-phen})_3{}^{2+} + Fe(o\text{-phen})_3{}^{3+} \quad (7)$$

$$Cr^{2+}aq + (NH_3)_5Co^{III}Cl^{2+} \rightleftharpoons CrCl^{2+}aq + Co^{2+}aq + 5NH_3 \quad (8)$$

The same two mechanisms are also operative in metal-catalysed oxidations of organic compounds (1).

10.2.1 *Outer-sphere or direct electron-transfer oxidation*

The essential feature of this mechanism is that there is no transfer of ligands between the reactants. Kinetically this means that the rate of reaction is faster than the rate of substitution of the ligands. This condition is fulfilled when the ligands attached to the metal ion are difficult to displace as in iron(III) tris-*o*-phenanthroline. Oxidation of cyclohexanone with this oxidant proceeds by an outer-sphere mechanism. Evidence for this comes from the quantitative reduction of the iron(III) complex to iron(II) tris-*o*-phenanthroline showing the absence of any ligand transfer. Further, the rate of oxidation is greater than the rate of hydrolysis of the iron(III) complex and is also independent of the concentration of free ligand. This is consistent with a transition state in which there is direct electron transfer from the oxidant to cyclohexanone, which is almost certainly in its enolic form (reaction 9) (2).

$$Fe^{III} (o\text{-phen})_3^{3+} + \overset{OH}{\underset{}{\bigcirc}} + H_2O \longrightarrow Fe^{II} (o\text{-phen})_3^{2+} + \overset{O}{\underset{}{\bigcirc}} + H_3O^+ \quad (9)$$

A number of other oxidants possessing exchange-inert ligands, such as hexachloroiridate(IV) and ferricyanide, also generally react by a non-bonded mechanism.

The outer-sphere mechanism is also operative when the organic substrate is unlikely to be capable of displacing the ligands of the oxidant even though these may be liable to exchange with solvent or free ligand. This mechanism is operative in the oxidation of alkylaromatics, e.g. toluene, with cobalt(III) acetate or manganese(III) acetate (3–5). The stability of cobalt(III) and manganese(III) compounds is largely dependent on the donor ability of attached ligands. The donor ability of hydrocarbons is very weak and hence it is unreasonable to postulate the displacement of acetate by hydrocarbon. Oxidation must consequently proceed by an outer-sphere process involving direct electron transfer giving the toluene radical cation which subsequently loses a proton (reactions *10* and *11*).

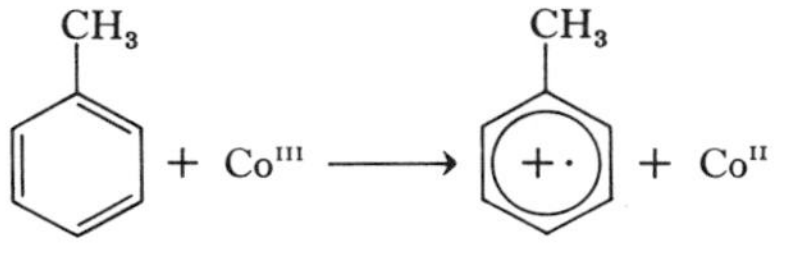

$$(10)$$

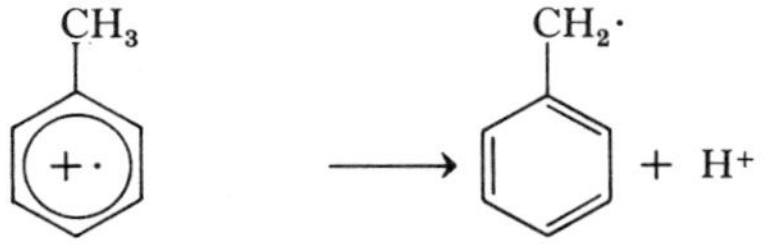

$$(11)$$

10.2.2   *Inner-sphere or bonded electron-transfer oxidation*

Inner-sphere or bonded electron transfer of organic compounds generally proceeds through a complex between the oxidant and the substrate, with electron transfer taking place within the complex. Evidence for this comes from kinetic studies rather than from the product composition, since in the vast majority of oxidations of organic compounds the final product is a free organic compound.

An example of an oxidation which proceeds by the bonded mechanism is provided by the oxidation of pinacol by managanese(III) pyrophosphate (6). The rate of oxidation is given by the expression:

$$\frac{-d[Mn^{III}]}{dt} = \frac{K[Mn^{III}][pinacol]}{[pinacol] + [pyrophosphate]}$$

This implies that a complex is formed between the pinacol and the manganese, and that this complex may then either revert to the reactants or undergo electron transfer to the products (reaction *12* on p. 308.)

The situation is different when a strong complex is formed between the oxidant and the substrate as occurs in the oxidation of carboxylic acids by cobalt(III) salts (7). The formation of such a complex effectively

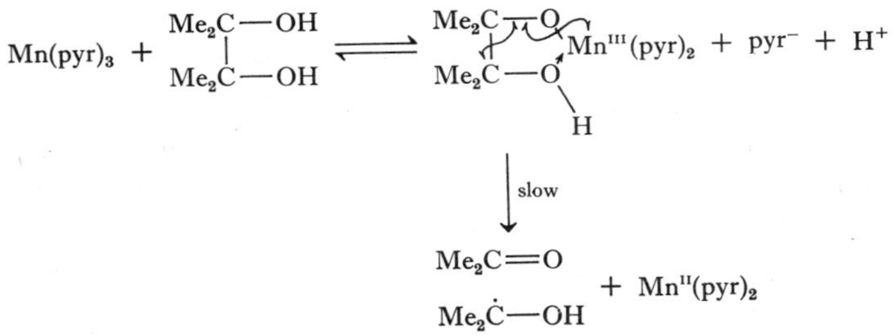

$$Mn(pyr)_3 + \quad \begin{matrix} Me_2C{-}OH \\ | \\ Me_2C{-}OH \end{matrix} \quad \rightleftharpoons \quad \begin{matrix} Me_2C{-}O \\ \diagdown \\ Me_2C{-}O \\ \diagup \\ H \end{matrix} Mn^{III}(pyr)_2 \; + \; pyr^- \; + \; H^+$$

$$\Big\downarrow \text{slow}$$

$$\begin{matrix} Me_2C{=}O \\ \\ Me_2\dot{C}{-}OH \end{matrix} \quad + \quad Mn^{II}(pyr)_2$$

(12)

removes a significant portion of the substrate from the solution. The complex then breaks down in a slow step (reactions *13* and *14*):

$$RCO_2H + Co^{3+}aq \rightleftharpoons RCO_2Co^{2+}aq + H^+aq \tag{13}$$

$$RCO_2Co^{2+}aq \longrightarrow R\cdot + CO_2 + Co^{2+}aq \tag{14}$$

This type of behaviour is revealed by the rate of reaction being no longer first-order in the substrate:

$$\frac{-d[Co^{III}]}{dt} = \frac{kK[Co^{III}][RCO_2H]}{K[RCO_2H]+1}$$

This behaviour is referred to as 'Michaelis–Menten' kinetics and is frequently observed in oxidations by cobalt(III), cerium(IV), manganese(III), and vanadium(V). Oxidations by these ions often proceed by a bonded mechanism, particularly when the substrates contain donor groups such as hydroxyl or carboxylate which readily form complexes with metal ions.

10.3 *MECHANISM OF THE OXIDATION OF RADICALS*

In the same way that there are two mechanisms whereby organic substrates undergo oxidation to radicals, there are two routes available for the oxidation of radicals. Kochi has designated these as electron-transfer oxidation (reaction *15*) and ligand-transfer oxidation (reaction *16*) (8):

$$R\cdot + Cu^{2+} \longrightarrow R^+ + Cu^+ \tag{15}$$

$$R\cdot + CuCl_2 \longrightarrow RCl + CuCl \tag{16}$$

The first of these corresponds to outer-sphere oxidation of organic compounds and the latter to inner-sphere or bonded oxidation.

10.3.1   *Electron-transfer oxidation*

This essentially involves the oxidation of a radical to a carbonium ion which undergoes subsequent reaction to give stable products. The product composition can be used to provide evidence for the intermediacy of carbonium ions. Thus, oxidation of cyclobutyl radicals by lead(IV) gives the same mixture of acetates as that obtained in solvolyses proceeding via the carbonium ion (reactions *17* and *18*) (9):

$$\square^{\bullet} \xrightarrow[\text{AcOH}]{\text{Pb}^{\text{IV}}} \square\text{--OAc} + \triangleright\text{--CH}_2\text{OAc} + \text{CH}_2\text{=CH(CH}_2)_2\text{OAc} \quad (17)$$
$$\qquad\qquad\qquad (52\%) \qquad\qquad (44\%) \qquad\qquad (4\%)$$

$$\square\text{--OTos} \xrightarrow{\text{OAc}^-} \square\text{--OAc} + \triangleright\text{--CH}_2\text{OAc} + \text{CH}_2\text{=CH(CH}_2)_2\text{OAc} \quad (18)$$
$$\qquad\qquad\qquad (47\%) \qquad\qquad (48\%) \qquad\qquad (5\%)$$

Cyclobutyl radicals unlike cyclobutyl carbonium ions do not undergo rearrangement.

The ease of oxidation of a radical to a carbonium ion will depend on the oxidation potential of both the radical and the oxidant. The oxidation potential of the radical will be controlled largely by the ability of the groups attached to the tervalent carbon to stabilize the developing positive charge. In accord with this, *p*-methoxybenzyl radicals are much more readily oxidized by lead(IV) than benzyl radicals, owing to the stabilizing effect of the methoxy group on the *p*-methoxybenzyl cation (10). Any stabilizing effect the methoxy group had on the radical would be minimal. Similarly, the radical $\dot{\text{C}}\text{Me}_2\text{OH}$ is more easily oxidized than the isomeric radical $\dot{\text{C}}\text{H}_2\text{CHMe(OH)}$ because of the ability of the hydroxyl group to stabilize the positive charge in the resultant carbonium ion. These results suggest that there is considerable development of carbonium-ion character in the transition state in electron-transfer oxidation of radicals.

Iron(III) is a much more efficient oxidant than titanium(IV) in the oxidation of $\dot{\text{C}}\text{Me}_2\text{OH}$ radicals (11). This is consistent with iron(III) having a higher oxidation potential than titanium(IV).

Somewhat surprisingly copper(II) is a very much more efficient oxidant than cobalt(III) or lead(IV) even though it has a lower oxidation potential. Thus copper(II) acetate oxidizes 2-phenylethyl radicals very efficiently, whereas lead(IV) acetate is very much less effective (10). Not only is copper(II) a more efficient oxidant but a different ratio of products is obtained with it than with lead(IV) (reactions *19* and *20*):

$$Ph\overset{\cdot}{C}HEt \xrightarrow[\text{AcOH}]{Pb(IV)} PhCH(OAc)Et + PhCH{=}CHMe \qquad (19)$$
$$(96\%) \qquad\qquad (4\%)$$

$$Ph\overset{\cdot}{C}HEt \xrightarrow[\text{AcOH}]{Cu(II)} PhCH(OAc)Et + PhCH{=}CHMe \qquad (20)$$
$$(80\%) \qquad\qquad (20\%)$$

Kochi has shown that, in general, oxidation with copper(II) acetate results in the formation of more alkene than does oxidation with other oxidants. This difference in behaviour is particularly noticeable in the case of cyclobutyl radicals which are oxidized by copper(II) acetate to give almost exclusively cyclobutene (9). In view of the increase of strain, loss of a β-hydrogen from the radical or a β-proton from the cation would not be expected nor is it observed in other reactions of cyclobutyl radicals or cations. To account for this, Kochi proposed that in the transition state leading to the formation of cyclobutene, there is some specific bonding between the β-hydrogen of the radical and the copper (II) species, and also that some of the driving force for elimination is derived from synergic bonding between the incipient alkene and the copper(II) species (reaction *21*).

$$(21)$$

Some specific involvement of copper in the oxidation process as envisaged above for the reaction of cyclobutyl radicals, which is not possible with other metal oxidants, could explain why copper(II) is generally a better oxidant of alkyl radicals than cobalt(III) or lead(IV). The only instances in which lead(IV) is as efficient an oxidant as copper(II) are when oxidative elimination cannot occur, as in the case of benzyl and neopentyl radicals. Oxidation then must be a pure electron-transfer process.

The carbonium-ion pathway in the oxidation of radicals with copper(II) is much more important when the oxidations are carried out in solvents capable of complexing with the copper such as acetonitrile or pyridine, or in the presence of complexing agents such as *o*-phenanthroline or α,α-bipyridyl. The complexes decrease the amount of interaction between the incipient double bond and copper(I), and at the same time increase the oxidation potential of the copper(II) because of the increased stability of the complexed copper(I) compound. Thus, oxidation of cyclobutyl radicals in acetonitrile or in the presence of *o*-phenanthroline or α,α-bipyridyl gives typical carbonium-ion product (9, 12):

$$\square^{\bullet} \xrightarrow[\text{CH}_3\text{CN/AcOH}]{\text{Cu(OAc)}_2} \square^{\text{--OAc}} + \triangleright\!\!-\!\text{CH}_2\text{OAc} + \text{CH}_2\!\!=\!\!\text{CH(CH}_2)_2\text{OAc} \quad (22)$$

(52%)   (44%)   (4%)

Similarly oxidation of octenyl radicals with copper(II) acetate in presence of pyridine gives the same mixture of octenyl acetates as that obtained using lead(IV) but different from that with copper(II) acetate in benzene (reactions *23–25*) (10):

$$C_5H_{11}\diagdown\diagup^{\bullet} \xrightarrow{\text{Cu(OAc)}_2} C_5H_{11}\diagdown\!\!\underset{\text{OAc}}{\diagup}\!\!\diagup + C_5H_{11}\diagdown\diagup\!\!\underset{\text{OAc}}{\diagdown} \quad (23)$$

(85%)   (15%)

$$C_5H_{11}\diagdown\diagup^{\bullet} \xrightarrow{\underset{C_5H_5N}{\text{Cu(OAc)}_2}} C_5H_{11}\diagdown\!\!\underset{\text{OAc}}{\diagup}\!\!\diagup + C_5H_{11}\diagdown\diagup\!\!\underset{\text{OAc}}{\diagdown} \quad (24)$$

(49%)   (51%)

$$C_5H_{11}\diagdown\diagup^{\bullet} \xrightarrow{\text{Pb}^{\text{IV}}} C_5H_{11}\diagdown\!\!\underset{\text{OAc}}{\diagup}\!\!\diagup + C_5H_{11}\diagdown\diagup\!\!\underset{\text{OAc}}{\diagdown} \quad (25)$$

(55%)   (45%)

Direct evidence for copper alkyl intermediates has been obtained by studying the e.s.r. spectra of 2-butyl radicals produced by reaction of 2-butyl lithium with copper(II) salts (13). The resultant radicals are not only much more stable than simple 2-butyl radicals but their hyperfine splitting is altered in such a way as to indicate that the unpaired electron in the π-orbital is coupled equally with two equivalent protons. This observation has led to the structure (1) being put forward for the complexed radical.

$$\text{Cu}^{2+} \longleftarrow \quad \begin{array}{c} \text{H}\diagdown\overset{+}{\underset{|}{C}}\diagup\text{R} \\ \cdot \\ \text{H}\diagup\underset{C}{}\diagdown\text{R}' \end{array}$$

(1)

Oxidations and reductions

There are few reported examples of electron-transfer reductions of organic radicals. Electron-transfer reductions will occur most readily when the resultant carbanion is relatively stable (reaction 26) (14):

$$\begin{array}{cc}
\underset{|}{CH_2\dot{C}HCO_2Me} & \underset{|}{CH_2\bar{C}HCO_2Me} \\
CH_2CH_2CO_2Me & CH_2CH_2CO_2Me
\end{array} \xrightarrow{\ Cu^+\ } \qquad (26)$$

10.3.2 Ligand-transfer oxidation

Ligand-transfer oxidations involve the direct transfer of groups from the metal salt to the radical. They are most commonly encountered in the case of the transfer of a halogen atom, but other groups such as azide, thiocyanate and xanthate can also act as ligands in ligand-transfer reactions (15, 15a).

The chief distinction between a ligand-transfer and an electron-transfer process is that the former but not the latter is relatively unaffected by electronic effects in the radical. This has been demonstrated in ligand-transfer reactions of substituted benzylic radicals with a chloro-lead(IV) complex which proceed readily in all cases to give the benzylic chlorides (reaction 27) (16):

$$X-C_6H_4CH_2\cdot + Pb^{IV}Cl \longrightarrow X-C_6H_4CH_2Cl + Pb^{III} \qquad (27)$$

There seems no reason to suppose that copper(II) halides do not behave similarly. This points to only a small degree of electron-transfer in the transition state for ligand transfer. The transition state can be represented as:

$$[R\cdot\ Cl-Cu^{II}-Cl \longleftrightarrow R-Cl\ Cu^{I}Cl]$$

The transition state is much more akin to that of an atom transfer in a typical free-radical reaction. This is borne out by the fact that the butenyl radical gives the same proportions of crotyl chloride and α-methylallyl chloride on reaction with both cupric chloride and t-butyl hypochlorite (equation 28). The proportions of these two products are quite different from those which would have been obtained in an electron-transfer reaction proceeding through a butenyl carbonium ion (8):

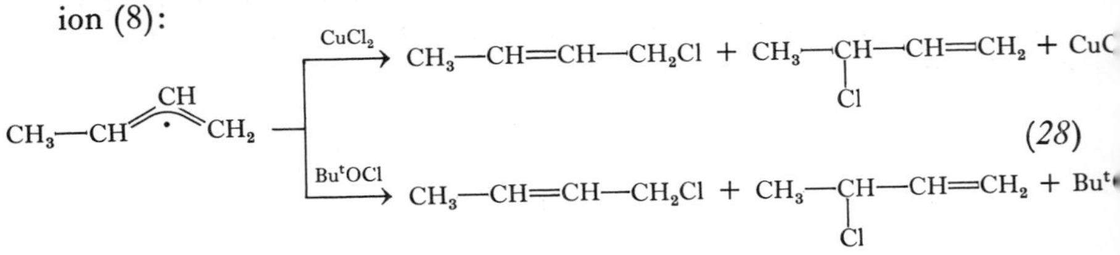

$$\qquad (28)$$

312

Much the same picture emerges for ligand-transfer reductions, e.g. the chromium(II) reductions of alkyl halides, in which a halogen atom is transferred from the alkyl halide to the chromium in the rate-determining step (reaction *29*) (16):

$$RX + Cr^{II}aq \longrightarrow R \cdot + Cr^{III}X \tag{29}$$

Substituents in a series of *p*-substituted benzyl halides had little effect on the overall rate of reaction, which is consistent with a ligand-transfer process.

A second characteristic feature of ligand-transfer oxidations is that they proceed without rearrangement of the alkyl group. Thus neopentyl radicals are oxidized by copper(II) chloride to give only neopentyl chloride (reaction *30*), whereas in electron-transfer oxidations using copper(II) acetate rearranged products are obtained (reaction *31*) (17):

$$Me_3CCH_2 \cdot \xrightarrow{CuCl_2} Me_3CCH_2Cl \tag{30}$$

$$Me_3CCH_2 \cdot \xrightarrow{Cu(OAc)_2} Me_3CCH_2^+ \xrightarrow{-e} Et\overset{+}{C}Me_2$$

$$\longrightarrow EtC(Me){=}CH_2 + MeCH{=}CMe_2 + EtCMe_2(OAc) \tag{31}$$

Cyclobutyl radicals are similarly converted by copper(II) chloride to cyclobutyl chloride without rearrangement (reaction *32*) in contrast to their behaviour with copper(II) acetate.

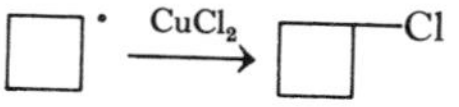

$$\square \cdot \xrightarrow{CuCl_2} \square{-}Cl \tag{32}$$

As a consequence of there being little development of charge in the transition state in ligand-transfer reactions, it is found that such reactions can take place with radicals containing electron-withdrawing substituents which would not be susceptible to electron-transfer oxidation. Thus the 2-cyano-2-propyl radical is inert to oxidation by copper(II) acetate but reacts readily with copper(II) chloride (reaction *33*) (18):

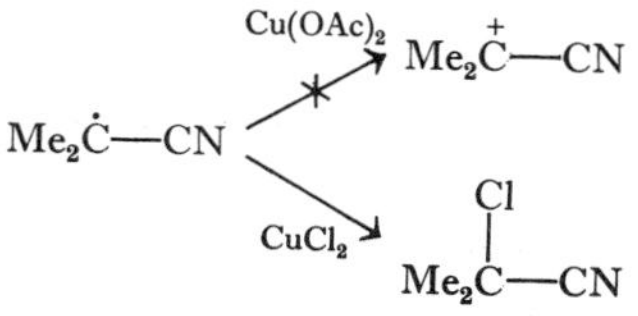

$$Me_2\dot{C}{-}CN \begin{cases} \xrightarrow[\times]{Cu(OAc)_2} Me_2\overset{+}{C}{-}CN \\ \xrightarrow{CuCl_2} Me_2\overset{\overset{\displaystyle Cl}{|}}{C}{-}CN \end{cases} \tag{33}$$

The rate of ligand-transfer oxidation approaches the diffusion-controlled limit, as evidenced from the ability of copper(II) chloride to trap the cyclopropylmethyl radical prior to its fragmentation to the but-3-en-1-yl radical:

$$\triangleright\!\!-CH_2\cdot + CuCl_2 \longrightarrow \triangleright\!\!-CH_2Cl + CuCl \qquad (34)$$

10.4 *OXIDATION OF ORGANIC COMPOUNDS WITH METAL-ION OXIDANTS*

The ability of an organic compound to undergo oxidation is dependent on three factors:

    (1) the oxidation potential of the oxidant,

    (2) the oxidation potential of the organic compound,

    (3) the availability of a suitable pathway whereby oxidation can occur.

These factors will be considered by reference to the oxidation of several classes of organic compound by one-electron oxidants. In the main, attention will be directed in this section to the first step in the oxidation process, i.e. the formation of the radical intermediate. The oxidation of phenols in which the emphasis lies more on the products of oxidation, i.e. on the subsequent reactions of the intermediate phenoxy radicals, will be considered separately.

10.4.1 *Alcohols*

Transition-metal ions of high oxidation potential oxidize monohydric alcohols by a bonded mechanism. The electron-transfer step can involve the transfer of one or two electrons. Only oxidations involving the transfer of a single electron will be discussed here, e.g. oxidations brought about by cerium(IV), cobalt(III), manganese(III) and vanadium(V).

Kinetic studies by Waters and his collaborators (19, 20) have indicated that the oxidation of alcohols proceeds through a complex between the alcohol and the oxidant. In some instances, notably oxidations with cerium(IV) and vanadium(V), colour changes reveal the presence of intermediate complexes.

Oxidations of primary and secondary alcohols involve the removal of an α-hydrogen from the alcohol in the rate-determining step, as evidenced by deuterium isotope effects in the oxidation of cyclohexanol and α-deuteriocyclohexanol with cerium(IV) (21), cobalt(III) (22) and vanadium(V) (23). Consequently, mechanisms have been proposed for these oxidations which involve cleavage of an α-C–H bond with concomitant reduction of the transition metal in a cyclic process (reaction *35*):

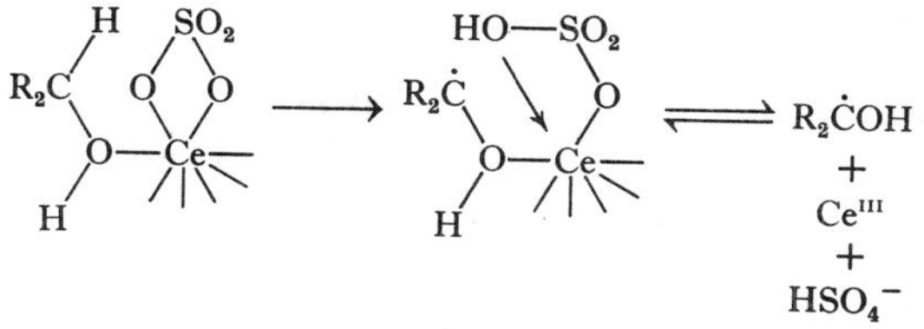

$$(35)$$

Some oxidations proceed with C–C bond fission rather than C–H bond fission. The former is favoured when the radical resulting from C–C bond fission is more stable than that resulting from C–H bond fission, as exemplified by the oxidation of 2-phenylethanol by vanadium(V) (reactions *36* and *37*) (24):

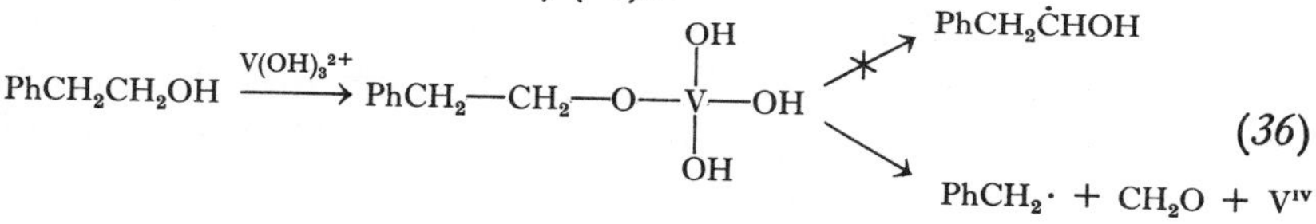

$$(36)$$

$$\text{PhCH}_2\cdot \xrightarrow{\text{V(OH)}_3{}^{2+}} \text{PhCH}_2\text{OH} \xrightarrow{\text{V(OH)}_3{}^{2+}} \text{PhCHO} \qquad (37)$$

This type of C–C bond fission provides a route for the oxidation of tertiary alcohols when a group can be split off as a stable radical (reaction *38*) (25):

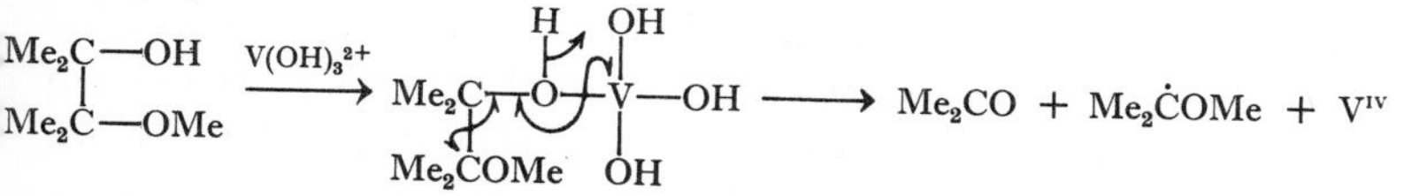

$$(38)$$

With stronger oxidants such as cobalt(III), the stability of the radical is of lesser importance and consequently cobalt(III) is able to oxidize tertiary alcohols which do not have structural features to stabilize the derived radical (reaction *39*) (26):

$$\text{Et}_3\text{C}\!-\!\text{OH} \xrightarrow{\text{Co}^{\text{III}}} \text{Et}_2\text{CO} + \text{Et}\cdot + \text{Co}^{\text{II}} \qquad (39)$$

The reaction products obtained from tertiary alcohols containing different alkyl groups are identical to those obtained from the homolysis of the analogous di-t-alkyl peroxides (reaction *40* on p. 316).

This suggests that these oxidations may involve the intermediacy of t-alkoxy radicals which undergo subsequent fragmentation. Alternatively, the alkyl radical may be extruded in the oxidation step since the nature of the alkyl groups attached to the tertiary carbon has a consider-able effect on the rate of reaction.

$$R_2\overset{\overset{\displaystyle R'}{|}}{C}\text{—OH}$$

$$R_2\overset{\overset{\displaystyle R'}{|}}{C}\text{—O—O—}\overset{\overset{\displaystyle R'}{|}}{C}R_2 \quad\xrightarrow[\Delta]{\text{Co}^{\text{III}}}\quad R\cdot + R'\cdot + R_2CO + RCOR' \qquad (40)$$

Oxidations of secondary alcohols with cobalt(III) also result in C–C bond fission in addition to the normal C–H bond fission (reaction *41*):

$$Et_2CHOH \longrightarrow EtCHO + Et_2CO \qquad (41)$$
$$ (85\%) \quad (15\%)$$

Glycols are readily oxidized by vanadium(V) and other transition-metal single-electron oxidants. Kinetic studies show that oxidations of di-tertiary but not di-secondary glycols with vanadium(V) and also with manganese(III) involve a cyclic chelate (reaction *42*) (27, 28):

$$V(OH)_3^{2+} + \begin{array}{l}-\overset{|}{C}\text{—OH}\\-\overset{|}{C}\text{—OH}\\\end{array} \underset{K}{\rightleftharpoons} \begin{array}{l}-\overset{|}{C}\text{—O}\\-\overset{|}{C}\text{—O}\end{array}\hspace{-0.5em}\overset{OH}{\underset{H}{V}}\hspace{-0.5em}OH \xrightarrow{k} \begin{array}{l}\diagdown C{=}O\\\diagup\\\diagup\overset{..}{C}\text{—OH}\\\diagup\end{array} + V^{\text{IV}} \qquad (42)$$

Further evidence for the cyclic intermediates comes from the observa-tion that pinacol is oxidized much more readily than its monomethyl ether. The overall rate of oxidation is determined not only by the rate of decomposition of the complex (k), but also by the stability of the complex with respect to the reactants. It is this second factor which probably accounts for the slowness of oxidation of *cis*- and *trans*-1,2-dimethyl-cyclohexane-1,2-diols relative to that of pinacol, since the chelate ring obtained from the dimethylcyclohexane-1,2-diols would be strained as a result of its fusion with a six-membered ring (28). The rates of de-composition of the complexes should be similar as one might reasonably expect the resultant radicals to have comparable stabilities. Oxidation of di-secondary glycols with vanadium(V) results in the conversion of the alcohol group to carbonyl, as in the oxidation of monohydric secondary alcohols.

Cyclic complexes do not appear to be involved in the oxidation of glycols, including di-tertiary glycols, with more powerful oxidants such as cerium(IV) (29). Thus, ethylene glycol and its monomethyl ether are oxidized at comparable rates. Oxidation of these substrates with cerium(IV) does not involve a simple oxidation of the $>$CHOH group to $>$C=O, since butane-2,3-diol is oxidized to acetaldehyde. The mechanism for this reaction must thus account for the conversion of the alcohol function to the carbonyl with concomitant C–C bond fission (reaction *43*):

$$
\begin{array}{c}
-\overset{|}{\underset{|}{C}}-OH \\
-\overset{|}{\underset{|}{C}}-OH
\end{array}
\xrightarrow{Ce^{IV}}
\begin{array}{c}
-\overset{|}{\underset{|}{C}}-O-\overset{\cdot}{Ce}^{IV} \\
-\overset{|}{\underset{|}{C}}-OH
\end{array}
\longrightarrow
\begin{array}{c}
\overset{\diagdown}{\diagup}C=O \\
\overset{\diagdown}{\underset{\diagup}{\overset{\cdot\cdot}{C}}}-OH
\end{array}
+ Ce^{III}
\qquad (43)
$$

In sharp contrast to the ease with which glycols are oxidized by vanadium(V) and other inner-sphere oxidants, pinacol was found to be resistant to oxidation by iron(III) tris-*o*-phenanthroline and hexachloroiridate(IV). These latter are both outer-sphere oxidants and the resistance to oxidation of pinacol can be attributed to the absence of low-energy outer-sphere oxidation routes (reaction *44*): the resultant radical cation would be of high energy:

$$
\begin{array}{c}
Me_2C-OH \\
\underset{|}{} \\
Me_2C-OH
\end{array}
+ IrCl_6^{2-} \xrightarrow{\quad\times\quad}
\begin{array}{c}
Me_2\overset{+\cdot}{C}-OH \\
\underset{|}{} \\
Me_2C-OH
\end{array}
+ IrCl_6^{3-}
\qquad (44)
$$

The products of inner-sphere oxidation with vanadium(V) are of lower energy enabling reaction to proceed (reaction *42*).

Similarly, alcohols are much less readily oxidized by outer-sphere oxidants. These results emphasize the importance of the availability of a suitable pathway for the oxidation process.

The oxidations discussed above have in the main little synthetic value due to the ease with which the products undergo further oxidation by the fairly powerful oxidants employed. This is not true for copper(II) oxidations of α-ketols to α-diketones. Copper(II) is a weak oxidant and thus the product is resistant to further oxidation under the conditions employed. Thus, copper(II) sulphate in pyridine oxidizes benzoin to benzil in 86% yield (reaction *45*) (30):

$$
PhCH(OH)COPh \xrightarrow[C_5H_5N]{CuSO_4} PhCOCOPh \qquad (45)
$$

The detailed mechanism of this reaction is as yet in dispute, though it probably involves a single-electron oxidation of a copper(I) chelate of the ene-diol followed by a second one-electron oxidation (reaction *46*) (31):

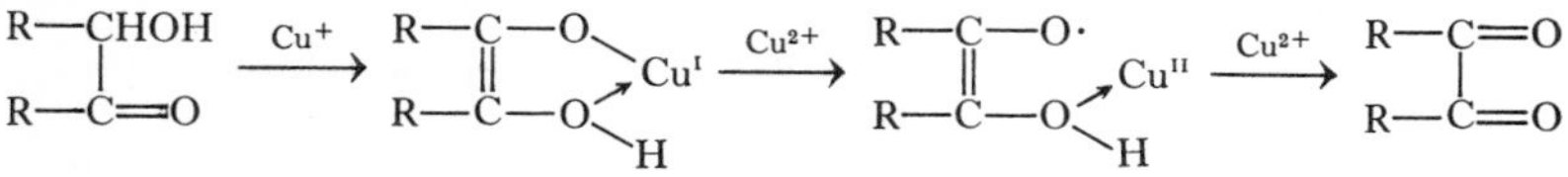

$$R-\underset{R-C=O}{\overset{CHOH}{|}} \xrightarrow{Cu^+} R-\underset{R-C-O}{\overset{C-O}{\underset{H}{\parallel}}}Cu^I \xrightarrow{Cu^{2+}} R-\underset{R-C-O}{\overset{C-O\cdot}{\underset{H}{\parallel}}}Cu^{II} \xrightarrow{Cu^{2+}} R-\underset{R-C=O}{\overset{C=O}{|}} \quad (46)$$

10.4.2 *Aldehydes and ketones*

The oxidations of ketones, particularly cyclohexanone, have been studied in considerable detail using a variety of oxidants. The rates of oxidation of cyclohexanone by cerium(IV), cobalt(III) and vanadium(V) are greater than its rate of enolization, indicating that the ketone is attacked rather than the enol. Isotope effects indicate that the α-C–H bond is broken in the rate-determining step (reaction *47*) (32):

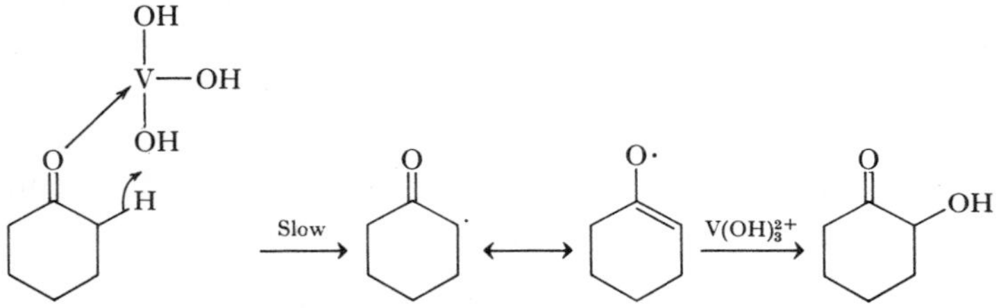

$$(47)$$

The intermediate radical has been identified in the oxidation of cyclopentanone with vanadium(V) by carrying out the oxidation in the presence of acrylonitrile. The infrared spectrum of the resultant polyacrylonitrile showed the typical absorption of a carbonyl group in a five-membered ring consistent with the proposed mechanism.

In contrast to the resistance to oxidation of alcohols by outer-sphere oxidants, ketones undergo facile oxidation by both iron(III) tris-*o*-phenanthroline and hexachloroiridate(IV) (2).

In general these oxidations have little or no synthetic value as the initial products undergo further oxidation. From a synthetic point of view the use of silver oxide for the oxidation of aldehydes to carboxylic acids is highly recommended. Detailed mechanistic studies on this have not been carried out though it is virtually certain that it is a homolytic process, as silver(I) is a one-electron oxidant.

10.4.3 *Carboxylic acids*

Extensive studies have been made by Waters and co-workers on the oxidation of a variety of carboxylic acids with cobalt(III) perchlorate in aqueous solution (7, 33). Kinetic studies point to an inner-sphere mechanism involving the rapid reversible formation of a cobalt(III) complex which breaks down slowly to free-alkyl radicals (reactions *48* and *49*):

$$RCO_2H + Co(H_2O)_6^{3+} \longrightarrow [RCO_2Co(H_2O)_5]^{2+} + H_3O^+ \quad (48)$$

$$[RCO_2Co(H_2O)_5]^{2+} \longrightarrow R\cdot + CO_2 + Co^{2+} \text{ aq} \quad (49)$$

The relative rates of oxidation of aliphatic carboxylic acids containing primary, secondary and tertiary alkyl groups, suggest that the breakdown of the cobalt(III) complex occurs directly to give alkyl radicals in accord with their relative stabilities rather than stepwise giving first acyloxy radicals (reaction *50*):

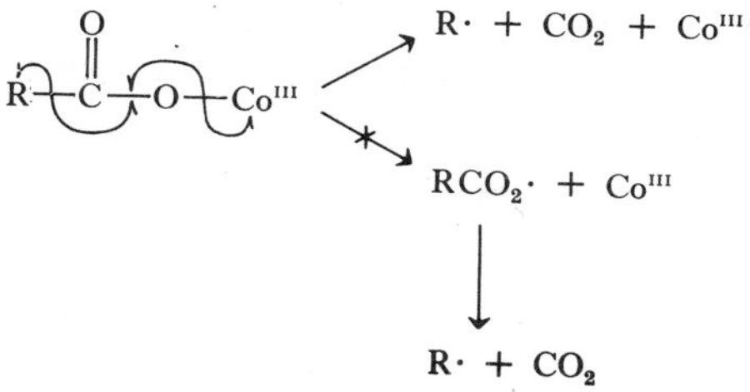

$$(50)$$

Evidence for the intermediacy of alkyl radicals comes from the detection of bibenzyl in the oxidation of phenylacetic acid, and from the formation of bromoalkanes in the oxidations of ω-phenylalkanoic acids in the presence of bromoform or bromotrichloromethane (33):

$$PhCH_2CO_2H \xrightarrow{Co^{III}} PhCH_2\cdot \quad (51)$$

$$2PhCH_2\cdot \longrightarrow PhCH_2CH_2Ph \quad (52)$$

$$Ph(CH_2)_{n+1}CO_2H \xrightarrow{Co^{III}} Ph(CH_2)_nCH_2\cdot \quad (53)$$

$$Ph(CH_2)_nCH_2\cdot + CHBr_3 \longrightarrow Ph(CH_2)_nCH_2Br + \dot{C}HBr_2 \quad (54)$$

$$Ph(CH_2)_nCH_2\cdot + CBrCl_3 \longrightarrow Ph(CH_2)_nCH_2Br + CCl_3\cdot \quad (55)$$

The alkyl radicals formed in these reactions undergo subsequent oxidation to aldehydes via the alcohol. The alcohol may be formed directly from the radical in a ligand-transfer process (reaction *56*) or via the carbonium ion (reactions *57* and *58*):

$$PhCH_2\cdot \xrightarrow{Co^{III}(OH)} PhCH_2OH \quad (56)$$

$$PhCH_2\cdot \xrightarrow{Co^{III}} PhCH_2^+ \quad (57)$$

$$PhCH_2^+ \xrightarrow[-H^+]{H_2O} PhCH_2OH \qquad (58)$$

$$PhCH_2OH \xrightarrow{Co^{III}} PhCHO \qquad (59)$$

Kochi has extended the study of the oxidative decarboxylation of carboxylic acids with single-electron oxidants to non-aqueous media. He has shown that lead salts of carboxylic acids undergo decarboxylation on heating or irradiation (34):

$$Pb(OAc)_4 + RCO_2H \longrightarrow RCO_2Pb(OAc)_3 \qquad (60)$$

$$RCO_2Pb(OAc)_3 \longrightarrow R\cdot + CO_2 + Pb^{III} \qquad (61)$$

$$R\cdot + Pb^{IV} \longrightarrow R^+ + Pb^{III} \qquad (62)$$

The oxidation of alkyl radicals can be brought about very much more efficiently by traces of copper(II) acetate. This brings about the oxidation of the alkyl radical to alkene in high yield.

$$R\cdot + Cu^{II} \longrightarrow R(-H) + Cu^I + H^+ \qquad (63)$$

This has been developed as an efficient method for the synthesis of alkenes (35).

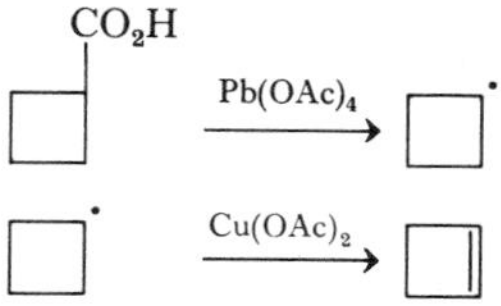

$$\qquad (64)$$

$$\qquad (65)$$

In the absence of copper(II) acetate, oxidation of alkyl radicals is slow and the main products are formed as a result of radical combinations or by reaction of the radical with solvent.

From observations on the relative ease of the photochemical decarboxylation of a series of aliphatic carboxylic acids by lead tetra-acetate, Kochi proposed that alkyl radicals are formed in the rate-determining step, i.e. cleavage of the Pb–O and C–O bonds is concerted (36). The same is true for decarboxylations of carboxylates of cobalt(III) (37) and manganese(III) but not for those of cerium(IV) (38). In the latter, the structure of the alkyl group has little influence on the rate of decarboxylation and it is consequently assumed that acyloxy radicals are formed initially, and that these undergo decarboxylation to alkyl radicals in a subsequent fast reaction.

$$RCO_2Ce^{IV} \longrightarrow RCO_2\cdot + Ce^{III} \qquad (66)$$

$$RCO_2\cdot \longrightarrow R\cdot + CO_2 \qquad (67)$$

10.4.4 *Aromatic compounds*

Aromatic compounds are also susceptible to oxidation by certain one-electron oxidants. Thus, manganese(III) acetate oxidizes p-methoxytoluene to p-methoxybenzyl acetate by the reaction scheme outlined in reactions (68–71) (3).

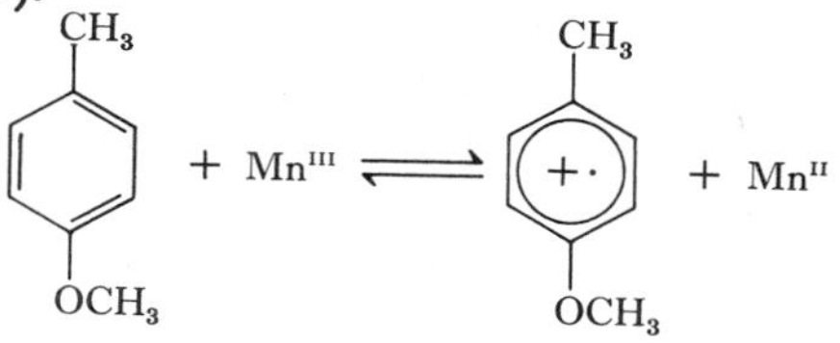

$$\text{(} p\text{-methoxytoluene)} + Mn^{III} \rightleftharpoons \text{(radical cation)}^{+\cdot} + Mn^{II} \tag{68}$$

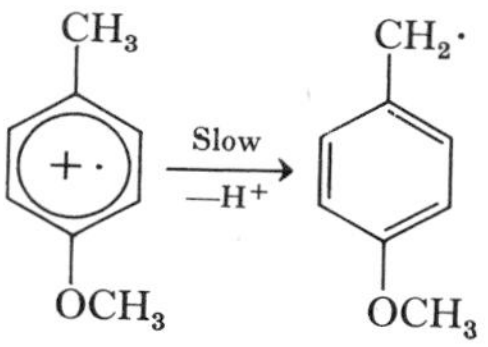

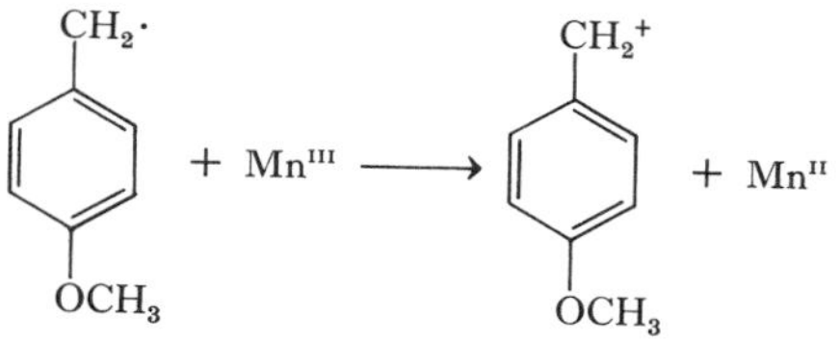

$$\text{(radical cation)}^{+\cdot} \xrightarrow[-H^+]{\text{Slow}} \text{(benzylic radical)} \tag{69}$$

$$\text{(CH}_2^{\cdot}\text{ radical)} + Mn^{III} \longrightarrow \text{(CH}_2^{+}\text{ cation)} + Mn^{II} \tag{70}$$

$$\text{(CH}_2^{+}\text{ cation)} + OAc^- \longrightarrow \text{(CH}_2OAc) \tag{71}$$

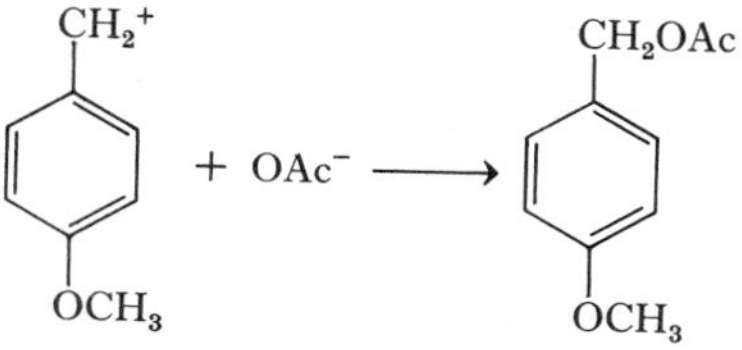

The essential feature of this scheme is the reversible electron-transfer giving a radical cation, followed by slow loss of a proton to give a benzylic radical.

The reactivity of aromatic hydrocarbons towards electron-transfer oxidation by manganese(III) acetate is dependent on the oxidation potential of the aromatic compound. This is greater for toluene than for p-methoxytoluene, and in agreement with this it has been shown that toluene is inert towards electron-transfer oxidation by manganese(III) acetate, though it is attacked by $\dot{C}H_2CO_2H$ radicals produced in the thermolysis of the manganese salt (4).

Oxidations and reductions

More powerful oxidants such as cobalt(III) acetate are capable of oxidizing toluene as well as p-methoxytoluene (5). The rate of reaction of p-substituted toluenes is considerably retarded by electron-withdrawing substituents, consistent with the electron-transfer mechanism. The observation of an isotope effect is consistent with the proposed reaction scheme in which the reversible electron-transfer step is followed by the slow loss of a proton.

Radical cations have been detected spectroscopically in the cobalt(III) acetate oxidations of polycyclic aromatic hydrocarbons having low ionization potentials, thus lending further support to the proposed mechanism:

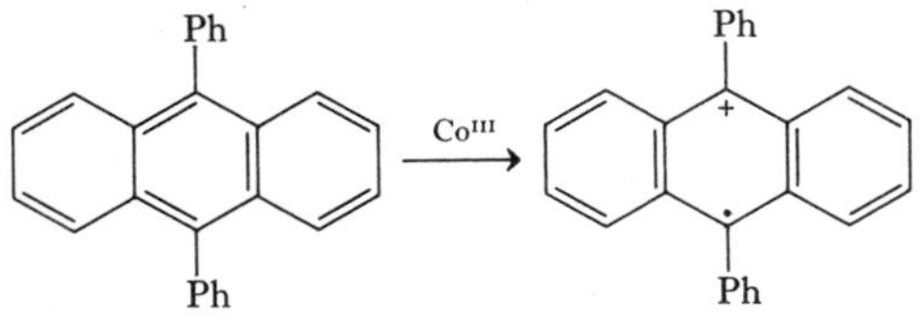

$$(72)$$

Nuclear acetoxylation can also occur as in the oxidation of 2-methylnaphthalene by manganese(III) or cobalt(III) acetates. This results from direct attack on the radical cation as shown in Scheme 1. This mechanism is similar to that proposed for the nuclear acetoxylation of aromatic compounds in the anodic oxidation in acetic acid.

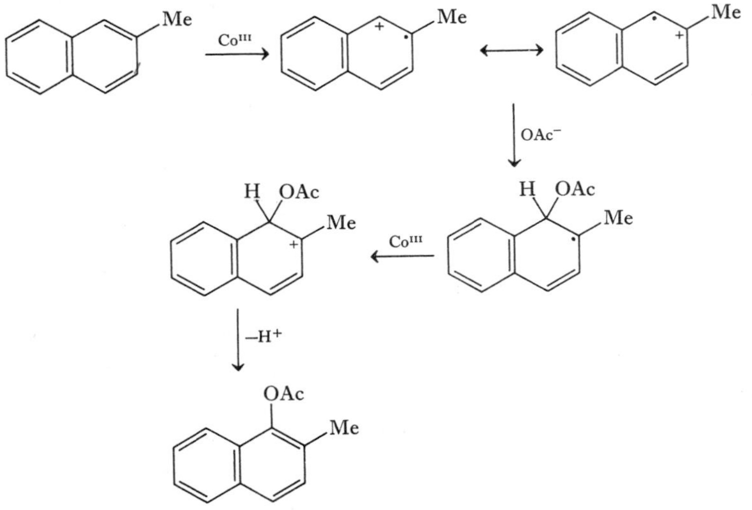

Scheme 1

322

In the presence of chloride, the rate of reaction is markedly enhanced as is the proportion of nuclear attack. This may well be as a result of a ligand-transfer reaction between a chlorocobalt(III) acetate and the radical cation (Scheme 2) which would be expected to be a more facile process than reaction with acetate followed by a second electron transfer.

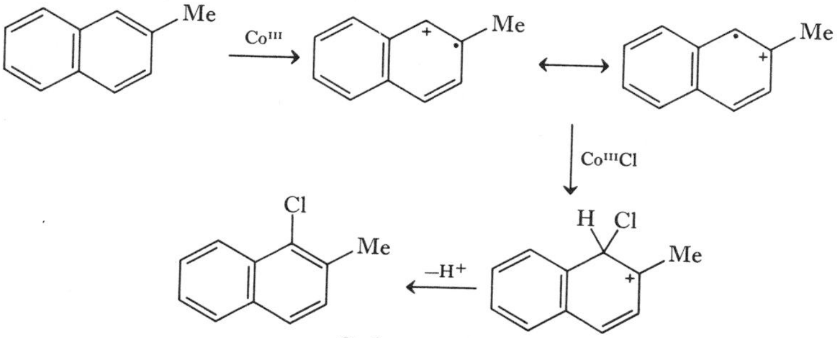

Scheme 2

Nuclear acetoxylation of reactive aromatics can also be effected by lead tetra-acetate in the presence of a suitable radical initiator, such as di-isopropyl peroxy carbonate (39) (see chapter 11, p. 446).

Copper(II) bromide and chloride have been shown to be very effective halogenating agents for anthracene, 9-alkyl-, 9-aryl-, and 9-halogenoanthracenes under heterogeneous conditions in non-polar solvents (40).

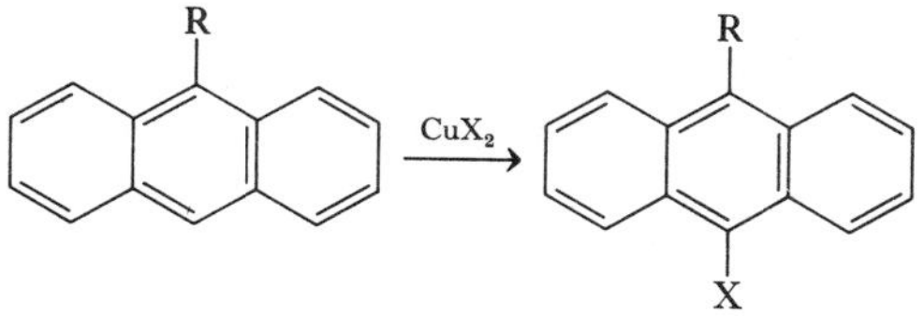

$$(73)$$

The reactions are undoubtedly those of copper(II) halides with the aromatic compound rather than of halogen, produced by dissociation of copper(II) halide:

$$CuX_2 \nrightarrow CuX + X\cdot \qquad (74)$$

That these reactions of copper(II) halides are confined to polycyclic aromatic compounds, has led to the suggestion that the aromatic system is complexed to the surface of the copper(II) halide thereby enabling reaction to occur. Water and ethanol, which break up the polymeric nature of the copper(II) halide, retard or inhibit the reaction.

The mechanism proposed for the reaction is outlined in Scheme 3. The essential feature of this scheme is that it involves a halogen atom transfer to give the radical (2), which reacts in a second atom-transfer reaction to give the halogenated anthracene. This scheme is consistent with the known propensity of copper(II) halides to react by a ligand-transfer route. Evidence for this scheme is based on the fact that chlorination of 9-bromoanthracene with chlorine gives a mixture of 9-bromo-10-chloroanthracene and 9,10-dichloroanthracene, whereas

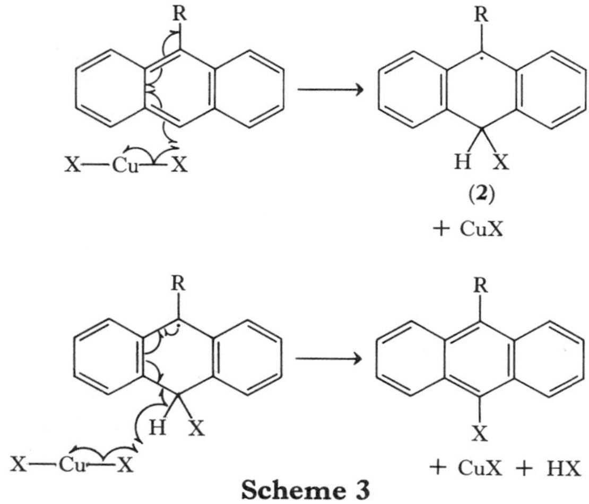

Scheme 3

chlorination with copper(II) chloride gives exclusively 9-bromo-10-chloroanthracene. Chlorination with chlorine proceeds through 9-bromo-9,10-dichloro-9,10-dihydroanthracene (3), which undergoes loss of both hydrogen bromide and hydrogen chloride to give the two products (reaction 75):

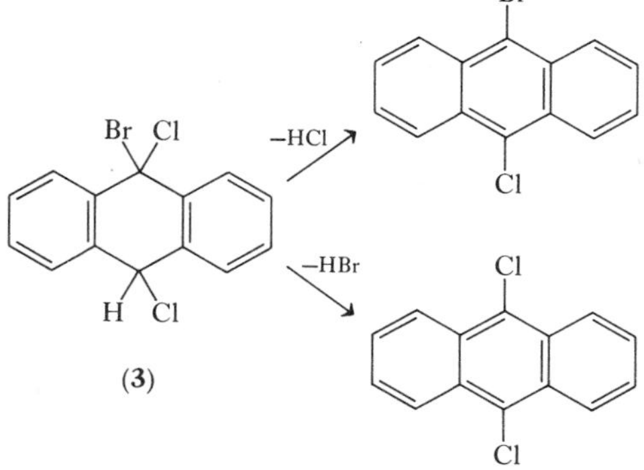

$$(75)$$

That no 9,10-dichloroanthracene is obtained in chlorination with copper(II) chloride indicates that a different route must be involved.

Further evidence for the participation of free radicals in halogenations with copper(II) halides comes from a study of their reactions with 9-methyl-10-phenylanthracene. The only products from these reactions are 9-halomethyl-10-phenylanthracenes which undoubtedly arise as a result of reaction of 9-anthrylmethyl radicals with copper(II) halide (Scheme 4). It has been suggested that the 9-anthrylmethyl radicals

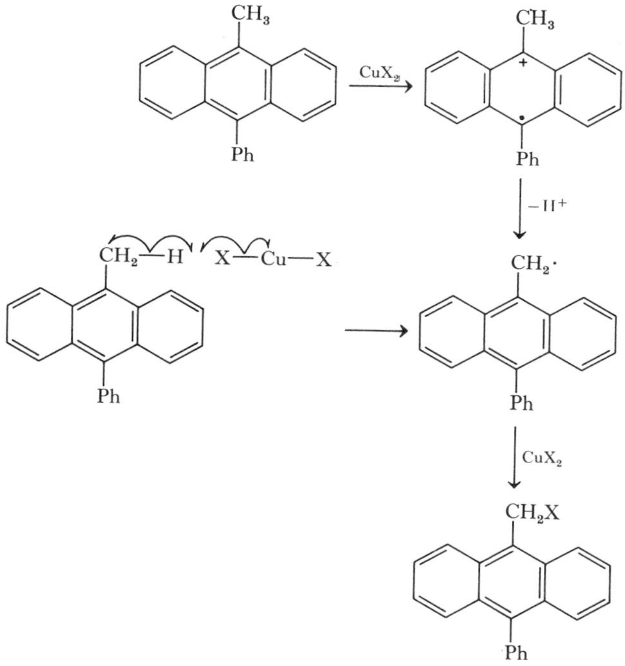

Scheme 4

arise either as a result of a direct reaction between the methyl group and the copper halide, or more probably via a radical cation formed in an electron-transfer reaction. No dimeric products were obtained in this reaction since the anthrylmethyl radicals would be generated on the surface of the copper(II) halide and never become free enough to escape into the solution and dimerize (40 a).

A further mode of reaction of copper(II) halides with anthracene derivatives is exemplified in the formation of bianthron-9-yl from reactions of 9-acyloxy- and 9-alkoxyanthracenes (41). Some of the

10-halogenated product is also formed. The route to bianthron-9-yl consists of a four-centre reaction between the 9-alkoxyanthracene and copper(II) halide to give the mesomeric 9-anthryloxy radical which dimerizes (Scheme 5). Dimerization occurs here rather than reaction with copper(II) halide because of the greater stability of the 9-anthryloxy radical compared to that of the 9-anthrylmethyl radical.

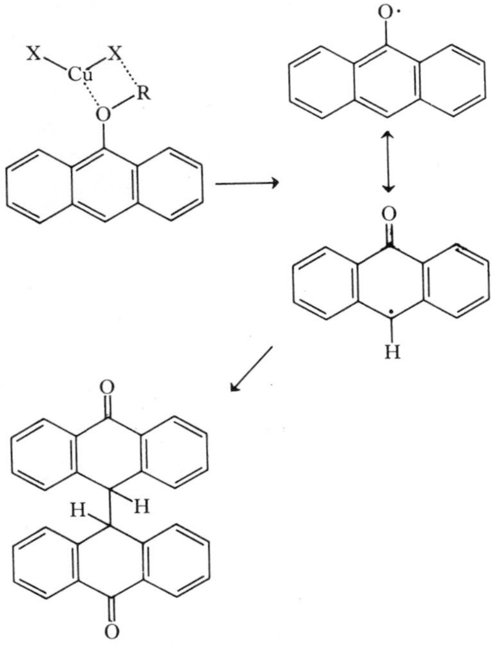

Scheme 5

10.5 *OXIDATION OF PHENOLS*

The oxidation of phenols is a topic of immense importance because of the widespread occurrence of phenolic coupling in the biosynthesis of phenolic natural products, and also because of the role of phenols as antioxidants. Consequently this topic merits examination in some detail. Interest lies not so much in the formation of the phenoxy radical but in its subsequent reactions–dimerization, disproportionation and oxidation.

10.5.1 *Formation of the phenoxy radical*

Oxidation of phenols can be brought about by a wide variety of single-electron oxidants including enzymes. Oxidation of the unionized phenol or of the phenolate anion may occur (reactions *76* and *77*).

$$ArOH \xrightarrow{-e} Ar\overset{+\cdot}{O}H \xrightarrow{-H^+} ArO\cdot \qquad (76)$$

$$ArO^- \xrightarrow{-e} ArO\cdot \qquad (77)$$

The latter is the major pathway in neutral or alkaline aqueous solution whereas the former is more important in strongly acidic solutions or in non-polar solvents. The formation of phenoxy radicals in the oxidation of a variety of phenols under different conditions has been amply confirmed by e.s.r. techniques (42).

10.5.2 *The mechanism of the coupling reaction*

No less than five distinct dimeric products have been isolated from the oxidation of phenol. These are formed by carbon–carbon (C–C) and carbon–oxygen (C–O) coupling of the mesomeric phenoxy radical (4) followed by enolization of the resultant dimer as shown (*78–82*):

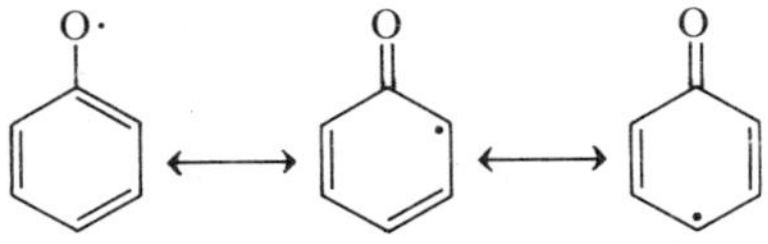

(4)

ortho–ortho C–C coupling

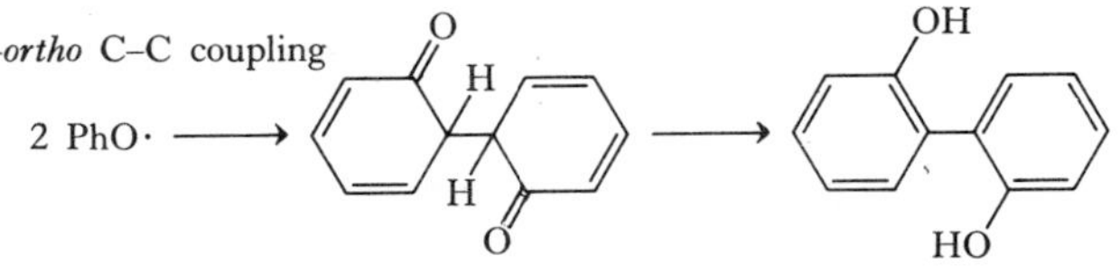

(78)

ortho–para C–C coupling

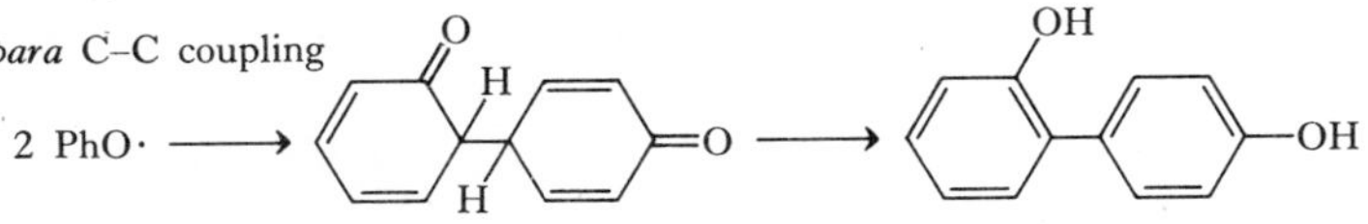

(79)

para–para C–C coupling

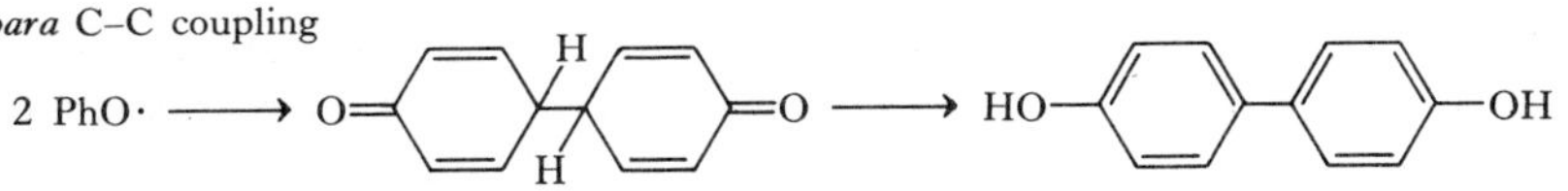

(80)

Oxidations and reductions

ortho C–O coupling

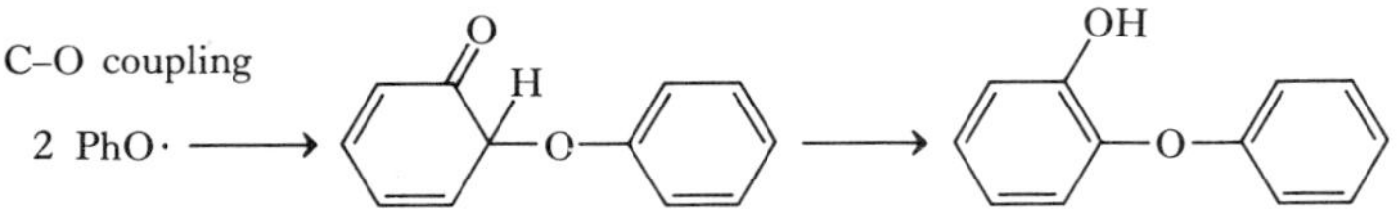

$$2 \text{ PhO} \cdot \longrightarrow \quad \longrightarrow \quad \tag{81}$$

para C–O coupling

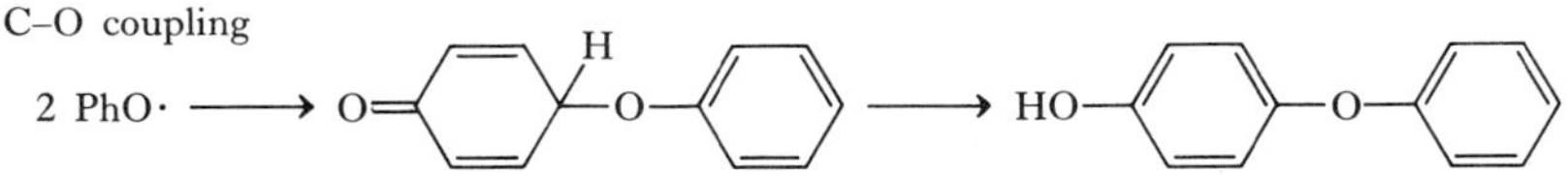

$$2 \text{ PhO} \cdot \longrightarrow \quad \longrightarrow \quad \text{(8}$$

The formation of dimers obtained from phenol has been formulated as involving radical coupling. Kinetic studies indicate that this is the predominant route to dimers in the case of the oxidation of simple phenols with weak oxidants.

Two other routes, which could also lead to dimeric products, have to be considered (43). The phenoxy radical could undergo further oxidation to the phenoxonium ion. This latter could then substitute a phenol molecule (reaction *83*):

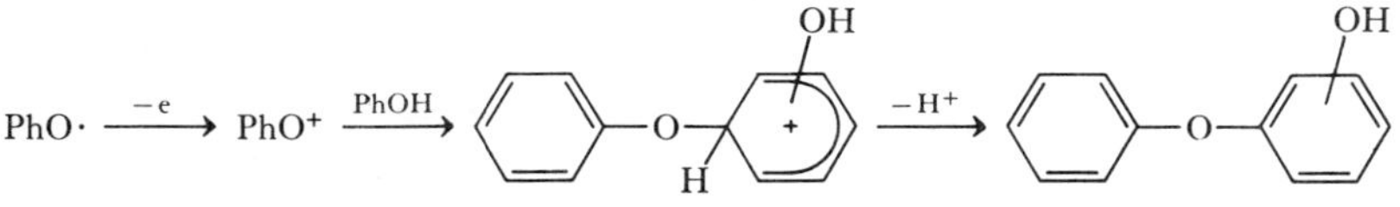

$$\text{PhO} \cdot \xrightarrow{-e} \text{PhO}^+ \xrightarrow{\text{PhOH}} \quad \xrightarrow{-\text{H}^+} \quad \tag{83}$$

The other plausible route involves radical substitution by the phenoxy radical on a phenol molecule followed by oxidation of the resultant cyclohexadienyl radical (reaction *84*):

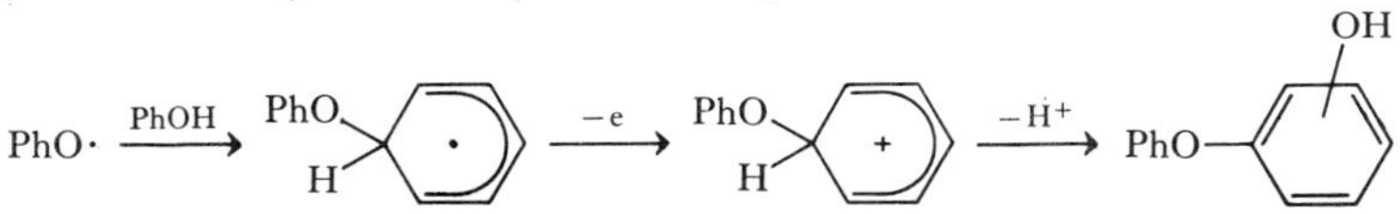

$$\text{PhO} \cdot \xrightarrow{\text{PhOH}} \quad \xrightarrow{-e} \quad \xrightarrow{-\text{H}^+} \text{PhO} \quad \tag{84}$$

Both these routes could in a similar way lead to C–C coupled products.

The phenoxonium ion is undoubtedly formed to a limited extent in phenol oxidations when a strong oxidant is used such as cerium(IV) or hexachloroiridate(IV). Weaker oxidants do not oxidize phenols beyond the phenoxy radical. The involvement of the phenoxonium ion in the hexachloroiridate(IV)-catalysed oxidation of 2,6-dimethylphenol is indicated by the formation of 2,6-dimethylbenzoquinone (*43 a*). This occurs to the extent of at least 20% (Scheme 6). 2,6-Dimethylbenzoquinone is also obtained quantitatively from the anodic oxidation of 2,6-dimethylphenol (*43 b*). Controlled anodic oxidation of a mixture of 2,6-di-t-butyl-4-methylphenol and phenol in

328

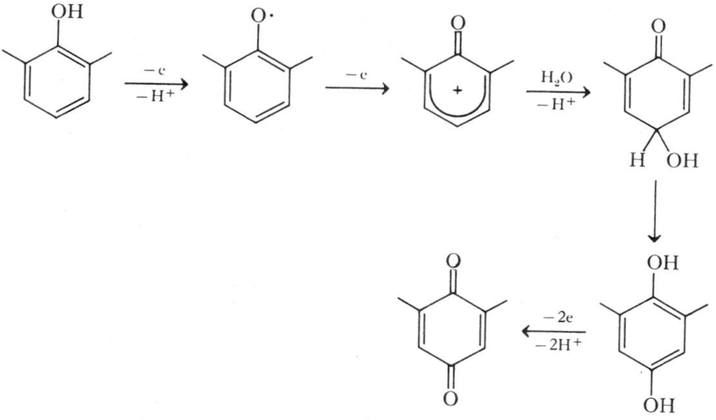

Scheme 6

aqueous solution at a potential at which only the 2,6-di-t-butyl-4-methylphenol is oxidized, leads to the coupled product (5) together with the hydroxycyclohexadienone (6) (Scheme 7). These result from reac-

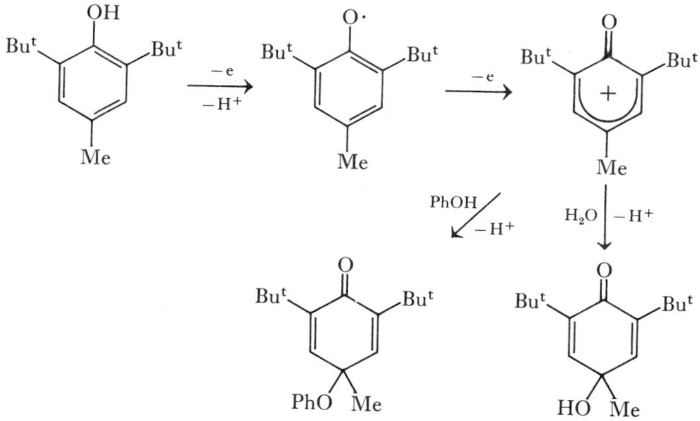

(5)   (6)

Scheme 7

tion of the phenoxonium ion with phenol and water respectively. These examples in which phenoxonium ions are involved, however, represent the exception rather than the general rule in phenol oxidation. Furthermore the products derived from phenoxonium ions are generally quite diagnostic of their involvement.

The absence of any intramolecularly coupled product in the oxidation

329

of the monoalkylated diphenol (**7**; R = Me, PhCH$_2$) has been interpreted as indicating that radical substitution does not occur. The diphenol (**7**; R = H) itself gives the intramolecularly coupled product (**8**) (44).

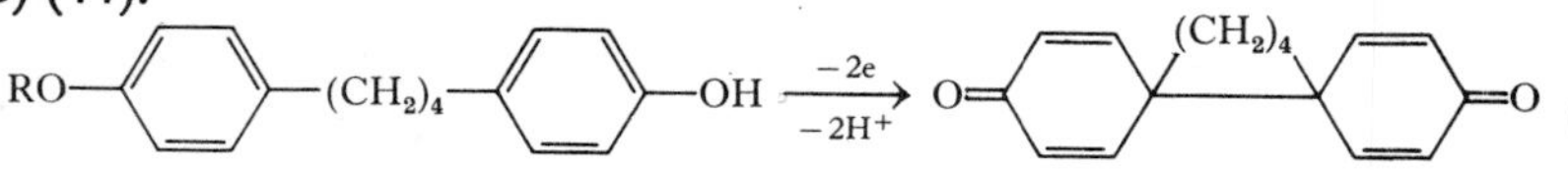

$$RO\text{—}\langle\ \rangle\text{—}(CH_2)_4\text{—}\langle\ \rangle\text{—}OH \xrightarrow[-2H^+]{-2e} O=\langle\ \rangle\overset{(CH_2)_4}{\underset{}{}}\langle\ \rangle=O \quad (85)$$

$$\qquad\qquad (7) \qquad\qquad\qquad\qquad\qquad\qquad (8)$$

A more cogent argument against radical substitution is that oxidation of *p*-cresol in a large excess of 1,2- or 1,3-dimethoxybenzene, or *p*-methylanisole gave no evidence of cross-coupling though these substrates would have been expected to have been at least as reactive as *p*-cresol towards radical substitution (45). Littler has, however, argued that if the first step of radical substitution is reversible, then the importance of this reaction sequence will be governed by the ease of oxidation of the intermediate cyclohexadienyl radical (43 a). This latter property may well be altered by replacement of OH by OCH$_3$. The absence of cross-coupled products can similarly be taken as evidence against the participation of phenoxonium ions.

If radical substitution were an important pathway to dimers, then one might expect to obtain *meta*-coupled products as well as *ortho*- and *para*-coupled products. Radical coupling is a much more feasible process when the phenolic hydroxyl and the aromatic nucleus are held in close proximity, especially when the aromatic nucleus is reactive towards radical substitution as with naphthalene derivatives. Thus cyclization of (**9**) gives (**10**) (46):

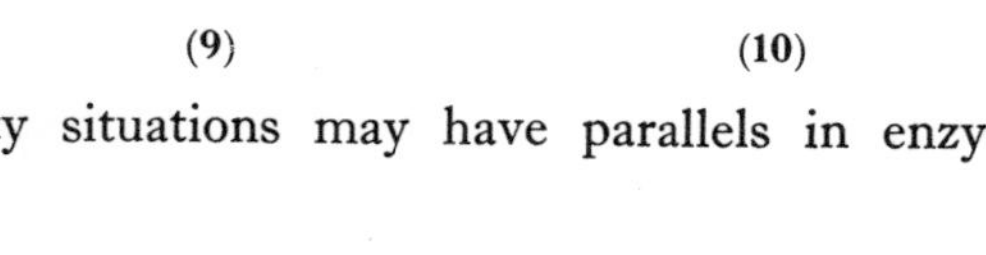

$$(9) \qquad\qquad\qquad\qquad (10) \qquad\qquad\qquad\qquad (86)$$

Close-proximity situations may have parallels in enzymic coupling reactions.

10.5.3 *Products of phenol oxidation*

The products obtained in the oxidation of phenols are greatly influenced by the structure of the phenol and also by the conditions used in the oxidation. These factors are best illustrated by reference to the oxidation of several different types of phenols.

(i) p-*Alkylphenols*. Oxidation of a p-alkylphenol leads to the typical C–C and C–O coupled products (**11** and **12**). In the case of p-cresol, these products are accompanied by a third product, Pummerer's ketone (**14**), formed by *ortho–para* C–C coupling followed by an intramolecular Michael addition of the phenolic hydroxyl to the cyclohexadienone (**13**).

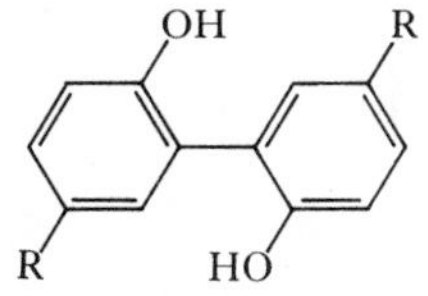

(11) 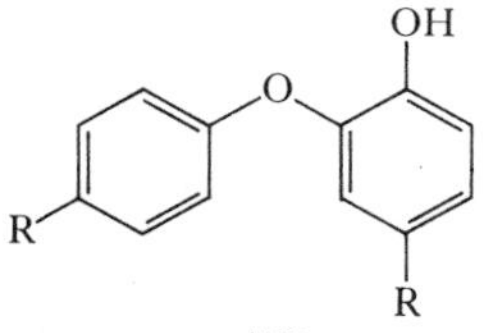   (12)

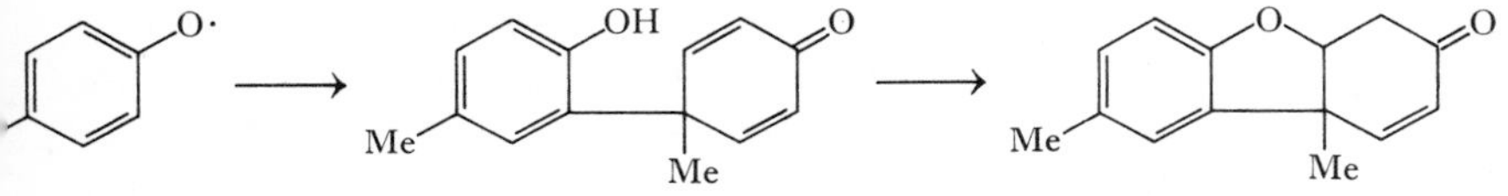

(13)   (14)   (87)

This type of product is also obtained in the oxidation of other p-alkyl-phenols with primary alkyl groups (48). p-t-Butylphenol does not form any of this type of product because of the severe steric interaction which would have to be overcome in the initial coupling.

One of the intriguing features of phenol oxidation in general, and of the oxidation of p-cresol in particular, is the variation in the yields of dimeric products obtained under different conditions. The whole picture is confused by the fact that the dimeric products are themselves readily oxidized to trimers and higher products. Thus in the oxidation of p-cresol the trimer (**15**) is formed by coupling of the radical derived

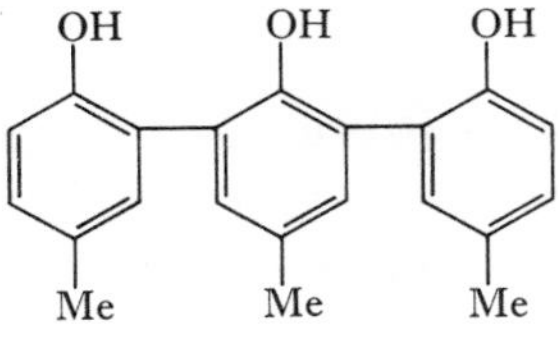

(15)

from the C–C dimer with that from p-cresol. If the oxidations are carried out in the presence of excess phenol, the extent of further oxidation is greatly cut down, thus enabling a more accurate understanding of the factors controlling the influence of the reaction conditions on the products to be obtained.

The yield of Pummerer's ketone is markedly influenced by the temperature of the reaction, decreasing to virtually zero when oxidations are carried out at elevated temperatures and increasing considerably when carried out at $-30\ °C$. This is interpreted as indicating that the initial coupling process (reaction *88*) is reversible (45):

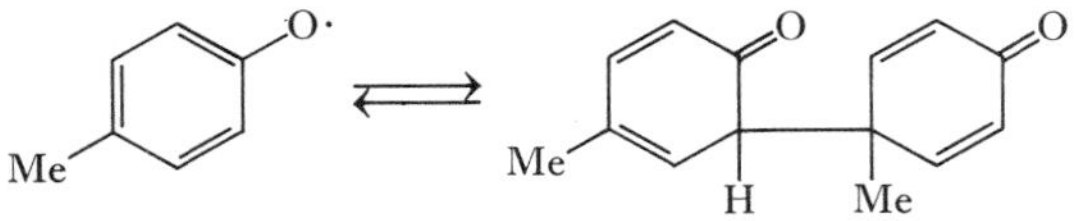

$$(88)$$

The bond formed in *ortho–para* C–C coupling would be expected to be weaker than that formed in *ortho–ortho* C–C or C–O coupling because of the steric interaction of the methyl group. Thus if the coupling reactions were reversible, it is reasonable that the importance of *ortho–para* C–C coupling would decrease at higher temperatures. Evidence for the reversibility of the coupling reaction comes from a study of the oxidation of 2,6-di-t-butylphenol in benzene to the diketo-dimer (**16**), enolization of which only occurs slowly because of steric hindrance. The

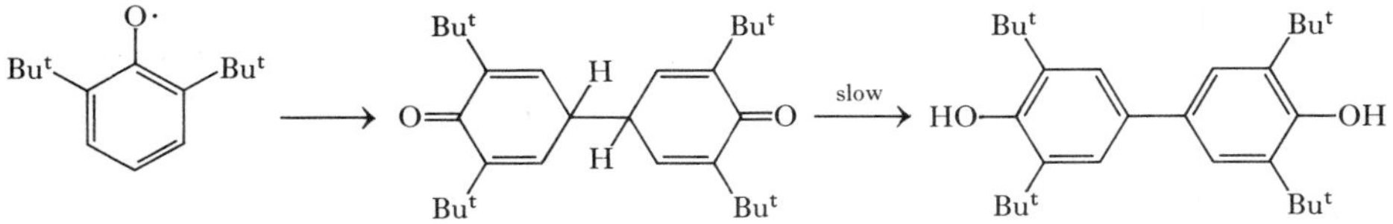

(16)   (17)

diketo-dimer (**16**) but not the diphenol (**17**) dissociates at elevated temperatures to the 2,6-di-t-butylphenoxy radical (Scheme 8). These results also point to the role of enolization in phenol oxidations. This would be affected by both the solvent and the pH of the reaction medium, and consequently both these factors might be expected to influence the product composition. There is evidence to suggest that this is so.

Waters has recently suggested (47) that the different proportions of C–C and C–O coupled products obtained under different conditions could result from a greater or lesser involvement of phenoxonium ions as reactive intermediates. Evidence against this is that there is little or no sign of substitution products, even when oxidations are carried out in solvents susceptible towards electrophilic substitution such as *p*-methoxytoluene and 1,3-dimethoxybenzene. It might also be expected

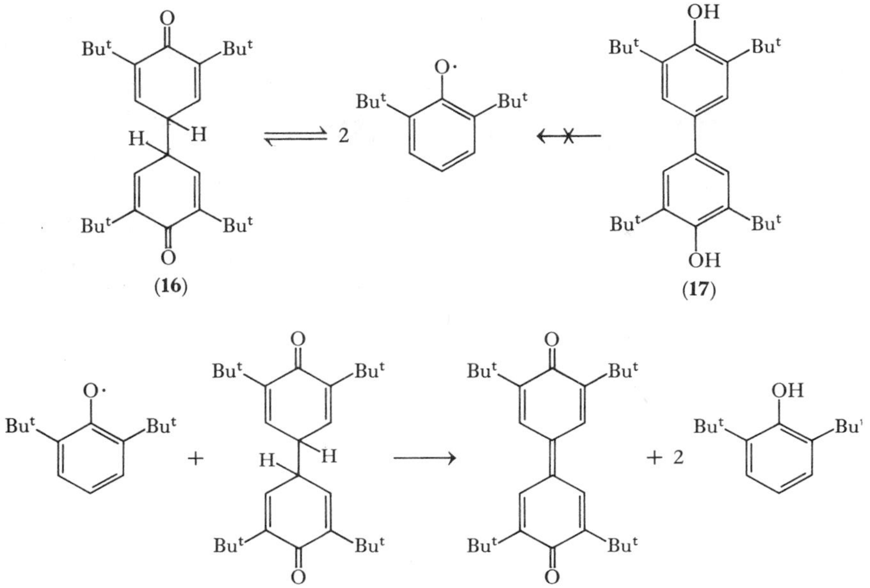

Scheme 8

that phenoxonium ions would play a more important role in oxidations using a strong oxidant, such as ceric ammonium nitrate, rather than a weak oxidant such as ferricyanide. In fact under the same reaction conditions both oxidants led to the same mixture of products in the case of *p*-cresol (45).

The product composition might also be expected to be influenced by the spin density distribution in the phenoxy radical. E.s.r. studies seem to indicate that there is greater spin density on oxygen in the *p*-ethyl-phenoxy radical than in the *p*-methylphenoxy radical (42). In accordance with this somewhat more of the C–O dimer is obtained from oxidations of *p*-ethylphenol than from *p*-cresol (45).

(ii) *Dialkylphenols*. Attention has been mainly directed to the study of 2,4- and 2,6-dialkylphenols. Products analogous to those obtained from *p*-alkylphenols could be derived from 2,4-dialkylphenols though in practice none of the C–O coupled products have thus far been

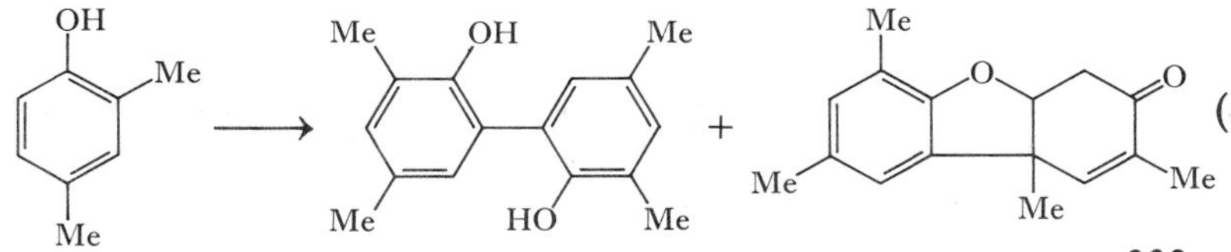

(*90*)

detected. 2,4-Dimethylphenol thus gives both the C–C dimer and a product analogous to Pummerer's ketone (reaction *90*). Replacement of the 4-methyl group by a t-butyl group prevents the formation of this type of product. This phenol and other 2-substituted 4-t-butylphenols, except 2,4-di-t-butylphenol, give, in addition to the C–C dimer, a trimeric spiroketal (**18**) (*4*), the formation of which is postulated to involve C–O coupling between radicals derived from the parent phenol and the C–C dimer (Scheme 9). The radical derived from the C–C dimer of 2,4-di-t-butylphenol is so hindered that it cannot couple with

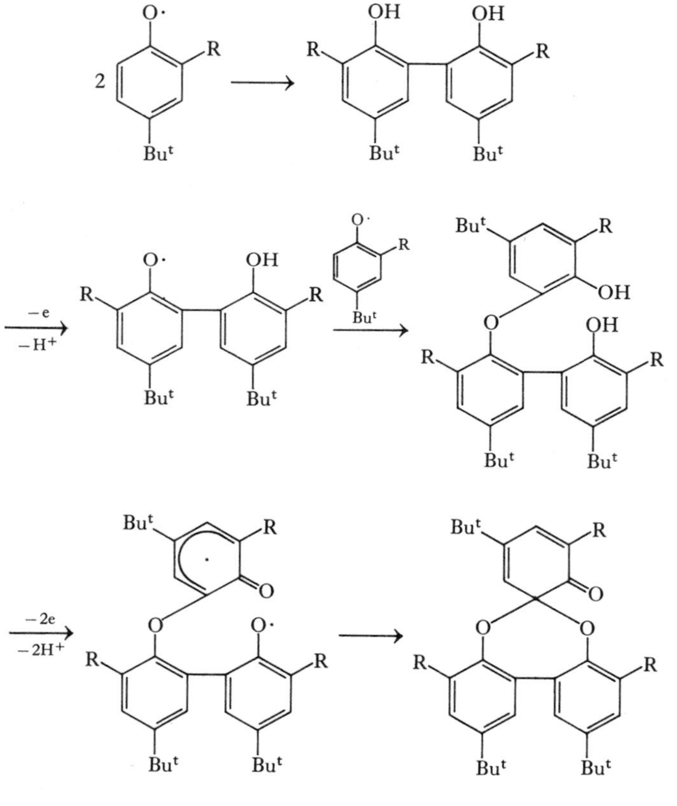

(R = Me, CH$_2$=CHCH$_2$, OMe)　　　　　(**18**)

Scheme 9

a 2,4-di-t-butylphenoxy radical. It can undergo further oxidation to the diradical, which then undergoes an intramolecular C–O coupling to give the spirocyclic quinolide ether (**19**) (Scheme 10) (*50*).

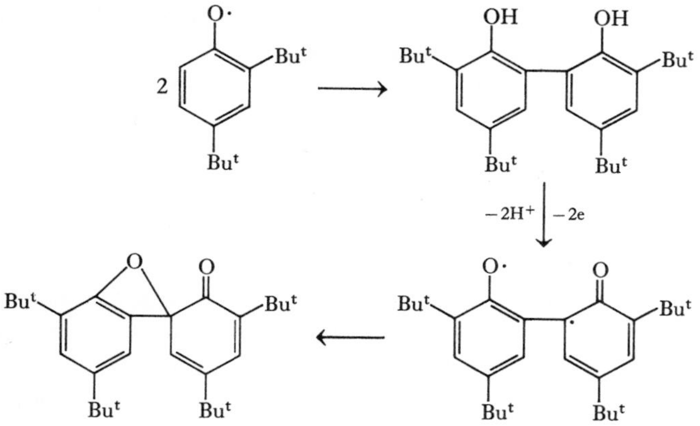

(19)   **Scheme 10**

Similarly, oxidation of 2,6-dialkylphenols can give rise to both C–C and C–O coupled products, the proportion of the latter decreasing with

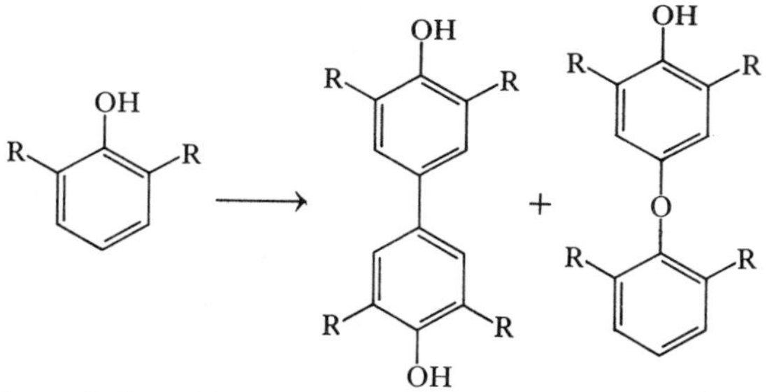

(91)

increasing size of the alkyl groups (51). In the case of 2,6-dimethyl-phenol, these products are accompanied by a ketonic compound (**20**) resulting from *ortho–para* C–C coupling (Scheme 11) (52). A significant

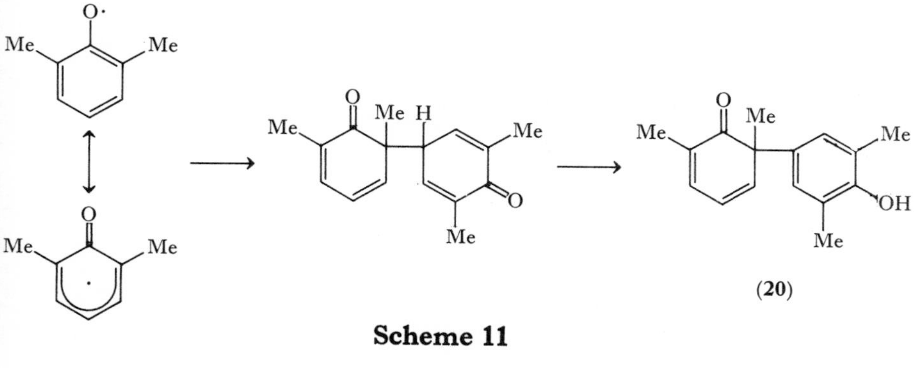

(20)

Scheme 11

335

feature of the products of oxidation of 2,6-dialkylphenols is that the C–C dimers are frequently oxidized further to the corresponding diphenoquinone (reaction *92*).

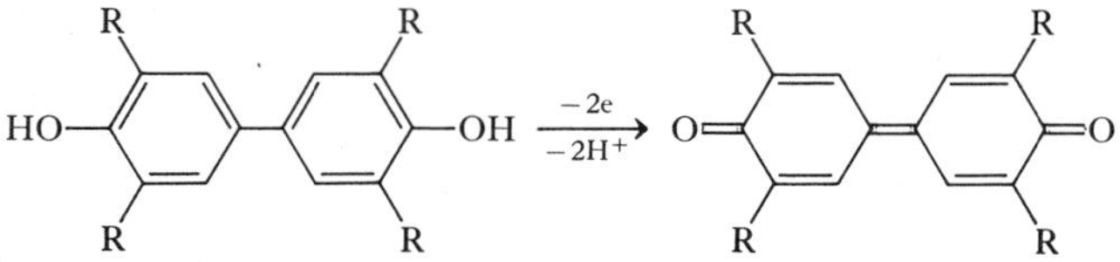

$$\text{HO} \quad \xrightarrow[-2H^+]{-2e} \quad \text{O} \qquad (92)$$

(iii) *2,4,6-Trialkylphenols.* 2,4,6-Trialkyl- or triaryl-phenols give rise to stable phenoxy radicals when the substituents are bulky groups possessing no α-C–H bonds. Such radicals do not undergo dimerization because of the strain which would be present in the dimer, and in the absence of any α-C–H bonds they cannot undergo disproportionation. 2,6-Di-t-butyl-4-alkylphenols, in which the alkyl group is primary or secondary, form radicals which disproportionate to the phenol and the quinone methide (**21**, R = H). The decay of the phenoxy radical has

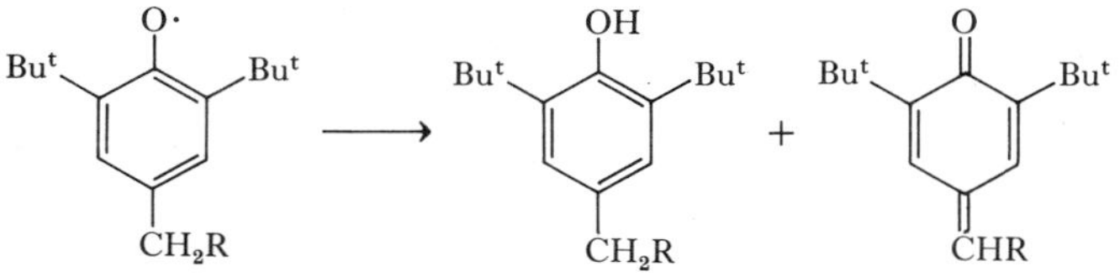

$$(93)$$

(21)

been shown to follow second-order kinetics (53). Quinone methides can be detected by u.v. spectroscopy, and when R $\neq$ H they can be isolated (54). They react in inert solvents to give equimolar amounts of the diphenol (**23**) and the stilbene quinone (**24**) (55). In this reaction the

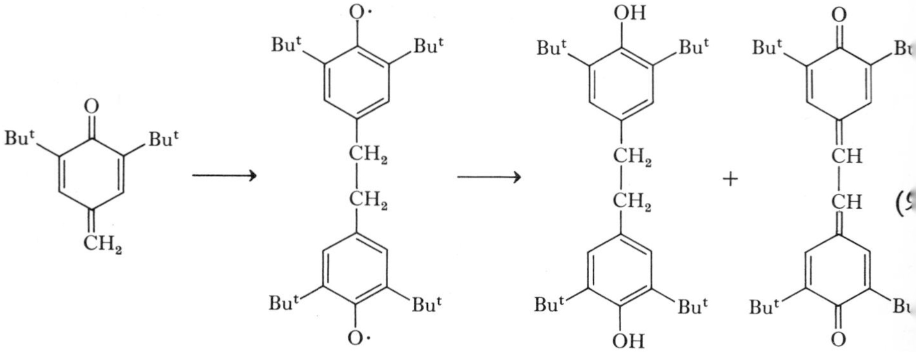

(22)   (23)   (24)

336

quinone methide behaves as if it were a diradical (22) which undergoes disproportionation (69). In aqueous solution, quinone methides undergo nucleophilic addition of water. Thus oxidation of 4-benzyl-2,6-di-t-butylphenol with aqueous potassium ferricyanide affords the alcohol (25) (48):

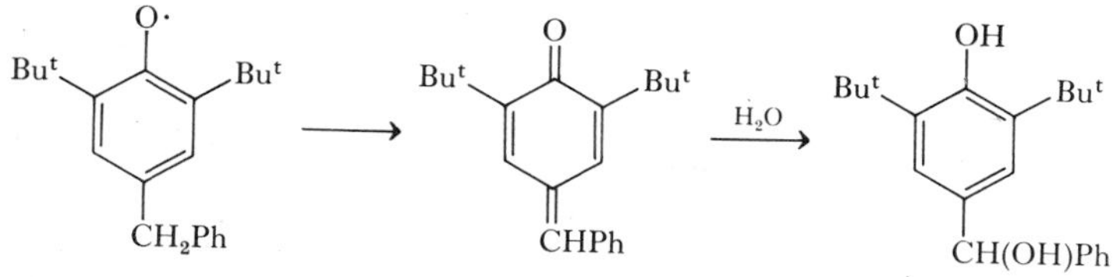

(95)

(25)

Both the dimeric products and also the alcohols could arise via benzylic radicals (26) resulting from hydrogen abstraction by the phenoxy radical. There is little evidence that this occurs in the oxidation of phenols though it could clearly occur particularly in the presence of excess phenol:

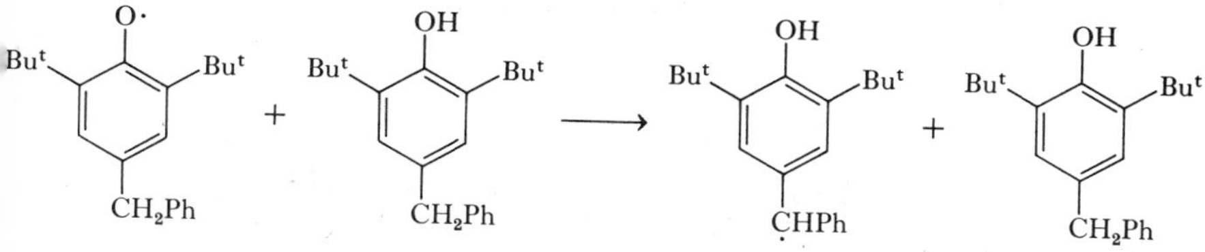

(96)

(26)

The resultant benzylic radicals could then dimerize or be further oxidized to the carbonium ion (27), hydration of which would afford the alcohol (28). It is certainly most improbable that intramolecular rearrangement of the phenoxy radical to a benzylic radical occurs, as has been proposed on several occasions (55a):

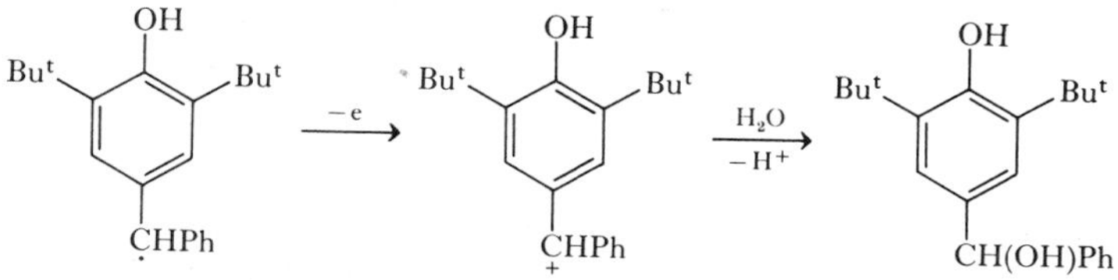

(97)

(26)   (27)   (28)

The unpaired electron in phenoxy radicals (29) obtained from

Oxidations and reductions

2,6-di-t-butyl-4-hydroxycinnamates can be further delocalized onto the side chain:

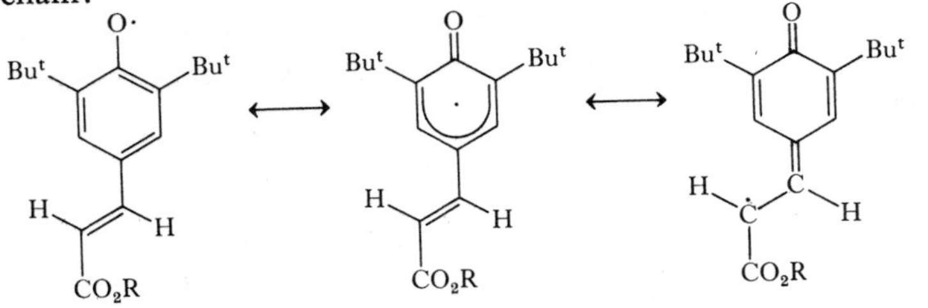

(29)

Dimerization of these radicals occurs at the α-carbon atom to give the bis-quinone methide (30) (56).

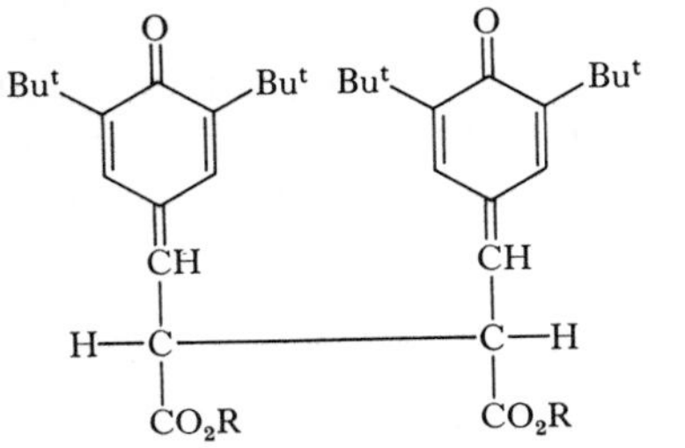

(30)

(iv) *Dihydric phenols*. Oxidation of catechols and hydroquinones yields initially the semiquinone radical anions. These are relatively

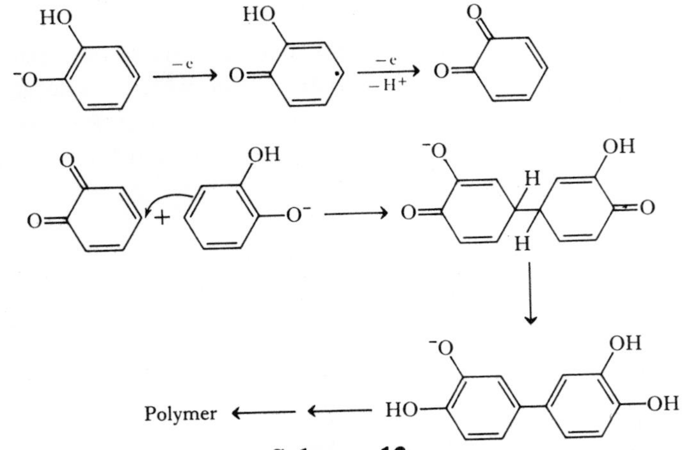

Scheme 12

stable and, rather than couple, are further oxidixed to *o*- and *p*-quinones respectively. The quinones thus obtained are very susceptible to attack by any nucleophilic species such as phenoxide anions, and hence give rise to coupled products (Scheme 12).

Resorcinols under anaerobic conditions give initially only C–C coupled products. Thus oxidation of orcinol (31) gives the C–C dimer (32), but none of the C–O dimer (57):

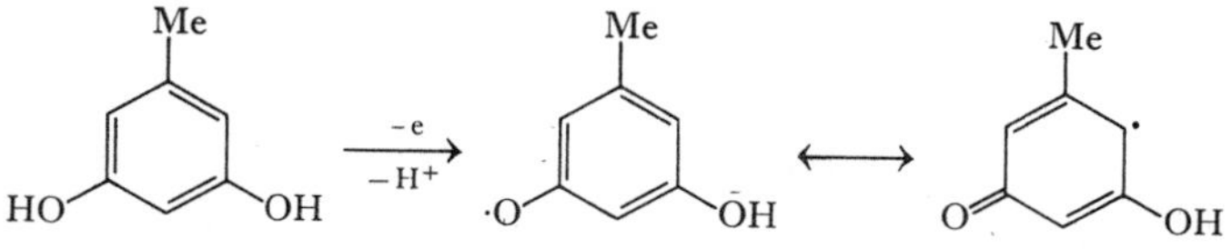

(31)

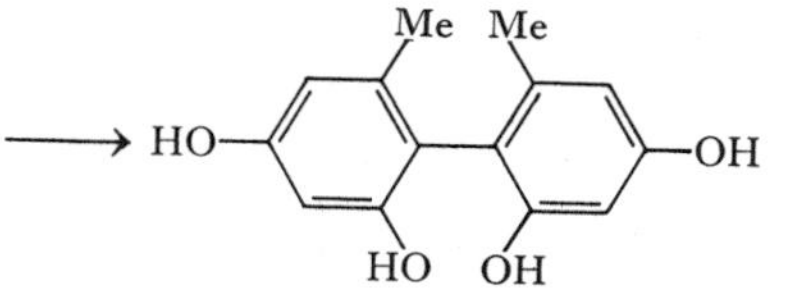

(32)

Further oxidation of the phenoxy radical does not occur as the formation of *m*-quinones is impossible.

Resorcinols behave differently on oxidation in air (58). The initial phenoxy radical (33) reacts with oxygen to give a hydroperoxy radical (34). The resultant hydroperoxide (35) on dehydration affords the quinone (36), which undergoes nucleophilic attack by the anion of orcinol to give the coupled product (37) (Scheme 13).

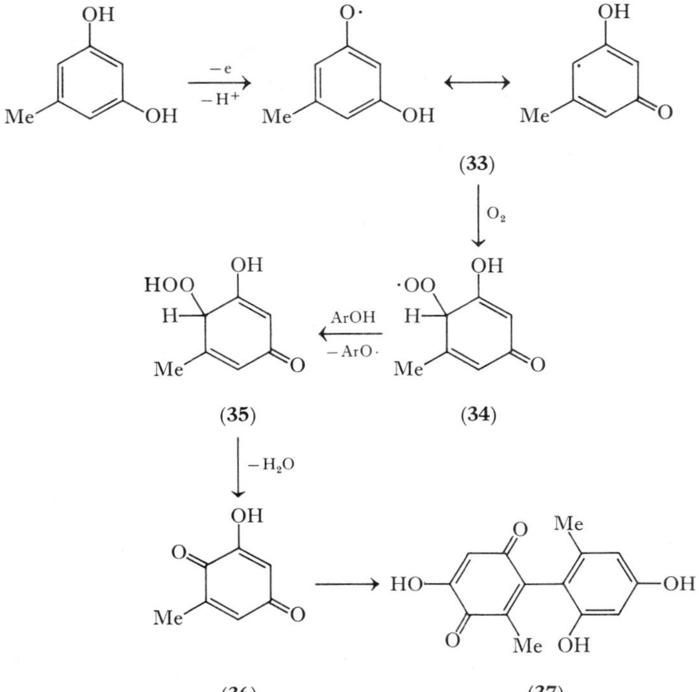

(33)

O₂

(35) ←ArOH / −ArO· (34)

−H₂O

(36) → (37)

Scheme 13

10.5.4 *Oxidation of phenols with nitroxides*

The foregoing discussion has considered metal-catalysed oxidations of phenols. Other one-electron oxidants, notably free radicals, can also be used to effect the homolytic oxidation of phenols. The use of nitroxides as oxidants has been studied in some detail (59). The chief difference between oxidation with metal ions and nitroxides is that the products of oxidation with the latter are quinones rather than the normal coupled products. This difference arises from the fact that the phenoxy radicals

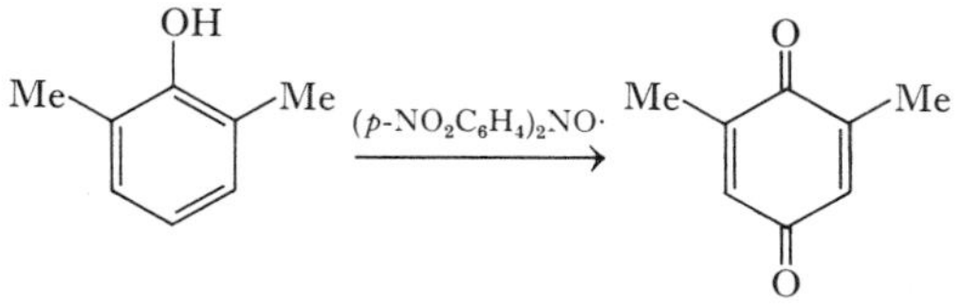

$$(p\text{-}NO_2C_6H_4)_2NO\cdot \tag{99}$$

(61%)

generated from the phenols couple with nitroxide radicals. The resultant product then breaks down to the quinone (Scheme 14).

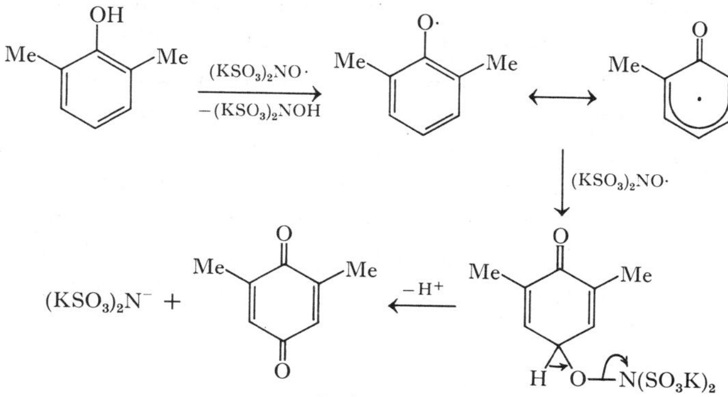

Scheme 14

Evidence supporting this mechanism comes from a study of the oxidation of 2,6-dimethylphenol with ^{17}O-labelled Fremy's salt (60). The labelled oxygen is incorporated to the extent of more than 97 % in the product, showing that the oxygen introduced into the phenol comes from the nitroxide and not from the solvent. An alternative mechanism for the formation of the quinone would involve the further oxidation of the phenoxy radical to the phenoxonium ion, solvation of which would have produced the quinone but without the incorporation of the oxygen of the nitroxide (Scheme 15).

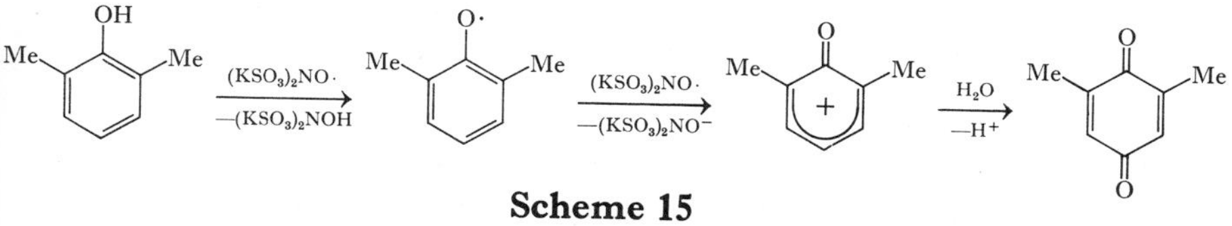

Scheme 15

10.5.5 *Oxidative coupling in the biosynthesis of natural products*

A crucial step in the biosynthesis of many alkaloids and other natural products, including lignins, aphid pigments and antibiotics, is now known to involve phenol oxidation by enzymes acting as one-electron oxidants (58, 61–63). Evidence supporting the theory of phenol oxidative coupling is largely based on experiments involving the feeding of labelled precursors. Such studies undertaken by Barton, Scott, Battersby and others have established the importance of phenolic coupling in the formation of C–C and C–O bonds in natural products. When analogous experiments have been carried out *in vitro* the yields of coupled product have invariably been very much lower than those obtained *in vivo*. This suggests that the enzymes in the plants or animals can fold the molecules being oxidized such that only one coupled product is obtained. In this connection it is notable that optically active materials are synthesized in the plant, whereas *in vitro* syntheses give racemic compounds.

The importance of oxidative phenol coupling in biosynthesis can be illustrated best by consideration of some specific examples. One of the best known of these is usnic acid which is clearly structurally related to Pummerer's ketone. Barton was able to synthesize the hydrate of usnic acid by the ferricyanide oxidation of *C*-methylphloroacetophenone (Scheme 16) (64).

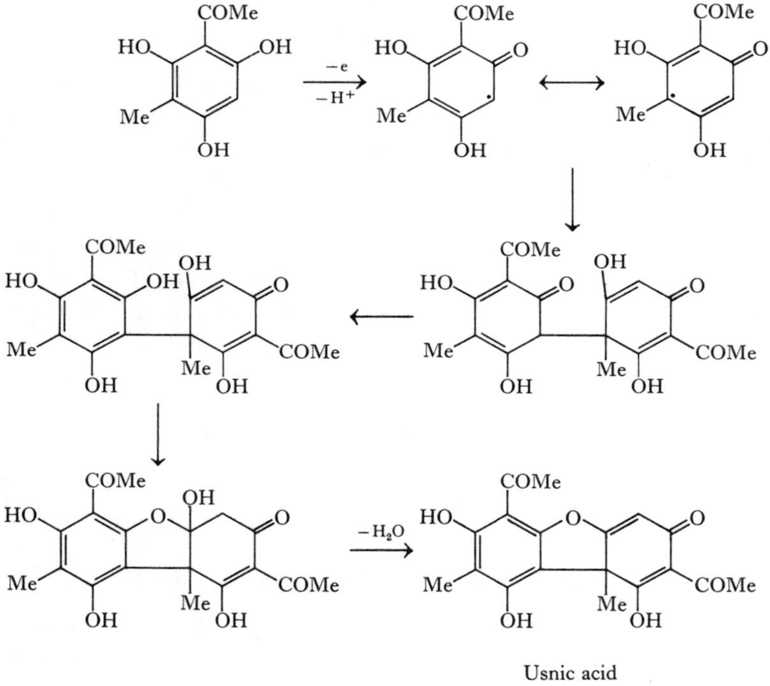

Usnic acid

Scheme 16

The biogenesis of more than 10% of known alkaloids is said to involve phenolic coupling. This is readily illustrated by reference to the benzylisoquinoline alkaloids. Reticuline has been shown by labelling experiments to be the precursor of both the aporphine and morphine alkaloids (Scheme 17) (65). Ferricyanide oxidation of reticuline has also been shown to produce a low yield of isoboldine, one of the former class of alkaloids (66).

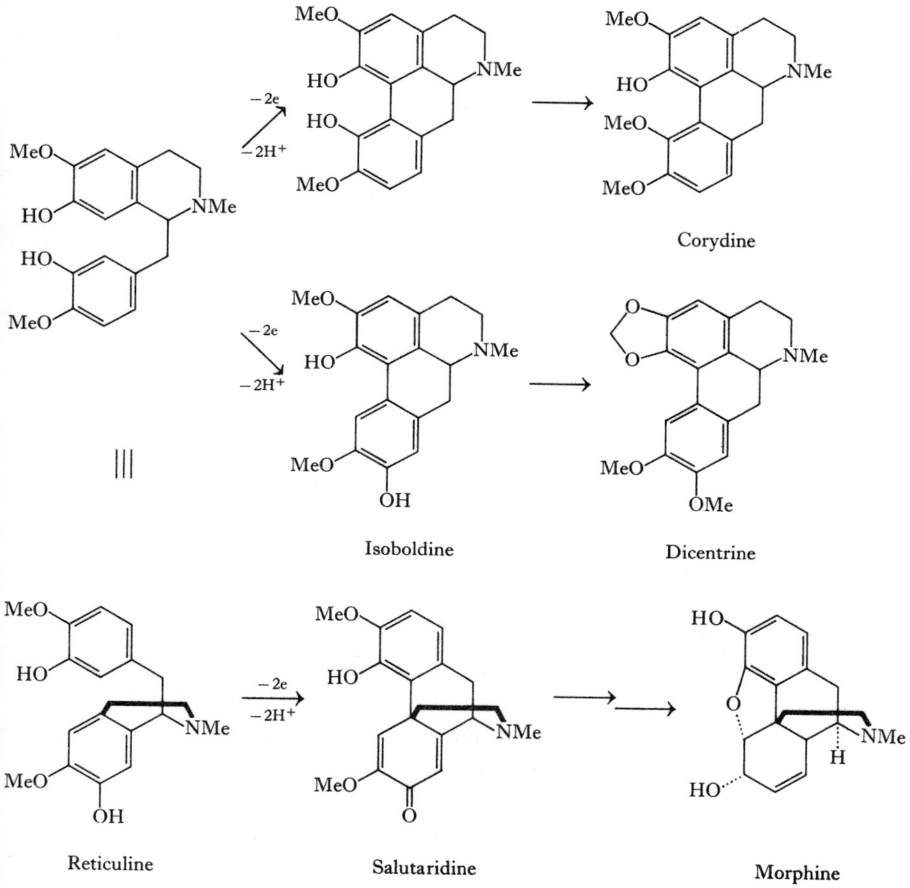

MeO
HO
HO
MeO
NMe
− 2e
− 2H⁺
Corydine
O
O
OH
OMe
Isoboldine
Dicentrine
Reticuline
Salutaridine
Morphine
H

Scheme 17

The biosynthesis of the culare alkaloids almost certainly involves C–O rather than C–C coupling of an isomer of reticuline (Scheme 18) (67).

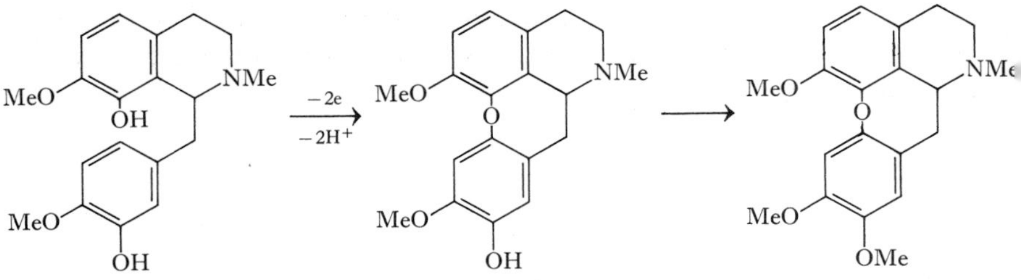

Cularine

Scheme 18

The fungal metabolites geodin (**39**, R = Me) and erdin (**39**, R = H) also exemplify C–O coupling. Their formation from the corresponding benzophenones (**38**) has been demonstrated *in vivo* and *in vitro* (66, 67):

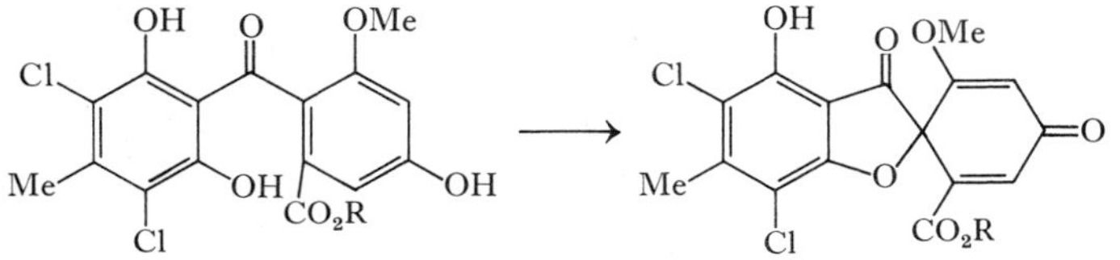

(38)   (39)

Oxidative coupling can also take place at the side chain if this is unsaturated as in coniferyl alcohol. Oxidative coupling of this gives rise to the lignans, such as guaiaretic acid (Scheme 19) (61).

The material making up lignin, which composes much of the woody tissues of plants, is similarly derived from coniferyl alcohol (and other alcohols of similar structure) by side-chain C–C, *ortho–ortho* C–C, side-chain C–O, and *ortho* C–O coupling reactions.

344

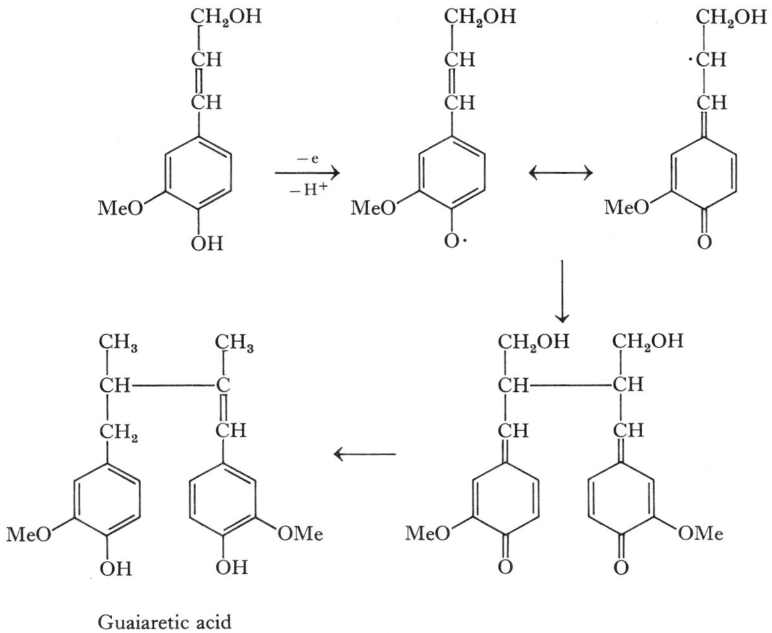

Guaiaretic acid

Scheme 19

10.6 *METAL AND METAL-ION CATALYSED REDUCTION OF ORGANIC COMPOUNDS*

Organic compounds can undergo homolytic reduction in much the same way as they undergo oxidation. The metal or metal ion acts as an electron source rather than as an electron acceptor. Rather fewer examples of this type of process are known, and consequently it will be discussed in less detail than oxidation. The metal-ion catalysed reductions of diazonium salts and peroxides will be discussed separately because, in both cases, interest lies more in the subsequent oxidation of the derived radicals.

10.6.1 *Dissolving-metal reductions*

One of the oldest and most useful reduction procedures is the dissolving-metal reduction or, as it is commonly called, Birch reduction (68–70). The technique involves the use of a solution of a strongly electropositive metal in an inert solvent, such as liquid ammonia, ethylamine, or other low-molecular-weight amine, containing a proton source. The proton source is frequently an alcohol. The metal functions as an electron

345

source resulting in the reduction of a double bond to a resonance-stabilized radical anion:

$$X{=}Y + M \longrightarrow [\ddot{X}{-}\dot{Y} \longleftrightarrow \dot{X}{=}\ddot{Y}] + M^+ \qquad (101)$$

The radical anion, thus generated, generally accepts a proton:

$$[\ddot{X}{-}\dot{Y} \longleftrightarrow \dot{X}{-}\ddot{Y}] + H^+ \longrightarrow \dot{X}{-}YH \qquad (102)$$

Further reduction of the resultant radical leads to the dihydro compound:

$$\dot{X}{-}YH + M \longrightarrow \ddot{X}{-}YH + M^+ \qquad (103)$$

$$\ddot{X}{-}YH + H^+ \longrightarrow HX{-}YH \qquad (104)$$

Extensive use has been made of this procedure for the reduction of a wide variety of organic compounds.

(i) *Aromatic compounds.* Benzene is reduced by lithium in liquid ammonia containing a little ethanol to 1,4-dihydrobenzene (71). Reduction of monosubstituted benzenes could lead to either the

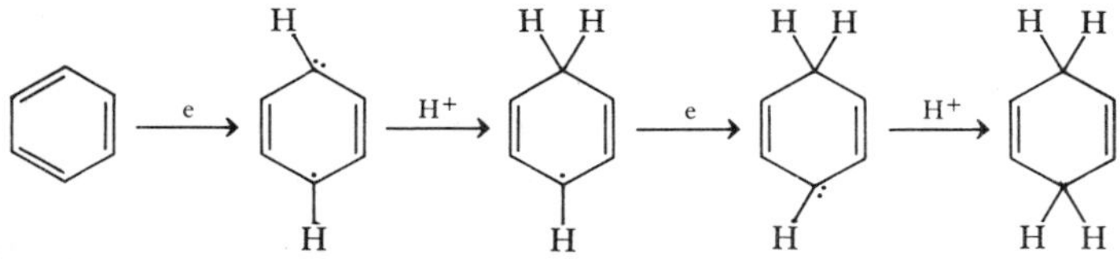

1-substituted-1,4-dihydrobenzene or the 1-substituted-2,5-dihydrobenzene. The electron adds to the aromatic system to give the most stable radical anion. In the case of toluene, molecular-orbital calculations show that the lowest vacant orbital has a node along the 1,4-axis and hence the methyl- and *para*-carbons should have approximately zero spin, whilst the *ortho*- and *meta*-carbons should have the same spin. (In agreement with this the e.s.r. spectrum of the toluene radical anion is a quintet (72).) Similarly, other monosubstituted benzenes with electron-donating substituents give rise to anions in which the electron density is greatest at the *ortho*- and *meta*-positions, and consequently reduction of anisole leads to the 1-methoxy-2,5-dihydrobenzene (73). The converse is true for benzenes containing electron-withdrawing substituents, when the 1-substituted-1,4-dihydrobenzene is obtained (74).

Controlled reduction of polycyclic aromatic hydrocarbons generally gives only the 9,10-dihydro compounds resulting from addition of

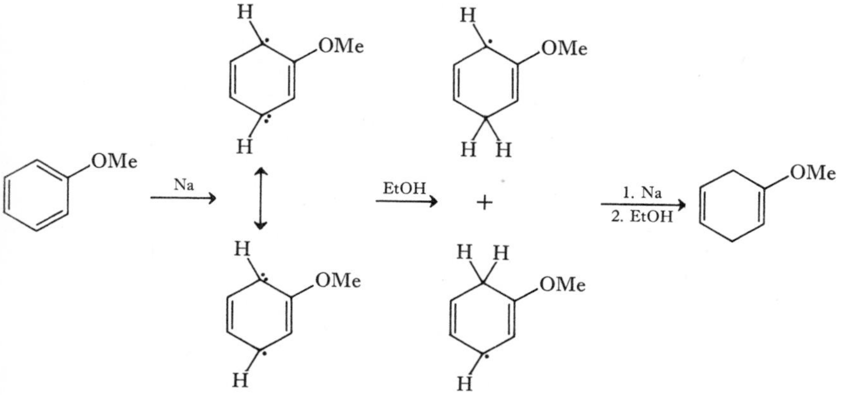

$$(106)$$

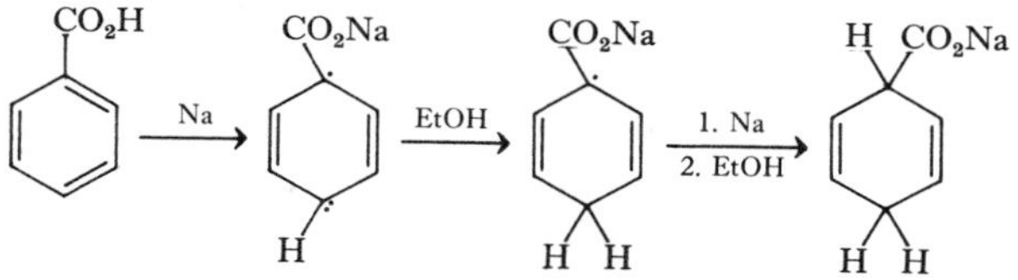

$$(107)$$

hydrogens at the positions predicted by molecular-orbital calculations to have the greatest charge density in the di-anion (75):

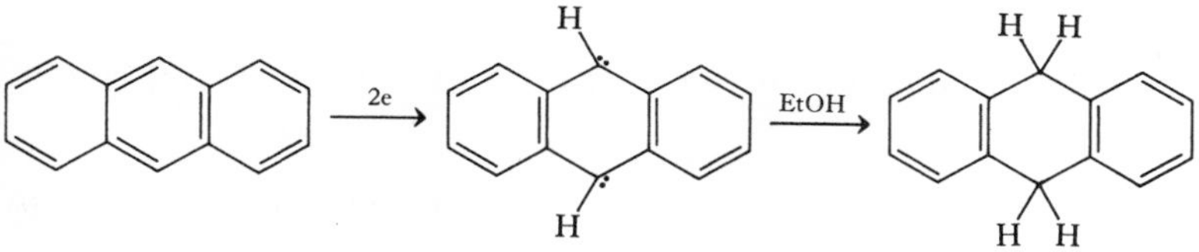

$$(108)$$

(ii) *Alkynes.* Reduction of alkynes with sodium in liquid ammonia is frequently the method of choice for the synthesis of *trans*-olefins (76):

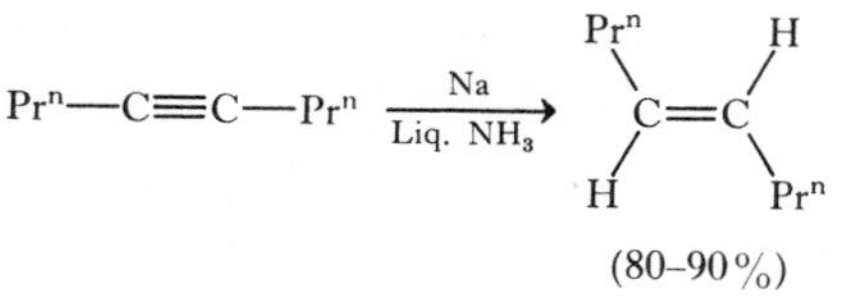

$$(109)$$

$$(80–90\%)$$

The reaction proceeds by addition of an electron to give a radical anion (**40**), protonation of which gives the *trans*-radical (**41**). The preference for the formation of the *trans*-radical can be considered analogous to the *trans*-addition of electrophilic reagents to alkynes. Subsequent reduction

347

and protonation of the derived anion (42) leads to the product (Scheme 20).

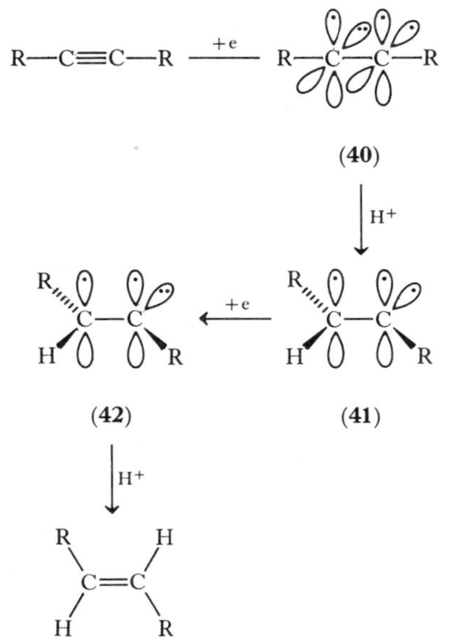

$$R-C{\equiv}C-R \xrightarrow{+e} R-C{\cdots}C-R \quad (40)$$

(40)

$$\downarrow H^+$$

(42) $\xleftarrow{+e}$ (41)

$$\downarrow H^+$$

$$\underset{H}{\overset{R}{>}}C{=}C\underset{R}{\overset{H}{<}}$$

Scheme 20

(iii) *Carbonyl groups.* Reduction of ketones with sodium (either in solution in liquid ammonia or the free metal) gives the corresponding alcohol (reaction *110*) (77):

$$\text{n-C}_5\text{H}_{11}\text{COCH}_3 \longrightarrow \text{n-C}_5\text{H}_{11}\text{CH(OH)CH}_3 \qquad (110)$$
$$(60\%)$$

The reaction proceeds by transfer of an electron to give the radical anion (43), protonation of which occurs from the less hindered side to give the alkoxy radical (44). Subsequent electron transfer leads to the alkoxide (45), and thence to the product:

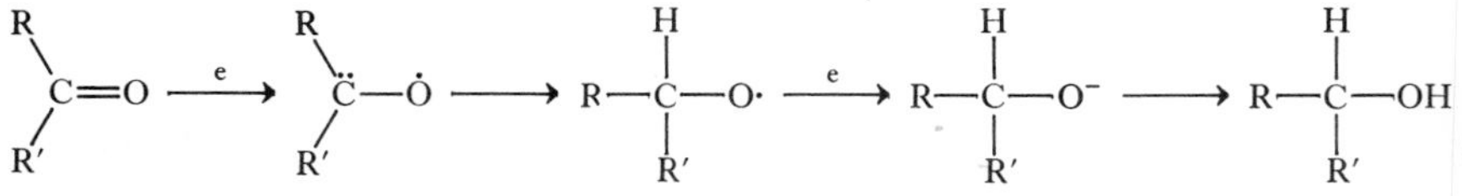

$$\underset{R'}{\overset{R}{>}}C{=}O \xrightarrow{e} \underset{R'}{\overset{R}{>}}\ddot{C}-\dot{O} \longrightarrow R-\underset{R'}{\overset{H}{\underset{|}{C}}}-O\cdot \xrightarrow{e} R-\underset{R'}{\overset{H}{\underset{|}{C}}}-O^- \longrightarrow R-\underset{R'}{\overset{H}{\underset{|}{C}}}-OH$$

(43)   (44)   (45)

If the reduction is carried out under anhydrous aprotic conditions,

348

dimerization of the radical anion occurs to give, after protonation in the work-up procedure, the pinacol (78):

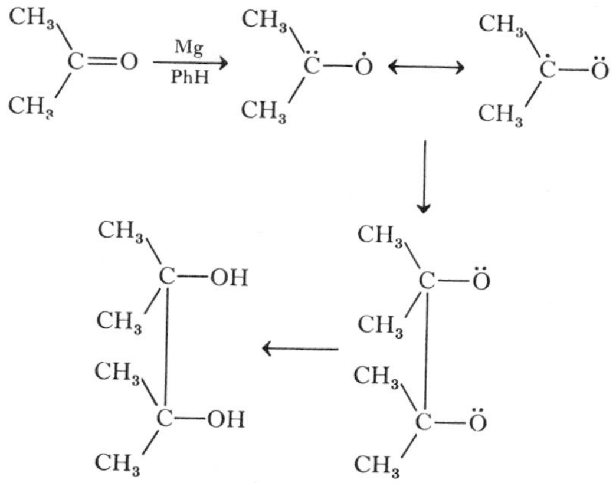

$$(112)$$

Reduction of esters with sodium in ethanol also results in the formation of alcohols (reaction *113*) (79) by the scheme outlined (Scheme 21).

$$\text{n-}C_{11}H_{23}CO_2Et \xrightarrow[\text{EtOH}]{\text{Na}} \text{n-}C_{11}H_{23}CH_2OH \qquad (113)$$

$$(65\text{--}75\,\%)$$

In the absence of a proton source, dimerization of the radical anion occurs to give eventually the acyloin (Scheme 21). This procedure is of considerable synthetic value particularly in the synthesis of medium and large-ring acyloins (reaction *114*) (80, 81).

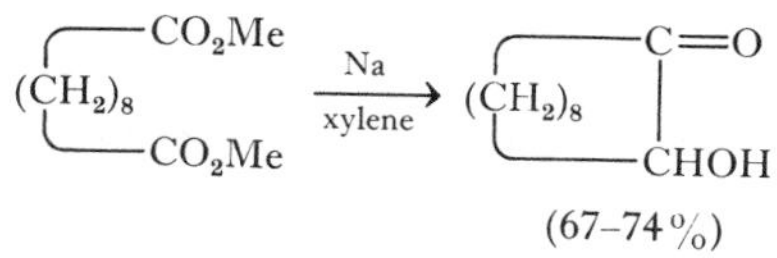

$$(114)$$

$$(67\text{--}74\,\%)$$

10.6.2   *Reductions with chromium(II) salts*

Chromium(II) salts have been successfully employed for the reduction of organic halides to hydrocarbons (82):

$$CH_2{=}CHCH_2Br \xrightarrow{CrSO_4} CH_2{=}CHCH_3 \qquad (115)$$

$$(100\,\%)$$

Simple chromium(II) salts are effective reductants for the reactive allylic, benzylic and α-carbonyl halides. This reactivity can be dramatic-

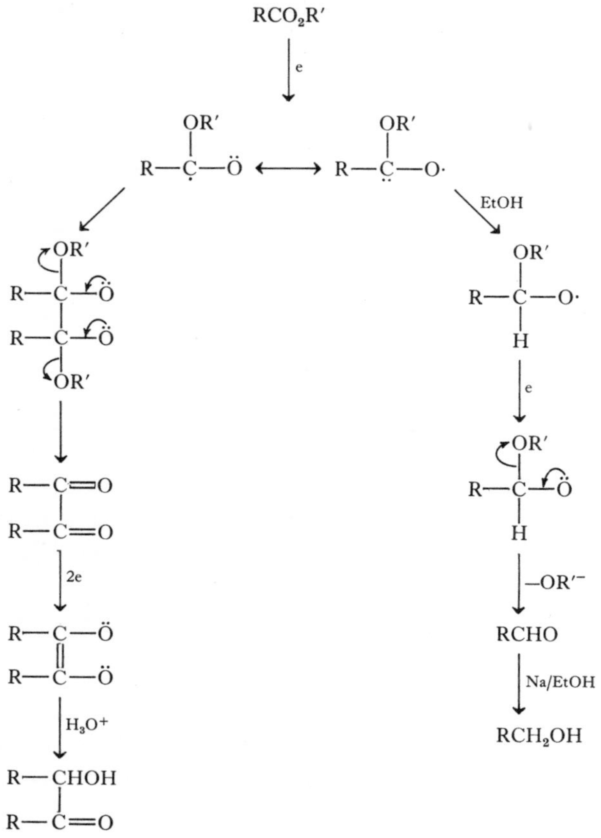

$$RCO_2R'$$

$\downarrow$ e

$$\underset{OR'}{R-\overset{|}{C}-\ddot{O}} \longleftrightarrow \underset{OR'}{R-\overset{|}{C}-O\cdot}$$

(left branch)

$$R-\overset{|OR'}{\underset{|}{C}}-\overset{\frown}{O}$$
$$R-\overset{|}{\underset{|OR'}{C}}-\overset{\frown}{O}$$

$\downarrow$

$$R-C=O$$
$$R-C=O$$

$\downarrow$ 2e

$$R-\overset{|}{C}-\ddot{O}$$
$$R-\overset{||}{C}-\ddot{O}$$

$\downarrow$ H_3O^+

$$R-CHOH$$
$$R-\overset{|}{C}=O$$

(right branch, EtOH)

$$\underset{H}{R-\overset{OR'}{\overset{|}{C}}-O\cdot}$$

$\downarrow$ e

$$\underset{H}{R-\overset{OR'}{\overset{|}{C}}-\overset{\frown}{\ddot{O}}}$$

$\downarrow$ $-OR'^-$

$$RCHO$$

$\downarrow$ Na/EtOH

$$RCH_2OH$$

Scheme 21

ally increased by complexing the chromium(II) ion with ligands such as ethylenediamine, and ethanolamine, such that these complexed compounds can readily effect the reduction of primary alkyl chlorides (83).

The mechanism of the reaction has been demonstrated both by Castro and Kray (84), and by Kochi and his co-workers (85) to be as shown in the following reactions:

$$PhCH_2X + Cr^{2+} \longrightarrow PhCH_2\cdot + CrX^{2+} \qquad (116)$$

$$PhCH_2\cdot + Cr^{2+} \longrightarrow PhCH_2Cr^{2+} \qquad (117)$$

$$PhCH_2Cr^{2+} + X^- \longrightarrow PhCH_2CrX^+ \qquad (118)$$

$$PhCH_2CrX^+ + H^+ \longrightarrow PhCH_3 + CrX^{2+} \qquad (119)$$

The first stage of the process is a ligand-transfer reaction between the organic halide and chromium(II) with generation of a radical. Reaction of this with chromium(II) gives the alkylchromium ions in a fast reaction. The benzylchromium(III) ion can be detected spectroscopically in the reaction between benzyl halides and chromium(II) perchlorate. The protonolysis of the alkylchromium species is a two-stage process. The first stage involves association of the alkylchromium ion with a nucleophile followed by a rate-determining protonolysis of this adduct. Evidence for this comes from the dependence of the rate of disappearance of the benzylchromium ion on the anion.

The reduction of benzyl halides and of other aralkyl halides leads to dimeric rather than reduced products. There is considerable evidence that these dimers arise not from radical coupling but from the reaction of free aralkyl halide with the aralkylchromium ion (reaction *120*) (86):

$$ArCH_2X + ArCH_2Cr^{2+} \longrightarrow ArCH_2CH_2Ar + CrX^{2+} \qquad (120)$$

β-Substituted bromoalkanes lead to both alkenes and the reduced compound, the proportion of alkene being greater the better the leaving group. Thus β-halo-, alkoxy-, acetoxy- and tosyloxy-alkyl bromides give exclusively alkenes, whereas when the leaving group is azido or cyano the reduced compound is the principal product (87):

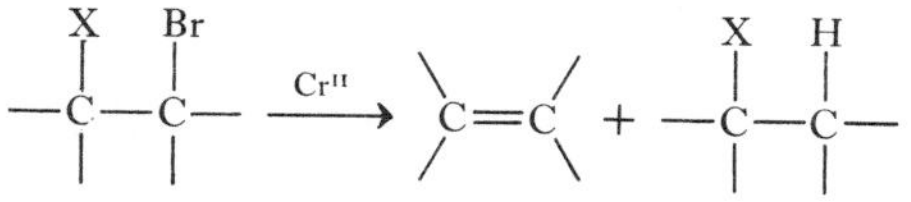

$$ \qquad (121)$$

Barton has shown that when the above reaction is carried out in the presence of a good hydrogen donor such as butane thiol, the initial alkyl radical undergoes a hydro-gentransfer reaction rather than form the alkylchromium ion and subsequently the elimination product (88):

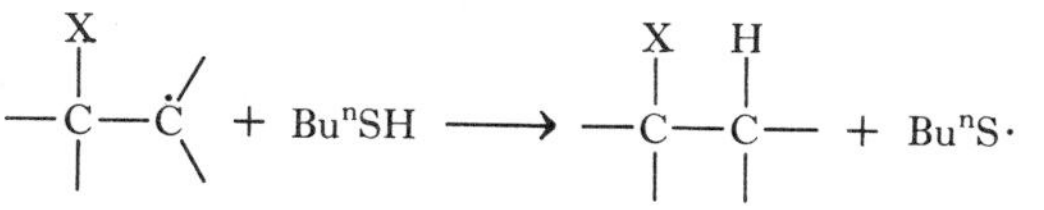

$$ \qquad (122)$$

$$2\ Bu^nS\cdot \longrightarrow Bu^nSSBu^n \qquad (123)$$

Other hydrogen donors have also been used, including 1,4-dihydrobenzene, hypophosphorous acid and triphenyltin hydride. This method has been used particularly effectively for the synthesis of β-hydroxy steroids (reaction *124*).

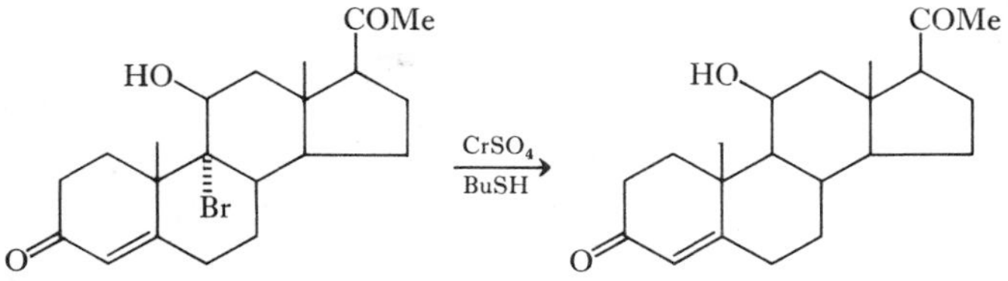

$$(124)$$

(78%)

In the absence of a suitable hydrogen donor, the major product was the diene and none of the reduced compound was obtained.

10.6.3 *Reductions with trialkyl and triphenyl tin hydrides*

Tri-n-butyl- and triphenyl-tin hydrides have been extensively used in recent years to accomplish the reduction of alkyl, cycloalkyl, acyl and aryl halides to the corresponding hydrocarbons (cf. p. 299):

$$RX + R_3'SnH \longrightarrow RH + R_3'SnH \qquad (125)$$

Kuivila (89) has clearly established the radical nature of the reaction for which he has proposed the mechanism outlined in the reactions below:

$$Q\cdot + R_3'SnH \longrightarrow HQ + R_3'Sn\cdot \qquad (126)$$

$$R_3'Sn\cdot + RX \longrightarrow R_3'SnX + R\cdot \qquad (127)$$

$$R\cdot + R_3'SnH \longrightarrow RH + R_3'Sn\cdot \qquad (128)$$

$$R_3'Sn\cdot + R_3'Sn\cdot \longrightarrow R_3'SnSnR_3' \qquad (129)$$

The initiation is brought about by adventitious traces of oxygen, though in some instances, with less reactive halides, it is necessary to add a radical initiator such as azoisobutyronitrile. Evidence for the intermediacy of free radicals includes the observations that optically active 1-phenylethyl chloride gives racemic 1-deuterioethylbenzene (reaction *130*), that α-methylallyl chloride and crotyl chloride both give the same mixture of butenes (reaction *131*), and that the reaction is retarded by radical inhibitors such as hydroquinone.

$$(+)\text{-PhCHClCH}_3 \xrightarrow{\text{Ph}_3\text{SnD}} (\pm)\text{-PhCHDCH}_3 \qquad (130)$$

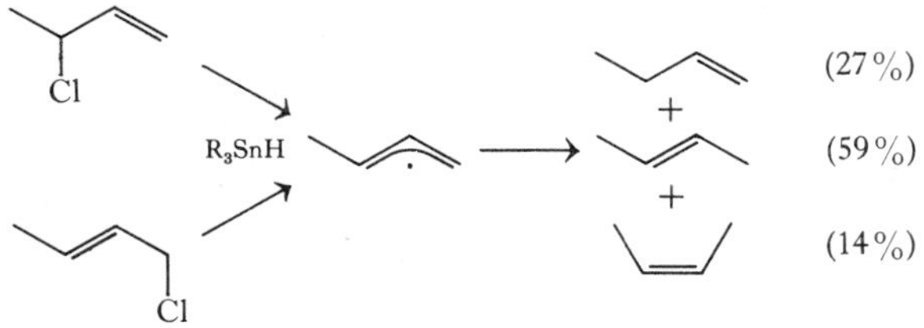

(27%)

+

(59%) $$(131)$$

+

(14%)

Carlsson and Ingold (90), from a kinetic study of the reduction of alkyl chlorides, concluded that the rate-determining step is the abstraction of chlorine by an organotin radical (reaction *127*), whilst for alkyl bromides the rate-controlling step is hydrogen abstraction from the organotin hydride (reaction *128*). The deuterium isotope effect for this step, k_H/k_D, is about 3.

The rate of chlorine abstraction from benzylic chlorides is slightly increased by electron-withdrawing substituents. The ability of tin to accommodate a positive charge enables one to consider polar contributions to the resonance hybrid for the transition state.

$$R \cdots X \cdots Sn \longleftrightarrow R^- \quad \cdot XSn^+ \longleftrightarrow R\cdot \quad X^-Sn^+$$

The effect is only small, as would be expected for a ligand-transfer reaction.

The rate of hydrogen abstraction by t-butyl radicals from tin hydrides is in the order: $Me_3SnH > Bu_3SnH > Ph_3SnH$

This has been attributed to the greater stability of the triphenyltin radical compared to the trimethyltin radical, and also to the weaker Sn–H bond in triphenyltin hydride. This arises from the decreased importance of the polar contribution $\overset{\delta+}{>}Sn-\overset{\delta-}{H}$ because of the $-I$ effect of the phenyl groups.

Reductions with triphenyl- and tri-n-butyltin hydrides have considerable synthetic importance (91) (reaction *132*):

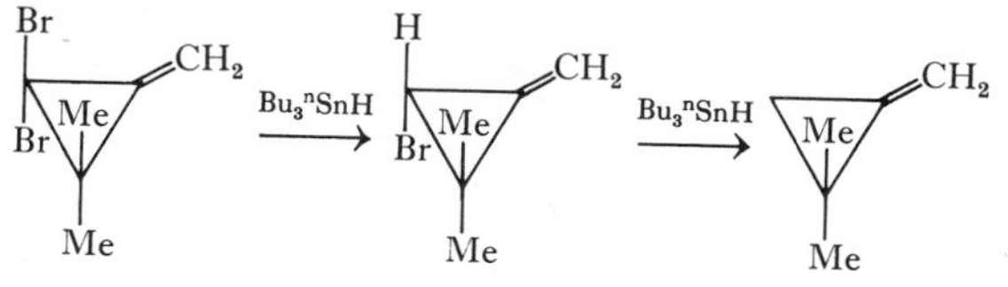

(68%)   (53%)   (*132*)

The different reactivity of iodides, bromides and chlorides enables the more reactive halogen to be removed specifically from dihalides (92):

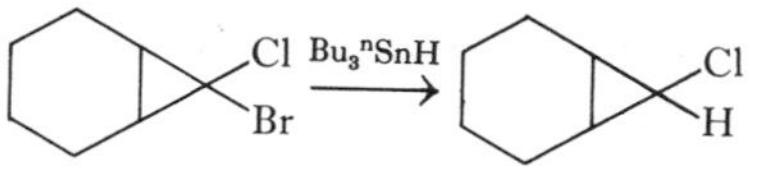

(97%)   (*133*)

A particularly useful procedure is to use a mixture of lithium aluminium hydride and an organotin halide for the *in situ* formation of organotin hydride provided the substrate has no groups which would be reduced by the lithium aluminium hydride. In such instances only catalytic amounts of the organotin compound need be used with one mole of lithium aluminium hydride (93):

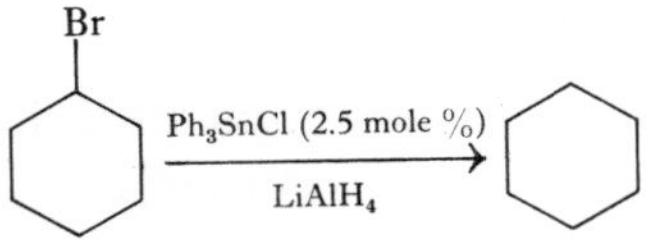

$$(134)$$

The use of organosilanes as alternatives to organotin hydrides has been advocated by Jackson in view of their greater ease of handling (94). There are a few reports in the literature of the reduction of organic halides by organosilanes (95) but they do not seem to have been generally used in synthesis.

10.7 REDOX REACTIONS OF DIAZONIUM SALTS

Many reactions of diazonium salts proceed by a radical mechanism. This is particularly true of those reactions which are catalysed by copper(I) salts or copper, including the Sandmeyer, Gattermann, Meerwein and Pschorr reactions (reactions *135–138*).

$$ArN_2Cl \xrightarrow{CuCl} ArCl + N_2 \qquad (135)$$

$$ArN_2Cl \xrightarrow{Cu} ArCl + N_2 \qquad (136)$$

$$ArN_2Cl + CH_2{=}CHX \xrightarrow{CuCl} ArCH_2CHClX + N_2 \qquad (137)$$

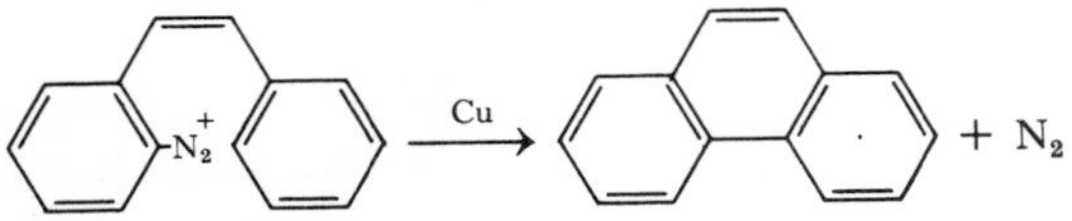

$$(138)$$

The key step in all these reactions is the electron-transfer reduction of the diazonium salt with the resultant generation of aryl radicals (reactions *139* and *140*) (96–98).

$$ArN_2^+ + CuCl_2^- \longrightarrow Ar\cdot + N_2 + CuCl_2 \qquad (139)$$

$$ArN_2^+ + Cu \longrightarrow Ar\cdot + N_2 + Cu^+ \qquad (140)$$

In the Sandmeyer reaction the aryl radicals then undergo a ligand-transfer reaction with copper(II) chloride, generated in the first stage of the reaction, or dimerize to biaryls (reaction *142*).

$$Ar\cdot + CuCl_2 \longrightarrow ArCl + CuCl \qquad (141)$$

$$Ar\cdot + Ar\cdot \longrightarrow Ar\text{---}Ar \qquad (142)$$

As Waters has pointed out, the salient feature of this mechanism is that it can only occur within closely defined redox potential limits (99). The redox potentials of the Cu^+/Cu^{2+} and Cu/Cu^+ systems in aqueous solution allow for both facile oxidation and reduction as envisaged in the two steps of the Sandmeyer and Gattermann reactions.

The catalyst in the Meerwein reaction is copper(II) chloride in aqueous acetone, which Kochi has shown gives copper(I) chloride and chloroacetone (100):

$$CH_3COCH_3 + 2CuCl_2 \longrightarrow CH_3COCH_2Cl + 2CuCl + HCl \quad (143)$$

The copper(I) chloride can then reduce the diazonium salt with the generation of aryl radicals. The aryl radicals thus produced may add to the unsaturated compound (reaction *144*), react with acetone to give the reduced product (reaction *146*), react with copper(II) chloride to give the Sandmeyer product (reaction *141*), or dimerize (reaction *142*). The relative concentrations of alkenes and copper(II) chloride determine the importance of these competing reactions:

$$Ar\cdot + CH_2{=}CHY \longrightarrow ArCH_2\dot{C}HY \qquad (144)$$

$$ArCH_2\dot{C}HY + CuCl_2 \longrightarrow ArCH_2CHClY + CuCl \qquad (145)$$

$$Ar\cdot + CH_3COCH_3 \longrightarrow ArH + CH_3COCH_2\cdot \qquad (146)$$

$$CH_3COCH_2\cdot + CuCl_2 \longrightarrow CH_3COCH_2Cl + CuCl \qquad (147)$$

The intervention of aryl radicals in the Sandmeyer and Gattermann reactions has been confirmed by their ability to promote polymerization of vinyl monomers (99). Further, under Meerwein and Gattermann conditions, it has been possible to arylate aromatic substrates to the same mixture of arylated products as that obtained with other sources of aryl radicals (101, 102).

The probable mechanism of the Pschorr reaction is outlined in Scheme 22. Cyclization of the intermediate radical occurs rather than reaction with solvent because of the close proximity of the radical centre to the second aromatic system. As is discussed later (p. 439), the

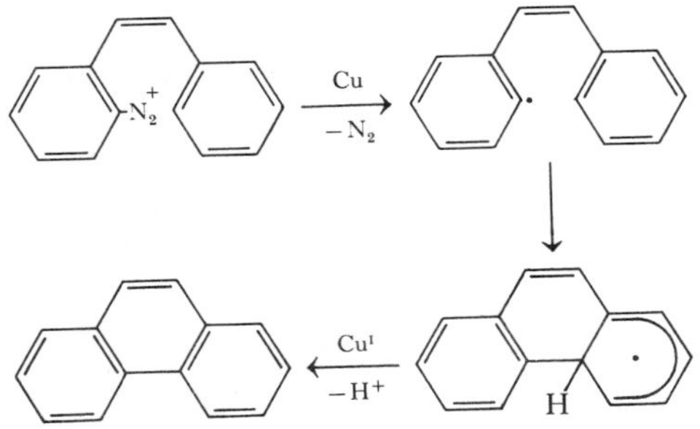

Scheme 22

Pschorr reaction only proceeds by a homolytic pathway when the cyclization step involves the formation of a planar six-membered ring (103).

Replacement of the diazonium group by nitro is a further example of a copper-catalysed reaction of a diazonium salt. The key step in this reaction is the aryl radical attack on a nitrite anion to give a nitro-aromatic radical anion (104), oxidation of which gives the products (reactions *148–150*):

$$ArN_2^+ + Cu \longrightarrow Ar\cdot + N_2 + Cu^+ \qquad (148)$$

$$Ar\cdot + NO_2^- \longrightarrow ArNO_2^{\overline{\cdot}} \qquad (149)$$

$$ArNO_2^{\overline{\cdot}} + Cu^+ \longrightarrow ArNO_2 + Cu \qquad (150)$$

The same reaction can be brought about using titanium(III) rather than copper as the reductant, and in that case the identity of the nitroaromatic radical anion has been confirmed by e.s.r. spectroscopy (105).

The one-electron reduction of diazonium salts with the consequent generation of aryl radicals can also be accomplished by other reducing agents having appropriate redox potentials, e.g. iodide ion.

$$ArN_2^+ + I^- \longrightarrow Ar\cdot + N_2 + I\cdot \qquad (151)$$

$$Ar\cdot + I\cdot \longrightarrow ArI \qquad (152)$$

Copper or copper(I) salts are unnecessary to promote this reaction.

The substitution of the diazonium group by hydrogen is a reaction of considerable synthetic utility (106). The reaction is best brought about using hypophosphorous acid as reductant. Again the key step in the reaction is the single-electron reduction of the diazonium salt. This is followed by a hydrogen transfer between the aryl radical and the

356

hypophosphorous acid (reactions *153–156*):

$$ArN_2^+ + H_2PO_2^- \longrightarrow Ar\cdot + N_2 + H_2PO_2\cdot \qquad (153)$$

$$ArN_2^+ + H_2PO_2\cdot \longrightarrow Ar\cdot + N_2 + H_2PO_2^+ \qquad (154)$$

$$Ar\cdot + H_3PO_2 \longrightarrow ArH + H_2PO_2\cdot \qquad (155)$$

$$H_2PO_2^+ + H_2O \longrightarrow H_3PO_3 + H^+ \qquad (156)$$

Supporting evidence for reaction (*155*) comes from the observation that there is very little incorporation of deuterium when the reaction is carried out using a fresh solution of hypophosphorous acid in heavy water (107). The extent of incorporation is much greater if the hypophosphorous acid is allowed to remain in contact with the heavy water for some time before the diazonium salt is added.

10.8 REDOX REACTIONS OF PEROXIDES

The thermolysis or pyrolysis of a variety of peroxide compounds, including hydroperoxides, dialkyl peroxides, diacyl peroxides and peresters, has been extensively used as a source of alkoxy or acyloxy radicals, and fragmentation of these radicals provides a source of alkyl radicals. The generation of radicals from these classes of compound is catalysed by single-electron reductants such as iron(II) and copper(I) salts. This section will be concerned with these reactions and with the subsequent reactions of the radicals produced, particularly when these involve further oxidation or reduction of the intermediate radical.

10.8.1 *Fenton's reaction*

One of the oldest and best known radical reactions is that between hydrogen peroxide and iron(II) sulphate discovered by Fenton in 1894:

$$Fe^{2+} + H_2O_2 \longrightarrow Fe^{3+} + OH^- + HO\cdot \qquad (157)$$

Clear evidence for the radical nature of this reaction came from the ability of free hydroxyl radicals to initiate polymerization of vinyl monomers. The resultant monomers were shown to contain hydroxyl end-groups (108):

$$HO\cdot + CH_2{=}CHCN \longrightarrow HOCH_2\dot{C}HCN$$

$$\xrightarrow{n(CH_2=CHCN)} HO(CH_2CHCN)_n CH_2\dot{C}HCN \qquad (158)$$

The hydroxyl radicals produced in Fenton's reaction may react with an organic substrate either by hydrogen abstraction (reaction *159*) or by addition to an unsaturated system (reaction *160*):

$$RH + HO\cdot \longrightarrow R\cdot + H_2O \qquad\qquad (159)$$

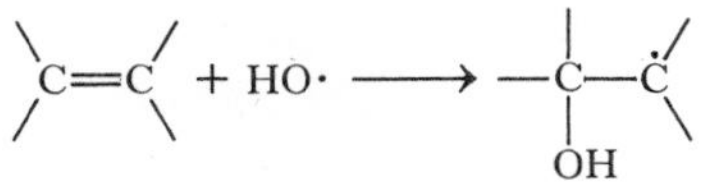

$$\ce{\overset{\textstyle\diagdown}{\underset{\textstyle\diagup}{C}}=\overset{\textstyle\diagup}{\underset{\textstyle\diagdown}{C}} + HO\cdot \longrightarrow -\overset{|}{\underset{\overset{|}{OH}}{C}}-\overset{\diagup}{\underset{\diagdown}{C}}} \qquad (160)$$

Coffman and his collaborators have used hydroxyl radicals to bring about oxidative coupling (109):

$$RH + HO\cdot \longrightarrow R\cdot + H_2O \qquad\qquad (161)$$

$$R\cdot + R\cdot \longrightarrow R\!-\!R \qquad\qquad (162)$$

Table 10-1 records details of some syntheses which have been accomplished in this way. The procedure is most effective for coupling of compounds with only one type of C–H bond. This is because hydroxyl radicals are very reactive and hence unselective abstraction takes place. The other limitation is that the reaction is confined to water-soluble compounds.

TABLE 10-1 *Preparation of dimeric products using Fenton's reagent* (109)

Substrate	Dimer	Yield (%)
Me_3COH	$HOCMe_2CH_2CH_2CMe_2OH$	36
Me_3CCO_2H	$HO_2CCMe_2CH_2CH_2CMe_2CO_2H$	37
CH_3CO_2H	$HO_2CCH_2CH_2CO_2H$	4
CH_3CN	$NCCH_2CH_2CN$	18
CH_3CH_2CN	$NCCH_2CH_2CH_2CH_2CN$	
	$NCCH(CH_3)CH(CH_3)CN$	60
	$NCCH(CH_3)CH_2CH_2CN$	

This procedure has been extended by carrying out the reactions in presence of a 1,3-diene (110). The alkyl radicals generated by hydrogen abstraction by hydroxyl radicals add to a 1,3-diene and the resultant radical dimerizes:

$$HO\cdot + Me_3COH \longrightarrow Me_2C(OH)CH_2\cdot + H_2O \qquad (163)$$

$$Me_2C(OH)CH_2\cdot + CH_2{=}CHCH{=}CH_2$$
$$\longrightarrow Me_2C(OH)CH_2CH_2CH{=}CHCH_2\cdot$$
$$\longrightarrow Me_2C(OH)CH_2CH_2CH{=}CHCH_2CH_2CH{=}CHCH_2CH_2C(OH)Me_2$$
$$(64\%) \qquad\qquad (164)$$

A certain amount of 1,2-addition also occurs.

There is little evidence for the formation of compounds with one or with more than two butadiene units. This is because the radical obtained by addition to the diene is resonance-stabilized and thus of much lower reactivity than the first alkyl radical.

The production of hydroxyl radicals from hydrogen peroxide can also be brought about by other single-electron reductants, e.g. cobalt(II), copper(I) and titanium(III). The last of these has been extensively used by Norman in his studies of the reactions of hydroxyl radicals with organic compounds (111). It possesses certain advantages over iron(II), namely that it can be used over a much wider pH range (c. 0.5–11) if a sequestering agent is used above pH 2, and also titanium(IV) is a weaker oxidant than iron(III). This latter is important since a radical, R·, produced from an organic compound, RH, may undergo oxidation to the corresponding cation particularly when the resultant cation is tertiary and/or is stabilized by electron-donating groups:

$$R· + Ti^{4+} \longrightarrow R^+ + Ti^{3+} \tag{165}$$

One might thus expect the titanium(III)–hydrogen peroxide system to be the preferred system for the synthesis of dimeric products. This does not appear to have been exploited at the present time.

Hydroxylation of aromatic compounds can also be effected by Fenton's reagent. Thus, benzene gives phenol and biphenyl by the mechanism outlined in Scheme 23 (112) (see also p. 461).

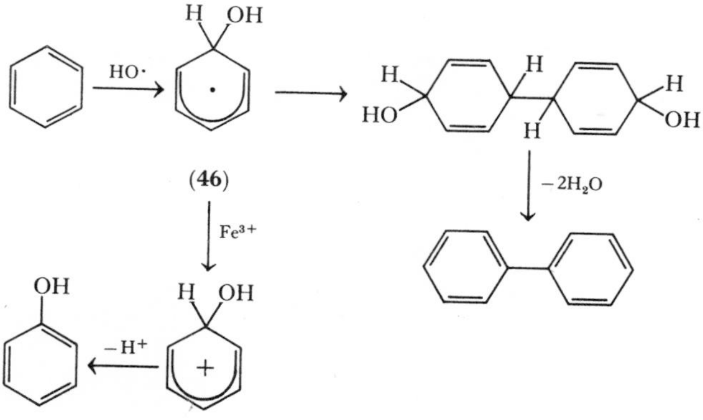

Scheme 23

Dixon and Norman (113) have observed the e.s.r. spectrum of the hydroxycyclohexadienyl radical (46). Oxidation of this followed by proton loss gives rise to phenol. Biphenyl arises from dimerization of the hydroxycyclohexadienyl radical followed by dehydration. The proportion of phenol increases with an increase in the concentration of iron(III), and is decreased by addition of fluoride which complexes with iron(III), thereby preventing oxidation of the radical (46). The formation of biphenyl is favoured by a high concentration of hydroxycyclohexadienyl radicals, whose formation is favoured by a high concentration of hydrogen peroxide.

10.8.2 *Reactions of hydroperoxides*

In the same manner as hydroxyl radicals are produced from hydrogen peroxide, alkoxy radicals are obtained from the reduction of hydroperoxides with iron(II) or other one-electron reductants (114, 115):

$$Bu^tOOH + Fe^{2+} \longrightarrow Fe^{3+} + OH^- + Bu^tO\cdot \qquad (166)$$

The behaviour of 1-substituted cycloalkyl hydroperoxides is interesting in that the derived alkoxy radicals undergo very ready fragmentation to give carbon radicals (reaction *167*) (116):

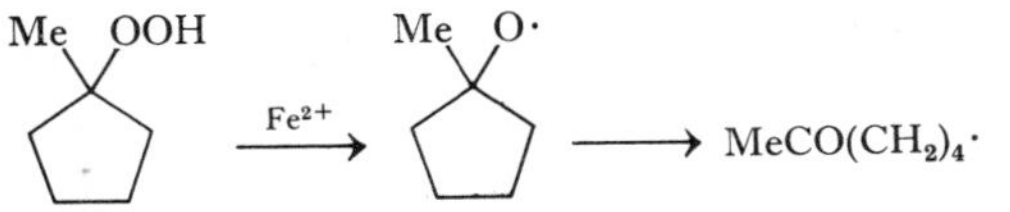

$$(167)$$

These latter may then either dimerize (reaction *168*), undergo electron-transfer oxidation (reaction *169*), or ligand-transfer oxidation (reactions *170–174*) according to the particular reaction conditions (117–119):

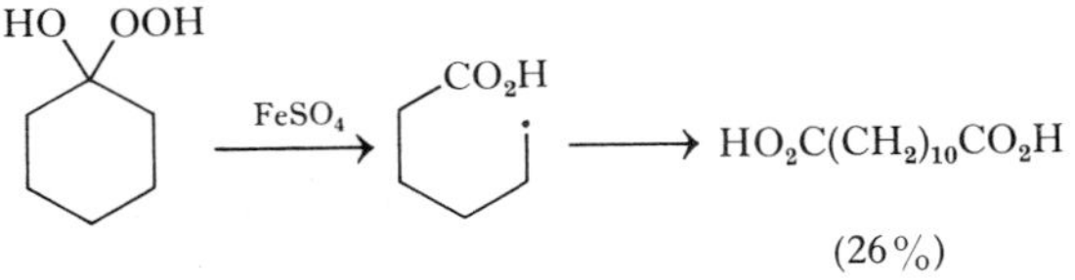

$$(168)$$

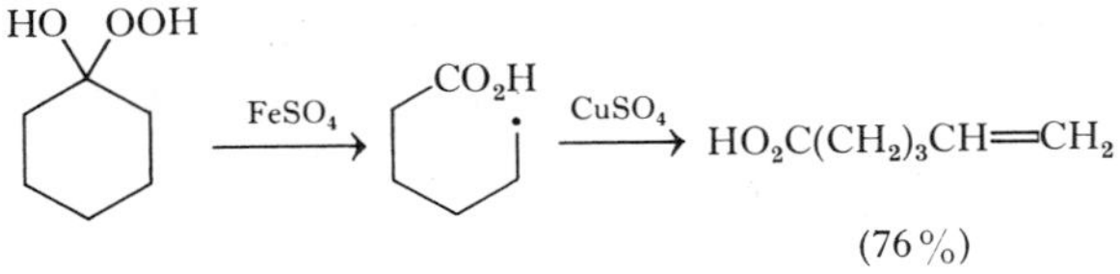

(26%)

$$(169)$$

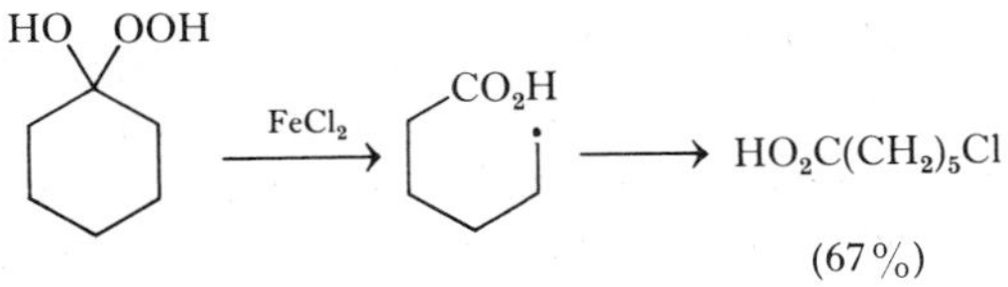

(76%)

$$(170)$$

(67%)

$$\text{(171)} \qquad \begin{array}{c} \text{HO} \quad \text{OOH} \\ \text{[cyclohexane ring]} \end{array} \xrightarrow{\text{CuBr}} \begin{array}{c} \text{CO}_2\text{H} \\ \text{[ring radical]} \end{array} \longrightarrow \text{HO}_2\text{C(CH}_2)_5\text{Br} \quad (49\%)$$

$$\text{(172)} \qquad \begin{array}{c} \text{HO} \quad \text{OOH} \\ \text{[cyclohexane ring]} \end{array} \xrightarrow{\text{Fe(SCN)}_6^{4-}} \begin{array}{c} \text{CO}_2\text{H} \\ \text{[ring radical]} \end{array} \longrightarrow \text{HO}_2\text{C(CH}_2)_5\text{SCN} \quad (64\%)$$

$$\text{(173)} \qquad \begin{array}{c} \text{HO} \quad \text{OOH} \\ \text{[cyclohexane ring]} \end{array} \xrightarrow{\text{FeSO}_4} \begin{array}{c} \text{CO}_2\text{H} \\ \text{[ring radical]} \end{array} \xrightarrow{\text{N}_3^-} \text{HO}_2\text{C(CH}_2)_5\text{N}_3 \quad (48\%)$$

$$\text{(174)} \qquad \begin{array}{c} \text{HO} \quad \text{OOH} \\ \text{[cyclohexane ring]} \end{array} \xrightarrow[\text{KCN}]{\text{CuCN}} \begin{array}{c} \text{CO}_2\text{H} \\ \text{[ring radical]} \end{array} \longrightarrow \text{HO}_2\text{C(CH}_2)_5\text{CN} \quad (28\%)$$

These equations give some indication of the synthetic possibilities of these reactions.

10.8.3 *Reactions of peresters*

The single-electron catalysed reduction of peresters is another useful route to alkoxy radicals (reaction *175*):

$$\text{RCO}_3\text{R}' + \text{Cu}^{\text{I}} \longrightarrow \text{RCO}_2\text{Cu}^{\text{II}} + \text{R}'\text{O}\cdot \qquad (175)$$

The reaction leading to alkoxy radicals is energetically more favourable than the alternative pathway which would give acyloxy radicals (reaction *176*):

$$\text{RCO}_3\text{R}' + \text{Cu}^{\text{I}} \longrightarrow \text{RCO}_2\cdot + \text{R}'\text{OCu}^{\text{II}} \qquad (176)$$

Copper(I) salts, however, do more than merely promote the production of radicals: they markedly influence the course of subsequent reactions (cf. reactions *177* and *178*):

$$\text{PhCHMe}_2 + \text{PhCO}_3\text{Bu}^t \xrightarrow{\Delta} \text{PhCMe}_2\text{CMe}_2\text{Ph} \qquad (177)$$

$$\text{PhCHMe}_2 + \text{PhCO}_3\text{Bu}^t \xrightarrow{\text{Cu}^{\text{I}}} \text{PhCMe}_2\text{OCOPh} \qquad (178)$$

Oxidations and reductions

This latter reaction, which results in the selective oxidation at carbon atoms, has been very extensively studied from both synthetic and mechanistic aspects by the groups of Kharasch (119), Sosnovsky (120), Lawesson (121), and Kochi (122). It has the advantage over the analogous reaction with hydroperoxides in that it is more selective, more versatile, and is of greater synthetic utility, since the products (esters) are much more readily converted into other derivatives than ethers. An example of the reaction is the synthesis of 3-benzoyloxycyclohexene from cyclohexene and t-butyl perbenzoate in the presence of copper(I) bromide in 71–80 % yield (reaction *179*) (123):

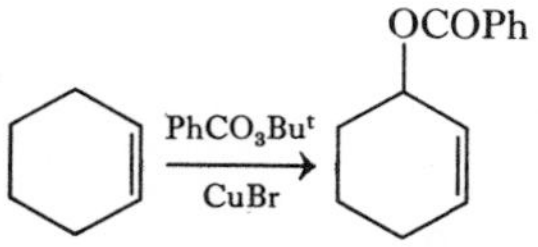

$$
\text{(179)}
$$

The mechanism of the reaction involves the generation of an alkoxy radical in the electron-transfer reduction of the perester (reaction *180*). This abstracts a hydrogen atom from the substrate (reaction *181*), and the resulting alkyl radical then undergoes a pseudo-ligand-transfer oxidation by the copper(II) carboxylate (reaction *182*):

$$RCO_3R' + Cu^I \longrightarrow RCO_2Cu^{II} + R'O\cdot \qquad (180)$$

$$R'O\cdot + R''H \longrightarrow R''\cdot + R'OH \qquad (181)$$

$$R''\cdot + RCO_2Cu^{II} \longrightarrow R''OCOR + Cu^I \qquad (182)$$

More detailed information on the final stage of this process has been obtained by studying the copper(I)-catalysed decomposition of t-butyl peracetate in butadiene (122). The proportions of the isomeric acetoxy-t-butoxybutenes (reaction *183*) differ from those obtained in the silver-catalysed acetolysis of 4-t-butoxy-1-chlorobut-2-ene (reaction *184*) which undoubtedly proceeds via a t-butoxybutenyl cation. The proportions of the two products also differ from those of t-butoxychlorobutenes obtained in the ligand-transfer oxidation of t-butoxybutenyl radicals by copper(II) chloride (reaction *185*):

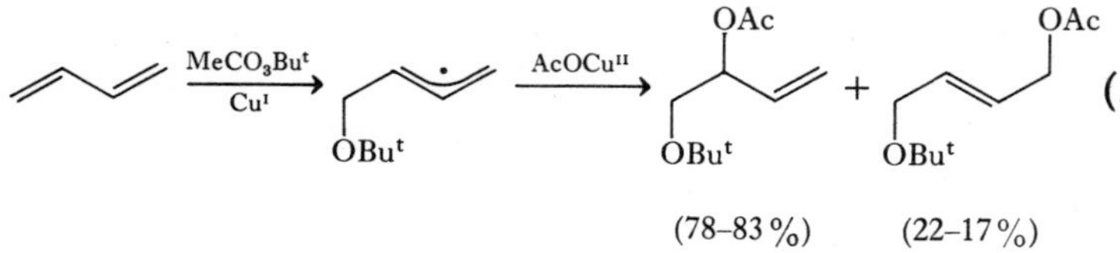

$$
\text{(183)}
$$

(78–83 %)   (22–17 %)

362

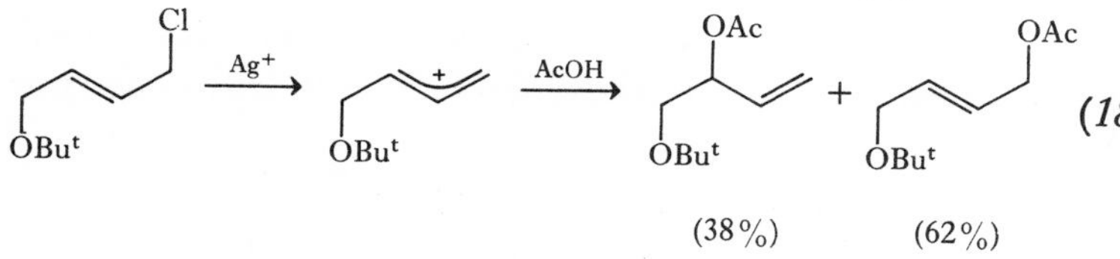

$$\text{(184)}$$

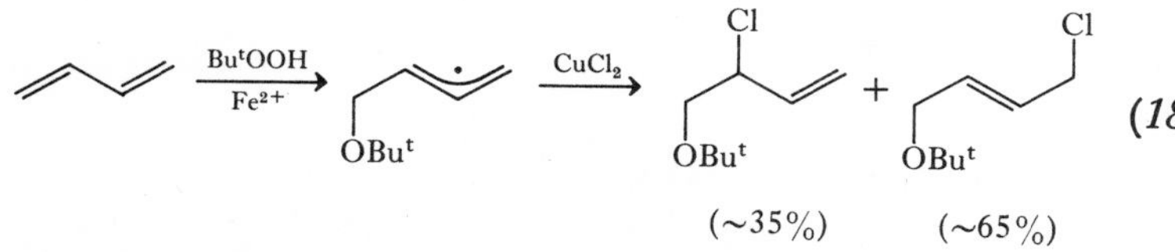

(38%)   (62%)

$$\text{(185)}$$

(~35%)   (~65%)

These results indicate that the oxidation of t-butoxybutenyl radicals in the copper(I)-catalysed decompositions of peresters is neither a pure electron-transfer nor a pure ligand-transfer process. The transition state can best be represented by (47) which has both electron- and ligand-transfer components.

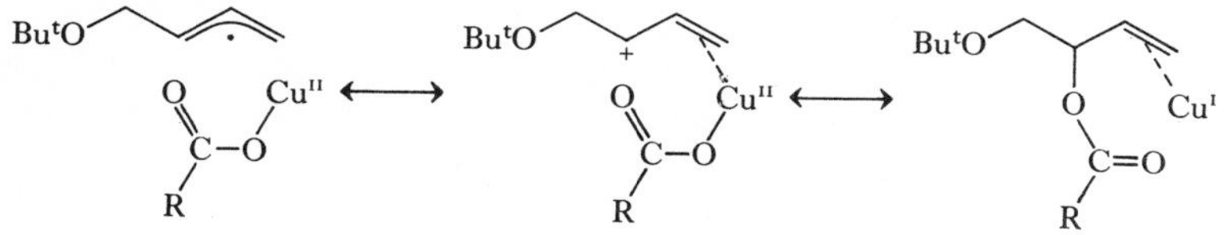

(47)

The reactions are remarkably insensitive to solvent changes. Even in protic solvents such as methanol and aqueous t-butanol, the proportions of butenyl ethers and butenols is very low. This would not be the case were free carbonium ions involved.

The formation of predominantly *trans*-2-benzoyloxy-4-t-butyl-1-methylenecyclohexane (49) from 4-t-butyl-1-methylenecyclohexane (48) is in accord with the above mechanism (124). Co-ordination of the copper to the double bond would be expected to occur preferentially but not exclusively on the least hindered side of the molecule, i.e. on the side of the molecule *trans* to the t-butyl group thereby directing attack onto the axial position at C-2.

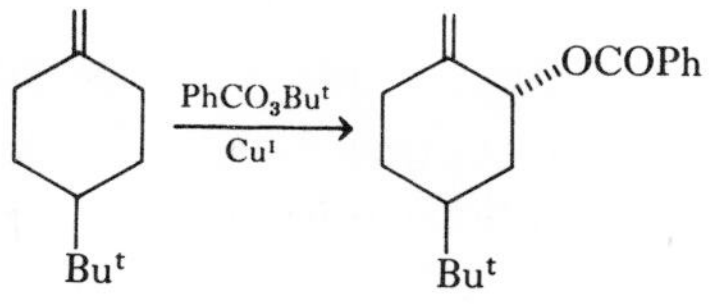

$$\text{(186)}$$

(48)   (49)

363

Oxidations and reductions

An unusual reaction is that between norbornadiene and t-butyl perbenzoate which leads to 7-t-butoxynorbornadiene (50) rather than the expected benzoate (125).

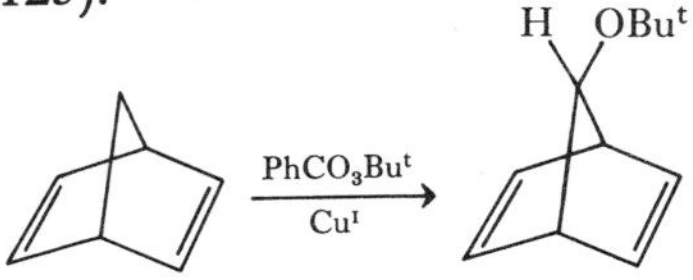

$$(187)$$

(50)

2-Deuterionorbornadiene (51) gave 7-t-butoxynorbornadiene in which deuterium was located at all skeletal positions in an approximately statistical distribution. This result can be explained by Scheme 24.

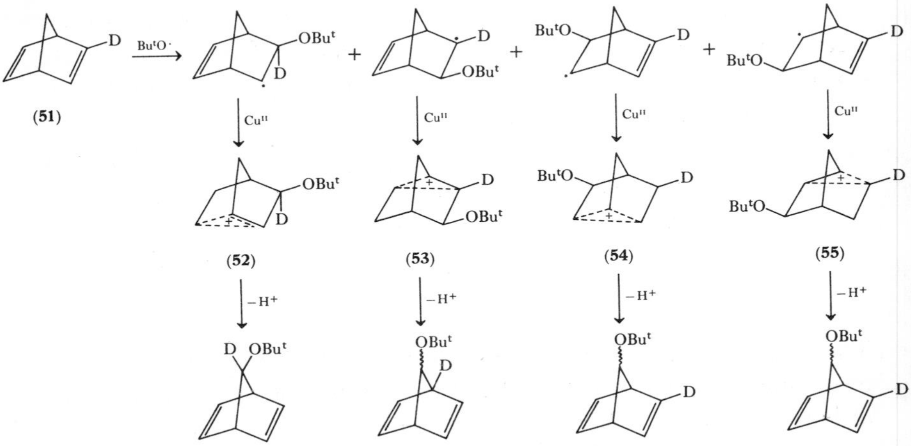

Scheme 24

Oxidation of the radicals, obtained by addition of the t-butoxy radical to norbornadiene, gives rise to the non-classical norbornenyl carbonium ions (52–55) which give the products by proton loss. This result establishes that hydrogen abstraction at C-7 does not occur. This result also indicates that under certain favourable circumstances further oxidation of the intermediate radical to the corresponding cation may occur.

Ethers, sulphides, alcohols and aldehydes also undergo acyloxylation with t-butyl peresters and copper(I) salts though in some instances further reaction occurs (reactions 188–191):

$$PhOCH_3 + PhCO_3Bu^t \xrightarrow{Cu^I} PhOCH_2OCOPh \qquad (188)$$

$$CH_3CH_2SCH_2CH_3 + PhCO_3Bu^t \xrightarrow{Cu^I} CH_3CH_2SCH(OCOPh)CH_3 \qquad (189)$$

$$PhCH_2OH + PhCO_3Bu^t \xrightarrow{Cu^I} Ph\overset{\overset{\displaystyle OCOPh}{|}}{C}HOH + PhCHO + PhCO_2H \qquad (190)$$

$$PhCHO + PhCO_3Bu^t \xrightarrow{Cu^I} (PhCO)_2O \qquad (191)$$

Single-electron reductants other than copper(I) will also bring about the catalysed decomposition of peresters. The oxidized species, e.g. iron(III) or chromium(III) are, however, very much less efficient oxidants for radicals than copper(II) and hence are much less suitable species in these reactions.

10.8.4 *Reactions of dialkyl peroxides*

Copper(I) salts and other single-electron reductants catalyse the decomposition of dialkyl peroxides (126, 127). The temperature required for the catalysed decomposition of di-t-amyl peroxide is about 20 °C less than that required for the uncatalysed decomposition:

$$EtCMe_2OOCMe_2Et \xrightarrow{Cu^I} EtCMe_2O\cdot + EtCMe_2OCu^{II}$$
$$\downarrow$$
$$Et\cdot + Me_2CO \qquad (192)$$

Unsymmetrical peroxides such as t-butyl 4-methyl-4-octyl peroxide undergo decomposition by both possible routes with the pathway leading to t-butoxy radicals predominating:

$$Bu^tOOR + Cu^I \left\{ \begin{array}{l} \rightarrow Bu^tO\cdot + ROCu^{II} \\ \rightarrow RO\cdot + Bu^tOCu^{II} \end{array} \right. \qquad (193)$$
$$(R = 4\text{-methyl-4-octyl})$$

The decomposition of di-t-butyl peroxide in benzaldehyde containing a trace of copper(I) chloride gives t-butyl benzoate in 83 % yield (128):

$$(Bu^tO)_2 + Cu^I \longrightarrow Bu^tO\cdot + Bu^tOCu^{II} \qquad (194)$$

$$Bu^tO\cdot + PhCHO \longrightarrow Ph\dot{C}{=}O + Bu^tOH \qquad (195)$$

$$Ph\dot{C}{=}O + Bu^tOCu^{II} \longrightarrow PhCOOBu^t + Cu^I \qquad (196)$$

In the absence of the copper(I) chloride none of this is obtained. This result emphasizes the similarity between the catalysed decompositions of dialkyl peroxides and peresters.

10.8.5 *Reactions of diacyl peroxides*

The fundamental processes in the copper(I)-catalysed decompositions of diacyl peroxides and peresters are similar. Such differences in product composition as do occur, for example in the decompositions in butenes, can be attributed to the different initiating radical in the two processes; alkoxy radicals from peresters (reaction *197*) and acyloxy radicals from diacyl peroxides (reaction *198*):

$$PhCO_3Bu^t + Cu^I \longrightarrow Bu^tO\cdot + PhCO_2Cu^{II} \qquad (197)$$

$$(PhCO_2)_2 + Cu^I \longrightarrow PhCO_2\cdot + PhCO_2Cu^{II} \qquad (198)$$

Alkoxy radicals react with the three isomeric butenes primarily by hydrogen abstraction with the formation of butenyl radicals (reaction *199*) which are subsequently oxidized to a mixture of 1-acyloxybut-2-ene and 3-acyloxybut-1-ene in which the latter predominates (reaction *200*) (129):

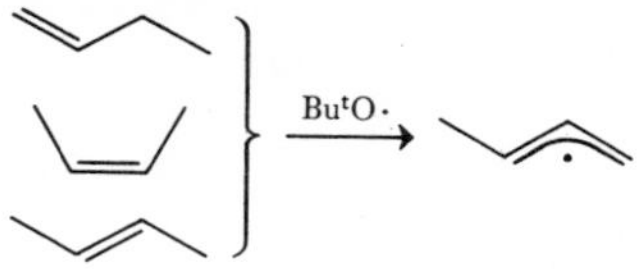

$$(199)$$

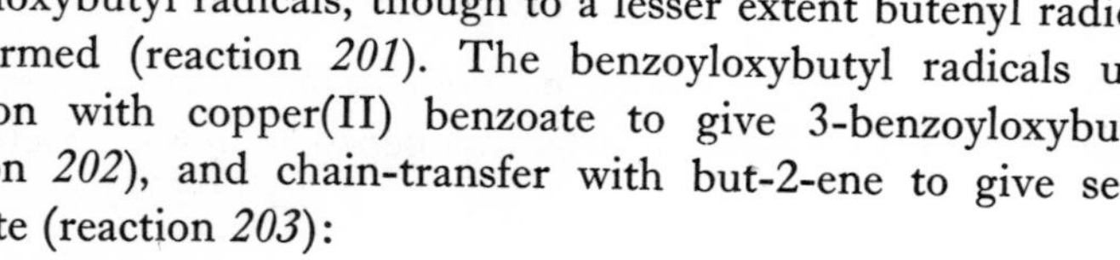

$$(200)$$

$$(\sim 10\%) \qquad (\sim 90\%)$$

Benzoyloxy radicals react with the but-2-enes mainly by addition to give benzoyloxybutyl radicals, though to a lesser extent butenyl radicals are also formed (reaction *201*). The benzoyloxybutyl radicals undergo oxidation with copper(II) benzoate to give 3-benzoyloxybut-1-ene (reaction *202*), and chain-transfer with but-2-ene to give sec-butyl benzoate (reaction *203*):

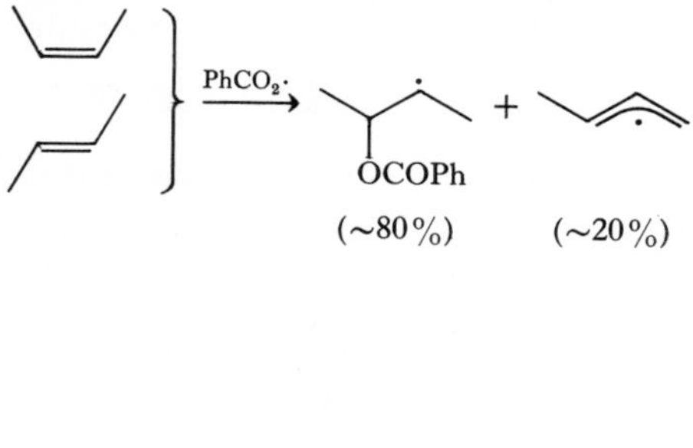

$$(201)$$

$$(\sim 80\%) \qquad (\sim 20\%)$$

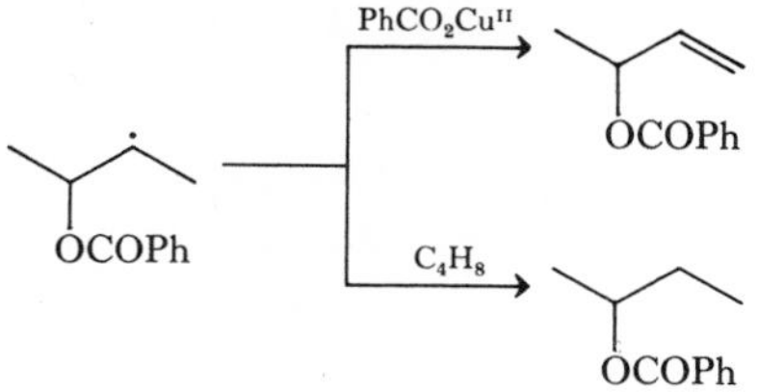

$$\xrightarrow{\text{PhCO}_2\text{Cu}^{\text{II}}} \quad \text{(structure)} \quad \text{OCOPh} \tag{202}$$

$$\xrightarrow{\text{C}_4\text{H}_8} \quad \text{(structure)} \quad \text{OCOPh} \tag{203}$$

The catalysed decomposition of dibenzoyl peroxide in the presence of but-1-ene affords a mixture of 1-benzoyloxybut-2-ene (reaction *205*) and n-butyl benzoate (reaction *206*). The formation of significant quantities of 3-benzoyloxybut-1-ene in this reaction is attributed to oxidation of butenyl radicals (reaction *200*). This latter process also accounts for the formation of some of the 1-benzoyloxybut-2-ene:

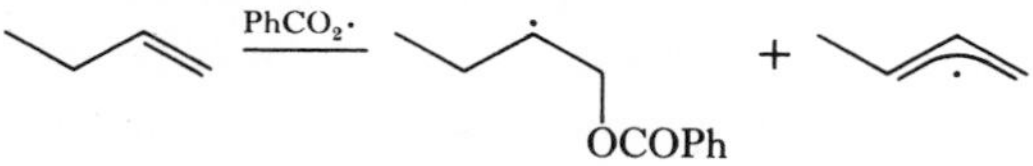

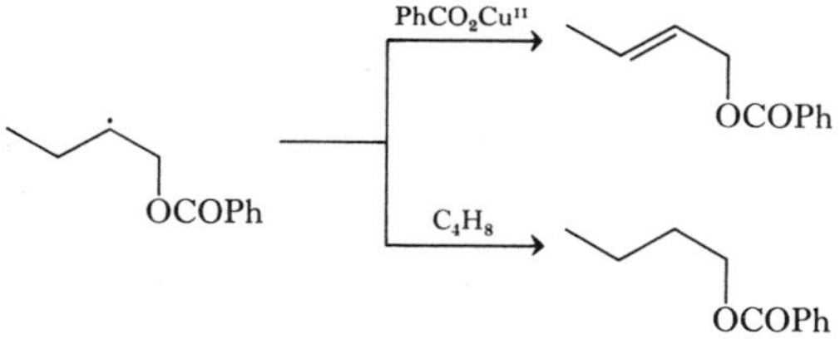

$$\xrightarrow{\text{PhCO}_2\cdot} \quad \text{OCOPh} \quad + \quad \text{(structure)} \tag{204}$$

$$\xrightarrow{\text{PhCO}_2\text{Cu}^{\text{II}}} \quad \text{OCOPh} \tag{205}$$

$$\xrightarrow{\text{C}_4\text{H}_8} \quad \text{OCOPh} \tag{206}$$

The copper(I)-catalysed decomposition of dibenzoyl peroxide in aromatic solvents gives very much higher yields of biaryls than are obtained in the absence of a catalyst (130):

$$\text{Ph}\cdot + \text{PhH} \longrightarrow [\text{Ph}\cdot\text{C}_6\text{H}_6]\cdot \tag{207}$$

$$[\text{Ph}\cdot\text{C}_6\text{H}_6]\cdot \xrightarrow{\text{Cu}^{\text{II}}} [\text{Ph}\cdot\text{C}_6\text{H}_6]^+ \tag{208}$$

$$[\text{Ph}\cdot\text{C}_6\text{H}_6]^+ \xrightarrow{-\text{H}^+} \text{Ph}-\text{Ph} \tag{209}$$

This arises as a result of efficient oxidation of the intermediate phenyl-cyclohexadienyl radicals (61) thereby preventing their dimerization. This result clearly has important synthetic implications. Further discussion of this topic is included in chapter 11 (p. 430).

10.9 ELECTROCHEMICAL OXIDATIONS AND REDUCTIONS

The foregoing sections in this chapter have considered oxidation and reductions in which the transfer of an electron has been brought about by metals or metal ions. Electron transfer can also be achieved electrochemically and it is such reactions which will be considered in this section. Electrodes can either act as an electron sink in anodic oxidation or as an electron source in cathodic reduction. If a single electron is transferred, a radical species will be produced which can undergo further oxidation or reduction, react further in the solution surrounding the electrode, or, in certain instances, accumulate in the solution. We shall consider here some common examples of electrochemical oxidation and reduction which involve single-electron transfers, and hence the intermediacy of radical species. Particular attention will be given to comparison of reactions of radicals produced electrochemically with those obtained by chemical methods.

10.9.1 *The Kolbe reaction*

The Kolbe reaction involving the oxidation of carboxylates is undoubtedly the best-known and most extensively studied example of anodic oxidation (reactions *210–211*) (131–134):

$$RCO_2^- \longrightarrow RCO_2 \cdot \longrightarrow R \cdot + CO_2 \qquad (210)$$

$$R \cdot + R \cdot \longrightarrow R-R \qquad (211)$$

Extensive use of the Kolbe reaction has been made in the synthesis of long-chain compounds (reaction *212*) (135):

$$MeO_2C(CH_2)_8CO_2H \longrightarrow MeO_2C(CH_2)_{16}CO_2Me \qquad (212)$$
$$(69–74\%)$$

Electrolysis of mixtures of two acids gives rise not only to the two symmetrical products but also to the unsymmetrical product:

$$RCO_2^- \xrightarrow[-CO_2]{-e} R \cdot \qquad (213)$$

$$R'CO_2^- \xrightarrow[-CO_2]{-e} R' \cdot \qquad (214)$$

$$R \cdot + R' \cdot \longrightarrow R-R + R-R' + R'-R' \qquad (215)$$

Use of one acid in excess favours the formation of the unsymmetrical product. This technique is particularly valuable if only one of the acids is readily available (reactions *216*, *217*) (136, 137):

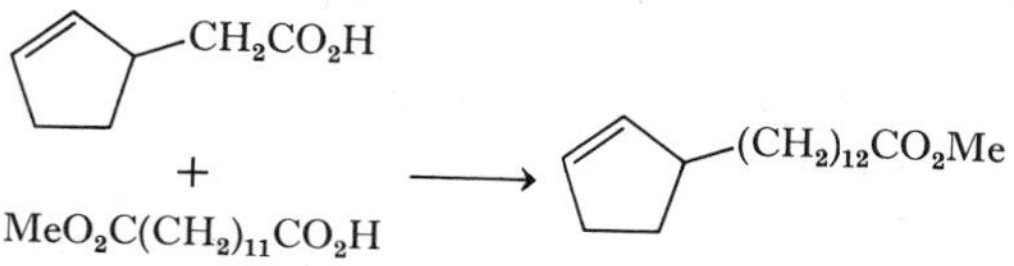

$$(216)$$

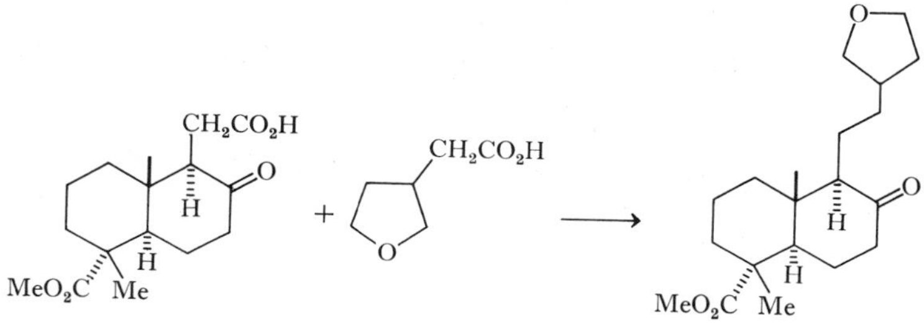

$$(217)$$

Electrolysis of β,γ-unsaturated acids gives rise to allyl radicals which can dimerize in several ways (138):

$$CH_3CH{=}CHCH_2CO_2H \xrightarrow[-CO_2]{-e,-H^+} CH_3CH{=}CHCH_2\cdot \longleftrightarrow CH_3\dot{C}HCH{=}CH_2$$

$$\begin{cases} CH_3CH{=}CHCH_2CH_2CH{=}CHCH_2 \\ CH_3CH{=}CHCH_2CH(CH_3)CH{=}CH_2 \\ CH_2{=}CHCH(CH_3)CH(CH_3)CH{=}CH_2 \end{cases} \quad (218)$$

The configuration of the disubstituted double bonds is retained in the products as expected from the configurational stability of allylic radicals (cf. chapter 4, p. 88).

In much the same way α-cyanoalkyl radicals generated from the electrolysis of α-cyanoalkanoic acids dimerize to give both C–C and C–N coupled products (139):

$$(CH_3)_2\overset{\overset{\displaystyle CN}{|}}{C}CO_2^- \xrightarrow[-CO_2]{-e} (CH_3)_2\dot{C}{-}C{\equiv}N \longleftrightarrow (CH_3)_2C{=}C{=}N\cdot$$

$$\longrightarrow (CH_3)_2\overset{\overset{\displaystyle CN}{|}}{C}{-}\overset{\overset{\displaystyle CN}{|}}{C}(CH_3)_2 + (CH_3)_2C{=}C{=}N{-}\overset{\overset{\displaystyle CN}{|}}{C}(CH_3)_2 \quad (219)$$

Oxidations and reductions

The dimeric products obtained from the Kolbe reaction are accompanied by a number of other products resulting from the disproportionation of alkyl radicals (reactions *220* and *221*) (140), and from attack by alkyl radicals on the solvent (reactions *222* and *223*) (140 and 141):

$$CH_3CD_2CO_2^- \xrightarrow[-CO_2]{-e} CH_3CD_2\cdot \longrightarrow CH_2{=}CD_2 + CH_3CHD_2 \quad (220)$$

$$CD_3CH_2CO_2^- \xrightarrow[-CO_2]{-e} CD_3CH_2\cdot \longrightarrow CH_2{=}CD_2 + CD_3CH_2D \quad (221)$$

$$CD_3CO_2^- \xrightarrow[-CO_2]{-e} CD_3\cdot \xrightarrow{CD_3CO_2H} CD_4 \quad (222)$$

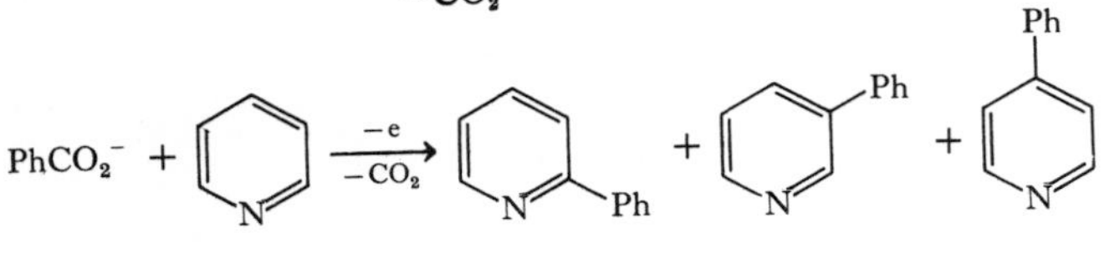

$$(56\%) \qquad (35\%) \qquad (9\%) \qquad (223)$$

The proportions of the isomeric phenylpyridines obtained from the electrolysis of benzoic acid in pyridine are the same as in the phenylation of pyridine using dibenzoyl peroxide as the source of phenyl radicals.

Alcohols and esters are also frequently obtained as by-products in the Kolbe reaction. The formation of these is best explained on the basis of further oxidation of the radical to the cation, followed by reaction with the solvent or substrate to give respectively alcohols and esters:

$$R\cdot \xrightarrow{-e} R^+ \qquad (224)$$

$$R^+ + H_2O \longrightarrow ROH + H^+ \qquad (225)$$

$$R^+ + RCO_2H \longrightarrow RCO_2R + H^+ \qquad (226)$$

In these products the alkyl group has frequently undergone rearrangement typical of the involvement of carbonium ions.

The nature of the electrode has been shown to have a considerable influence on the type of products obtained on electrolysis. Smooth platinum anodes favour the production of radical products, whereas carbon anodes result in the almost exclusive formation of products derived from carbonium ions. A comparative study of the products of electrolysis of 1-methylcyclohexaneacetic acid using platinum and carbon anodes illustrates this point very effectively (143). At a platinum anode 74% of the observed products are formed from radical intermediates (Scheme 25 a), compared to only 2% at a carbon anode (Scheme

370

25*b*). The carbonium ion products are easily identified by the fact that rearrangement has occurred. This contrasts with the radical products which are completely unrearranged. This is in agreement with the propensity that this type of carbonium ion has to undergo rearrangement.

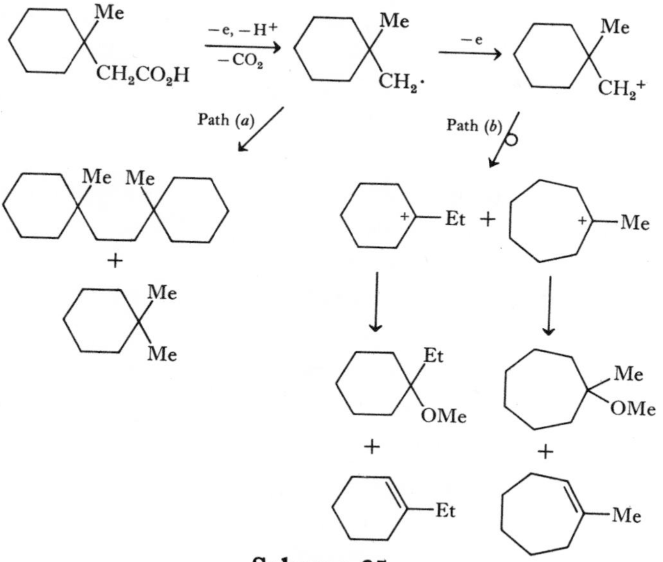
Me
CH2CO2H
−e, −H+
−CO2
Me
CH2·
−e
Me
CH2+
Path (a)
Path (b)
Me Me
+
Me
Me
Et +
Me
Et
OMe
Me
OMe
+
+
Et
Me

Scheme 25

The Kolbe reaction is limited to acids which give alkyl radicals that are not readily oxidized to the corresponding cation. Effectively this has been shown by Eberson to include alkyl radicals with ionization potentials greater than 8 eV (131). This includes primary alkyl radicals, cycloalkyl radicals, and radicals containing electron-withdrawing substituents (CN, CO_2R, $CONH_2$). Alkyl radicals with ionization potentials less than 8 eV undergo further oxidation to the cation and thence give typical carbonium-ion products. Radicals which come into this category include secondary and tertiary alkyl radicals as well as primary alkyl radicals containing α-substituents capable of stabilizing the carbonium ion centre, e.g. NHCOR, OR. Acids of this type rarely give the Kolbe products in yields of greater than 10%. Reaction sequences (*227–229*)

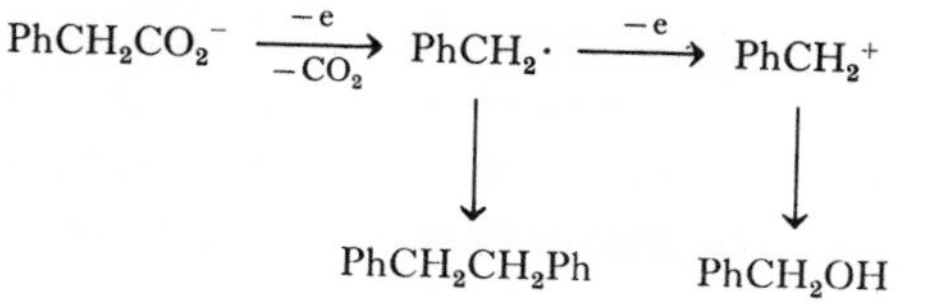
PhCH2CO2−
−e
−CO2
PhCH2·
−e
PhCH2+
PhCH2CH2Ph
PhCH2OH

(*227*)

Oxidations and reductions

$$\text{MeO}-\langle\bigcirc\rangle-\text{CH}_2\text{CO}_2^- \xrightarrow[-\text{CO}_2]{-e} \text{MeO}-\langle\bigcirc\rangle-\text{CH}_2\cdot \longrightarrow \text{MeO}-\langle\bigcirc\rangle-\text{CH}_2^+$$

$$\text{MeO}-\langle\bigcirc\rangle-\text{CH}_2\text{CH}_2-\langle\bigcirc\rangle-\text{OMe} \qquad \text{MeO}-\langle\bigcirc\rangle-\text{CH}_2\text{O} \quad \xrightarrow[\text{MeOH}]{-\text{H}^+}$$

(2

$$\text{Ph}_2\text{CHCO}_2^- \xrightarrow[-\text{CO}_2]{-e} \text{Ph}_2\text{CH}\cdot \longrightarrow \text{Ph}_2\text{CH}^+ \xrightarrow[\text{MeOH}]{-\text{H}^+} \text{Ph}_2\text{CHOMe}$$

(2

illustrate the influence of the alkyl group on the nature of the products of electrolysis (144–146) (see also Table 10-2):

TABLE 10-2 *Products from the electrolysis of arylacetic acids* (146)

	Products (%)			
Aryl group	$\text{ArCH}_2\text{CH}_2\text{Ar}$	ArCH_2OMe	ArCHO	Others
C_6F_5	74	7	14	5
C_6H_5	68	13	8	11
$p\text{-CF}_3\text{C}_6\text{H}_4$	37	18	24	11
$p\text{-MeC}_6\text{H}_4$	38	46	8	8
$p\text{-Bu}^t\text{C}_6\text{H}_4$	13	78	0	9
$p\text{-ClC}_6\text{H}_4$	4	91	0	5
$p\text{-MeOC}_6\text{H}_4$	1	99	0	1

It is thus evident that the involvement of radical species in the Kolbe reaction is beyond doubt. It is, however, of considerable interest to relate the behaviour of radicals produced under Kolbe conditions to those obtained from the decomposition of diacyl peroxides. The products obtained from the electrolysis of potassium propionate in propionic acid at 100 °C have been compared with those derived from the thermolysis of dipropionyl peroxide also in propionic acid at 100 °C (Table 10-3) (147). The most notable difference is that less ethane and more butane is formed in the Kolbe reaction. This arises because under electrolytic conditions a relatively high concentration of radicals is produced in the neighbourhood of the anode, thereby favouring dimerization rather than reaction with solvent. In the thermolysis of the peroxide there will be an even distribution of radicals throughout the solution. A second difference concerns the fate of the $\text{CH}_3\dot{\text{C}}\text{HCO}_2\text{H}$

372

TABLE 10-3   *Products formed by electrolysis of propionic acid*
and in the decomposition of dipropionyl peroxide (147)

Product	Moles per 2 moles of EtCO$_2$H	Moles per mole of (EtCO$_2$)$_2$
C$_2$H$_4$	5.6[a]	3.0[b]
C$_2$H$_6$	14.0[a]	28.1[b]
C$_4$H$_{10}$	15.8[a]	6.9[b]
EtCO$_2$H	0.0	0.28
EtCHMeCO$_2$H	0.002–0.014	0.038
MeCHCO$_2$H | MeCHCO$_2$H	0.002	0.106

[a] Percentage of anode gases.
[b] Percentage of total gaseous products.

radicals generated by hydrogen abstraction. In the Kolbe reaction they tend to couple with ethyl radicals (reaction *232*) because of the availability of the latter in the vicinity of the anode, whereas in the peroxide decomposition dimerization is preferred (reaction *233*):

$$CH_3CH_2CO_2^- \xrightarrow[-CO_2]{-e} CH_3CH_2 \cdot \qquad (230)$$

$$CH_3CH_2 \cdot + CH_3CH_2CO_2H \longrightarrow CH_3CH_3 + CH_3\dot{C}HCO_2H \qquad (231)$$

$$CH_3\dot{C}HCO_2H + CH_3CH_2 \cdot \longrightarrow CH_3CH_2\overset{\overset{\displaystyle CH_3}{\displaystyle |}}{C}HCO_2H \qquad (232)$$

$$CH_3\dot{C}HCO_2H + CH_3\dot{C}HCO_2H \longrightarrow \overset{\displaystyle CH_3CHCO_2H}{\underset{\displaystyle CH_3CHCO_2H}{|}} \qquad (233)$$

The above discussion indicates that qualitatively, and sometimes quantitatively, alkyl radicals generated in the Kolbe reaction behave in the same way as those generated by other methods. Such differences as do occur are generally explicable on the basis of a higher radical concentration in the region of the anode than is customarily obtained in solution.

Much less is understood about the precise mode of formation of the alkyl radicals from the carboxylate anion. This could either be a concerted one-stage process (reaction *234*) or a two-stage process proceeding via the acyloxy radical (reaction *235*):

$$RCO_2^- \longrightarrow [R \cdots\cdots CO_2 \cdots\cdots e(\text{anode})] \longrightarrow R\cdot + CO_2 + e \quad (234)$$

$$RCO_2^- \xrightarrow{\;-e\;} RCO_2\cdot \longrightarrow R\cdot + CO_2 \quad\quad\quad (235)$$

Skell has argued in favour of the two-stage process on the basis of the relative insensitivity to the nature of the alkyl group of the ease of electrolysis of a series of carboxylates (148). In the concerted process the transition ·state would resemble the radical, and hence the ease of reaction should be susceptible to normal substituent effects. This was not the cáse. Eberson has pointed out that the experiments were conducted under conditions likely to favour carbonium ions and also that factors such as adsorption on the electrode surface might be the rate-controlling factor (149). It has also been argued that the formation of 4-phenylcoumarin from the electrolysis of 3,3-diphenylacrylic acid arises from the intramolecular attack of the acyloxy radical on one of the phenyl groups (Scheme 26) (150). Alternatively this could arise, as

$$Ph_2C\!\!=\!\!CHCO_2^- \xrightarrow{\;-e\;} Ph_2C\!\!=\!\!CHCO_2\cdot$$

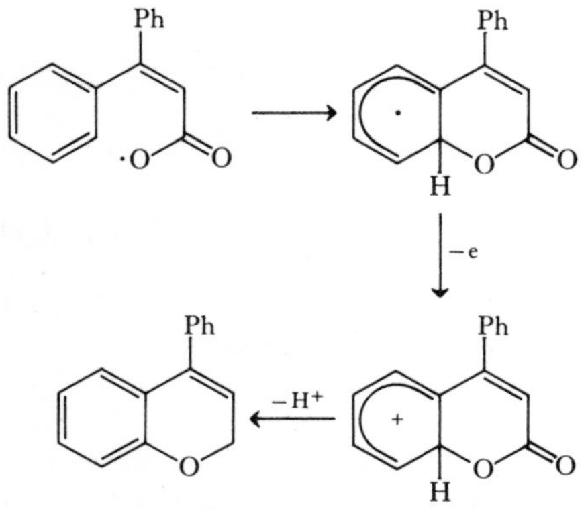

Scheme 26

Eberson has pointed out (131), by oxidation of the aromatic nucleus to a radical cation followed by nucleophilic attack by the carboxylate group (Scheme 27).

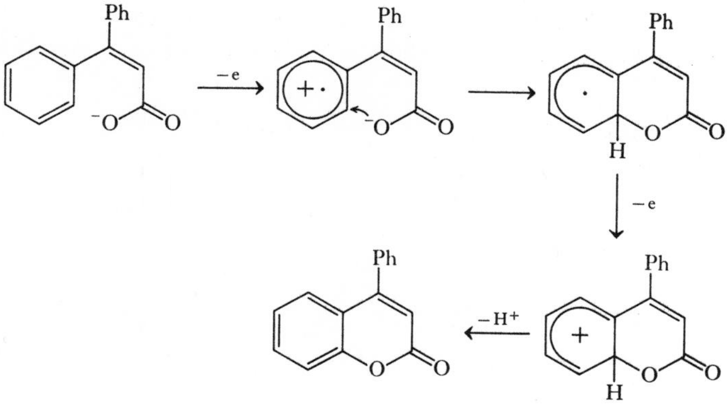

Scheme 27

10.9.2 *Anodic oxidation of other organic anions*

Anodic oxidation of phenols can lead to C–C coupled products of the type encountered in chemical oxidations (Scheme 28) (151). Oxidation of 2,4,6-trisubstituted phenols which give rise to stable phenoxy radicals proceeds further to give the corresponding phenoxonium ion (Scheme 29) (43*b*, 152).

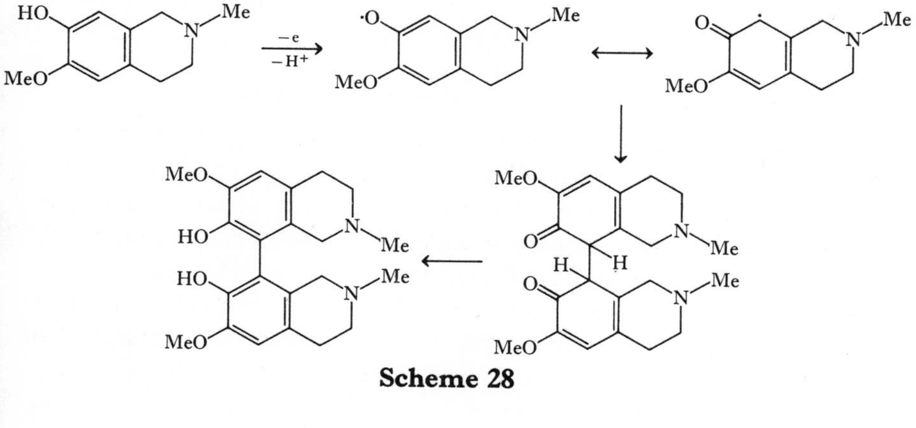

Scheme 28

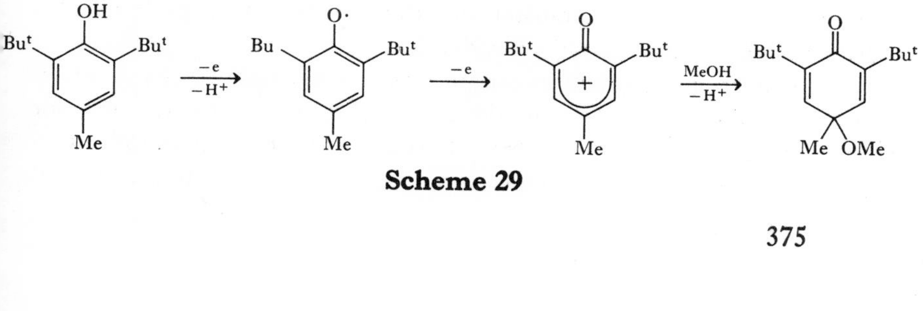

Scheme 29

375

Stabilized carbanions likewise undergo electrochemical oxidative coupling (Scheme 30) (153). Thus if the oxidation of 2-nitrobutane is

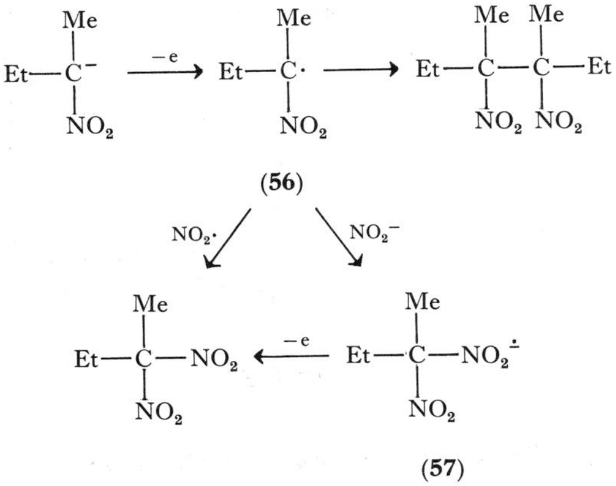

$$
\text{Et}-\underset{\underset{NO_2}{|}}{\overset{\overset{Me}{|}}{C}}{}^{-} \xrightarrow{-e} \text{Et}-\underset{\underset{NO_2}{|}}{\overset{\overset{Me}{|}}{C}}\cdot \longrightarrow \text{Et}-\underset{\underset{NO_2}{|}}{\overset{\overset{Me}{|}}{C}}-\underset{\underset{NO_2}{|}}{\overset{\overset{Me}{|}}{C}}-\text{Et}
$$

(56)

$NO_2\cdot \swarrow \qquad \searrow NO_2^-$

$$
\text{Et}-\underset{\underset{NO_2}{|}}{\overset{\overset{Me}{|}}{C}}-NO_2 \xleftarrow{-e} \text{Et}-\underset{\underset{NO_2}{|}}{\overset{\overset{Me}{|}}{C}}-NO_2^{\cdot-}
$$

(57)

Scheme 30

carried out in the presence of nitrite, the chief product is 2,2-dinitro-butane, which is probably formed by reaction of the radical **(56)** with nitrite to give the radical anion **(57)**, further oxidation of which affords the product. Alternatively, this could arise from coupling of the radical **(56)** with NO_2.

10.9.3 *Anodic oxidation of aromatic hydrocarbons*

The anodic oxidation of aromatic hydrocarbons leads initially to radical cations as a result of the transfer of one electron. If the derived radical cation is relatively stable, as is the case in that derived from 9,10-diphenylanthracene, it can be identified from its e.s.r. spectrum (154).

When only one of the reactive sites in anthracene is blocked then the resultant radical cations are less stable. This is so in the case of 9-phenyl-anthracene, the radical cation **(58)** of which can dimerize (Scheme 31 a) or undergo reaction with added nucleophiles (Scheme 31 b) (155). When this latter process occurs, reaction continues as indicated by a series of steps involving electron transfer and attack of the nucleophile to give, for example, 10-acetoxy-10-phenylanthrone.

Acetoxylation of aromatic compounds has been shown by Eberson and Nyberg to occur at a potential considerably below that required for the discharge of acetate ions (156). This demonstrates that acetoxylation of aromatic compounds involves nucleophilic attack on the radical

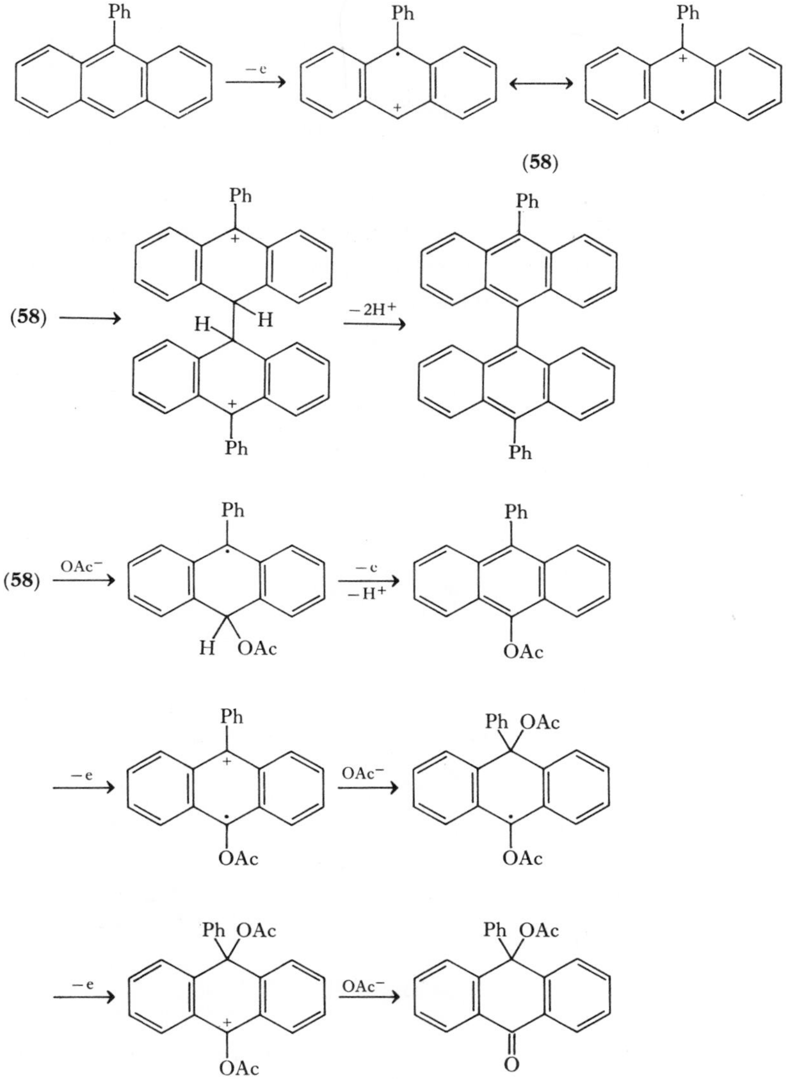

(58)

(58) ⟶

(58) $\xrightarrow{\text{OAc}^-}$

Scheme 31

cation (reaction *236*), rather than discharge of acetate followed by attack of the resultant acetoxy radical on the aromatic compound (reaction *237*).

Furthermore, the exclusive formation of *ortho*- and *para*-acetoxybiphenyls from biphenyl argues against a mechanism involving attack by

377

$$\text{ArH} \xrightarrow{-e} \text{ArH}^{+\cdot} \xrightarrow{\text{OAc}^-} \text{Ar}\overset{\cdot}{\underset{\text{OAc}}{<}}\overset{H}{} \xrightarrow{-e} \text{Ar}\overset{+}{\underset{\text{OAc}}{<}}\overset{H}{} \xrightarrow{-H^+} \text{ArOAc}$$

(236)

$$\text{AcO}^- \xrightarrow{-e} \text{AcO}\cdot \xrightarrow{\text{ArH}} \text{Ar}\overset{\cdot}{\underset{\text{OAc}}{<}}\overset{H}{} \xrightarrow[-\text{AcOH}]{\text{AcO}^\cdot} \text{ArOAc} \qquad (237)$$

acetoxy radicals but is consistent with attack by acetate on the radical
cation (Scheme 32). The reaction closely resembles the nuclear acetoxylation of anisole with lead tetra-acetate, which also leads to a mixture of
ortho- and *para*-acetoxyanisoles (39 a) (cf. p. 446).

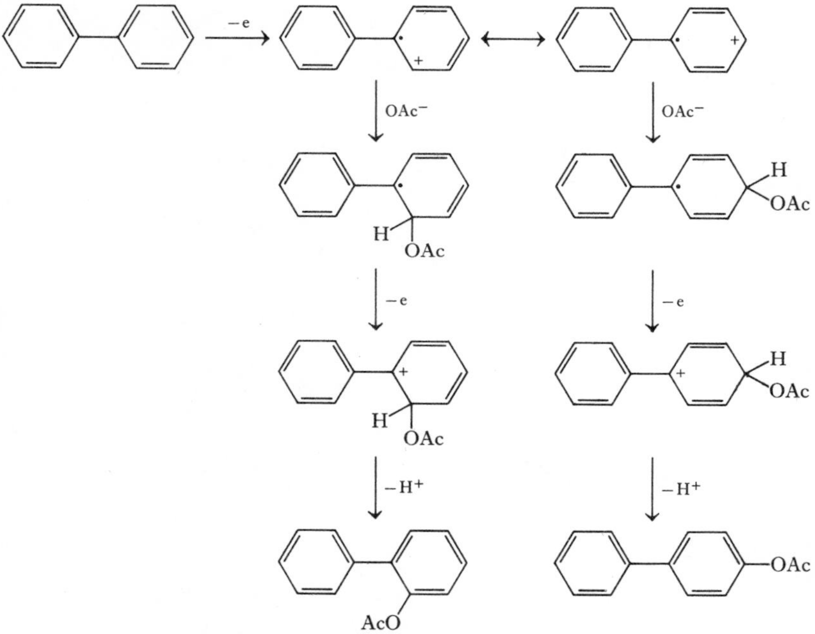

Scheme 32

The radical cation of 9,10-dimethylanthracene (**59**) undergoes nucleophilic attack to give the radical (**60**), which on dimerization gives (**61**),
rather than undergoing attack at the 9- or 10-positions (157, 158). The
radical (**60**) also attacks a second 9,10-dimethylanthracene radical cation
to give, after reaction with acetate, the acetoxy compound (**62**) (Scheme
33).

If the reaction is carried out using a somewhat higher electrode potential, the radical cation (**59**) undergoes further oxidation to the dication

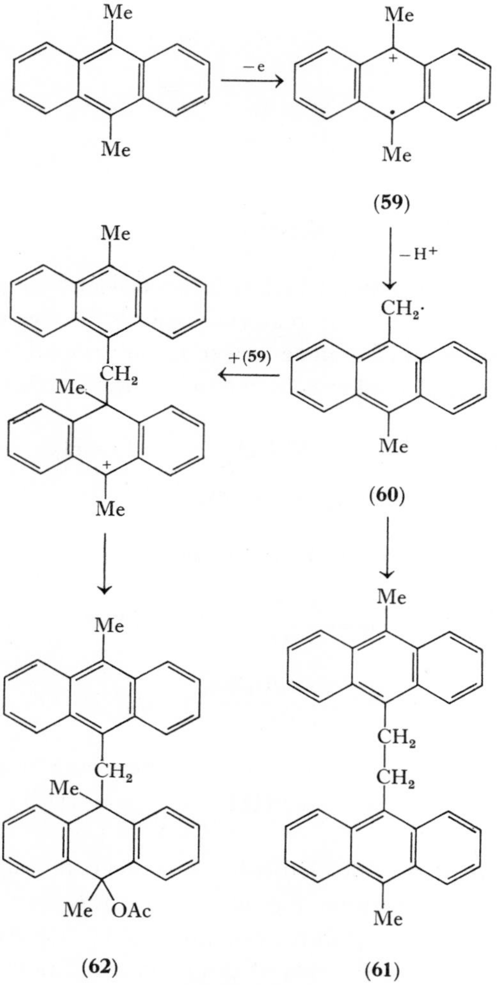

Scheme 33

(63) which, after reaction with acetate, gives 9,10-diacetoxy-9,10-dimethylanthracene (Scheme 34).

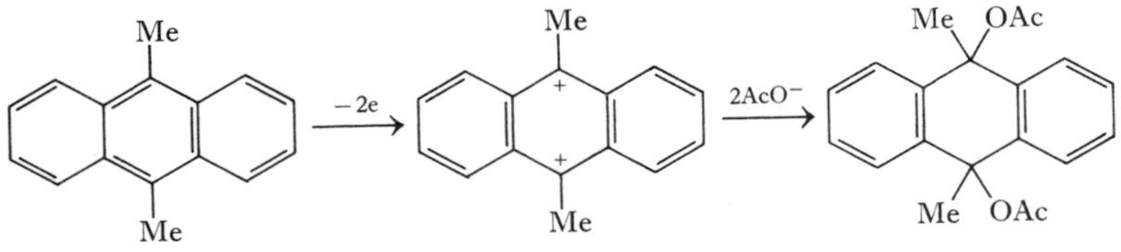

Oxidations and reductions

(63)

Scheme 34

A further mode of reaction which is frequently encountered in anodic oxidation is the oxidation of benzylic radicals to the corresponding cation. Thus oxidation of toluene in wet acetonitrile affords in addition to bibenzyl, benzyl alcohol and *N*-benzylacetamide (reactions *238–241*) (159):

$$\text{PhCH}_3 \xrightarrow{-e} \text{PhCH}_3^{+\cdot} \xrightarrow{-\text{H}^+} \text{PhCH}_2\cdot \qquad (238)$$

$$2\,\text{PhCH}_2\cdot \longrightarrow \text{PhCH}_2\text{CH}_2\text{Ph} \qquad (239)$$

$$\text{PhCH}_2\cdot \xrightarrow{-e} \text{PhCH}_2^+ \qquad (240)$$

$$\text{PhCH}_2^+ - \left\lfloor \begin{array}{l} \xrightarrow[-\text{H}^+]{\text{H}_2\text{O}} \text{PhCH}_2\text{OH} \\[2ex] \xrightarrow[-\text{H}^+]{\text{CH}_3\text{CN}/\text{H}_2\text{O}} \text{PhCH}_2\text{NHCOCH}_3 \end{array} \right. \qquad (241)$$

These anodic oxidations of aromatic hydrocarbons resemble in many ways the oxidations by cobalt(III), manganese(III) and lead(IV) acetates.

The anodic oxidation of olefins has also been investigated in some detail. The products from the electrochemical oxidation of 4,4′-dimethoxystilbene depend on the potential used (160). At low potentials oxidation occurs to give the radical cation which dimerizes and cyclizes to give **(64)** (Scheme 35*a*), whilst at higher potentials further oxidation of the radical cation to the dication takes place. This subsequently forms the hydroxyacetate **(65)** (Scheme 35*b*). More highly substituted olefins such as tetra-*p*-anisylethylene only give the dication.

Electrolysis of aromatic hydrocarbons (and also alkenes and amines) in methanol can lead to the formation of methoxy compounds in moderate yield. Thus electrolysis of tetralin in methanol gives 1-methoxytetralin (161). The oxidation takes place at a potential at which the tetralin is not converted into its radical cation, and is thus believed to

380

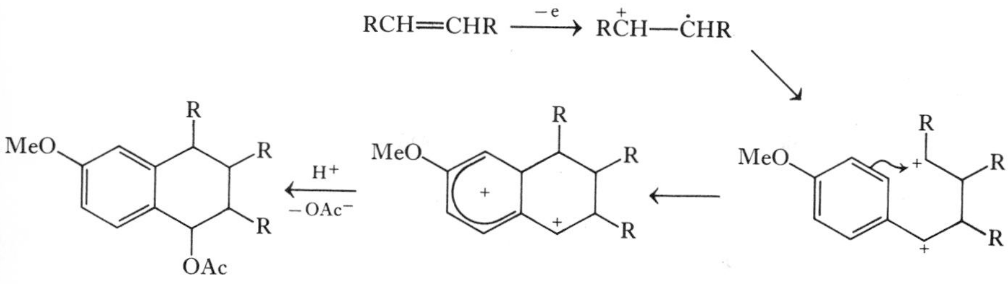

$$RCH{=}CHR \xrightarrow{-e} R\overset{+}{C}H{-}\overset{\cdot}{C}HR$$

(64)

Scheme 35 a

$$RCH{=}CHR \xrightarrow{-2e} R\overset{+}{C}H{-}\overset{+}{C}HR \longrightarrow \underset{\overset{|}{R}CH}{\overset{OAc}{|}}{-}\underset{\overset{|}{C}HR}{\overset{OH}{|}}$$

(65)

$[R = p\text{-}MeOC_6H_4]$

Scheme 35 b

involve methoxy radicals obtained by discharge of methoxide (reactions *242–243*).

$$MeO^- \xrightarrow{-e} MeO\cdot \qquad (242)$$

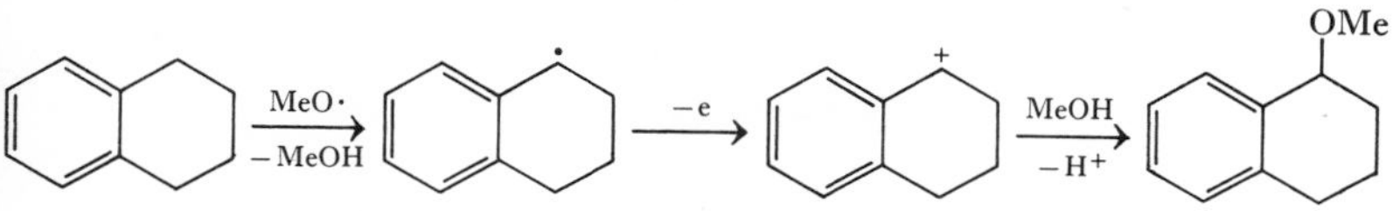

(243)

10.9.4 *Anodic oxidation of amines*

Amines, like hydrocarbons, undergo initial oxidation to radical cations. The radical cation from a secondary or tertiary aromatic amine undergoes dimerization to give a substituted benzidine provided that there is a free *para*-position (reactions *244* to *245*) (162):

$$Ph_3N \xrightarrow{-e} Ph_3\overset{+\cdot}{N} \qquad (244)$$

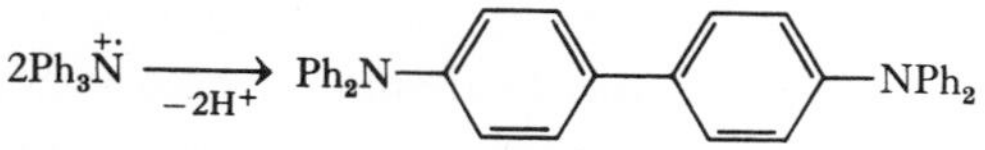

$$2Ph_3\overset{+\cdot}{N} \xrightarrow{-2H^+} Ph_2N{-}\!\!\!\bigcirc\!\!\!{-}\!\!\!\bigcirc\!\!\!{-}NPh_2 \qquad (245)$$

The oxidation in methanol of 2,4,6-tri-t-butylaniline, which has no free *ortho*- or *para*-positions at which coupling may occur, gives 2,6-di-t-butyl-4-methoxyaniline (Scheme 36) (163). The radical cation (**66**) and the radical (**67**) have both been detected by e.s.r. spectroscopy (164).

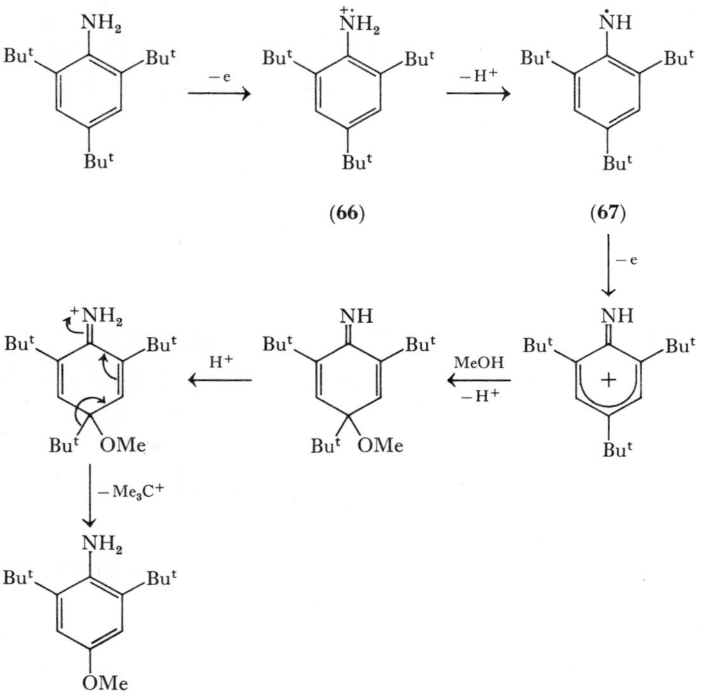

Scheme 36

The radical cations derived from primary amines give rise to *p*-amino-diphenylamines in their oxidized forms (**68** and **69**) even when the parent amine contains a *para*-substituent (165). This is true whether the *para*-substituent leaves as an anion (e.g. Cl, OMe) or as a cation (e.g. H). In the latter case the product is initially obtained in its reduced form (Scheme 37).

10.9.5 *Cathodic reduction of carbonyl compounds*

In many respects cathodic reduction of organic compounds closely resembles dissolving metal reductions. Cathodic reduction results initially in the formation of a radical anion, which is generally followed or accompanied by rapid protonation. The resultant radical then either dimerizes or undergoes further reduction to the carbanion which is subsequently protonated. Whether dimerization or further reduction occurs depends on the nature of the initial radical and also on the potential used. Thus reduction of ketones in acid solution gives the

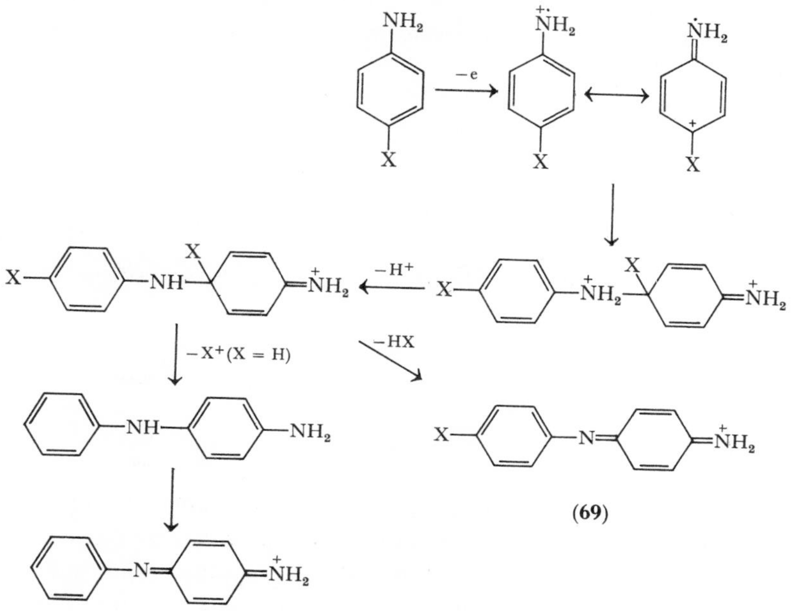

Scheme 37

pinacol, whereas at a more negative potential further reduction occurs, resulting in the formation of the carbinol (166).

$$R_2C{=}O \xrightarrow[H^+]{e} R_2\overset{\cdot}{C}{-}OH \qquad (246)$$

$$2R_2\overset{\cdot}{C}{-}OH \longrightarrow \begin{array}{c} R_2C{-}OH \\ | \\ R_2C{-}OH \end{array} \qquad (247)$$

$$R_2\overset{\cdot}{C}OH \xrightarrow[H^+]{e} R_2CHOH \qquad (248)$$

The stereochemistry of the pinacols produced in the reduction of unsymmetrical ketones and aldehydes has been extensively studied. Hydroxy-substituted benzaldehydes undergo reduction in alkaline solution to give exclusively the *meso*-hydrobenzoins (167). This is attributed to the electrostatic repulsion between the phenoxide anions present in the dimerizing radicals.

Oxidations and reductions

The cathodic reduction of acetophenone leads to a mixture of approximately 1:1 of *racemic*:*meso* products in acidic media, but of

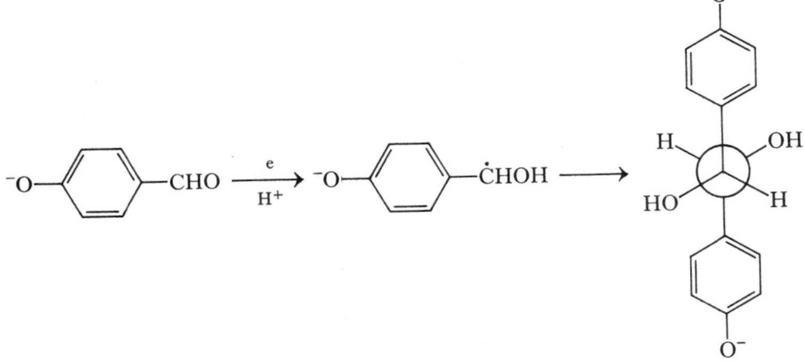

meso

(249)

about 3:1 in strongly basic media (168). The two factors which have the most influence on the proportions of these two products are steric factors which favour the formation of the *meso*-compound, and hydrogen bonding which favours the formation of the racemic mixture. This can be seen from consideration of the transition states (**70** and **71**) leading to the *meso*- and (±)-compounds. In acid solution these factors are

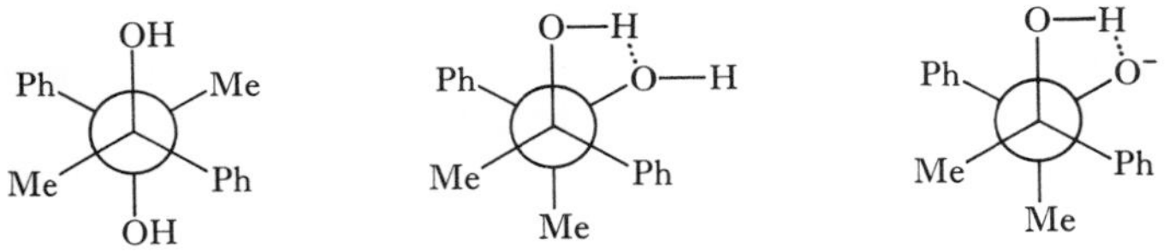

(**70**)   (**71**)   (**72**)

meso   (±)   (±)

more or less balanced, whereas in strongly alkaline solution protonation of the radical anions will be incomplete, and hence some product will arise through radical–radical anion coupling in which hydrogen bonding will be stronger (cf. **72**).

The radicals produced in the cathodic reduction of ketones can escape from the vicinity of the electrode and react with other organic substrates. Thus the reduction of acetone in the presence of acrylonitrile leads to γ-hydroxy-γ-methylvaleronitrile (169). In this process it is the

384

acetone rather than the acrylonitrile which accepts an electron (reactions *250–252*).

$$Me_2C{=}O \xrightarrow[H^+]{e} Me_2\dot{C}OH \qquad (250)$$

$$Me_2\dot{C}OH + CH_2{=}CHCN \longrightarrow Me_2C(OH)CH_2\dot{C}HCN \quad (251)$$

$$Me_2C(OH)CH_2\dot{C}HCN \xrightarrow[H^+]{e} Me_2C(OH)CH_2CH_2CN \quad (252)$$

10.9.6 *Cathodic reduction of alkenes*

Simple olefins are not reduced cathodically. However, when the double bond is conjugated with an electron-withdrawing group, reduction of the double bond to the saturated compound may occur (170):

$$CH_2{=}CHX \xrightarrow{e} [\dot{C}H_2{-}\bar{C}HX \longleftrightarrow \bar{C}H_2{-}\dot{C}HX] \xrightarrow{H^+} CH_3\dot{C}HX \qquad (253)$$

$$CH_3\dot{C}HX \xrightarrow[H^+]{e} CH_3CH_2X \qquad (254)$$

$$PhCH{=}CHCO_2^- \xrightarrow[2H^+]{2e} PhCH_2CH_2CO_2^- \qquad (255)$$
$$(80{-}90\,\%)$$

Alternatively the intermediate radical anion may add in a Michael-type reaction to give a dimeric radical anion, reduction of which followed by protonation leads to the hydrodimer (171):

$$CH_2{=}CHCO_2Et \longrightarrow \bar{C}H_2{-}\dot{C}HCO_2Et \qquad (256)$$

$$\bar{C}H_2{-}\dot{C}HCO_2Et + CH_2{=}CHCO_2Et \longrightarrow EtO_2C\dot{C}HCH_2CH_2\bar{C}HCO_2Et \qquad (257)$$

$$EtO_2C\dot{C}HCH_2CH_2\bar{C}HCO_2Et \xrightarrow{e} EtO_2C(CH_2)_4CO_2Et \quad (258)$$

This type of behaviour is favoured by carrying out the electrolysis using a high concentration of monomer. When proton availability is low, cathodic reduction of a mixture of two olefins at a potential at which only one of them is reduced leads to only two products, the homodimer from the more readily reducible olefin and the mixed dimer (cf. reactions *259–261*). This latter results from Michael addition of the radical anion, derived from the more readily reducible olefin, with the other olefin. This result is in complete accord with the proposed mechanism. Thus, electrolysis of a mixture of diethyl maleate and acrylonitrile at about -1.4 V gives no adiponitrile (172):

$$EtO_2CCH=CHCO_2Et \xrightarrow{e} EtO_2C\dot{C}H-\bar{C}HCO_2Et \quad (259)$$

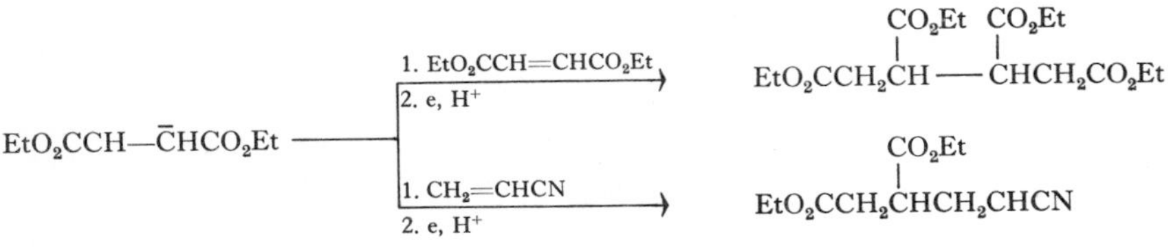

$$EtO_2CCH-\bar{C}HCO_2Et \begin{cases} \xrightarrow[\text{2. e, H}^+]{\text{1. EtO}_2\text{CCH=CHCO}_2\text{Et}} & EtO_2CCH_2\overset{\overset{\displaystyle CO_2Et}{|}}{CH}-\overset{\overset{\displaystyle CO_2Et}{|}}{CH}CH_2CO_2Et \quad (2\ldots) \\[3em] \xrightarrow[\text{2. e, H}^+]{\text{1. CH}_2\text{=CHCN}} & EtO_2CCH_2\overset{\overset{\displaystyle CO_2Et}{|}}{CH}CH_2CHCN \quad (2\ldots) \end{cases}$$

Simple molecular-orbital calculations have indicated that for acrylonitrile β,β coupling should occur predominantly, as is found (reaction *262*) (173):

$$CH_2=CHCN \xrightarrow[2H^+]{2e} NC(CH_2)_4CN \quad (262)$$

These calculations also predict that some of the α,β-coupled product, in addition to the β,β-coupled product, should be formed from α-methylacrylonitrile. This was found to be the case (reaction *263*) (174):

$$CH_2=CMeCN \longrightarrow NCCHMeCH_2CH_2CHMeCN$$
$$+ NCCMe_2CH_2CHMeCN \quad (263)$$

Electrochemical hydrodimerization with bis-deactivated olefins leads to cyclic compounds (reaction *264*) (175).

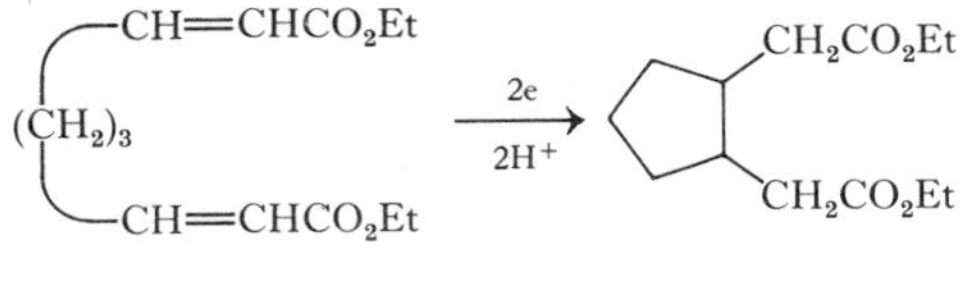

$$(CH_2)_3 \begin{cases} -CH=CHCO_2Et \\ \\ -CH=CHCO_2Et \end{cases} \xrightarrow[2H^+]{2e} \begin{array}{c} CH_2CO_2Et \\ \bigtriangleup \\ CH_2CO_2Et \end{array} \quad (264)$$

$$(\sim 100\%)$$

10.10 *RADICALS AS OXIDIZING AND REDUCING AGENTS*

Any discussion of homolytic oxidation and reduction would not be complete without reference to the role of radicals as oxidizing and reducing species. Waters (176) has pointed out that it should be possible to assign *two* oxidation potentials to radicals, E_{-e} for the oxidation to carbonium ions (reaction *265*), and E_{+e} for reduction to carbanions (reaction *266*).

$$R\cdot - e \xrightarrow{E_{-e}} R^+ \tag{265}$$

$$R\cdot + e \xrightarrow{E_{+e}} R^- \tag{266}$$

These potentials should be directly comparable to those for inorganic ions. They are not, however, nearly so readily obtained because it is exceedingly rare that free radicals and either of their related ions are stable in solution. One can get a measure of the values of E_{-e} and E_{+e} in polarography using concentrations such that electron transfer between the radical and the ion occurs more rapidly than competing radical–radical reactions. For cases where this is not possible, one can get some measure of their oxidation potentials according to whether they can be oxidized (or reduced) by a metal ion of known oxidation potential. Thus the oxidation potential for the reaction:

$$PhN_2\cdot \longrightarrow PhN_2^+ + e$$

must lie between that for $Cu^+ \longrightarrow Cu^{2+}$ and $Fe^{2+} \longrightarrow Fe^{3+}$, since reduction of the diazonium ion can be effected by copper(I) but not by iron(II).

Using this type of consideration it is possible to place E_{-e} values of radicals in the order:

$$CH_3\cdot < CR_3\cdot < \dot{C}H_2OH < \dot{C}R_2OR < \dot{C}R_2NH_2 < \dot{C}O_2H$$

This is the order of increasing stability of the resultant carbonium ion. In agreement with this, it is found that $\dot{C}H_2OH$ radicals but not methyl radicals can transfer an electron to biacetyl (*177*):

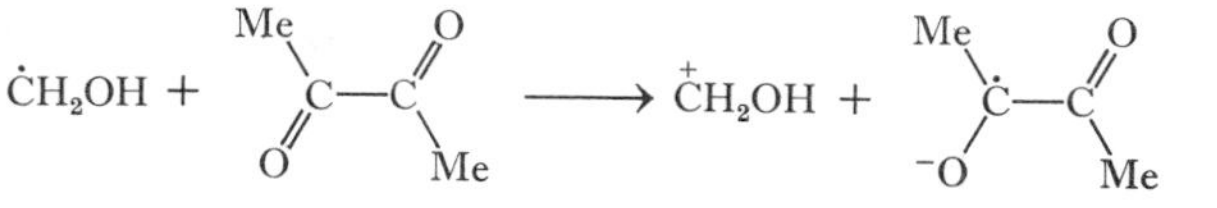

Fieser, from a study of critical oxidation potentials, has estimated E_{-e} values for a series of phenols and mercaptans. Electron-withdrawing substituents result in E_{-e} becoming more negative, indicating that the radicals are more readily reduced, or alternatively that the phenoxides are less readily oxidized.

In general, most organic carbon radicals are more powerful oxidants than reductants, reflecting the greater stability of carbonium ions as compared to carbanions. This is consistent with the nucleophilic character which they display in radical abstraction and in homolytic aromatic substitution. One of the few carbon radicals which is readily reduced is the radical (73) (178):

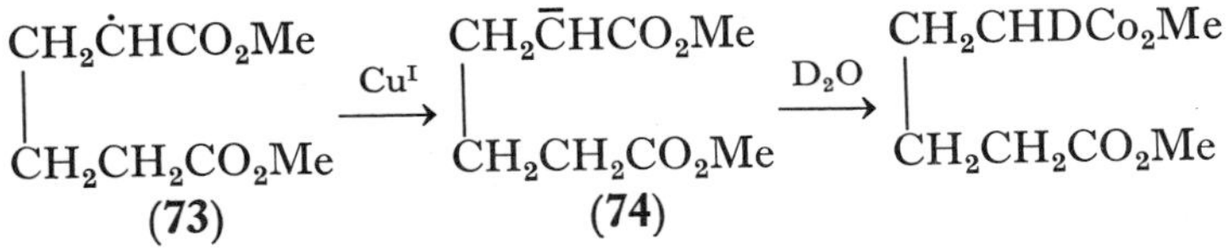

$$\begin{array}{c} CH_2\dot{C}HCO_2Me \\ | \\ CH_2CH_2CO_2Me \\ (73) \end{array} \xrightarrow{Cu^I} \begin{array}{c} CH_2\bar{C}HCO_2Me \\ | \\ CH_2CH_2CO_2Me \\ (74) \end{array} \xrightarrow{D_2O} \begin{array}{c} CH_2CHDCo_2Me \\ | \\ CH_2CH_2CO_2Me \end{array} \quad (268)$$

The oxidizing ability of this radical is due to the electron-withdrawing influence of the methoxycarbonyl group which stabilizes the resultant carbanion (74). Evidence for this rests on the incorporation of deuterium when the reaction is carried out in deuterium oxide.

An appreciation of the above considerations enables one to rationalize the different products obtained via related radicals. Thus the persulphate oxidation of *p*-methoxyphenylacetic acid which proceeds via *p*-methoxybenzyl radicals gives only a small amount of 4,4'-dimethoxybibenzyl, whereas bibenzyl is obtained in good yield from phenylacetic acid (179). The difference in behaviour is due to the more facile oxidation of *p*-methoxybenzyl radicals:

$$ArCH_2CO_2H \xrightarrow{S_2O_8^{2-}} ArCH_2\cdot \quad (269)$$

$$2ArCH_2\cdot \longrightarrow ArCH_2CH_2Ar \quad (270)$$

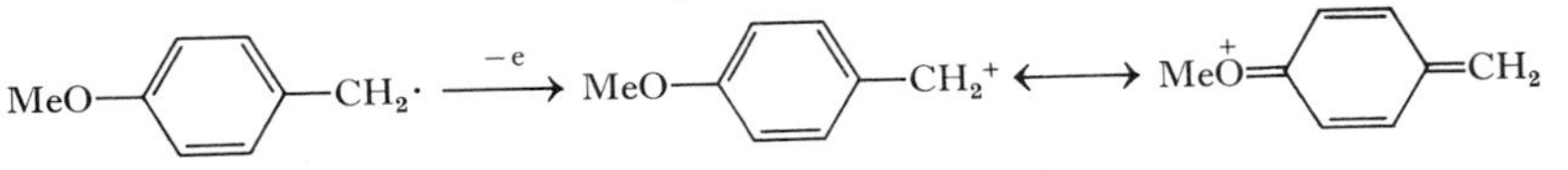

$$MeO-\!\!\!\bigcirc\!\!\!-CH_2\cdot \xrightarrow{-e} MeO-\!\!\!\bigcirc\!\!\!-CH_2^+ \longleftrightarrow Me\overset{+}{O}=\!\!\!\bigcirc\!\!\!=CH_2 \quad (27.$$

$$MeO-\!\!\!\bigcirc\!\!\!-CH_2^+ \xrightarrow[-H^+]{H_2O} MeO-\!\!\!\bigcirc\!\!\!-CH_2OH \quad (27.$$

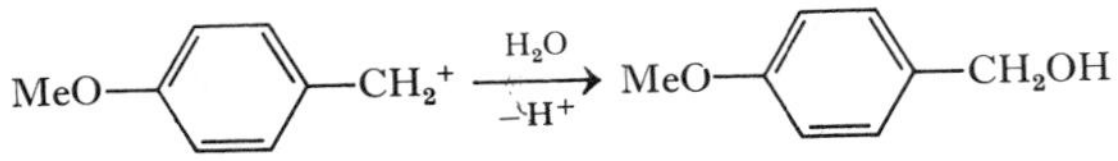

10.11 ELECTRON-TRANSFER SUBSTITUTION REACTIONS

Substitution reactions of alkyl halides are commonly held to involve attack by a nucleophile at a saturated carbon atom. This is undoubtedly the case in solvolyses of simple alkyl halides (reaction *273*):

$$CH_3I + OH^- \longrightarrow CH_3OH + I^- \quad (273)$$

That such a mechanism is not always operative in reactions which are formally of this type is indicated by the fact that it is possible to observe an e.s.r. signal in the reaction between triphenylmethyl perchlorate and potassium t-butoxide (180). This implies that the reaction is a homolytic electron-transfer process and not an ionic nucleophilic aliphatic substitution.

$$Ph_3C^+ClO_4^- + Bu^tO^-K^+ \longrightarrow Ph_3C\cdot + Bu^tO\cdot + K^+ + ClO_4^- \quad (274)$$

$$Ph_3C\cdot + Bu^tO\cdot \longrightarrow Ph_3COBu^t \quad (275)$$

There is an increasing amount of experimental evidence to indicate that this is a fairly general reaction pathway, being particularly common in reactions involving organometallic compounds and nitro compounds. Such reactions are properly included in a chapter on homolytic oxidation and reduction and will consequently be discussed in this section.

10.11.1 *Electron-transfer substitution reactions of* p-*nitrobenzyl and* p-*nitrocumyl chlorides*

Reaction of the salts of nitro compounds with aliphatic, allylic and benzylic halides has been developed as an effective method for the synthesis of aldehydes (181). The reaction proceeds by *O*-alkylation followed by a base-catalysed cleavage of the *O*-alkylated compound (reactions *276* and *277*):

$$PhCH_2Cl + Me_2\bar{C}NO_2 \longrightarrow Me_2C{=}\overset{\overset{\displaystyle O}{\uparrow}}{N}{-}OCH_2Ph + Cl^- \quad (276)$$

$$Me_2C{=}\overset{\overset{\displaystyle O}{\uparrow}}{N}{-}OCH_2Ph \xrightarrow{\text{base}} Me_2C{=}NOH + PhCHO \quad (277)$$

The reaction takes a very different course when the benzylic halide is substituted in the *ortho-* or *para-*positions by a nitro group. Under these circumstances only the *C*-alkylated product is obtained (182).

The reaction is retarded or inhibited by oxygen, *p*-dinitrobenzene or triphenylmethyl radicals (182, 183), all of which tends to suggest that a different mechanism is involved, and that this is an example of an electron-transfer reaction (Scheme 38). Initially electron-transfer occurs to give the radical anion (**75**), heterolytic fragmentation of which gives the *p*-nitrobenzyl radical (cf. chapter 12, p. 487). The *m*-nitrobenzyl radical anion would not be expected to undergo this type of fragmentation and in agreement with this it behaves 'normally', giving the *O*-alkylated compound. Coupling of the resultant *p*-nitrobenzyl radical with a 2-nitro-2-propyl anion gives the radical anion (**76**). This reacts

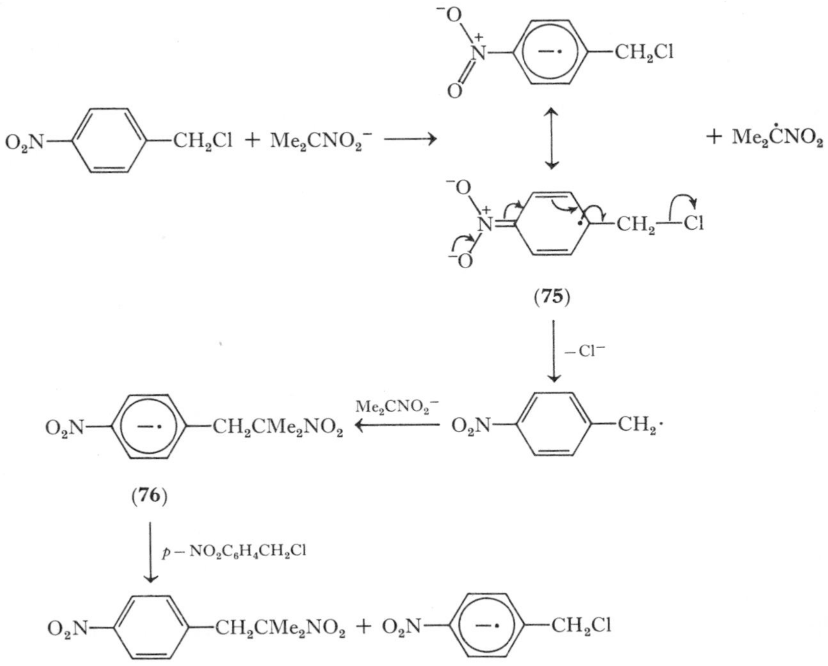

Oxidations and reductions

Scheme 38

with p-nitrobenzyl chloride in an electron-transfer process to give the product. p-Dinitrobenzene inhibits the reaction either by its ability to react with p-nitrobenzyl radicals (cf. 184), or because it competes effectively with the electron-transfer step in the reaction. Oxygen inhibits the reaction because of its ability to intercept carbon radicals with the resultant formation of peroxy radicals:

$$R \cdot + O_2 \longrightarrow RO_2 \cdot \qquad (278)$$

These reactions are also catalysed by light, consistent with the facile formation of nitroaromatic radical anions by the photolysis of nitro-aromatics in basic organic solvents (183).

p-Nitrocumyl chloride and α,p-dinitrocumene behave in an analogous manner, giving excellent yields of the C-alkylated products not only with the 2-nitro-2-propyl anion but also with other nucleophiles, e.g. thiophenoxide, n-butylmalonate anion and t-amines (185).

These reactions are particularly novel in that they constitute examples of very facile substitution at a tertiary carbon atom. The products are

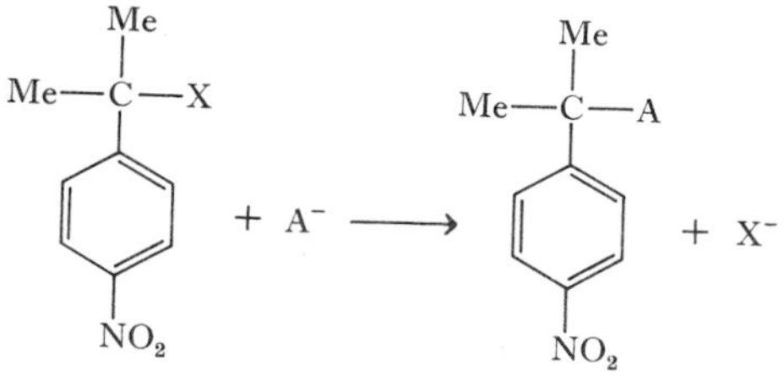

$$(279)$$

$$(X = Cl, NO_2)$$

different from those obtained in the corresponding reactions of cumyl chloride (reaction *280*). The primary substituted product (**77**) arises by an elimination–addition pathway:

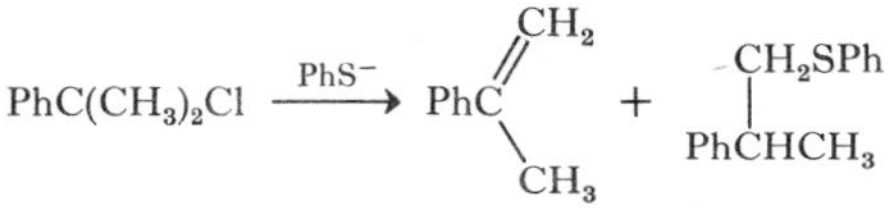

$$(280)$$

(**77**)

An interesting development of this procedure involves the use of catalytic amounts of 2-nitro-2-propyl anion to effect the substitution of *p*-nitrocumyl chloride and α,*p*-dinitrocumene by nucleophiles, which are themselves not sufficiently powerful reductants to bring about the formation of the initial radical anion (Scheme 39). The chain-

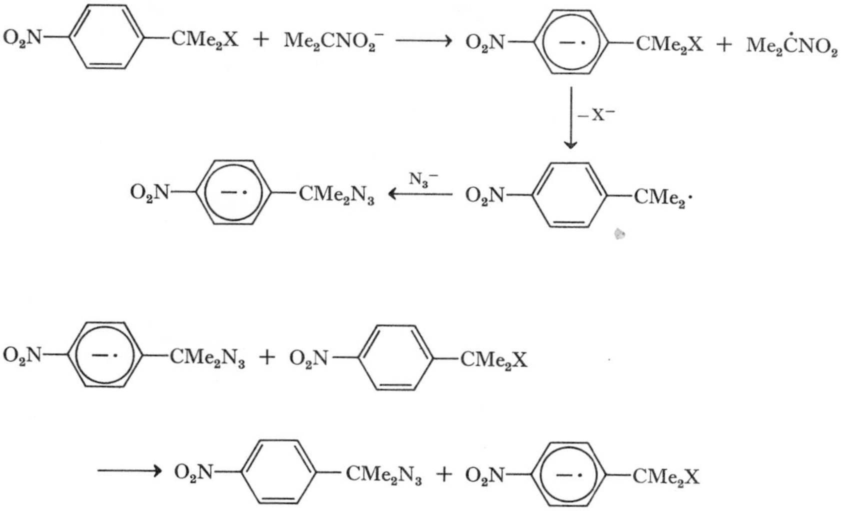

$$(X = Cl, NO_2)$$

Scheme 39

propagating sequence involves the addition of the anion to the *p*-nitrocumyl radical.

This type of process is not confined to reactions of *p*-nitrobenzylic halides. Thus, 2-chloro-2-nitropropane forms coupled products on reaction with anions (reactions *281* and *282*) (187):

$$\overset{\text{Cl}}{\underset{|}{\text{Me}_2\text{CNO}_2}} + \text{Me}_2\bar{\text{C}}\text{NO}_2 \longrightarrow \text{O}_2\text{NCMe}_2\text{CMe}_2\text{NO}_2 + \text{Cl}^- \quad (281)$$

$$\overset{\text{Cl}}{\underset{|}{\text{Me}_2\text{CNO}_2}} + \text{Et}\bar{\text{C}}(\text{CO}_2\text{Et})_2 \longrightarrow \text{O}_2\text{NCMe}_2\overset{\text{Et}}{\underset{|}{\text{C}}}(\text{CO}_2\text{Et})_2 + \text{Cl}^- \ (282)$$

These reactions are also inhibited by *p*-dinitrobenzene but catalysed by light.

10.11.2 *Electron-transfer reactions between alkyl halides and naphthalene radical anions*

Alkyl halides react with aromatic hydrocarbon radical anions to give, *inter alia*, alkylaromatics and dialkyl-dihydro-aromatics. The system which has been most extensively studied is that involving the naphthalene radical anion (NpH$^{\doteq}$) and particularly its reactions with 5-hexenyl halides (188–190). The alkylaromatic products are accompanied by reduction products, RH, and also by dimeric products, R—R.

$$\text{RX} + \text{NpH}^{\doteq} \longrightarrow \text{NpR} + \text{RH} + \text{R—R} \quad (283)$$

Garst has established that these products are formed as indicated in Scheme 40 (188). The salient feature of this scheme is the formation of

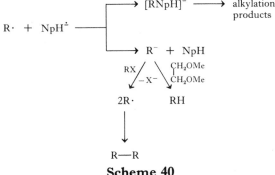

Scheme 40

alkyl radicals in an electron-transfer step. These radicals then either couple with naphthalene radical anion or undergo a second electron transfer to the carbanion. This latter either reacts with solvent (1,2-dimethoxyethane) to give the reduced compound, or reacts with alkyl halide in a further electron-transfer step to give a pair of alkyl radicals which combine to form dimers.

5-Hexenyl radicals undergo very rapid cyclization to cyclopentyl-methyl radicals:

$$(284)$$

Thus, formation of methylcyclopentane in the products can be taken as good evidence of radicals. The extent of such rearrangement is much less than that encountered in reactions of 5-hexenyl radicals in toluene. This is consistent with reduction of the radical to the carbanion competing fairly effectively with its cyclization. In accord with this, the extent of rearrangement decreases with increasing concentration of naphthalene radical anion. Radicals have thus little discrete existence in these reactions though they can be detected by use of ClDNP (184).

10.12   *AUTOXIDATION*

Autoxidation can be defined as a radical-chain reaction between molecular oxygen and organic compounds at low or moderate temperatures. This reaction almost invariably results in the formation of hydroperoxides:

$$RH + O_2 \longrightarrow ROOH \qquad (285)$$

These hydroperoxides then frequently undergo further decomposition.

The autoxidation reaction may be beneficial or deleterious according to the particular circumstances under which it occurs. Thus, it is important in the drying of paints and oils as it leads to the formation of protective films, though these films ultimately break down as the result of further autoxidation. The oils used in paints contain unsaturated esters which give rise to hydroperoxides. Decomposition of these to alkoxy radicals by traces of added metal salts results in the formation of a protective polymer film by reaction of the alkoxy radical with further olefin.

Autoxidation is widely used in the petrochemical industry. Particularly important is the production of cumene hydroperoxide, acid-catalysed decomposition of which provides phenol and acetone:

$$PhCHMe_2 \xrightarrow{O_2} \underset{\underset{OOH}{|}}{PhCMe_2} \xrightarrow{H^+} PhOH + Me_2CO \qquad (286)$$

The effects of autoxidation are highly undesirable in the rancidification of edible oils and the perishing of rubber. Petroleum and petroleum products are also susceptible to autoxidation giving rise to gummy products. In these cases it is necessary to find ways of controlling, or better inhibiting, autoxidation by suitable additives. These antioxidants are frequently phenols or amines.

10.12.1 *Kinetic features of autoxidation*

Alkanes, alkenes, alcohols, aldehydes and ethers are all susceptible to autoxidation by the same general process which involves the formation of a hydroperoxide (191):

$$RH + O_2 \xrightarrow{\text{Initiator}} ROOH \qquad (287)$$

The initiation, propagation and termination steps of this process are given by reactions (288–294).

$$\text{Initiation} \begin{cases} \text{Initiator} \longrightarrow 2In\cdot & (288) \\ In\cdot + RH \longrightarrow R\cdot + InH & (289) \end{cases}$$

$$\text{Propagation} \begin{cases} R\cdot + O_2 \longrightarrow ROO\cdot & (290) \\ ROO\cdot + RH \longrightarrow R\cdot + ROOH & (291) \end{cases}$$

$$\text{Termination} \begin{cases} ROO\cdot + ROO\cdot \longrightarrow \text{Products} & (292) \\ R\cdot + ROO\cdot \longrightarrow ROOR & (293) \\ R\cdot + R\cdot \longrightarrow R-R & (294) \end{cases}$$

(i) *Initiation*. Initiation can be achieved by addition of typical radical initiators such as azonitriles and peroxides. The overall rate of autoxidation when a radical initiator is employed is proportional to the square root of the initiator concentration. In the absence of radical initiators, autoxidations are frequently subject to long induction periods. This induction period arises from the time needed for a build-up of hydroperoxide, thermolysis or photolysis of which produces radicals:

$$ROOH \longrightarrow RO\cdot + HO\cdot \qquad (295)$$

This hydroperoxide may arise as a result of interaction of singlet molecular oxygen with the organic substrate.

(ii) *Propagation.* The overall rate of autoxidation is independent of the oxygen pressure at moderate pressures of oxygen but is proportional to the concentration of the organic substrate RH. This is consistent with reaction (290) being fast and reaction (291) being rate-limiting. One might expect reaction (290) to be fast since it is essentially a radical coupling process as molecular oxygen is a diradical. This step is probably diffusion controlled ($k \sim 10^9$ l mol^{-1} s^{-1}). The rate of reaction (291) is very susceptible to the chemical environment of the hydrogen being abstracted (Table 10-4). Examination of this table reveals that the reaction is faster, the more stable is the incipient radical R·. Polar and steric factors also influence the rate of this reaction.

TABLE 10-4 *Relative reactivities of hydrocarbons towards hydrogen abstraction by peroxy radicals at 30 °C* (192)

Hydrocarbon	Relative reactivity[a]
PhCH$_3$	1.00[b]
PhCH$_2$Ph	30
PhCH$_2$CH=CH$_2$	63
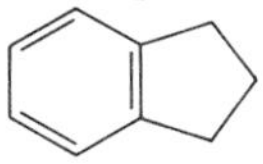 (indane structure)	15
PhCHMe$_2$	2.3
PhCH$_2$CH$_2$Ph	2.0
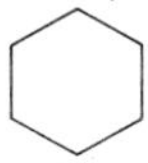 (cyclohexane structure)	0.014
 (cyclohexene structure)	18
 (cyclopentene structure)	21
CH$_2$=CHCH$_2$CH=CH$_2$	88

[a] Reactivity per active hydrogen relative to toluene.
[b] By definition.

Benson (193) has estimated the ROO–H bond strength to be about 375 kJ mol^{-1}, which means that this bond is somewhat stronger than an allylic, benzylic or aldehydic C–H bond, but comparable to a tertiary C–H bond. This helps one to understand the observed selectivity of

different sites in a molecule towards autoxidation. Reactions (*296–300*)
illustrate this:

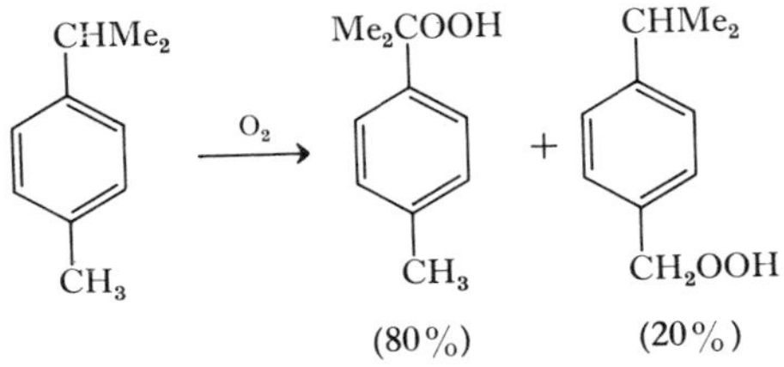

(*296*)

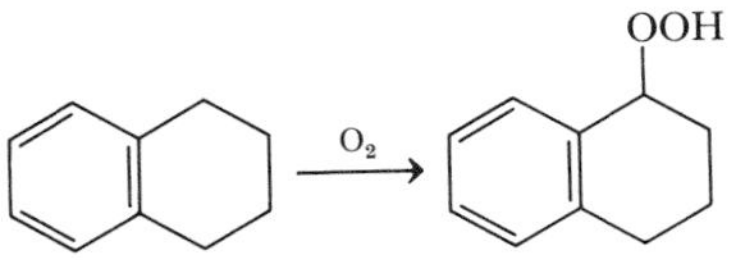

(80%) (20%)

(*297*)

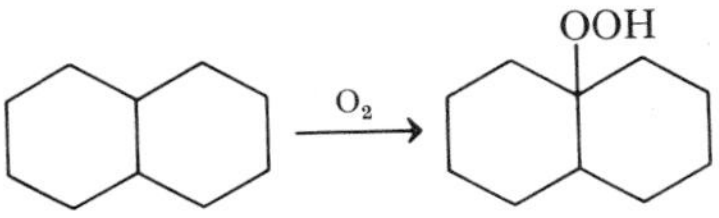

(*298*)

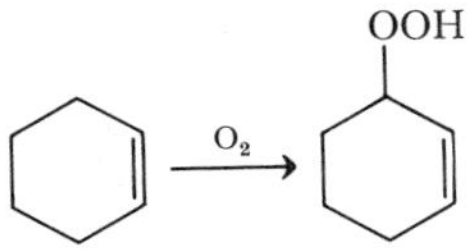

(*299*)

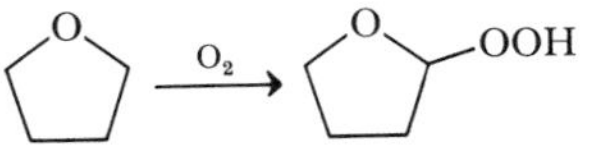

(*300*)

These examples illustrate that reaction occurs preferentially at allylic or
benzylic positions, which give rise to a resonance-stabilized allylic or
benzylic radical. Where no such position exists as in decalin, the tertiary
C–H bond is attacked, whilst in tetrahydrofuran the stabilizing effect of
the oxygen on the resultant radical promotes attack at the α-C–H bond.
Polar factors are also important here. The peroxy radical is an electrophilic radical, and hence in the transition state (**78**) the α-carbon will
bear a partial positive charge, which can be stabilized by the adjacent
oxygen atom.

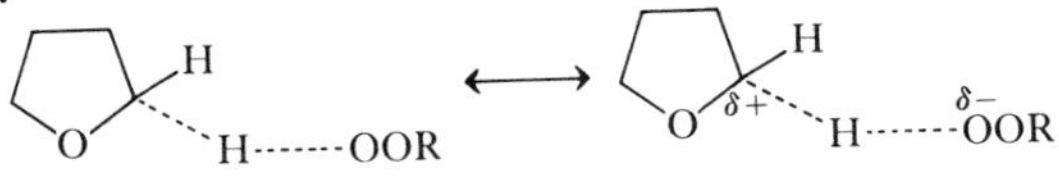

(**78**)

Groups which increase the electrophilicity of the peroxy radical enhance its reactivity. Thus, the radical (**79**) is much more reactive than the radical (**80**).

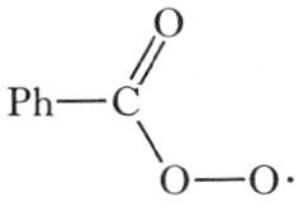

$$Ph-C\underset{O-O\cdot}{\overset{O}{\lesseqgtr}}\qquad\qquad Ph-\underset{H}{\overset{OR}{\underset{|}{\overset{|}{C}}}}-O-O\cdot$$

(**79**) 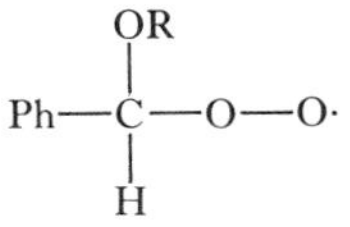   (**80**)

Compounds which can give rise to more than one allylic radical such as methyl oleate, give mixtures of all the possible hydroperoxides:

$$CH_3(CH_2)_6CH_2CH=CHCH_2(CH_2)_6CO_2Me$$

$$\xrightarrow{O_2} CH_3(CH_2)_6\overset{\overset{\textstyle OOH}{|}}{C}HCH=CH(CH_2)_7CO_2Me$$

$$+ CH_3(CH_2)_6CH=CH\overset{\overset{\textstyle OOH}{|}}{C}H(CH_2)_7CO_2Me$$

$$+ CH_3(CH_2)_7\overset{\overset{\textstyle OOH}{|}}{C}HCH=CH(CH_2)_6CO_2Me$$

$$+ CH_3(CH_2)_7CH=CH\overset{\overset{\textstyle OOH}{|}}{C}H(CH_2)_6CO_2Me \qquad (301)$$

(iii) *Termination.* The only termination step which need be considered at moderate pressures of oxygen (above 100 Torr) is reaction (*302*). Reactions (*303* and *304*) can be effectively ignored, since the concentration of R· is so much lower than that of RO$_2$·:

$$RO_2\cdot + RO_2\cdot \longrightarrow Products \qquad (302)$$

$$R\cdot + RO_2\cdot \longrightarrow ROOR \qquad (303)$$

$$R\cdot + R\cdot \longrightarrow R-R \qquad (304)$$

At lower pressures of oxygen (~ 10 Torr) reaction (*303*) becomes important, particularly when the alkyl radical R· is fairly stable. When this is so, the reaction of R· with oxygen is only moderately fast, and reaction (*303*) can compete with reaction (*302*).

The rate constant for the reaction between peroxy radicals is markedly dependent on the nature of the alkyl group of the peroxy radical. In

TABLE 10-5   *Chain termination constants for peroxy radicals*

Radical	Chain termination constant (194, 195) $(2k_t, \text{l mol}^{-1}\text{s}^{-1})$
$Bu^nOO\cdot$	0.4×10^8
$Bu^sOO\cdot$	1.5×10^6
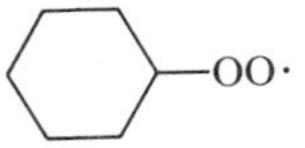 —OO·	2.0×10^6
$(CH_2{=}CH)_2CHOO\cdot$	11×10^6
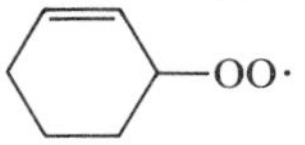 —OO·	5.6×10^6
$Bu^tOO\cdot$	0.4×10^3
$PhCMe_2OO\cdot$	2.6×10^3
$Ph_2CMeOO\cdot$	2.7×10^4

general, the chain-termination rate constants in the series tertiary alkyl-peroxy, secondary alkylperoxy and primary alkylperoxy are of the order of 10^3, 10^6 and 10^8 respectively (Table 10-5). The reaction involves the reversible formation of a dimer; which is generally formulated as a tetroxide. This subsequently decomposes to give products, the nature of which depend on the type of peroxy radical:

$$2ROO\cdot \overset{K}{\rightleftharpoons} ROOOOR \longrightarrow \text{Products} \qquad (305)$$

Russell has established that the decomposition of 1-phenylethylperoxy radicals leads to acetophenone, 1-phenylethanol and oxygen (196). He interprets this as proceeding via a tetroxide which decomposes through a cyclic transition state to give the observed products.

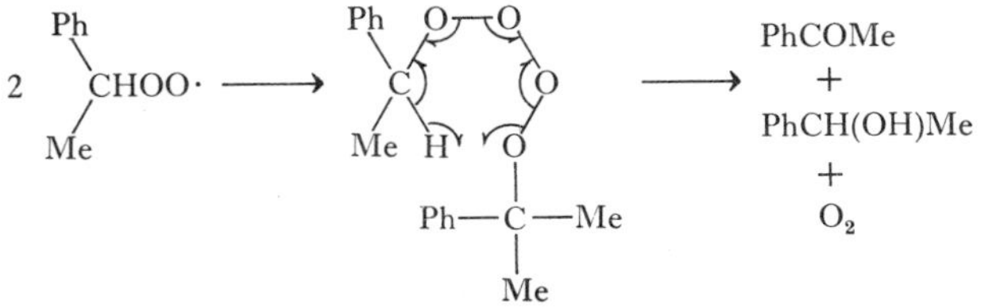

$$\qquad (306)$$

Confirmation of this scheme was obtained by studying the combination of 1-deuteriophenylethylperoxy radicals for which a deuterium isotope effect, $k_H/k_D = 1.9$, was observed. This is consistent with the involvement of the α-hydrogen in the transition state for the decomposition of the tetroxide:

$$2 \quad \begin{matrix} Ph \\ \diagdown \\ CDOO\cdot \\ \diagup \\ Me \end{matrix} \longrightarrow PhCOMe + PhCD(OD)Me + O_2 \qquad (307)$$

The Wigner spin-conversation rule demands that the oxygen liberated in the self-reaction of secondary peroxy radicals should either involve the formation of a singlet or triplet oxygen. This latter may be produced if the carbonyl compound is also produced in the excited triplet state, provided that the tetroxide is an intermediate and not merely a transition state (197). Kellogg has shown that the reaction does indeed give rise to triplet oxygen and the excited carbonyl compound, but that these are mutually quenched within the solvent cage (198). The resultant singlet oxygen has been trapped by 9,10-diphenylanthracene with the formation of the transannular peroxide (81) (198).

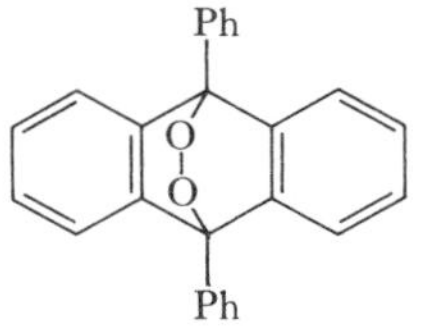

(81)

The greater termination efficiency for peroxy radicals in which the peroxy group is adjacent to a π-electron system may reflect either activation of the α-hydrogen, or alternatively stabilization of the incipient conjugated carbonyl compound.

The decomposition of the tetroxide derived from t-alkylperoxy radicals is much slower than that derived from primary or secondary peroxy radicals, because of the much greater energy of activation for this process (~ 640–60 kJ mol^{-1}). This undoubtedly arises from the absence of the cyclic pathway available for the decomposition of primary and secondary tetroxides. In the case of di-t-alkyltetroxides, decomposition occurs involving O–O cleavage with formation of t-alkoxy radicals and oxygen. The t-alkoxy radicals either dimerize within the solvent cage or diffuse away and start further radical chains, thereby complicating the kinetics of the overall autoxidation process. The oxygen generated in this process is likely to be in the triplet ground state and consistent with this no transannular peroxide was formed from 9,10-diphenylanthracene (197).

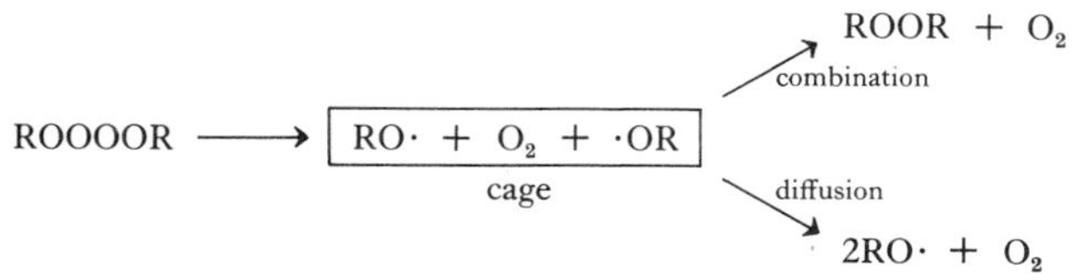

$$ROOOOR \longrightarrow \boxed{RO\cdot + O_2 + \cdot OR} \nearrow^{\text{combination}} ROOR + O_2 \qquad (308)$$
$$\text{cage} \searrow_{\text{diffusion}} 2RO\cdot + O_2$$

The rate constants for chain termination reactions of acylperoxy radicals, which do not have a hydrogen on the α-carbon atom, are very high (cf. Table 10-5). They may well proceed through cyclic six-centre transition states (reaction *309*):

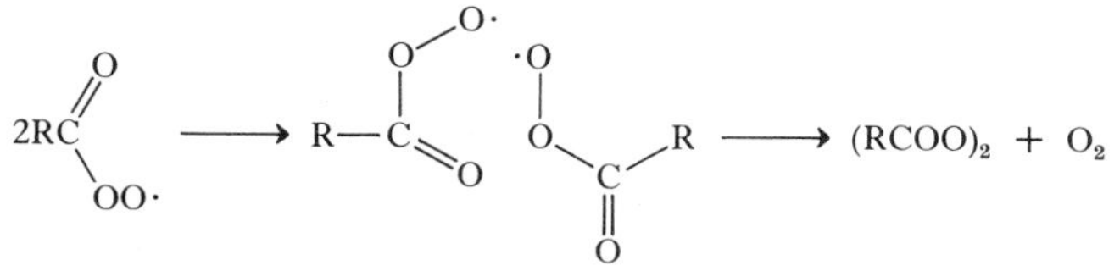

$$2R\overset{O}{\underset{OO\cdot}{\overset{\|}{C}}} \longrightarrow R-\overset{\cdot O}{\underset{O}{\overset{O-O\cdot}{C}}} \overset{\cdot O}{\underset{C}{\overset{O}{\longrightarrow}}} R \longrightarrow (RCOO)_2 + O_2 \quad (309)$$

This agrees with the observation that the molecular oxygen evolved appears to come from both radicals.

10.12.2 *Catalysis of autoxidation by metal salts*

Metal-catalysed decompositions of hydroperoxides can take place in two different ways according to the oxidation state of the metal (199, 200):

$$ROOH + M^{n+} \longrightarrow HOM^{n+} + RO\cdot \qquad (310)$$
$$ROOH + M^{(n+1)+} \longrightarrow ROO\cdot + M^{n+} + H^+ \qquad (311)$$

The effect of added metal salts is thus to initiate further radical chains, thereby increasing the overall rate of autoxidation. Use of larger quantities can, however, sometimes inhibit autoxidation:

$$ROO\cdot + M^{n+} \longrightarrow ROO^- + M^{(n+1)+} \qquad (312)$$

Metal salts can also initiate the autoxidation process in the absence of hydroperoxide:

$$Co^{3+} + RH \longrightarrow R\cdot + Co^{2+} + H^+ \qquad (313)$$

Thus, the rate of chain initiation of linoleic acid in benzene has been shown to be proportional to the concentration of added cobalt(III) stearate. This reaction is somewhat analogous to the oxidation of aromatic compounds by cobalt(III) acetate discussed earlier.

10.12.3 *Antioxidants*

The essential feature of an antioxidant is that it interrupts the autoxidation radical chain, thereby retarding the autoxidation process. It is thus
necessary that compounds used as antioxidants shall have a readily
abstractable hydrogen. This condition is satisfied by phenols and
aromatic amines, both of which are extensively employed as antioxidants
on account of their ability to act as efficient chain-transfer agents.

$$ROO\cdot + ArOH \longrightarrow ROOH + ArO\cdot \qquad (314)$$

$$ROO\cdot + ArNH_2 \longrightarrow ROOH + ArNH\cdot \qquad (315)$$

Ingold has shown that replacement of hydrogen by deuterium on the
phenolic hydroxyl group results in a considerable reduction in the
efficiency of this reaction, $k_H/k_D \approx 10$ for 2,6-di-t-butyl-4-methylphenol
(201). High values (10–15) were also obtained for other hindered and
unhindered phenols. Thus, the O–H bond of the antioxidant is broken
in the slow step of the chain-transfer process. Ingold has also shown that
the relative inhibiting efficiencies of a series of *meta-* and *para-*substituted phenols can be correlated by means of the Hammett equation
using σ^+ constants, i.e.

$$\log k_{314}(\text{rel.}) = \rho\sigma^+$$

The value of ρ varies from -1.12 to -1.49 according to the particular
series of phenols being investigated. The negative sign of ρ indicates
that the reaction is accelerated by electron-releasing substituents in the
phenol, showing that polar contributions to the transition state (82) for
hydrogen abstraction have to be considered. The correlation with σ^+
rather than σ emphasizes the importance of the dipolar structure (82c)
to the transition state, since *para-*substituents would be in direct conjugation with the oxygen. The small magnitude of ρ shows that there is
only a small degree of charge separation in the transition state.

(82)

Oxidations and reductions

The products obtained in the phenol-inhibited autoxidation of cumene are the typical products obtained in the oxidation of phenols as a result of the coupling of phenoxy radicals (e.g. reaction *316*) (202):

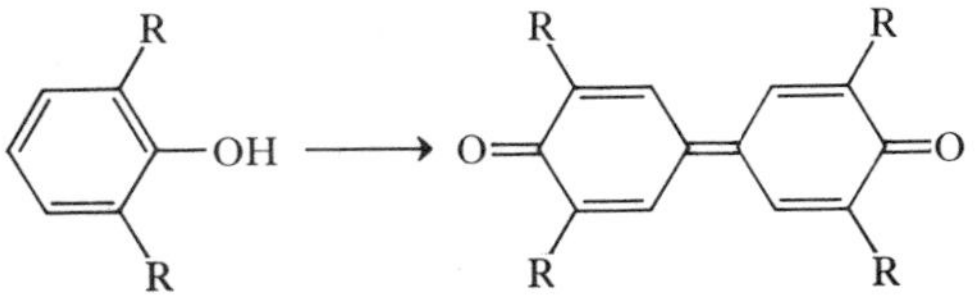

(316)

In the presence of a high concentration of peroxy radicals, 2,6-di-t-butyl-4-methylphenol gives the peroxy-cyclohexadienone (**83**) from coupling of the phenoxy radical with the peroxy radical (reaction *317*) (203):

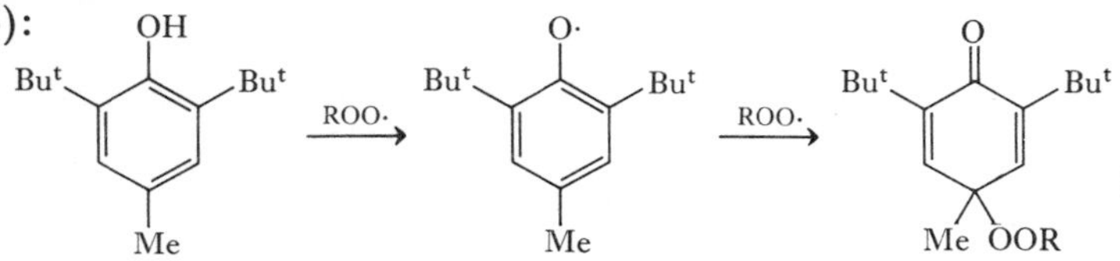

(317)

(**83**)

(R = Me$_2$CCN, Bu^t, α-tetralyl)

The kinetic scheme for autoxidation has to be modified in the presence of an antioxidant AH, by inclusion of reactions (*318–321*).

$$ROO\cdot + AH \longrightarrow ROOH + A\cdot \qquad (318)$$

$$A\cdot + ROO\cdot \longrightarrow ROOA \qquad (319)$$

$$2A\cdot \longrightarrow A\!-\!A \qquad (320)$$

$$A\cdot + RH \longrightarrow R\cdot + AH \qquad (321)$$

The rate of reaction (*318*) is strongly dependent not only on polar factors but also on steric factors. 2,6-Dialkylphenols react much less rapidly than simple phenols because of the steric protection of the reaction centre, and hence they are less efficient antioxidants (Fig. 10-1) (204). For hindered phenols reactions (*319*) and (*320*) are much faster than reactions (*318*) and (*321*), whereas for simple phenols all four reactions are competitive. Thus, for hindered phenols the chain carrier is the peroxy radical but for non-hindered phenols the chain is perpetuated by both peroxy and phenoxy radicals. Consequently, non-hindered phenols can only retard but not inhibit autoxidation reactions.

A mixture of a hindered and unhindered phenol is, however, a much

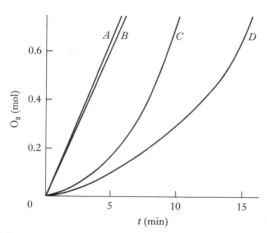

Fig. 10-1. The oxygen absorption of a chlorobenzene solution of 148 mmol 9,10-dihydroanthracene at 60 °C containing 124 mmol 2,2,3,3-tetraphenylbutane and curve A: no inhibitor; curve B: 0.1 mmol 2,6-di-t-butyl-4-methylphenol; curve C: 0.1 mmol 4-methoxyphenol; and curve D: 0.1 mmol 2,6-di-t-butyl-4-methylphenol and 0.1 mmol 4-methoxyphenol.

more efficient antioxidant than either alone (cf. Fig. 10-1). This is exemplified in reactions (*322–324*) for a mixture of 2,6-di-t-butyl-4-methylphenol and *p*-methoxyphenol (204):

$$\text{ROO} \cdot + \text{(phenol, OH/OMe)} \longrightarrow \text{ROOH} + \text{(phenoxyl, O} \cdot \text{/OMe)} \tag{322}$$

$$\text{(O} \cdot \text{/OMe)} + \text{(OH, Bu}^t\text{/Bu}^t\text{/Me)} \longrightarrow \text{(O} \cdot \text{, Bu}^t\text{/Bu}^t\text{/Me)} + \text{(OH/OMe)} \tag{323}$$

$$\text{ROO} \cdot + \text{(O} \cdot \text{, Bu}^t\text{/Bu}^t\text{/Me)} \longrightarrow \text{(O, Bu}^t\text{/Bu}^t\text{/ROO, Me)} \tag{324}$$

403

Oxidations and reductions

The reason for this synergistic behaviour of the two phenols arises from the regeneration of p-methoxyphenol in reaction (*323*). This reaction occurs as shown and not in the reverse direction, because of the relief of strain accompanying the formation of 2,6-disubstituted phenoxy radicals. Mahoney has estimated that this is of the order of 40 kJ mol^{-1} (*204*). In the absence of a hindered phenol reaction (*322*) would be reversible and consequently retardation would be less marked.

10.12.4 *Autoxidation of aldehydes*

Aldehydes are very susceptible to air oxidation. The autoxidation proceeds by the radical-chain mechanism outlined in reactions (*325–327*):

$$PhCHO + R \cdot \longrightarrow Ph\dot{C}O + RH \qquad (325)$$

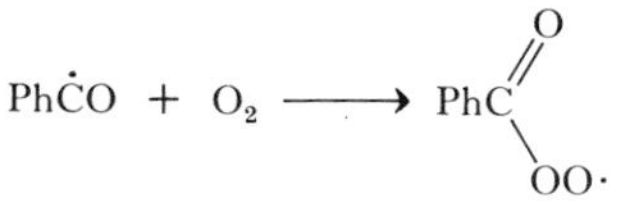

$$(326)$$

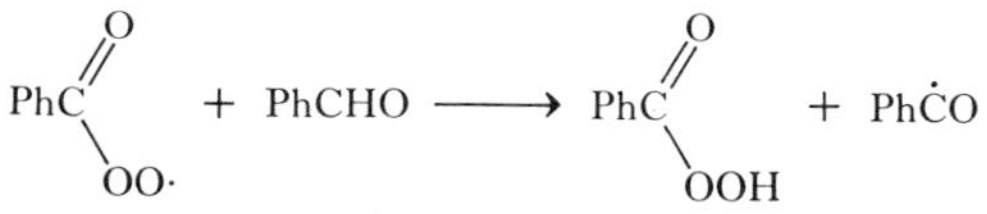

$$(327)$$

The final product obtained from these reactions is generally the carboxylic acid and not the peracid. This latter reacts with aldehyde in an acid-catalysed heterolytic process giving two molecules of carboxylic acid:

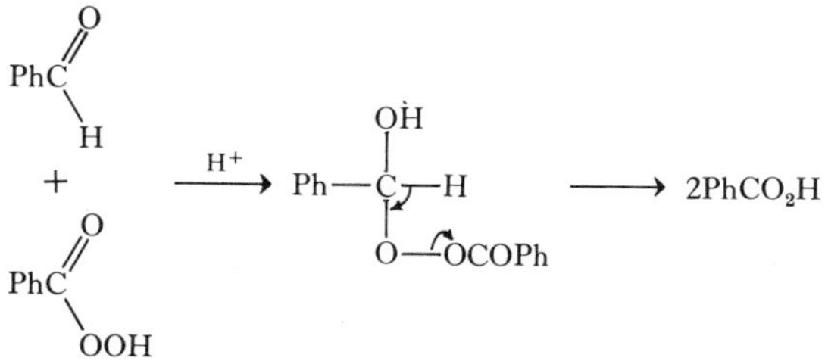

$$(328)$$

Evidence for the intermediacy of acyl radicals in the reaction scheme is the formation, at elevated temperatures ($\sim 100\,^{\circ}$C) and low oxygen concentrations, of carbon monoxide (*204a*). This results from the fragmentation of the intermediate acyl radicals:

$$R\dot{C}O \longrightarrow R \cdot + CO \qquad (329)$$

More carbon monoxide is evolved from aldehydes possessing secondary

404

alkyl groups than from those containing primary alkyl groups because of the higher stability of secondary alkyl radicals.

Initiation of the autoxidation may be brought about photochemically, by radical initiators, or by metal ions:

$$RCHO \xrightarrow{h\nu} R\dot{C}O + H\cdot \tag{330}$$

$$RCHO + R'\cdot \longrightarrow R\dot{C}O + R'H \tag{331}$$

$$RCHO + Co^{3+} \longrightarrow R\dot{C}O + Co^{2+} + H^+ \tag{332}$$

10.12.5   *Autoxidation of alkenes*

The propagation step in the autoxidations so far considered has been assumed to involve hydrogen abstractions from the substrate by the peroxy radical. In the case of alkenes this reaction (*333*) is accompanied by addition of a peroxy radical to the double bond (reaction *334*):

$$RO_2\cdot \; + \; \overset{|}{\underset{H}{-C}}-\overset{|}{C}=C\diagdown \; \longrightarrow \; ROOH \; + \; \overset{\cdot}{\underset{\diagup}{C}}-\overset{|}{C}=C\diagdown \tag{333}$$

$$RO_2\cdot \; + \; \diagdown_{\diagup}C=C_{\diagdown}^{\diagup} \; \longrightarrow \; ROO-\overset{|}{\underset{|}{C}}-\overset{\cdot}{C}_{\diagdown}^{\diagup} \tag{334}$$

Mayo has established that the abstraction mechanism occurs almost exclusively for cyclopentene, cyclohexene, and open-chain alkenes with a tertiary allylic hydrogen, whilst the addition pathway predominates for olefins with no reactive allylic hydrogens such as styrene and norbornene (205). The vast majority of alkenes react by both mechanisms to give mixtures of products.

The radical formed as a result of addition of a peroxy radical to a double bond, reacts further by addition of oxygen or by fragmentation to give an epoxide and an alkoxy radical:

$$RO_2\overset{|}{\underset{|}{C}}-\overset{\cdot}{C}_{\diagdown}^{\diagup} \; \xrightarrow{O_2} \; RO_2-\overset{|}{\underset{|}{C}}-\overset{|}{\underset{|}{C}}-OO\cdot \tag{335}$$

$$RO_2\overset{|}{\underset{|}{C}}-\overset{\cdot}{C}_{\diagdown}^{\diagup} \; \longrightarrow \; RO\cdot \; + \; \overset{O}{\underset{|}{-C}\diagup}\diagdown\underset{|}{C-} \tag{336}$$

The greater the stability of the β-peroxyalkyl radical, the greater is its tendency to react with oxygen rather than fragment.

Oxidations and reductions

The presence of oxygen during the polymerization of styrene has deleterious effects on the resultant polystyrene. The growing polymer radical may react with oxygen to give a polystyrylperoxy radical which reacts further to give, in addition to other products, a polyperoxide:

$$\text{\textasciitilde CH}_2\dot{\text{C}}\text{H} + \text{O}_2 \longrightarrow \text{\textasciitilde CH}_2\text{CHOO} \cdot \qquad (337)$$
$$\underset{\text{Ph}}{|} \qquad\qquad\qquad \underset{\text{Ph}}{|}$$

$$\xrightarrow{\text{PhCH=CH}_2} \text{\textasciitilde CH}_2\text{CHOOCH}_2\dot{\text{C}}\text{H} \xrightarrow{\text{O}_2} \text{\textasciitilde CH}_2\text{CHOOCH}_2\text{CHOO} \cdot \quad (338)$$
$$\qquad\qquad \underset{\text{Ph}}{|}\qquad\underset{\text{Ph}}{|} \qquad\qquad\qquad \underset{\text{Ph}}{|}\qquad\underset{\text{Ph}}{|}$$

This polyperoxy radical can break down to give styrene oxide and a polyalkoxy radical which 'unzips' to give benzaldehyde and formaldehyde:

$$\text{\textasciitilde}\left(\underset{\underset{\text{Ph}}{|}}{\text{OOCH}_2\text{CH}}\right)_n \underset{\underset{\text{Ph}}{|}}{\text{OOCH}_2\dot{\text{C}}\text{H}} \longrightarrow \text{\textasciitilde}\left(\underset{\underset{\text{Ph}}{|}}{\text{OOCH}_2\text{CH}}\right)_n \text{O}\cdot + \overset{\text{O}}{\overset{/\backslash}{\text{CH}_2\text{—CHPh}}}$$

$$(339)$$

$$\downarrow$$

$$n\text{PhCHO} + n\text{CH}_2\text{O}$$

10.12.6 Autoxidation of organoboron compounds

The autoxidation of organoboron compounds to organoperoxyboron compounds has been studied in considerable detail, particularly by Davies and Roberts (206):

$$\text{R—B}\overset{/}{\underset{\backslash}{}} + \text{O}_2 \longrightarrow \text{ROOB}\overset{/}{\underset{\backslash}{}} \qquad (340)$$

The importance of this reaction is that it is a model for the autoxidation of organometallic compounds in which the metal is cadmium, zinc, aluminium or magnesium (207).

The mechanism of the reaction was initially believed to involve coordination of oxygen to boron followed by a polar insertion of oxygen into the carbon–boron bond:

$$\text{R}_3\text{B} + \text{O}_2 \longrightarrow \text{R}_2\underline{\text{B}}\overset{\overset{\text{R} \quad \text{O}^+}{\quad\downarrow\quad|}}{\text{—O}} \longrightarrow \text{R}_2\text{B}\overset{\overset{\text{R—O}}{|}}{\text{—O}} \qquad (341)$$

Radicals were said not to be involved because common phenolic anti-oxidants had no influence on the rate of the reaction. Subsequently, Davies and Roberts showed that the reaction could be inhibited by the stable galvinoxyl radical and consequently proposed that autoxidation proceeded by a radical-chain mechanism (206):

Initiation: Initiator $\longrightarrow$ 2R· $\qquad\qquad$ (*342*)

Propagation: R· + O$_2$ $\longrightarrow$ RO$_2$· $\qquad$ (*343*)

$$\text{RO}_2\cdot + \quad \backslash \text{B—R} \quad \longrightarrow \text{ROOB} + \text{R·} \qquad (342)$$

$$\text{RO}_2\cdot + \overset{\diagdown}{\underset{\diagup}{\text{B}}}\text{—R} \longrightarrow \text{ROOB}^{\diagup}_{\diagdown} + \text{R·} \qquad (344)$$

Termination: 2ROO· $\longrightarrow$ Inactive products $\qquad$ (*345*)

Further confirmation for this scheme was obtained by isolation of racemic 1-phenylethyl hydroperoxide from the autoxidation of 1-phenyl-ethane boronic acid (206):

$$\underset{Ph}{\overset{H\,Me}{\diagdown\diagup}}\text{C—B(OH)}_2 \xrightarrow{\text{O}_2} \text{Me}\overset{H}{\underset{Ph}{\vert}}\text{C—O—O—B(OH)}_2 \xrightarrow{\text{H}_2\text{O}} \text{Me}\overset{H}{\underset{Ph}{\vert}}\text{C—OOH} \quad (346)$$

The key step in the reaction involves an S_H2 attack of the alkylperoxy radical at the boron atom. The inefficiency of phenols as antioxidants arises from the great speed of this process which Ingold has estimated to be some 10^7 times faster than abstraction of a hydrogen atom by a peroxy radical (208).

The autoxidation of organoboranes has been exploited by Brown *et al.* for the synthesis of alcohols (210) and hydroperoxides (211). Low-temperature oxidation of a trialkylborane, prepared *in situ* from an alkene, in tetrahydrofuran leads to a diperoxyborane (RO$_2$)$_2$BR (reactions *347–350*). Treatment of this with sodium hydroxide gives the alcohol in excellent yield (reaction *351*), whilst reaction with hydrogen peroxide affords the hydroperoxide (reaction *352*):

$$R_3B + O_2 \longrightarrow R\cdot + R_2BO_2\cdot \qquad (347)$$
$$R\cdot + O_2 \longrightarrow RO_2\cdot \qquad (348)$$
$$RO_2\cdot + R_3B \longrightarrow RO_2BR_2 + R\cdot \qquad (349)$$
$$RO_2\cdot + RO_2BR_2 \longrightarrow (RO_2)_2BR + R\cdot \qquad (350)$$
$$(RO_2)_2BR + OH^- + 2H_2O \longrightarrow 2ROH + ROO^- + B(OH)_3 \quad (351)$$
$$(RO_2)_2BR + H_2O_2 + 2H_2O \longrightarrow ROH + RO_2H + B(OH)_3 \quad (352)$$

Oxidations and reductions

Separation of the hydroperoxide from the alcohol is readily achieved by treating the reaction solution with excess alkali: the hydroperoxide dissolves in the aqueous phase leaving the alcohol in the organic phase.

10.12.7 Electron-transfer autoxidations

The autoxidations which we have considered thus far have all involved the two basic propagation steps (reactions *353* and *354*):

$$R \cdot + O_2 \longrightarrow RO_2 \cdot \tag{353}$$

$$RO_2 \cdot + RH \longrightarrow R \cdot + RO_2H \tag{354}$$

A rather different scheme has to be considered for the autoxidation of 2-nitropropane in basic solution. This reaction is subject to catalysis by iron(III) and also by nitrobenzene. To account for this Russell has proposed that the initiation step involves the generation of a 2-nitro-2-propyl radical in an electron-transfer step (183, 187). This reacts with oxygen and the resultant peroxy radical undergoes a second electron-transfer reaction with the 2-nitro-2-propyl anion:

$$Me_2CNO_2^- + Fe^{3+} \longrightarrow Me_2\dot{C}NO_2 + Fe^{2+} \tag{355}$$

$$Me_2CNO_2^- + PhNO_2 \longrightarrow Me_2\dot{C}NO_2 + PhNO_2^{\overset{.}{-}} \tag{356}$$

$$Me_2\dot{C}NO_2 + O_2 \longrightarrow \underset{\underset{NO_2}{|}}{Me_2COO \cdot} \tag{357}$$

$$\underset{\underset{NO_2}{|}}{Me_2COO \cdot} + Me_2CNO_2^- \longrightarrow Me_2\dot{C}NO_2 + \underset{\underset{NO_2}{|}}{Me_2COO^-} \tag{358}$$

$$\underset{\underset{NO_2}{|}}{Me_2COO^-} + Me_2CNO_2^- \longrightarrow 2\underset{\underset{NO_2}{|}}{Me_2CO^-} \tag{359}$$

$$\underset{\underset{NO_2}{|}}{Me_2CO^-} \longrightarrow Me_2CO + NO_2^- \tag{360}$$

The nature of the termination step which results in the eventual formation of acetone and nitrite ion is not fully understood.

The oxidations of triphenylmethane and diphenylmethane in basic dimethyl sulphoxide proceed by a different mechanism from the above process (209). The rate of oxidation is proportional to the rate of

408

ionization of the hydrocarbon, implying that the triphenylmethyl anion reacts with oxygen:

$$Ph_3CH + B: \xrightarrow{\text{slow}} Ph_3C^- + BH^+ \tag{361}$$

$$Ph_3C^- + O_2 \longrightarrow Ph_3COO^- \longrightarrow Ph_3COOH \tag{362}$$

$$Ph_3COO^- + Ph_3C^- \longrightarrow 2Ph_3CO^- \longrightarrow 2Ph_3COH \tag{363}$$

The reaction of the triphenylmethyl anion with oxygen contravenes multiplicity rules and hence it has been suggested that the reaction of the anion with oxygen is a two-stage process (reaction *364*):

$$Ph_3C^- + \uparrow \cdot O\!-\!O \cdot \uparrow \longrightarrow Ph_3C \cdot \uparrow \uparrow \cdot O_2^-$$

$$\longrightarrow Ph_3C \cdot \uparrow \downarrow \cdot O_2^- \longrightarrow Ph_3COO^- \tag{364}$$

10.12.8 *Autoxidation in biological systems*

The presence of unsaturated compounds in edible oils and fats clearly renders them liable to autoxidative breakdown. That this does not occur to a significant extent in bacteria-free systems suggests that such materials also contain antioxidants. This has been confirmed by the isolation of tocopherols (vitamin E) from the lipid-soluble portion of corn oil and other foodstuffs. The most potent of these tocopherols in terms of its efficiency as an antioxidant is α-tocopherol (**84**). The other tocopherols are also active, though to a lesser degree.

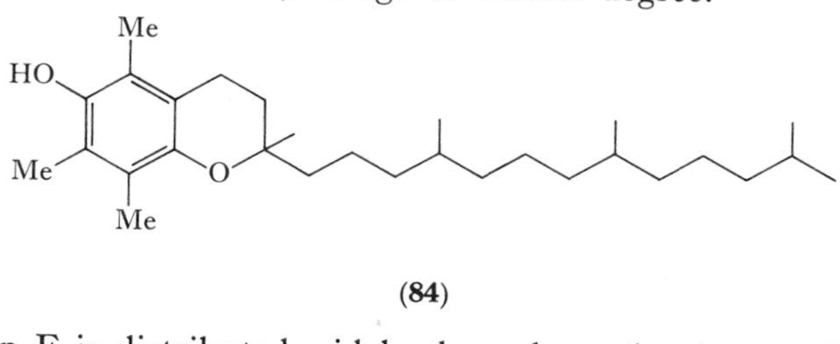

(**84**)

Vitamin E is distributed widely throughout the tissues of man and animals. The absence of this from their diet has been claimed to induce a wide variety of disorders, including the appearance of muscular dystrophy, atherosclerosis (the clogging of blood vessels by fatty deposits), diabetes, ulcers, and possibly infertility and abortion. Equally, these conditions can be treated by giving moderate doses of vitamin E or other antioxidants such as co-enzyme Q, 2,6-di-t-butyl-4-methoxy-phenol and diphenyl-*p*-phenylenediamine. All these compounds are radical scavengers and it seems likely that vitamin E also acts as a radical scavenger. In this role it protects the unsaturated lipids in cell membranes from oxidation though it may have other beneficial functions as well. Oxidation of this type would vitiate the ability of the membrane to allow the diffusion of selected biochemical species, thus adversely affecting the health of the cell.

Ozone and nitrogen dioxide which are present in relatively high concentrations in smog can induce disorders attributed to vitamin E deficiency. Both of these compounds can react readily with vitamin E, thus reducing the amount available as an antioxidant in cells.

There is little doubt that the oxidation of lipids causes the destruction of membranes. Much less is known about the repair mechanism for the damaged membranes though it seems likely that such repair processes become less efficient with increasing age. Membranes are also more susceptible to oxidation when they are damaged. It thus seems possible that if the amount of damage to membranes can be decreased by vitamin E the ageing process will be retarded. There is some evidence in support of this in that the average life span of mice is lengthened by including antioxidants in their diet. There are, however, other plausible explanations as to why antioxidants increase the life span. The radical theory of ageing is only one of several theories which have been formulated. It is still controversial and speculative. There is no concrete evidence whatsoever that the addition of vitamin E or antioxidants to the diet increases the life span of humans.

REFERENCES

1. J. S. Littler, *Essays in Free-Radical Chemistry*, Chem. Soc. Special Publ. no. 24, 1970, 383.
2. J. S. Littler and I. G. Sayce, *J. Chem. Soc.* 1964, 2545.
3. P. J. Andrulis, M. J. S. Dewar, R. Dietz and R. L. Hunt, *J. Amer. Chem. Soc.* 1966, **88**, 5473.
4. E. I. Heiba, R. M. Dessau and W. J. Koehl, *J. Amer. Chem. Soc.* 1969, **91**, 138.
5. E. I. Heiba, R. M. Dessau and W. J. Koehl, *J. Amer. Chem. Soc.* 1969, **91**, 6830.
6. A. Y. Drummond and W. A. Waters, *J. Chem. Soc.* 1953, 3119.
7. A. A. Clifford and W. A. Waters, *J. Chem. Soc.* 1965, 2796.
8. J. K. Kochi, *Science*, 1967, **155**, 415.
9. J. K. Kochi and J. D. Bacha, *J. Org. Chem.* 1968, **33**, 2746.
10. J. D. Bacha and J. K. Kochi, *J. Org. Chem.* 1968, **33**, 83.
11. R. O. C. Norman and P. R. West, *J. Chem. Soc.* B, 1969, 389.
12. J. K. Kochi and A. Bemis, *J. Amer. Chem. Soc.* 1968, **90**, 4038; J. K. Kochi, A. Bemis and C. L. Jenkins, *J. Amer. Chem. Soc.* 1968, **90**, 4616.

References

13. H. J. M. Bartelink, H. K. Ostendorf, B. C. Roest and H. A. J. Schepers, *Chem. Comm.* 1971, 878.
14. S. E. Schaafsma, E. J. F. Molenaar, H. Steinberg and Th. J. De Boer, *Rec. Trav. chim.* 1967, **86**, 1301.
15. F. Minisci, *Gazzetta*, 1959, **89**, 626, 1910, 2428; F. Minisci and A, Porblani, *ibid.* 1959, **89**, 1941; F. Minisci and U. Pallini, *ibid.* 1959, **89**, 2438.
15a. C. L. Jenkins and J. K. Kochi, *J. Org. Chem.* 1971, **36**, 3095, 3103.
16. J. K. Kochi and D. D. Davis, *J. Amer. Chem. Soc.* 1964, **86**, 5264.
17. J. K. Kochi, *J. Org. Chem.* 1965, **30**, 3265.
18. J. K. Kochi and D. Mog, *J. Amer. Chem. Soc.* 1965, **87**, 522.
19. W. A. Waters, *Mechanism of Oxidations of Organic Compounds*, Methuen, London, 1964.
20. W. A. Waters and J. S. Littler, *Oxidation in Organic Chemistry*, part A, ed. K. B. Wiberg, Academic Press, New York, 1965.
21. J. S. Littler, *J. Chem. Soc.* 1959, 4135.
22. D. G. Hoare and W. A. Waters, *J. Chem. Soc.* 1962, 965.
23. J. S. Littler and W. A. Waters, *J. Chem. Soc.* 1959, 4046.
24. J. R. Jones and W. A. Waters, *J. Chem. Soc.* 1960, 2772.
25. J. R. Jones, J. S. Littler and W. A. Waters, *J. Chem. Soc.* 1961, 630.
26. D. G. Hoare and W. A. Waters, *J. Chem. Soc.* 1964, 2552, 2560.
27. J. S. Littler, A. I. Mallet and W. A. Waters, *J. Chem. Soc.* 1960, 2761.
28. J. S. Littler and W. A. Waters, *J. Chem. Soc.* 1960, 2767.
29. J. R. Jones, J. S. Littler and W. A. Waters, *J. Chem. Soc.* 1961, 630.
29a. R. Cecil, J. S. Littler and G. Easton, *J. Chem. Soc.* B, 1970, 626.
30. H. T. Clarke and E. E. Dreger, *Org. Synth.* coll. vol. i, 1941, 87.
31. B. A. Marshall and W. A. Waters, *J. Chem. Soc.* 1960, 2392; 1961, 1579; K. B. Wibert and W. G. Nigh, *J. Amer. Chem. Soc.* 1965, **87**, 3849.
32. J. S. Littler, *J. Chem. Soc.* 1962, 832.
33. P. R. Sharan, P. Smith and W. A. Waters, *J. Chem. Soc.* B, 1968, 1322.
34. J. K. Kochi, *J. Amer. Chem. Soc.* 1965, **87**, 3609.
35. J. D. Bacha and J. K. Kochi, *Tetrahedron*, 1968, **24**, 2215.
36. J. K. Kochi, *Tetrahedron*, 1969, **25**, 1197.
37. S. S. Lande and J. K. Kochi, *J. Amer. Chem. Soc.* 1968, **90**, 5196.
38. R. A. Sheldon and J. K. Kochi, *J. Amer. Chem. Soc.* 1968, **90**, 6688.
39. R. A. McClelland, R. O. C. Norman and C. B. Thomas, *J. Chem. Soc.* C, 1972, 562.
40. D. C. Nonhebel, *Essays in Free-Radical Chemistry*, *Chem. Soc. Special Publ.* no. 24, 1970, 409; D. C. Nonhebel, *J. Chem. Soc.* 1963, 1216; D. Mosnaim and D. C. Nonhebel, *Tetrahedron*, 1969, **25**, 1591; D. Mosnaim, D. C. Nonhebel and J. A. Russell, *Tetrahedron*, 1969, **25**, 3485.
40a. S. Gibson, A. D. Mosnaim, D. C. Nonhebel and J. A. Russell, *Tetrahedron*, 1969, **25**, 5047.
41. D. C. Nonhebel and J. A. Russell, *Tetrahedron*, 1969, **25**, 3493.
42. T. J. Stone and W. A. Waters, *J. Chem. Soc.* 1964, 213, 4302.
43. D. H. R. Barton, *Chem. in Britain*, 1967, **3**, 330.
43a. R. Cecil and J. S. Littler, *J. Chem. Soc.* B, 1968, 1420.
43b. V. Ronlán, *Chem. Comm.* 1971, 1643.

44. D. H. R. Barton, *Proc. Chem. Soc.* 1963, 293.
45. D. T. Dalgleish, D. C. Nonhebel and P. L. Pauson, unpublished results.
46. A. Rieker, N. Zeller, K. Schurr and E. Müller, *Annalen*, 1966, **697**, 1.
47. W. A. Waters, *J. Chem. Soc.* B, 1971, 2026.
48. C. G. Haynes, A. H. Turner and W. A. Waters, *J. Chem. Soc.* 1956, 2523.
49. F. R. Hewgill, *J. Chem. Soc.* 1962, 4987; F. R. Hewgill and B. S. Middleton, *ibid.* 1965, 2914; D. F. Bowman and F. R. Hewgill, *J. Chem. Soc.* C, 1971, 1777.
50. E. Müller, R. Mayer, B. Narr, A. Rieker and K. Scheffler, *Annalen*, 1961, **645**, 25.
51. A. S. Hay, H. S. Blanchard, G. F. Endres and J. W. Eustance, *J. Amer. Chem. Soc.* 1959, **81**, 6335.
52. W. J. van Mijs, J. M. van Dijk, W. G. B. Huysmans and J. G. Westra, *Tetrahedron*, 1969, **25**, 4233.
53. R. Stebbins and F. Sicilio, *Tetrahedron*, 1970, **26**, 291.
54. C. D. Cook and B. E. Norcross, *J. Amer. Chem. Soc.* 1956, **78**, 3797; 1959, **81**, 1176.
55. C. D. Cook, N. G. Nash and H. R. Flanagan, *J. Amer. Chem. Soc.* 1955, **77**, 1783.
55a. A. I. Brodskii, V. D. Pokhodenko, V. A. Khizhnyi and N. N. Kalibabchuk, *Dokl. Akad. Nauk SSSR*, 1966, **169**, 339.
56. E. Müller, R. Mayer, H.-D. Spanagel and K. Scheffler, *Annalen*, 1961, **645**, 53.
57. H. Musso, U. v. Gizycki, H. Krämer and H. Döpp, *Chem. Ber.* 1965, **98**, 3952.
58. H. Musso, *Angew Chem. Internat. Edn*, 1963, **2**, 723.
59. A. R. Forrester and R. H. Thomson, *J. Chem. Soc.* C, 1965, 1844.
60. H.-J. Teuber and K. H. Dietz, *Angew Chem. Internat. Edn*, 1965, **4**, 871.
61. W. I. Taylor and A. R. Battersby, *Oxidative Coupling of Phenols*, Marcel Dekker, New York, 1967.
62. A. I. Scott, *Quart. Rev.* 1965, **19**, 1.
63. D. H. R. Barton, *Chem. in Britain*, 1967, **3**, 330.
64. D. H. R. Barton, A. M. Deflorin and O. E. Edwards, *J. Chem. Soc.* 1956, 530.
65. D. H. R. Barton, G. W. Kirby, W. Steglich, R. M. Thomas, A. R. Battersby, T. A. Dobson and H. Ramuz, *J. Chem. Soc.* 1965, 2423.
66. A. H. Jackson and J. A. Martin, *Chem. Comm.* 1965, 142, 420.
67. D. H. R. Barton and T. Cohen, *Festschrift Arthur Stoll*, Birkhauser, Basle, 1957, p. 117.
68. A. J. Birch, *Quart. Rev.* 1950, **4**, 69.
69. A. J. Birch and H. Smith, *Quart. Rev.* 1958, **12**, 1.
70. H. O. House, *Modern Synthetic Methods*, Benjamin, New York, 1965, chapter 3.
71. A. P. Krapcho and A. A. Bothner-By, *J. Amer. Chem. Soc.* 1960, **82**, 751.
72. T. R. Tuttle and P. I. Weissman, *J. Amer. Chem. Soc.* 1958, **80**, 5342.
73. A. L. Wilds and N. A. Nelson, *J. Amer. Chem. Soc.* 1953, **75**, 5360, 5366.
74. M. E. Kuehne and B. F. Lambert, *Org. Synth.* 1963, **43**, 22.
75. R. G. Harvey and K. Urberg, *J. Org. Chem.* 1968, **33**, 2510.

76. A. L. Henne and K. W. Greenlee, *J. Amer. Chem. Soc.* 1943, **65**, 2020.
77. F. C. Whitemore and T. Otterbacher, *Org. Synth.* coll. vol. II, 1943, 317.
78. R. Adams and E. W. Adams, *Org. Synth.* coll. vol. I, 1941, 459.
79. S. G. Ford and C. S. Marvel, *Org. Synth.* coll. vol. II, 1943, 372.
80. S. M. McElvain, *Org. Reactions*, 1948, **4**, 256.
81. N. L. Allinger, *Org. Synth.* coll. vol. IV, 1963, 840.
82. L. F. Fieser and M. Fieser, *Reagents for Organic Synthesis*, Wiley, New York, 1966, pp. 147–151.
83. J. K. Kochi and P. E. Mocadlo, *J. Amer. Chem. Soc.* 1966, **88**, 4094.
84. C. E. Castro and W. C. Kray, *J. Amer. Chem. Soc.* 1963, **85**, 2768.
85. J. K. Kochi and D. D. Davis, *J. Amer. Chem. Soc.* 1964, **88**, 5264.
86. J. K. Kochi and D. Buchanan, *J. Amer. Chem. Soc.* 1965, **87**, 853.
87. J. K. Kochi and D. M. Singleton, *J. Amer. Chem. Soc.* 1967, **89**, 6547; 1968, **90**, 1582; J. K. Kochi, D. M. Singleton and L. J. Andrews, *Tetrahedron*, 1968, **24**, 3503.
88. D. H. R. Barton, N. K. Basu, R. H. Hesse, R. H. Morehouse and M. M. Pechet, *J. Amer. Chem. Soc.* 1966, **88**, 3016.
89. H. G. Kuivila, *Accounts Chem. Res.* 1968, **1**, 299; L. W. Menapace and H. G. Kuivila, *J. Amer. Chem. Soc.* 1964, **86**, 3047.
90. D. J. Carlsson and K. U. Ingold, *J. Amer. Chem. Soc.* 1968, **90**, 7047.
91. W. Raman and H. G. Kuivila, *J. Org. Chem.* 1966, **31**, 772.
92. D. Seyferth, H. Yamazaki and D. L. Alleston, *J. Org. Chem.* 1963, **28**, 703.
93. H. G. Kuivila and L. W. Menapace, *J. Org. Chem.* 1963, **28**, 2165.
94. R. A. Jackson, *Essays in Free-Radical Chemistry, Chem. Soc. Special Publ.* no. 24, 1970, 295.
95. Y. Nagai, K. Yamazaki, I. Shiojima, N. Kobori and M. Hayashi, *J. Organometallic Chem.* 1967, **9**, 21.
96. D. C. Nonhebel and W. A. Waters, *Proc. Roy. Soc.* A, 1957, **247**, 16.
97. J. K. Kochi, *J. Amer. Chem. Soc.* 1956, **78**, 1228, 4815.
98. J. K. Kochi, *J. Amer. Chem. Soc.* 1957, **79**, 2942.
99. W. A. Waters, *Chemistry of Free Radicals*, Oxford, 2nd edn, 1948, p. 162.
100. J. K. Kochi, *J. Amer. Chem. Soc.* 1955, **77**, 5090, 5274.
101. S. C. Dickerman and W. Weiss, *J. Org. Chem.* 1957, **22**, 1070.
102. J. I. G. Cadogan, *Pure Appl. Chem.* 1967, **15**, 153.
103. D. F. DeTar, *Org. Reactions*, 1957, **9**, 409.
104. G. A. Russell, *Essays in Free-Radical Chemistry, Chem. Soc. Special Publ.* no. 24, 1970, 271.
105. A. L. J. Beckwith and R. O. C. Norman, *J. Chem. Soc.* B, 1969, 403.
106. N. Kornblum, *Org. Reactions*, 1944, **2**, 262.
107. G. P. Miklukhin and A. F. Rekasheva, *Dokl. Akad. Nauk SSSR*, 1952, **85**, 827.
108. J. H. Baxendale, M. G. Evans and G. S. Park, *Trans. Faraday Soc.* 1946, **42**, 155.
109. D. D. Coffman, E. L. Jenner and R. D. Lipscomb, *J. Amer. Chem. Soc.* 1958, **80**, 2864.
110. D. D. Coffman and E. L. Jenner, *J. Amer. Chem. Soc.* 1958, **80**, 2872.
111. R. O. C. Norman, *Essays in Free-Radical Chemistry, Chem. Soc. Special Publ.* no. 24, 1970, 117.

112. J. R. Lindsay Smith and R. O. C. Norman, *J. Chem. Soc.* 1963, 2897.
113. W. T. Dixon and R. O. C. Norman, *J. Chem. Soc.* 1964, 4857.
114. W. T. Dixon and R. O. C. Norman, *J. Chem. Soc.* 1963, 3119.
115. J. K. Kochi, *J. Amer. Chem. Soc.* 1962, **84**, 2785.
116. E. G. E. Hawkins and D. P. Young, *J. Chem. Soc.* 1950, 2804.
117. J. K. Kochi and F. F. Rust, *J. Amer. Chem. Soc.* 1962, **84**, 3946; H. E. De La Mare, J. K. Kochi and F. F. Rust, *J. Amer. Chem. Soc.* 1963, **85**, 1437.
118. F. Minisci and G. Belvedere, *Gazzetta*, 1960, **90**, 1299, 1306.
119. M. S. Kharasch, G. Sosnovsky and N. C. Yang, *J. Amer. Chem. Soc.* 1959, **81**, 5819.
120. G. Sosnovsky and S.-O. Lawesson, *Angew. Chem. Internat. Edn*, 1963, **3**, 269; G. Sosnovsky and D. J. Rawlinson, in *Organic Peroxides*, ed. D. Swern, Wiley Interscience, New York, 1969, vol. 1, p. 585.
121. S.-O. Lawesson and G. Schroll, in *The Chemistry of Carboxylic Acids and Esters*, ed. S. Patai, Interscience, London, 1969, p. 669.
122. J. K. Kochi, *J. Amer. Chem. Soc.* 1962, **84**, 774, 2785; J. K. Kochi and H. E. Mains, *J. Org. Chem.* 1965, **30**, 1862.
123. K. Pederson, P. Jakobsen and S.-O. Lawesson, *Org. Synth.* 1968, **48**, 18.
124. B. Cross and G. H. Whitham, *J. Chem. Soc.* 1961, 1650.
125. P. R. Story, *Tetrahedron Letters*, 1962, 401.
126. J. K. Kochi, *Tetrahedron*, 1962, **18**, 483; *J. Amer. Chem. Soc.* 1963, **85**, 1958.
127. G. Sosnovsky and D. J. Rawlinson, *Organic Peroxides*, ed. D. Swern, Wiley Interscience, New York, 1969, vol. 1, p. 561.
128. M. S. Kharasch and A. Fono, *J. Org. Chem.* 1959, **24**, 606.
129. J. K. Kochi, *J. Amer. Chem. Soc.* 1962, **84**, 1572.
130. D. H. Hey, K. S. Y. Liang and M. J. Perkins, *Tetrahedron Letters*, 1967, 1477.
131. L. Eberson in *The Chemistry of Carboxylic Acids and Esters*, ed. S. Patai, Interscience, London, 1969, p. 53.
132. B. C. L. Weedon, *Adv. Org. Chem.* 1960, **1**, 1.
133. J. H. P. Uttley, *Ann. Reports* B, 1968, 231; 1969, 217.
134. R. Brettle in *Modern Reactions in Organic Synthesis*, ed. C. J. Timmons, Van Nostrand-Reinhold, London, 1970, p. 155.
135. S. Swann and W. E. Garrison, *Org. Synth.* 1961, **41**, 33.
136. K. Mislow and I. V. Steinberg, *J. Amer. Chem. Soc.* 1955, **77**, 3807.
137. S. W. Pelletier, L. B. Hawley and K. W. Gopinath, *Chem. Comm.* 1967, 96.
138. R. F. Garwood, C. J. Scott and B. C. L. Weedon, *Chem. Comm.* 1965, 14.
139. M. Talât-Erben and A. N. Isfendiyaroğlu, *Canad. J. Chem.* 1958, **36**, 1156.
140. K. Clusius and W. Schanzer, *Z. phys. Chem. (Leipzig)*, 1943, **192A**, 273.
141. D. H. Hey and P. J. Bunyan, *J. Chem. Soc.* 1960, 3787.
142. W. J. Koehl, *J. Amer. Chem. Soc.* 1964, **86**, 4686.
143. S. D. Ross and M. Finkelstein, *J. Org. Chem.* 1969, **34**, 2923.
144. R. P. Linstead, B. R. Shepherd and B. C. L. Weedon, *J. Chem. Soc.* 1952, 2624.
145. B. Wladislaw and H. Viertler, *Chem. and Ind.* 1965, 39.

145a. M. Finkelstein and R. C. Petersen, *J. Org. Chem.* 1960, **25**, 136.

146. J. P. Coleman, J. H. P. Uttley and B. C. L. Weedon, *Chem. Comm.* 1971, 438.

147. S. Goldschmidt, W. Leicher and H. Haas, *Annalen*, 1952, **577**, 153.

148. P. H. Reichenbacher, M. Y.-C. Liu and P. S. Skell, *J. Amer. Chem. Soc.* 1968, **90**, 1816.

149. L. Eberson, *J. Amer. Chem. Soc.* 1969, **91**, 2402.

150. W. J. Koehl, *J. Org. Chem.* 1967, **32**, 614.

151. J. M. Bobbitt, J. T. Stock, A. Marchand and K. H. Weisgraber, *Chem. and Ind.* 1966, 2127.

152. F. J. Vermillion and J. A. Pearl, *J. Electrochem. Soc.* 1964, **111**, 1392.

153. C. T. Bahner, *Ind. Eng. Chem.* 1952, **44**, 317.

154. L. S. Marcoux, J. M. Fritsch and R. N. Adams, *J. Amer. Chem. Soc.* 1967, **89**, 5766.

155. V. D. Parker, *Acta Chem. Scand.* 1970, **24**, 3151, 3171.

156. L. Eberson and K. Nyberg, *J. Amer. Chem. Soc.* 1966, **88**, 1686.

157. V. D. Parker and L. Eberson, *Tetrahedron Letters*, 1969, 2839; *Acta Chem. Scand.* 1970, **24**, 3452.

158. V. D. Parker, *Chem. Comm.* 1969, 848.

159. V. D. Parker and B. E. Burgert, *Tetrahedron Letters*, 1968, 2411; K. Nyberg, *Chem. Comm.* 1969, 774.

160. V. D. Parker and L. Eberson, *Acta Chem. Scand.* 1970, **24**, 3553.

161. V. D. Parker and L. Eberson, *Tetrahedron Letters*, 1968, 2415.

162. R. N. Adams, *Accounts Chem. Res.* 1969, **2**, 175.

163. G. Cauquis, G. Fauvelot and J. Rigaudy, *Bull. Soc. chim. France*, 1968, 4928.

164. G. Cauquis and M. Genies, *Compt. rend.* 1967, **265**, 1340.

165. J. Bacon and R. N. Adams, *J. Amer. Chem. Soc.* 1968, **90**, 6596.

166. P. J. Elving and J. T. Leone, *J. Amer. Chem. Soc.* 1958, **80**, 1021.

167. J. Grimshaw and J. S. Ramsey, *J. Chem. Soc.* C, 1966, 653.

168. J. H. Stocker and R. M. Jenevein, *J. Org. Chem.* 1968, **33**, 294; J. H. Stocker, R. M. Jenevein and D. H. Kern, *J. Org. Chem.* 1969, **34**, 2810.

169. O. R. Brown and K. Lister, *Discuss. Faraday Soc.* 1968, **45**, 106.

170. A. W. Ingersoll, *Org. Synth.* coll. vol. I, 1941, 311.

171. M. M. Baizer and J. D. Anderson, *J. Electrochem. Soc.* 1964, **111**, 223, 226.

172. M. M. Baizer, *J. Org. Chem.* 1964, **29**, 1670; *Tetrahedron Letters*, 1963, 973.

173. M. Figeys and H. P. Figeys, *Tetrahedron*, 1968, **24**, 1097.

174. G. C. Jones and T. H. Ledford, *Tetrahedron Letters*, 1967, 615.

175. J. D. Anderson, M. M. Baizer and J. P. Petrovich, *J. Org. Chem.* 1966, **31**, 3890, 3897.

176. W. A. Waters, *Vistas in Free Radical Chemistry*, Pergamon Press, London, 1959, p. 151.

177. R. O. C. Norman and R. J. Pritchett, *J. Chem. Soc.* B, 1967, 378.

178. S. E. Schaafsma, H. Steinberg and Th. J. de Boer, *Rec. Trav. chim.* 1966, **85**, 70.

179. R. O. C. Norman and P. M. Storey, *J. Chem. Soc.* B, 1970, 1099.

180. K. A. Bilevitch, N. N. Bubnov and O. Yu. Okhlobystin, *Tetrahedron Letters*, 1968, 3465.

181. H. B. Hass and M. L. Bender, *J. Amer. Chem. Soc.* 1949, **71**, 1767.
182. R. C. Kerber, G. W. Urry and N. Kornblum, *J. Amer. Chem. Soc.* 1965, **87**, 4520.
183. G. A. Russell and W. C. Danen, *J. Amer. Chem. Soc.* 1968, **90**, 34.
184. R. A. Jackson and W. A. Waters, *J. Chem. Soc.* 1960, 1653.
185. N. Kornblum, T. M. Davies, G. W. Earl, N. L. Holy, R. C. Kerber, M. T. Musser and D. H. Snow, *J. Amer. Chem. Soc.* 1967, **89**, 725; N. Kornblum, G. W. Earl, N. L. Holy, J. W. Manthey, M. T. Musser, D. H. Snow and R. T. Swiger, *J. Amer. Chem. Soc.* 1968, **90**, 6221; N. Kornblum and F. W. Stuchal, *J. Amer. Chem. Soc.* 1970, **92**, 1804.
186. N. Kornblum, R. T. Swiger, G. W. Earl, H. W. Pinnick and F. W. Stuchal, *J. Amer. Chem. Soc.* 1970, **92**, 5513.
187. G. A. Russell, *Essays in Free-Radical Chemistry, Chem. Soc. Special Publ.* no. 24, 1970, 271.
188. J. F. Garst, *Accounts Chem. Res.* 1971, **4**, 400.
189. S. J. Cristol and R. V. Barbour, *J. Amer. Chem. Soc.* 1968, **90**, 2832.
190. G. D. Sargent and G. A. Lux, *J. Amer. Chem. Soc.* 1968, **90**, 7160.
191. K. U. Ingold, *Accounts Chem. Res.* 1969, **2**, 1; J. A. Howard, *Adv. Free-Radical Chem.* 1971, **4**, 49.
192. J. A. Howard and K. U. Ingold, *Canad. J. Chem.* 1967, **45**, 793.
193. S. W. Benson, *J. Amer. Chem. Soc.* 1965, **87**, 972.
194. J. A. Howard and K. U. Ingold, *J. Amer. Chem. Soc.* 1968, **90**, 1058.
195. J. A. Howard, K. Adamic and K. U. Ingold, *Canad. J. Chem.* 1969, **47**, 3793.
196. G. A. Russell, *J. Amer. Chem. Soc.* 1957, **79**, 3871.
197. J. A. Howard and K. U. Ingold, *J. Amer. Chem. Soc.* 1968, **90**, 1056.
198. R. E. Kellogg, *J. Amer. Chem. Soc.* 1969, **91**, 5433.
199. N. M. Emanuel, Z. K. Maizus, and I. P. Skibida, *Angew. Chem. Internat. Edn*, 1969, **8**, 97.
200. R. Hiatt, K. C. Irwin and C. W. Gould, *J. Org. Chem.* 1968, **33**, 1430.
201. K. U. Ingold, *Essays in Free-Radical Chemistry, Chem. Soc. Special Publ.* no. 24, 1970, 285.
202. R. F. Moore and W. A. Waters, *J. Chem. Soc.* 1952, 2432; 1954, 243.
203. C. E. Boozer, G. S. Hammond, C. E. Hamilton and J. N. Sen, *J. Amer. Chem. Soc.* 1955, **77**, 3233.
204. L. R. Mahoney, *Angew. Chem. Internat. Edn*, 1969, **8**, 547.
204a. P. Thüring and A. Perret, *Helv. Chim. Acta*, 1953, **36**, 13.
205. F. R. Mayo, *Accounts Chem. Res.* 1968, **1**, 193.
206. A. G. Davies and B. P. Roberts, *J. Chem. Soc. B*, 1967, 17.
207. A. G. Davies and B. P. Roberts, *J. Chem. Soc. B*, 1968, 1074.
208. K. U. Ingold and B. P. Roberts, *Free Radical Substitution Reactions*, Wiley Interscience, New York, 1971, p. 54.
209. G. A. Russell and A. G. Bemis, *J. Amer. Chem. Soc.* 1966, **88**, 5491.
210. H. C. Brown, M. M. Midland and G. W. Kabalka, *J. Amer. Chem. Soc.* 1971, **93**, 1024.
211. H. C. Brown and M. M. Midland, *J. Amer. Chem. Soc.* 1971, **93**, 4078.

11

Homolytic aromatic substitution

11.1 *INTRODUCTION*

This chapter considers substitution reactions of aromatic compounds in which the attacking species is a radical. This class of reaction has been very extensively studied, particularly by Hey and his co-workers. Arylation will thus be considered in most detail because of the large amount of quantitative data pertaining to this reaction and also because of its synthetic importance.

The first clear piece of evidence for the presence of reactive radicals in solution was obtained by Hey in 1934, from a study of the Gomberg reaction, which involved the decomposition of diazonium salts in a range of aromatic substrates (1). He observed that the *para*-substituted biaryl was always obtained irrespective of the electronic characteristic of the substituent already present in the aromatic compound. This was in marked contrast to the effects of substituents on aromatic nitration. It was to account for these and other anomalies that Hey and Waters in 1937 propounded the idea of short-lived neutral free radicals (2).

11.2 *ARYLATION*

Arylation will be considered in more detail than other homolytic aromatic substitutions, firstly because of its importance in giving an understanding of the basic principles of this reaction. Secondly, it is a procedure of considerable synthetic utility.

11.2.1 *Sources of aryl radicals*

The products of homolytic aromatic arylation are essentially independent of the source of aryl radicals (Table 11-1), though there are some slight exceptions to this, particularly as it applies to the proportion of substitution and hydrogen abstraction in reactions with alkylbenzenes:

$$\text{PhMe} \xrightarrow{\text{Ph·}} \text{[biphenyl structure with Me]} + \text{PhCH}_2\text{CH}_2\text{Ph} \qquad (1)$$

TABLE 11-1 *Isomer distributions with different sources of phenyl radicals*

Radical source	Substrate	% o	% m	% p	Reference
Dibenzoyl peroxide	PhCl	50	32	18	3
Gomberg	PhCl	64	22	14	4
AgI(OCOPh)$_2$	PhCl	60	24	16	5
PhN=NCPh$_3$	PhCl	58	28	14	6
PhN(NO)Ac	PhCl	56	28	16	6
PhNHNH$_2$/Ag$_2$O	PhCl	65	22	13	7
PhN=NNHPh	PhCl	60	24	16	8
(PhCO$_2$)$_2$	PhNO$_2$	56	12	32	9
Pb(OCOPh)$_4$	PhNO$_2$	45	18	37	10
PhI(OCOPh)$_2$	PhNO$_2$	57.5	14	28.5	11
PhN$_2^+$BF$_4^-$/C$_5$H$_5$N	PhNO$_2$	59	14	27	12
PhN$_2^+$BF$_4^-$/NO$_2^-$	PhNO$_2$	68	5	27	13
PhN=NCPh$_3$	PhNO$_2$	79	7	14	14
(PhCO$_2$)$_2$	C$_5$H$_5$N	54	32	14	15
Pb(OCOPh)$_4$	C$_5$H$_5$N	52	32.5	15.5	15
PhI(OCOPh)$_2$	C$_5$H$_5$N	58	28	14	15
PhN=NCPh$_3$	C$_5$H$_5$N	53	31	16	15
PhCO$_2$H (electrolysis)	C$_5$H$_5$N	56	35	9	16

(i) *Aryl radicals from diaroyl peroxides and other sources of aroyloxy radicals.* The most common source of aryl radicals is a diaroyl peroxide; thermolysis or photolysis generates aroyloxy radicals which subsequently lose carbon dioxide to give aryl radicals:

$$(\text{ArCO}_2)_2 \longrightarrow 2\,\text{ArCO}_2\cdot \qquad (2)$$

$$\text{ArCO}_2\cdot \longrightarrow \text{Ar}\cdot + \text{CO}_2 \qquad (3)$$

The production of aryl radicals from peroxides generally provides a cleaner method of arylation than the methods based on azo and diazo compounds, and in the case of benzenoid compounds better yields of products are obtained (18). The efficiency of this method as a means of synthesizing biaryls can be greatly improved by carrying out the reactions in oxygen (19) or with small amounts of nitrobenzene (20) or iron(III) benzoate (21) (cf. p. 430). Disadvantages of this method are that diaroyl peroxides are less accessible than amines, and also that aroyloxylation of reactive aromatic compounds, such as anisole and naphthalene, accompanies arylation.

Electrolysis of benzoic acid (the Kolbe reaction) provides a further source of benzoyloxy radical (16):

$$ArCO_2{}^- \xrightarrow{-e} ArCO_2\cdot \qquad (4)$$

This source of radicals is of little synthetic use as it is necessary to carry out the electrolysis in aqueous media, in which few aromatic compounds are soluble. Pyridine, however, has been successfully phenylated in this way. It suffers from the further disadvantage that aromatic compounds tend to be more easily oxidized than the benzoic acid (cf. chapter 10, p. 376). Any quantitative data thus obtained should be considered with some suspicion.

Closely related to the decomposition of aroyl peroxides are those of lead tetrabenzoate (30, 17):

$$Pb(OCOPh)_4 \xrightarrow{125\,°C} Pb(OCOPh)_2 + 2PhCOO\cdot \qquad (5)$$

Benzoyloxy radicals are also generated by the decomposition of silver halide dibenzoates (5):

$$AgX(OCOPh)_2 \longrightarrow 2PhCOO\cdot + AgX \qquad (6)$$

This method is of little general use as, in addition to phenylation and benzoyloxylation of the aromatic substrate, halogenation also occurs and with some substrates this is the principal reaction.

(ii) *Aryl radicals from diazo, azo and related compounds.* One of the earliest sources of aryl radicals used was *N*-nitrosoacetanilide, decomposition of which in benzene was found by Bamberger in 1897 to give biphenyl:

$$PhN(NO)Ac + PhH \longrightarrow Ph{-}Ph + N_2 + AcOH \qquad (7)$$

Extensive use has been made of this procedure in the synthesis of biaryls (22).

The mechanism whereby *N*-nitrosoacetanilide breaks down to give phenyl radicals has been the subject of extensive study over many years. Originally it was thought to rearrange to the diazoacetate, which subsequently decomposed to phenyl and acetoxy radicals (23):

$$PhN(NO)Ac \longrightarrow Ph{-}N{=}N{-}OAc \xrightarrow{slow} Ph\cdot + N_2 + AcO\cdot \qquad (8)$$

Later, it was suggested that the rearrangement was the rate-determining step and this was followed by rapid dissociation of the diazoacetate (24):

$$PhN(NO)Ac \xrightarrow{slow} Ph{-}N{=}N{-}OAc \xrightarrow{fast} Ph\cdot + N_2 + AcO\cdot \qquad (9)$$

Homolytic substitution

These mechanisms failed to take into account the absence of any evidence for the existence of acetoxy radicals, though there was excellent evidence that N-nitrosoacetanilide gave phenyl radicals. This difficulty was overcome by the mechanism proposed by Rüchardt (25) (Scheme 1),

$$PhN(NO)Ac \longrightarrow Ph-N=N-OAc \rightleftharpoons PhN_2^+ + AcO^-$$

$$PhN(NO)Ac + AcO^- \longrightarrow Ph-N=N-O^- + Ac_2O$$

$$PhN_2^+ + Ph-N=N-O^- \longrightarrow (PhN=N)_2O$$

$$\longrightarrow Ph\cdot + N_2 + Ph-N=N-O\cdot \qquad (1)$$

$$Ph\cdot + PhH \longrightarrow$$

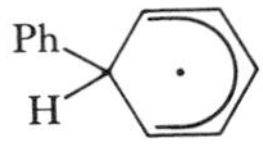

$$[Ph\cdot C_6H_6]\cdot + Ph-N=N-O\cdot \longrightarrow Ph-Ph + Ph-N=N-OH$$

Scheme 1

and substantiated by Cadogan, Paton and Thomson (26). The latter observed an e.s.r. signal corresponding to the radical (1) when the decomposition was carried out in the cavity of an e.s.r. spectrometer.

Aryl radicals can also be generated from diazonium salts by use of the Gomberg reaction, in which sodium hydroxide is added to a vigorously stirred solution of the cold diazonium salt and the aromatic substrate (22). This reaction was believed to involve formation of the covalent diazo hydroxide which decomposed to give aryl and hydroxyl radicals:

$$ArN_2^+ + OH^- \longrightarrow Ar-N=N-OH \longrightarrow Ar\cdot + N_2 + HO\cdot \qquad (10)$$

Rüchardt and Merz have suggested, however, that the reaction follows an analogous sequence to that proposed for the decomposition of N-nitrosoacetanilide (27). A somewhat cleaner modification of the Gomberg reaction, developed by Hey, involves the use of sodium acetate instead of sodium hydroxide (22). The Gomberg and Gomberg–Hey procedures suffer from the disadvantage that a heterogeneous system is used. This can be overcome by diazotizing the aromatic amine *in situ* in an organic solvent with amyl nitrite at 60–80 °C (28).

A similar mechanism may also be operative in arylations brought about by the acid-catalysed decompositions of 1-aryl-3,3-dialkyltri-azenes (29, 30) or diazoaminobenzenes (8):

$$Ar-N=N-NR_2 + H^+ \rightleftharpoons Ar-N=N-\overset{+}{N}HR_2 \qquad (11)$$

$$Ar-N=N-\overset{+}{N}HR_2 \longrightarrow ArN_2^+ + R_2NH \qquad (12)$$

$$ArN_2^+ + Ar\!-\!N\!=\!N\!-\!NR_2 \longrightarrow (ArN\!=\!N)_2\overset{+}{N}R_2 \qquad (13)$$

$$(ArN\!=\!N)_2\overset{+}{N}R_2 \longrightarrow Ar\!\cdot + N_2 + Ar\!-\!N\!=\!N\!-\!\overset{+\cdot}{N}R_2 \qquad (14)$$

$$Ar\!\cdot + PhH \longrightarrow [Ar\!\cdot\!PhH]\!\cdot \qquad (15)$$

$$[Ar\!\cdot\!PhH]\!\cdot + Ar\!-\!N\!=\!N\!-\!\overset{+\cdot}{N}R_2 \longrightarrow Ar\!-\!Ph + Ar\!-\!N\!=\!N\!-\!\overset{+}{N}HR_2 \qquad (16)$$

In arylations under all the above conditions, the normal chain-propagating step for the formation of aryl radicals may be reduction of the diazonium cation by an arylcyclohexadienyl radical, the resultant arylcyclohexadienyl cation forming the biaryl by loss of a proton (29):

$$ArN_2^+ + [Ar\!\cdot\!PhH]\!\cdot \longrightarrow Ar\!\cdot + N_2 + [Ar\!\cdot\!PhH]^+ \qquad (17)$$

$$[Ar\!\cdot\!PhH]^+ \longrightarrow Ar\!-\!Ph + H^+ \qquad (18)$$

Electron-transfer reduction of diazonium cations is an effective way of generating aryl radicals and is brought about by one-electron reductants:

$$ArN_2^+ + M^{n+} \longrightarrow Ar\!\cdot + N_2 + M^{(n+1)+} \qquad (19)$$

Thus, Norman and Waters have generated phenyl radicals from the stable benzene diazonium zincichloride with zinc in acetone (32), whilst Cadogan employed the diazonium fluoroborate and copper powder in moist acetone (33). The Meerwein reaction, which employs copper(II) chloride in aqueous acetone, proceeds by a radical mechanism. If the reaction is carried out in the presence of an aromatic substrate, arylation of this takes place (34):

$$2CuCl_2 + Me_2CO \longrightarrow 2CuCl + MeCOCH_2Cl + HCl \qquad (20)$$

$$ArN_2^+ + CuCl_2^- \longrightarrow Ar\!\cdot + N_2 + CuCl_2 \qquad (21)$$

Similarly aryl radicals are produced by reduction of diazonium salts with ferrocene (34a):

$$ArN_2^+ + FcH \longrightarrow Ar\!\cdot + N_2 + FcH^+ \qquad (22)$$

Electrolytic reduction of diazonium salts has recently been developed as a source of aryl radicals (36):

$$ArN_2^+ + e \longrightarrow Ar\!\cdot + N_2 \qquad (23)$$

The same mechanism probably operates in the generation of aryl radicals from treatment of diazonium salts with nitrite in dimethyl sulphoxide (13) or pyridine in sulpholan (12):

$$ArN_2^+ + NO_2^- \longrightarrow Ar\cdot + N_2 + NO_2 \qquad (24)$$

$$ArN_2^+ + C_5H_5N \longrightarrow Ar\cdot + N_2 + C_5H_5N^+\cdot \qquad (25)$$

Aryl radicals can also be generated by thermolysis of certain covalent diazo compounds such as arylazotriarylmethanes. These undergo decomposition at 80 °C to aryl and triarylmethyl radicals (35), the driving force for the reaction being the stability of the triarylmethyl radicals:

$$ArN{=}NCAr_3 \longrightarrow Ar\cdot + N_2 + Ar_3C\cdot \qquad (26)$$

It is probably more accurate to represent this process as a two-step reaction (cf. chapter 2, p. 25):

$$ArN{=}NCAr_3 \longrightarrow ArN_2\cdot + Ar_3C\cdot \qquad (27)$$

$$ArN_2\cdot \longrightarrow Ar\cdot + N_2 \qquad (28)$$

Phenylazoisobutane is much more stable and only undergoes decomposition at 300 °C (35 a):

$$PhN{=}NCMe_3 \longrightarrow Ph\cdot + N_2 + Me_3C\cdot \qquad (29)$$

It thus seems most unlikely that diazohydroxides and related compounds would undergo facile decomposition at room temperature to aryl radicals (reaction 10), since a t-butyl radical is very much more stable than a hydroxyl radical.

(iii) *Aryl radicals produced by photolysis of haloarenes and organometallic compounds.* Photolysis of aryl iodides and also aryl bromides in aromatic substrates has been shown by Bryce-Smith (37) and Kharasch (38) to generate aryl radicals:

$$ArI \longrightarrow Ar\cdot + I\cdot \qquad (30)$$

Wolf and Kharasch (39) used this technique to prepare *p*-terphenyl from 4-iodobiphenyl in a yield of 91 %:

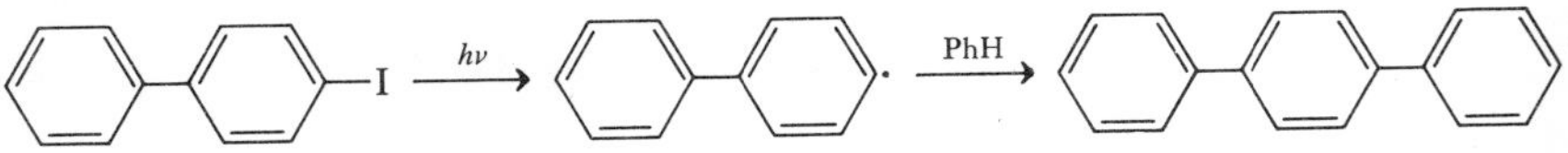

In this reaction the yield of *p*-terphenyl was significantly increased by the presence of oxygen. Photolyses of diphenylmercury, tetraphenyllead, and triphenylbismuth have all been employed as sources of phenyl radicals (37). Support for the radical nature of these reactions comes from the isomer distribution of the phenylated products.

(iv) *Other sources of aryl radicals.* Phenyl iodosobenzoate, on heating to 125 °C, breaks down to give phenyl radicals. The phenyl radicals come from the phenyl group of the iodobenzene moiety rather than from the benzoate groups, as was believed at one time (40).

Aryl radicals are also generated in reactions between aryl halides and Grignard reagents in the presence of catalytic quantities of cobalt(II) salts (41).

11.2.2 *The mechanism of the arylation reaction*

Three arylation pathways can be considered:

$$\text{(i)} \quad \text{Ar·} + \text{Ar'H} \longrightarrow \text{Ar—Ar'} + \text{H·} \qquad (32)$$

$$\text{(ii)} \quad \text{Ar'·} + \text{ArH} \longrightarrow \text{Ar'—H} + \text{Ar·} \qquad (33)$$

$$\text{Ar·} + \text{Ar'·} \longrightarrow \text{Ar—Ar'} \qquad (34)$$

(iii)

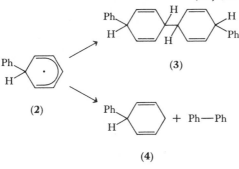

$$\text{Ar·} + \text{PhH} \longrightarrow \quad\quad \xrightarrow{\;R·\;} \text{Ar—Ph} + \text{RH} \qquad (35)$$

Of the three possibilities, the first is energetically most unlikely as it would be considerably endothermic. There is no sign of the reduction of nitro groups in the phenylation of nitro compounds as would be expected with such a powerful reductant as atomic hydrogen. The second route can be eliminated because of the absence of any of the symmetrical biaryls, Ar—Ar and Ar'—Ar', in addition to the unsymmetrical coupled product Ar—Ar'. Strong support for the third route comes from the isolation of 1',1'',4',4''-tetrahydroquaterphenyl (3) and dihydrobiphenyl (4) in addition to biphenyl from the decomposition of dibenzoyl peroxide in benzene in dilute solution (42) (Scheme 2). These

Scheme 2

products clearly arise from the dimerization and disproportionation of phenylcyclohexadienyl radicals (**2**).

The second step in homolytic aromatic arylation invariably involves a bimolecular reaction between two radical species and is consequently very fast. The first step, i.e. the formation of the phenylcyclohexadienyl radical, is thus likely to be the rate-limiting step on energetic considerations. Further evidence for this conclusion has been obtained from the absence of a significant primary isotope effect (43, 44, 45).

The failure to observe a large isotope effect has also been interpreted as indicating that the formation of the phenylcyclohexadienyl radicals is an irreversible process (44, 45). Recently it has been demonstrated by Perkins that this is not so, at least at high temperatures (46). Thus phenylcyclohexadienyl radicals, generated by the thermolysis of 1′,1″,4′,4″-tetrahydroquaterphenyl (**3**) undergo fragmentation to give phenyl radicals. This is indicated by the formation of chlorobiphenyls when the reaction is carried out in chlorobenzene as solvent (Scheme 3). Further indication of the reversibility of this step is that there is a definite

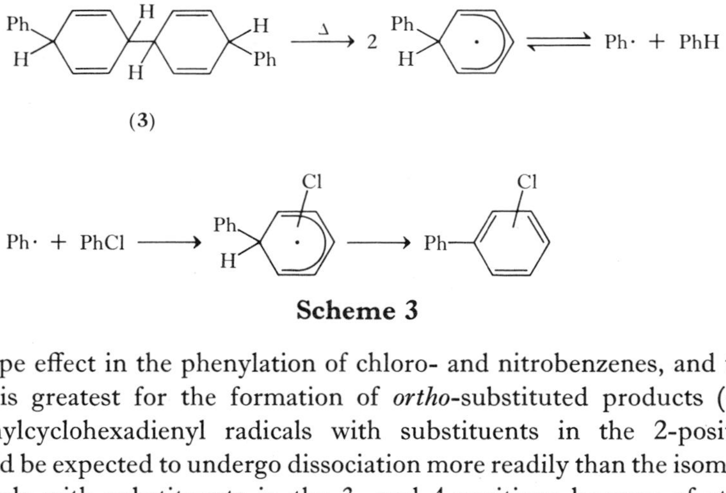

(**3**)

Scheme 3

isotope effect in the phenylation of chloro- and nitrobenzenes, and that this is greatest for the formation of *ortho*-substituted products (47). Phenylcyclohexadienyl radicals with substituents in the 2-position would be expected to undergo dissociation more readily than the isomeric radicals with substituents in the 3- and 4-positions because of steric effects. Consequently, the observation of an isotope effect in these examples is not altogether unexpected.

11.2.3   *Quantitative aspects of the arylation reaction*

A great number of systematic studies of the phenylation of different benzenoid compounds using a variety of different sources of phenyl radicals have been carried out. These studies have measured the isomer distributions of the products and the reactivities of the compounds with respect to benzene, $^{PhH}_{PhX}K$. This value is known as the rate factor of the compound. Use of these enables calculations to be made of the partial rate factors F_o, F_m and F_p. These give a measurement of the relative reactivity of any one position compared to that of a position in benzene which has unit reactivity, thus for a monosubstituted benzene:

$$6^{PhX}_{PhH}K = 2F_o + 2F_m + F_p$$

where

$$F_o = \tfrac{1}{2} \times 6 \times {}^{PhX}_{PhH}K \times \frac{\% \ \textit{ortho}\text{-product}}{100}$$

$$F_m = \tfrac{1}{2} \times 6 \times {}^{PhX}_{PhH}K \times \frac{\% \ \textit{meta}\text{-product}}{100}$$

and

$$F_p = 6 \times {}^{PhX}_{PhH}K \times \frac{\% \ \textit{para}\text{-product}}{100}$$

In order to obtain precise values of partial rate factors, it is necessary to be able to measure accurately the isomer distribution of the phenylated compound. In recent years this has been accomplished by use of g.l.c. The earlier results, including some quoted in Table 11-1, were obtained by i.r. methods and are probably rather less accurate. This may account for some of the differences in the values in Table 11-1. It is also necessary to measure accurately the reactivity of the particular compound relative to a standard. The standard frequently chosen was nitrobenzene, since the nitrobiphenyls present in the total phenylated product could be estimated quantitatively by titration with titanium(III) chloride (9). Thus, the value of $^{PhX}_{PhNO_2}K$ could be obtained and knowing the value of $^{PhNO_2}_{PhH}K$, $^{PhX}_{PhH}K$ could be calculated:

$$^{PhX}_{PhH}K = {}^{PhX}_{PhNO_2}K \times {}^{PhNO_2}_{PhH}K$$

Recently Davies, Hey and Summers have used *p*-dichlorobenzene as a standard reference compound rather than nitrobenzene (48). This has the advantage that on phenylation it yields only 2,5-dichlorobiphenyl which can be conveniently estimated by g.l.c. The partial rate factors using *p*-dichlorobenzene as standard are slightly higher than those obtained using nitrobenzene as standard, except in the case of toluene when appreciably higher values are obtained with *p*-dichlorobenzene.

425

The phenylation of toluene is complicated by the fact that side-chain attack leading to bibenzyl is a competing process. This reaction is, however, inhibited by nitrobenzene. No explanation has been put forward as to the reason for the variation of total and partial rate factors with solvent.

TABLE 11-2 *Total and partial rate factors for the phenylation of monosubstituted benzenes, PhX (48)*

Aromatic substrate	Standard solvent	Radical source	Total rate factor	Partial rate factor		
				F_o	F_m	F_p
PhNO$_2$	PhH	(PhCO$_2$)$_2$	2.94	5.50	0.86	4.90
PhMe	PhNO$_2$	(PhCO$_2$)$_2$	1.46	2.68	0.70	2.02
PhMe	C$_6$H$_4$Cl$_2$	(PhCO$_2$)$_2$	2.58	4.70	1.24	3.55
PhBu^t	PhNO$_2$	(PhCO$_2$)$_2$	1.03	0.73	1.53	1.63
PhBu^t	C$_6$H$_4$Cl$_2$	(PhCO$_2$)$_2$	1.09	0.70	1.64	1.81
PhCl	PhNO$_2$	(PhCO$_2$)$_2$	2.06	3.65	1.55	1.98
PhCl	C$_6$H$_4$Cl$_2$	(PhCO$_2$)$_2$	2.20	3.90	1.65	2.12
PhBr	PhNO$_2$	(PhCO$_2$)$_2$	1.82	2.95	1.64	1.78
PhBr	C$_6$H$_4$Cl$_2$	(PhCO$_2$)$_2$	1.90	3.05	1.70	1.92
PhBr	PhH	PhN$_2$$^+$ BF$_4$$^-$	1.75	2.94	1.42	1.85
PhOMe	C$_6$H$_4$Cl$_2$	(PhCO$_2$)$_2$	2.71	5.6	1.23	2.31
PhOMe	PhH	PhN$_2$$^+$ BF$_4$$^-$	1.96	3.9	1.0	1.92
Ph—Ph	PhNO$_2$	(PhCO$_2$)$_2$	2.94	2.1	1.0	2.5

Table 11-2 quotes some partial rate factors. Some of the discrepancies between values for a particular substrate are probably due to difficulties in analysis. It is not possible to say whether any of the differences can be attributed to a different radical source. This should be tested by one group using a standardized analytical procedure. Such an investigation may well provide results of interest.

It can be readily seen from Table 11-2 that the rate of homolytic arylation of all the substrates used is greater than that of benzene, and *ortho*- and *para*-substitution is more favoured than *meta*-substitution. The only exception to this is when steric effects are important, as in t-butylbenzene. The reason for this behaviour is that both electron-releasing and electron-attracting groups can delocalize the unpaired electron onto the group when attack occurs *ortho* or *para* to that group (cf. **5** and **6**).

No completely satisfactory explanation has been proposed for the predominance of attack at the *ortho*-position. One theory is that the

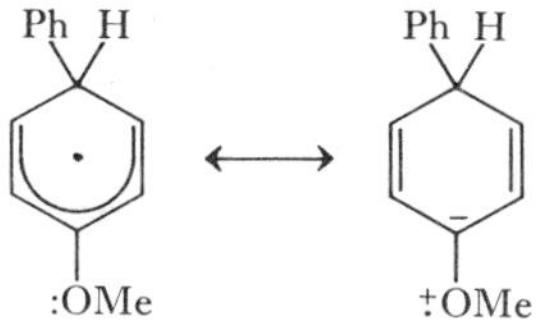

(5)

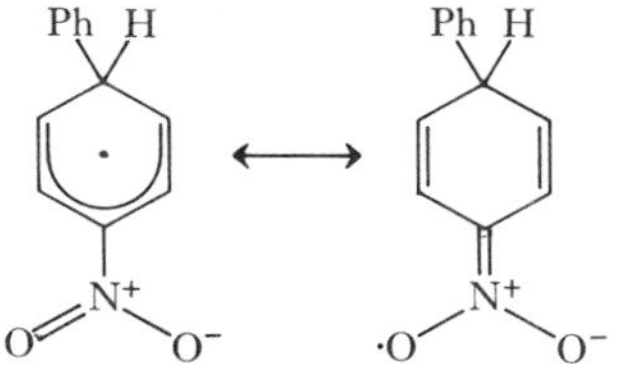

(6)

radical first forms a complex with the aromatic compound at the point of greatest electron density (49). This is invariably at the substituent or *ortho* to it, depending on whether the substituent is electron-attracting or electron-releasing. When this complex collapses to the σ-complex, the new bond is most likely to be formed at the *ortho*-position. There is no satisfactory explanation as to why the *meta*-position in mono-substituted benzenes is normally more reactive than one position in benzene.

Similar studies have been made on *p*-disubstituted benzenes (48). The results again show that all substituents except t-butyl are more efficient than hydrogen in promoting phenylation. When both sub-stituents have the same electronic characteristics, the effect of the two substituents is additive. Thus, the total rate factor of *p*-chloronitro-benzene (4·50) is significantly greater than the values for chlorobenzene or nitrobenzene. When the substituents have opposing electronic effects, the reactivity is less than that of the more reactive of the two mono-substituted benzene compounds.

11.2.4   *Arylation with diaroyl peroxides*

The discussion of the mechanism of the arylation reaction has been directed to the formation of the phenylcyclohexadienyl radical. The fate of this depends on the environment in which it is generated, in particular on the presence of any other species in the reaction medium with which it may react. This is clearly going to depend largely on the source of aryl radical used, and it is for this reason that the detailed mechanism of the reaction will be considered in the light of the radical source.

There are essentially three ways in which the phenylcyclohexadienyl radical may react:

(i) it can disproportionate to give dihydrobiphenyls and biphenyl (reaction *36*),

(ii) it can dimerize to give tetrahydroquaterphenyls (reaction *37*),

(iii) it can be oxidized to biphenyl (reactions *38* and *39*):

$$\text{Ph} \underset{H}{\diagdown}\!\!\bigcirc\!\!\cdot \longrightarrow \text{Ph}\underset{H}{\diagdown}\!\!\bigcirc \left(\text{and } \text{Ph}\underset{H}{\diagdown}\!\!\bigcirc \right) + \text{Ph—Ph} \qquad (36)$$

$$\text{Ph}\underset{H}{\diagdown}\!\!\bigcirc\!\!\cdot \longrightarrow \text{Ph}\underset{H}{\diagdown}\!\!\bigcirc\!\!\underset{H}{\overset{H}{\bigcirc}}\!\!\bigcirc\!\!\underset{H}{\diagup}\text{Ph} \qquad (37)$$

(+ other isomers)

$$\text{Ph}\underset{H}{\diagdown}\!\!\bigcirc\!\!\cdot + \text{R}\!\cdot \longrightarrow \text{Ph—Ph} + \text{RH} \qquad (38)$$

$$\text{Ph}\underset{H}{\diagdown}\!\!\bigcirc\!\!\cdot \longrightarrow \text{Ph}\underset{H}{\diagdown}\!\!\overset{+}{\bigcirc} \longrightarrow \text{Ph—Ph} + \text{H}^+ \qquad (39)$$

Decomposition of dibenzoyl peroxide in benzene and substituted benzenes generally results in the formation of a large amount of a high-boiling residue. This has been shown to consist largely of tetrahydro-quaterphenyls. Thus, the residue from the reaction in benzene on treatment with *o*-chloroanil gave a mixture of 2,2'-, 2,4'-, and 4,4'-quaterphenyls, as would be expected on the basis of dimerization of phenylcyclohexadienyl radicals (50). Decomposition of unsymmetrical diaroyl peroxides in benzene followed by dehydrogenation of the non-volatile residue showed the presence of both possible symmetrical quaterphenyls as well as the unsymmetrical one, thereby showing that arylcyclohexadienyl radicals have a definite existence outside the solvent cage.

A study of the variation in yields of dihydrobiphenyl, biphenyl and benzoic acid with peroxide concentration from the decomposition of

428

dibenzoyl peroxide in benzene, indicates that in very dilute solution equimolar amounts of biphenyl and dihydrobiphenyl are obtained (50). This is consistent with disproportionation of phenylcyclohexadienyl radicals being the only route to biphenyl at infinite dilution. The yield of dihydrobiphenyl falls, whilst that of biphenyl increases at higher peroxide concentrations, showing that some phenylcyclohexadienyl radicals react by a different route. Gill and Williams (52) examined the rate of decomposition of dibenzoyl peroxide in benzene, and showed that the rate of decomposition of the peroxide followed the rate law

$$\frac{-d[P]}{dt} = k_1[P] + k_2[P]^{\frac{3}{2}}$$

where P represents the peroxide. The term in $[P]^{\frac{3}{2}}$ arises from the induced decomposition of the peroxide (cf. chapter 2, p. 15). It has been established that this involves reaction of the phenylcyclohexadienyl radicals with the peroxide:

$$[Ph \cdot C_6H_6]\cdot + (PhCO_2)_2 \longrightarrow Ph\!-\!Ph + PhCO_2\cdot + PhCO_2H \quad (40)$$

The mechanism of this reaction probably involves electron transfer from the σ-radical to the lowest anti-bonding orbital of the peroxidic oxygen (Scheme 4). It was also shown that the reactions of cyclohexa-

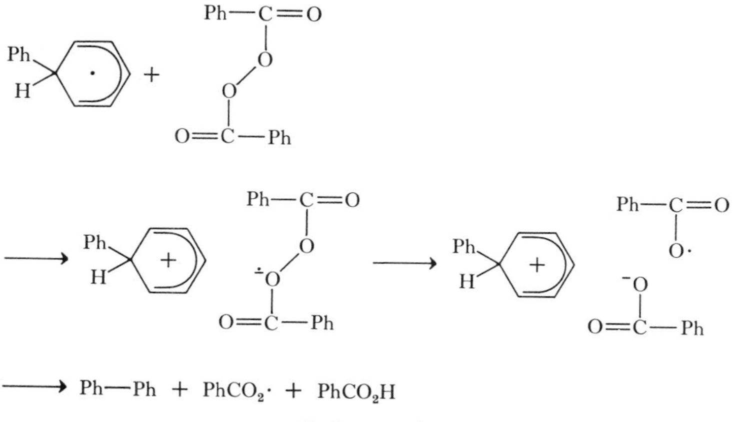

$$\longrightarrow Ph\!-\!Ph + PhCO_2\cdot + PhCO_2H$$

Scheme 4

dienyl radicals with phenyl radicals (reaction *41*) or with benzoyloxy radicals (reaction *42*) did not occur to any significant extent (less than 5%):

$$[Ph \cdot C_6H_6]\cdot + Ph\cdot \longrightarrow Ph\!-\!Ph + PhH \quad (41)$$

$$[Ph \cdot C_6H_6]\cdot + PhCO_2\cdot \longrightarrow Ph\!-\!Ph + PhCO_2H \quad (42)$$

11.2.5 *The effect of additives on the peroxide reaction*

The decomposition of diaroyl peroxides in aromatic substrates has in its simplest form only limited synthetic utility because of the low yields of biaryl produced. The method would be of much greater value if some means were available of preventing the dimerization of the intermediate σ-radicals. This has been achieved in several ways (Table 11-3): notably by addition of nitro compounds (20, 53, 54), addition of iron(III) or copper(II) salts (21, 55), or by carrying out the reactions in the presence of oxygen (19). In each case the σ-radical is oxidized before its dimerization can take place.

TABLE 11-3   *Effect of additives on the yield of biaryl produced in the decomposition of diaroyl peroxides in aromatic substrates*

Peroxide	Substrate	Additive	Yield[a]	Reference
$(PhCO_2)_2$	PhH	—	40	53
$(PhCO_2)_2$	PhH	$m\text{-}C_6H_4(NO_2)_2$	85	53
$(p\text{-}ClC_6H_4CO_2)_2$	PhH	—	47	53
$(p\text{-}ClC_6H_4CO_2)_2$	PhH	$m\text{-}C_6H_4(NO_2)_2$	90	54
$(3,5\text{-}Cl_2C_6H_3CO_2)_2$	PhH	—	50	53
$(3,5\text{-}Cl_2C_6H_3CO_2)_2$	PhH	$m\text{-}C_6H_4(NO_2)_2$	80	54
$(PhCO_2)_2$	PhF	—	48	20
$(PhCO_2)_2$	PhF	$m\text{-}C_6H_4(NO_2)_2$	80	20
$(PhCO_2)_2$	PhH	—	36	55
$(PhCO_2)_2$	PhH	$Cu(OCOPh)_2$	88	55
$(PhCO_2)_2$	PhCl	—	45	55
$(PhCO_2)_2$	PhCl	$Cu(OCOPh)_2$	78	55
$(PhCO_2)_2$	PhCl	air	100	55
$(PhCO_2)_2$	PhH	—	38	21
$(PhCO_2)_2$	PhH	$Fe(OCOPh)_3$	90	21
$(PhCO_2)_2$	PhH	—	38–47	19
$(PhCO_2)_2$	PhH	O_2	126–151	19

[a] Yield of biaryl based on 1 mole of product per mole of peroxide.

Hey, Perkins, and co-workers, have established that the action of nitro compounds is explicable on the basis of their reduction to nitroso compounds (56). The reduction of the nitro group to nitroso is probably effected by the phenylcyclohexadienyl radicals. Addition of phenyl radicals to the nitroso compound would give the stable nitroxide:

$$PhNO_2 \longrightarrow PhNO \tag{43}$$

$$Ph\cdot + PhNO \longrightarrow Ph_2NO\cdot \tag{44}$$

The e.s.r. spectrum of the decomposing dibenzoyl peroxide in benzene in the presence of nitrobenzene shows a signal characteristic of diphenyl nitroxide. This should be capable of oxidizing phenylcyclohexadienyl radicals to biphenyl efficiently:

$$[\text{Ph·C}_6\text{H}_6]\cdot + \text{Ph}_2\text{NO·} \longrightarrow \text{Ph—Ph} + \text{Ph}_2\text{NOH} \qquad (45)$$

The addition of copper(II) or iron(III) salts effects the conversion of the σ-radical to the corresponding cation, which loses a proton to give the biaryl (reactions *46* and *47*). The copper(I) or iron(II) thus generated would promote the catalyzed decomposition of the dibenzoyl peroxide (*58*):

$$(46)$$

$$(47)$$

$$(\text{PhCO}_2)_2 + \text{Cu}^+ \longrightarrow \text{PhCO}_2\cdot + \text{PhCO}_2^- + \text{Cu}^{2+} \qquad (48)$$

The work of Kochi on electron-transfer oxidations of radicals by copper(II) salts would indicate that the oxidation of the radical to the biaryl may well be a concerted process (59, 60) (cf. chapter 10, p. 310). A feature about the influence of metal salt additives is that equimolar yields of biaryl and carboxylic acid are obtained, consistent with the proposed scheme.

The addition of metal salts, at least in some instances, changes the isomer distribution of products. Thus, the yield of *m*-methoxybiphenyl from the reaction of dibenzoyl peroxide with anisole is greater in the presence of copper(II) benzoate (62). In the absence of copper(II) salts, biaryls arise as a result of disproportionation of the σ-radicals or from the induced decomposition of dibenzoyl peroxide. The copper(II) salts oxidize the σ-radicals more effectively than dibenzoyl peroxide. This is particularly true for the radical (7) which would have a higher oxidation potential than (8) or (9). Consequently, it may well be that oxidation of (7) by dibenzoyl peroxide in which there is at least a degree of electron-transfer in the transition state (Scheme 4), does not proceed as readily as that of (8) and (9), and hence dimerization is favoured. Copper(II) salts, being stronger oxidants, may act indiscriminately for all three

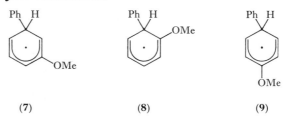

| (7) | (8) | (9) |

radicals. If the radicals (**8**) and (**9**) are more readily oxidized than (**7**) by dibenzoyl peroxide but not by copper(II) salts, one would expect that the proportion of *m*-methoxybiphenyl would be increased by the addition of copper(II) salts. If the above reasoning is correct, one might expect that the amount of *meta*-phenylation of monosubstituted benzene derivatives with electron-withdrawing substituents would be *decreased* by the addition of copper(II) salts. This has only been examined in the case of nitrobenzene, which, as we shall see later, is atypical. It would be of value to examine this effect with, for example, methyl benzoate. The influence of iron(III) benzoate on the isomer distribution of biaryls, resulting from the phenylation with dibenzoyl peroxide of chlorobenzene and toluene, was small (21). The only significant effect was an increase in the amount of *ortho*-substitution, probably arising from oxidation of the *ortho* σ-radical before it could dissociate back to the substrate and phenyl radicals. The role of electronic effects on the relative ease of oxidation of the isomeric σ-radicals would be much less than with substituted benzenes, containing more powerful electron-attracting or electron-donating groups.

The mode of action of oxygen in effecting the oxidation of the σ-radicals is to remove the hydrogen atom presumably with the formation of a hydroperoxide. This latter can then bring about the oxidation of a second σ-radical:

$$[Ph \cdot C_6H_6] \cdot + O_2 \longrightarrow Ph\text{---}Ph + HO_2 \cdot \qquad (49)$$

$$[Ph \cdot C_6H_6] \cdot + HO_2 \cdot \longrightarrow Ph\text{---}Ph + H_2O_2 \qquad (50)$$

Hydrogen peroxide has been detected in these reactions (63). In the presence of oxygen two moles of biaryl could theoretically be produced per mole of diaroyl peroxide, thereby accounting for the figures quoted in Table 11-3. Consistent with the proposed scheme, the amount of benzoic acid produced in the presence of oxygen is less than in the absence of oxygen and much less than in the presence of the other additives. At lower temperatures some phenol is also obtained, resulting from reaction of phenyl radicals with oxygen:

$$Ph\cdot + O_2 \longrightarrow PhO_2\cdot \longrightarrow PhOH \qquad (51)$$

That this does not occur at higher temperatures is due to the lower solubility of oxygen.

Both the relative reactivities of a wide range of aromatic substrates and the isomer distributions of derived biaryls have been shown to be unaffected by the presence of oxygen (64). This is in spite of the almost threefold increase in some instances in the yield of biaryl brought about by the presence of oxygen. These results have been taken to indicate that the various reactions which the σ-radicals may undergo (dimerization, disproportionation, or reaction with peroxide) are essentially non-selective. The non-polar character of the hydrogen-abstraction reaction with oxygen compared with copper(II) salts must account for the lack of influence of oxygen on the isomer distribution.

11.2.6 *Arylation with arylazotriarylmethanes*

Arylation with arylazotriarylmethanes differs from arylations with most other sources of aryl radicals in that the stable triarylmethyl radical is formed along with the aryl radical (35):

$$ArN{=}NCAr_3 \longrightarrow Ar\cdot + N_2 + Ar_3C\cdot \qquad (52)$$

As a consequence of the relatively high concentration of triarylmethyl radicals, certain differences in the reaction products are observed. Firstly, no dihydrobiaryls are produced as the σ-radicals, which are also relatively long-lived, suffer hydrogen abstraction by the triarylmethyl radicals before they can disproportionate or dimerize. The second difference is that coupled products formed by reaction of the σ-radicals with triarylmethyl radicals can be isolated (35) and in some instances are the major products (65).

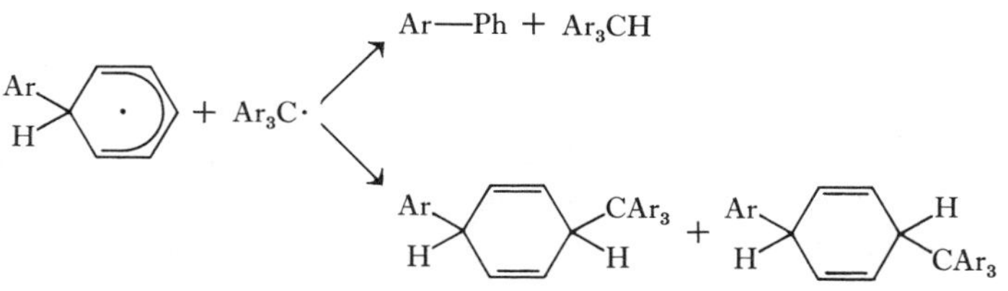

$$Ar{-}Ph + Ar_3CH \qquad (53)$$

$$(54)$$

Homolytic substitution

Isomer distributions found in arylations using arylazotriphenylmethanes are sometimes significantly different from those obtained with other sources of aryl radicals. This arises because of differences in the ratio of hydrogen abstraction and coupling (reactions *53–54*) of the isomeric σ-radicals (65).

11.2.7 *Arylation with* N-*nitrosoacetanilides*

No dihydrobiaryls are produced in the arylation of aromatic substrates with *N*-nitrosoacetanilides. This indicates that the intermediate σ-radicals must be rapidly oxidized. This could be effected by the stable radical, $PhN{=}NO\cdot$:

$$[Ph{\cdot}C_6H_6]{\cdot} + PhN{=}NO{\cdot} \longrightarrow Ph{-}Ph + PhN{=}NOH \qquad (55)$$

This radical might be expected to have comparable efficiency to diphenylnitroxide in the abstraction of a hydrogen from the phenylcyclohexadienyl radical. Alternatively, oxidation could be brought about by the diazonium ion in an electron-transfer process (66) (reactions *56* and *57*):

$$[Ph{\cdot}C_6H_6]{\cdot} + PhN_2{}^+ \longrightarrow [Ph{\cdot}C_6H_6]^+ + Ph{\cdot} + N_2 \qquad (56)$$

$$[Ph{\cdot}C_6H_6]^+ + OAc^- \longrightarrow Ph{-}Ph + AcOH \qquad (57)$$

The scheme would satisfactorily explain all the products from this reaction, particularly the formation of acetic acid.

11.2.8 *The influence of polar effects in the aryl radical*

No consideration has so far been given to the polar character, if any, of the aryl radicals in homolytic arylation. Clearly such effects with the phenyl radical are secondary to the ability of the substituent in the substituted phenylcyclohexadienyl radical to delocalize the unpaired electron, since the electronic nature of the substituent in the aromatic substrate has little influence on the isomer distribution. Investigations using substituted phenyl radicals have shown that the isomer distribution is altered in a predictable manner by the electronic and steric nature of the substituents (Tables 11-4, 11-5 and 11-6). Thus, *p*-nitrophenyl radicals react less readily with nitrobenzene than with anisole and give more of the *meta*-arylated product, in line with their being somewhat electrophilic. The effect is even more pronounced with perfluorophenyl radicals as judged from the orientation of the products. Conversely,

434

TABLE 11-4 *Relative rates and partial rate factors for arylation with substituted aryl radicals at 20 °C (67)*

Substrate	Partial rate factor	X = NO$_2$	Cl	H	Me	MeO
			\multicolumn{4}{c}{Attacking radical (p — XC$_6$H$_4$·)}			
PhNO$_2$	F_o	0.93	4.35	9.38	10.73	6.45
	F_m	0.35	0.61	1.16	1.07	1.19
	F_p	1.53	6.18	9.05	8.93	8.36
	$\frac{\text{PhNO}_2}{\text{PhH}}K$	0.68	2.69	5.02	5.50	3.94
PhCl	F_o	1.53	2.70	3.09	3.10	3.08
	F_m	0.65	0.87	1.01	1.04	1.82
	F_p	1.01	1.33	1.48	1.16	1.24
	$\frac{\text{PhCl}}{\text{PhH}}K$	0.89	1.41	1.61	1.57	1.92
PhOMe	F_o	5.17	3.93	3.56	3.69	3.68
	F_m	0.84	0.94	0.93	1.09	1.03
	F_p	2.30	1.54	1.29	1.52	1.31
	$\frac{\text{PhOMe}}{\text{PhH}}K$	2.39	1.88	1.71	1.85	1.79
PhMe	F_o	3.28	2.97	3.30	3.27	3.09
	F_m	1.36	1.07	1.09	1.00	1.00
	F_p	1.51	1.32	1.27	1.33	1.18
	$\frac{\text{PhMe}}{\text{PhH}}K$	1.80	1.56	1.68	1.65	1.56

TABLE 11-5 *Isomer distributions in the phenylation and pentafluorophenylation of aromatic substrates at 80 °C (68)*

Substrate	Radical	ortho-	meta-	para-	Ratio $(o+p)/m$
		\multicolumn{3}{c}{Isomer distribution (%)}			
PhCl	Ph·	56.9	25.6	17.5	2.9
PhCl	C$_6$F$_5$·	64.7	20.6	14.7	3.9
PhNO$_2$	Ph·	62.5	9.8	27.7	9.2
PhNO$_2$	C$_6$F$_5$·	20.8	53.4	25.8	0.87

p-tolyl radicals are slightly nucleophilic and consequently react more rapidly with nitrobenzene than with anisole and give less of the *meta*-substituted product. Substituents in the *ortho*-position in the attacking radical decrease the overall reactivity of the radical, as seen in the total rate factor, as well as decreasing the amount of *ortho*-substitution (Table 11-6). These effects can be explained as a result of steric factors. The differences in partial rate factors in the various tables arise from the different reaction conditions employed.

A careful analysis of the results quoted in Table 11-4 has shown that

TABLE 11-6 *Relative rates and partial rate factors for the arylation of nitrobenzene* (69)

		Partial rate factors		
Radical	$\frac{PhNO_2}{PhH}K$	F_o	F_m	F_p
o-$NO_2C_6H_4\cdot$	0.26	0.42	0.14	0.42
m-$NO_2C_6H_4\cdot$	0.43	0.68	0.73	0.75
p-$NO_2C_6H_4\cdot$	0.94	1.64	0.43	1.6
o-$ClC_6H_4\cdot$	0.82	0.88	0.66	2.0
m-$ClC_6H_4\cdot$	1.3	2.2	0.58	2.2
$C_6H_5\cdot$	2.9	5.5	0.86	4.9
o-$MeC_6H_4\cdot$	2.2	2.7	1.2	5.2
m-$MeC_6H_4\cdot$	3.0	5.5	1.2	4.7
p-$MeC_6H_4\cdot$	3.4	6.1	1.2	5.8

the partial rate factors at the *meta*-position, F_m, can be related to the Hammett σ constant for the substituent in the aromatic substrate:

$$\log F_m = \rho\sigma_m$$

This indicates that it is primarily the inductive effects of the substituents which determine the polar characteristic of the aryl radical. Table 11-7 gives the values for the reaction constant ρ for arylation for the series of aryl radicals.

TABLE 11-7 *Hammett reaction constants for arylation at 20 °C* (67)

Radical	Reaction constant (ρ)
p-$NO_2C_6H_4\cdot$	-0.81
p-$ClC_6H_4\cdot$	-0.27
$C_6H_5\cdot$	0.05
p-$MeC_6H_4\cdot$	0.03
p-$MeOC_6H_4\cdot$	0.09

11.2.9 *Side-chain attack in reactions of alkylbenzenes*

In addition to undergoing arylation in the nucleus, alkylbenzenes also suffer hydrogen abstraction, the resultant benzylic radicals dimerizing:

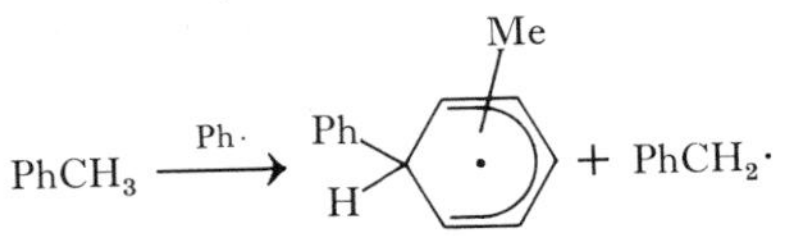

$$PhCH_3 \xrightarrow{Ph\cdot} \quad + PhCH_2\cdot \tag{58}$$

$$2PhCH_2\cdot \longrightarrow PhCH_2CH_2Ph \tag{59}$$

TABLE 11-8   *Amount of side-chain attack in the*
phenylation of alkylbenzenes (70)

Alkylbenzene	Side-chain attack (mole %)
PhMe	12
PhEt	51
PhPr^i	55.5
PhBu^t	0

TABLE 11-9   *The influence of electronic effects in the aryl radical*
on the amount of side-chain attack in alkylbenzenes (71)

Radical	Side-chain attack (mole %)	
	PhMe	PhPr^i
p-NO$_2$C$_6$H$_4$·	0	0
p-ClC$_6$H$_4$·	11	37.5
p-BrC$_6$H$_4$·	—	50
C$_6$H$_5$·	12	55.5
p-MeC$_6$H$_4$·	42	—
p-MeOC$_6$H$_4$·	47	88

The extent of side-chain attack is determined by the stability of the resultant benzylic radical (Table 11-8), i.e. on the degree of substitution at the radical centre. The amount of side-chain attack is also sensitive to electronic effects in the attacking radical (Table 11-9), being greater with more nucleophilic radicals (71). A nucleophilic species would be less prone to attack the electron-rich π-system than the methyl group. Substitution of the benzene ring by electron-withdrawing substituents might be expected to reduce the degree of side-chain attack, as has been observed (72).

Perusal of the literature indicates that the amount of side-chain attack varies with the source of aryl radical. It is consistently greatest when peroxides are used as the aryl radical source, suggesting that at least some of the side-chain attack is brought about by aroyloxy radicals and not by aryl radicals (73):

$$ArCH_3 + Ar'CO_2· \longrightarrow ArCH_2· + Ar'CO_2H \qquad (60)$$

It has also been shown that additives such as nitrobenzene very markedly reduce the amount of side-chain attack (Table 11-10). We

TABLE 11-10   *Effect of additives on the degree of side-chain attack by phenyl radicals* (46)

Substrate	Additive	Side-chain attack (%)
PhMe	—	13
PhMe	PhNO$_2$	0
p-Me$_2$C$_6$H$_4$	—	55
p-Me$_2$C$_6$H$_4$	PhNO$_2$	4.6

have already seen that nitrobenzene is converted into a nitroxide under the reaction conditions and that the nitroxide, because of its fairly high concentration, is an effective hydrogen-abstracting agent (p. 430). It is difficult to account for the role of nitrobenzene on the proportion of side-chain attack unless one considers that the initial σ-radical is formed reversibly (reaction *61*), but that the formation of the benzylic radical by hydrogen abstraction is an irreversible process (reaction *62*). In this case the nitrobenzene would effect the efficient oxidation of the σ-radicals, thereby preventing their dissociation, and consequently the degree of side-chain attack would be reduced:

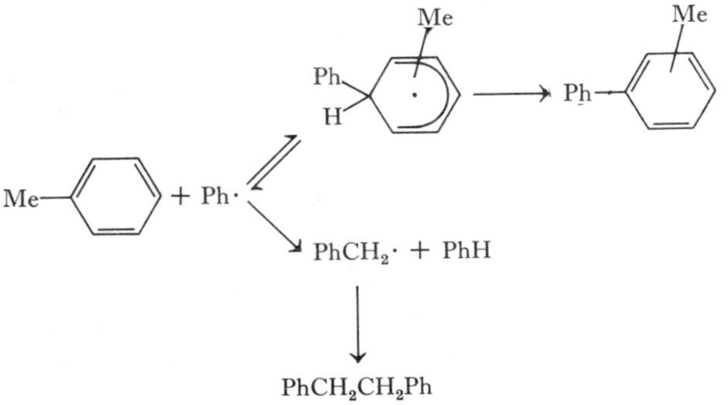

$$(61)$$

$$(62)$$

The correctness of this postulate should be capable of verification by carrying out phenylations at higher temperatures when dissociation of the σ-radical may occur more readily. Results indicate that there is indeed an increase in the amount of bibenzyl obtained in the reaction of toluene with dibenzoyl peroxide on increasing the temperature from 80 to 110 °C (74). Further confirmation of the reversible nature of the addition of phenyl radicals has recently been obtained by Perkins, who showed that phenylcyclohexadienyl radicals, derived by dissociation of tetrahydroquaterphenyl, can act as a source of phenyl radicals (46) (see Scheme 3) (p. 424).

11.2.10 *Intramolecular arylation*

The Pschorr synthesis of phenanthrene derivatives, which in its original form involved the copper-catalysed cyclization of the diazonium salt of *trans*-2-amino-α-phenylcinnamic acid to phenanthrene-9-carboxylic acid (reaction *63*), is the most-studied intramolecular arylation reaction (75):

$$\text{[structure: CO}_2\text{H diazonium compound]} \longrightarrow \text{[phenanthrene-9-carboxylic acid]} \qquad (63)$$

The conditions most frequently employed in these intramolecular arylations involve the copper-catalysed decomposition of the diazonium fluoroborate in acetone or of a diazonium salt prepared *in situ* in acid solution (76). More recently cuprous oxide has been used in place of copper (77). A recent development, which is claimed to give better yields, involves diazotization of the amine salt in acetone with isopentyl nitrite, followed by decomposition of the diazonium salt with sodium iodide (78). Photolysis of *o*-iodo-*cis*-stilbenes has also been used for Pschorr-type cyclizations (reaction *64*) (79):

$$\text{[o-iodo-cis-stilbene structure]} \xrightarrow{h\nu} \text{[phenanthrene]} \qquad (64)$$

These methods all involve procedures which have been shown for simpler systems to result in the generation of aryl radicals. Consequently the Pschorr reaction has been formulated as an intramolecular arylation (Scheme 5). Further, if an ionic mechanism were operative, one would predict that the yield of cyclized product would be markedly dependent on the electronic nature of substituents in the nucleus undergoing attack. This is not the case.

Good evidence that the reaction proceeds by a homolytic mechanism comes from a study of reactions of diazotized 2-amino-*cis*-stilbene and the corresponding *trans*-stilbene. The radical generated from the *cis*-stilbene affords phenanthrene with both copper powder in benzene and with hypophosphorous acid, whereas the *trans*-compound arylates the benzene and undergoes reduction in the two reactions (*65* and *66*) (33).

439

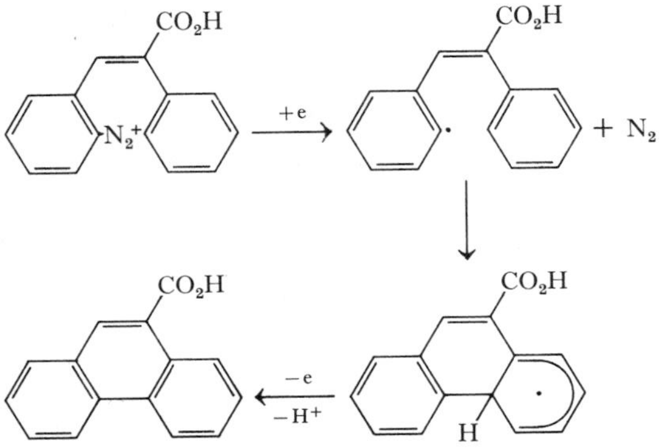

Scheme 5

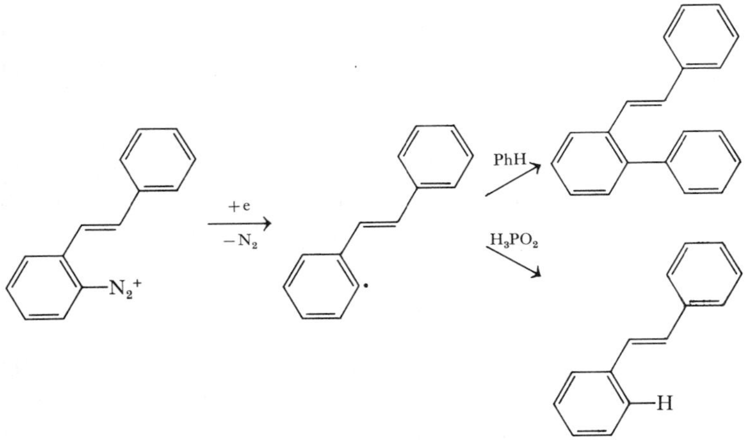

$$(65)$$

$$(66)$$

Even more convincing evidence for the involvement of radicals comes from the isolation of the dimer (**13**) which was formed along with *N*-methylphenanthridone (**15**) and the spiro-dienone (**14**) when the diazonium salt (**10**) was treated with either copper powder (80) or sodium iodide (81) (Scheme 6). The dimer can only reasonably be envisaged as arising from coupling of the radicals (**12**), which are formed from cyclization of the aryl radical (**11**). Oxidation of (**12**) leads to the spiro-dienone whereas rearrangement followed by oxidation gives the phenanthridone. That the mechanism proceeds as outlined, is shown by the fact that pyrolysis of the dimer at 200 °C gives *N*-methylphenanthridone almost quantitatively.

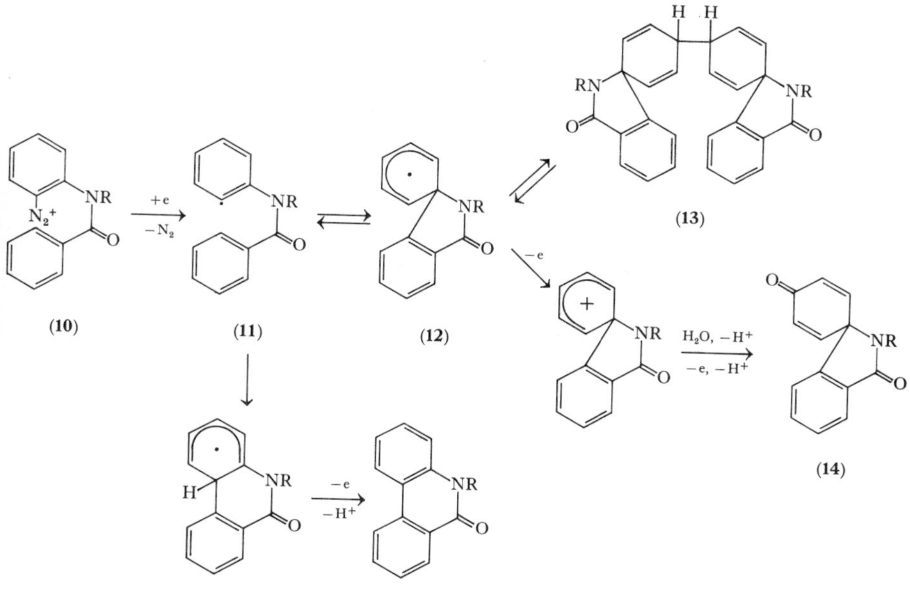

(13)

(10)   (11)   (12)

(14)

(15)

Scheme 6

The examples mentioned thus far have all involved the formation of a planar or near-planar six-membered ring. The homolytic route seems to be particularly favoured by such conditions. Competitive reactions, such as reduction in the presence of hypophosphorous acid or arylation of the solvent in the presence of benzene, do not intrude. In a cyclization involving the formation of a five-membered ring these side reactions become dominant (reactions *67–69*) (82, 83). In these circumstances

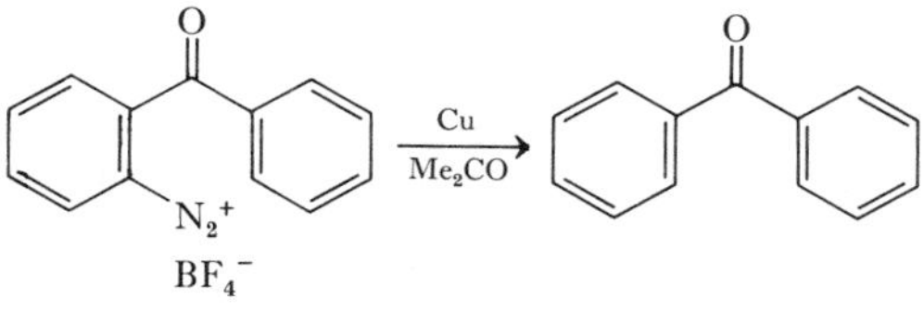

(67)

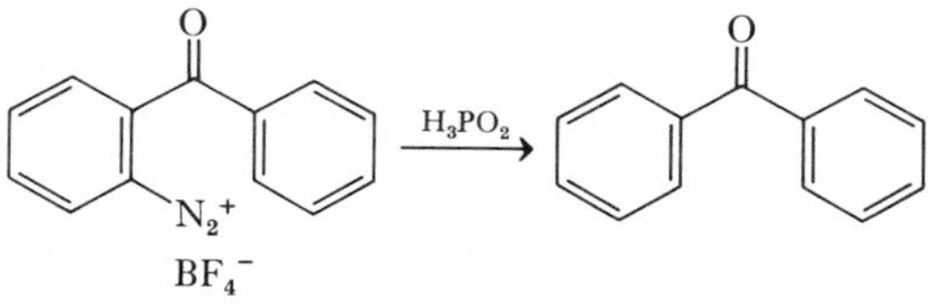

(68)

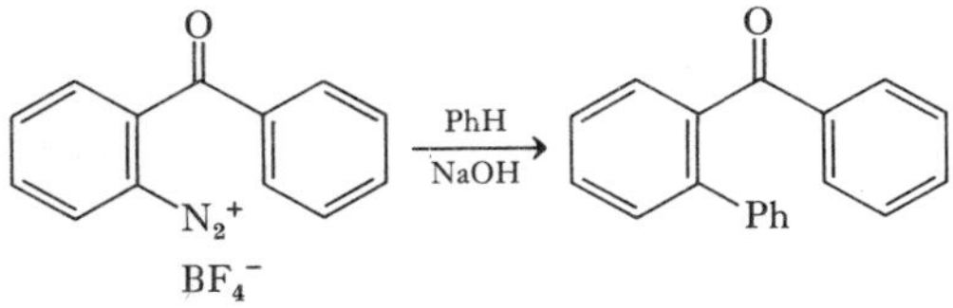

$$(69)$$

cyclization is more likely to proceed via a carbonium ion intermediate (83). Thus, diazotized 2-aminobenzophenone on treatment with copper in acetone gives no fluorenone, though this is obtained in 70% yield on heating an acidified solution of the diazonium salt. Hey has suggested that radical cyclization only occurs when the distance between the two

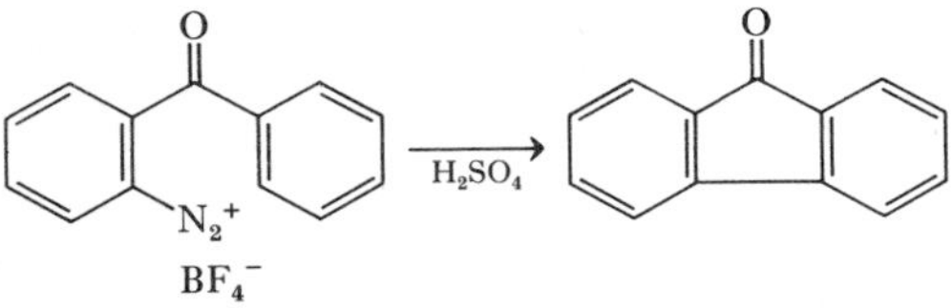

$$(70)$$

positions involved in the formation of the new bond is short (83). This distance has been calculated to be about 1.5 Å for cyclization to fluorenes. Cyclization by a heterolytic route would appear to be less critically controlled by this factor.

11.3   *ACYLOXYLATION*

Esters always accompany biaryls in the arylation of aromatic substrates with diaroyl peroxides. Combination of aryl and aroyloxy radicals within the solvent cage to give esters which are derived from the peroxide does not occur widely, particularly in thermolyses, when loss of carbon dioxide occurs readily:

$$(PhCO_2)_2 \longrightarrow Ph \cdot + CO_2 + PhCO_2 \cdot \longrightarrow PhCO_2Ph \qquad (71)$$

The isolation of approximately 10% of phenyl benzoate from the photolysis of dibenzoyl peroxide in chlorobenzene, however, indicates that this can occur under photolytic conditions (84) (cf. p. 21).

Attack of aromatic substrates by aroyloxy radicals also occurs, especially in the case of reactive aromatic substrates such as naphthalene (85), anthracene (86) and anisole (87, 88). Thus reaction of dibenzoyl peroxide with naphthalene gives α- and β-benzoyloxynaphthalenes in addition to α- and β-phenylnaphthalenes.

There is a lot of good evidence that the formation of the σ-benzoyl-oxycyclohexadienyl radical is reversible. First, the yield of phenyl benzoate obtained in the photochemical decomposition of dibenzoyl peroxide in benzene is increased in the presence of oxygen (74). The σ-radicals are oxidized to the ester. This competes effectively with dissociation of benzoyloxy radicals to phenyl radicals (Scheme 7).

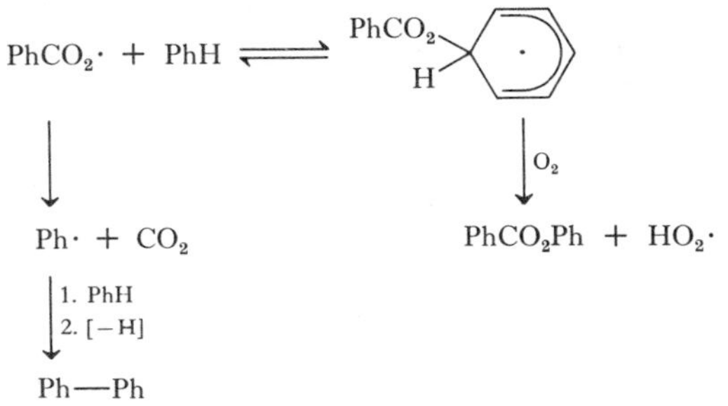

Scheme 7

Secondly, the yield of bibenzyl from the photolysis of dibenzoyl peroxide in toluene is increased by the addition of acid. The benzoyloxycyclo-hexadienyl radical undergoes acid-catalysed fragmentation to the benzyl radical (reaction 72). Otherwise the benzoyloxy radical undergoes decarboxylation to give the phenyl radical and thence biphenyl by reaction with benzene.

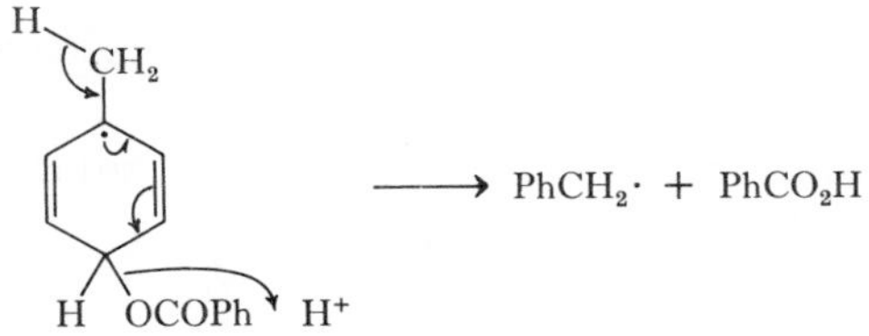

$$PhCH_2\cdot + PhCO_2H \qquad (72)$$

The extent of benzoyloxylation of anisole and other methoxybenzenes is far greater than would be expected on the basis of the activating influence of the methoxy group. Thus, the reaction of dibenzoyl peroxide with *p*-methylanisole gave neither biaryl- nor bibenzyl-type products: the major product was 2-methoxy-5-methylphenyl benzoate (72):

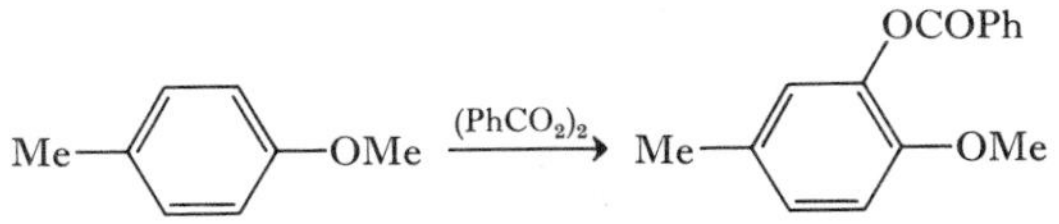

$$(73)$$

Similarly, *p*-dimethoxybenzene gave predominantly 2,5-dimethoxy-phenyl benzoate. *p*-Substituted anisoles in which the substituent is electron-attracting undergo normal phenylation. We consider that these results can best be interpreted by Scheme 8, in which the key step

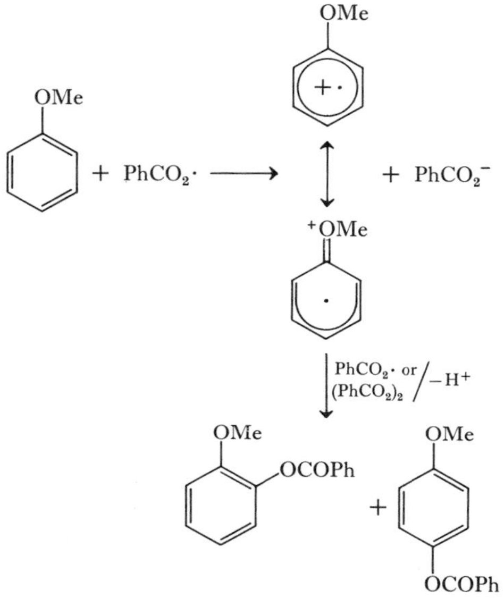

Scheme 8

involves electron transfer to give the radical cation. Electron-withdrawing groups would raise the oxidation potential of the substituted anisole, thereby disfavouring formation of the radical cation. This scheme

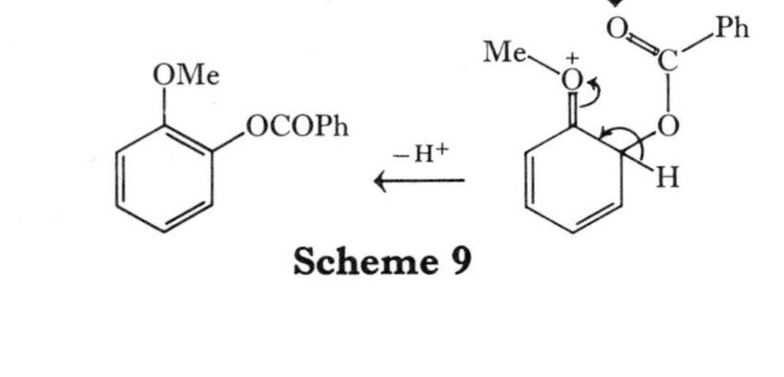

Scheme 9

444

appears more satisfactory than the alternative one proposed by Lynch and Moore (Scheme 9) (87). Consistent with Scheme 8, only *ortho*- and *para*-benzoyloxyanisoles are obtained from the benzoyloxylation of anisole. Scheme 9 only satisfactorily accounts for the formation of the *ortho*-isomer. Had the reaction been a 'normal' homolytic substitution process, some of the *meta*-isomer would undoubtedly have been obtained.

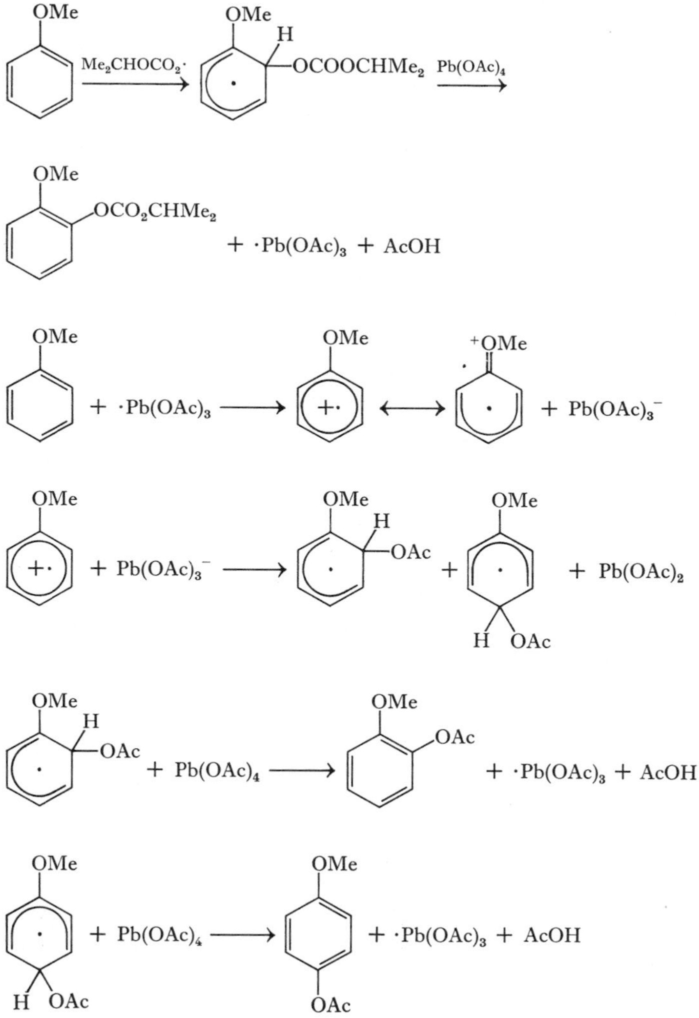

Scheme 10

A similar route involving acyloxylation of the radical cation seems to be involved in the acyloxylation of anisole by lead tetra-acetate and lead tetrabenzoate. Norman and his co-workers have shown that these reactions are catalysed by radical initiators such as di-isopropyl peroxy-dicarbonate, thus providing good evidence that such processes are radical-chain reactions (62). The mechanism for the acyloxylation is shown in Scheme 10. A particularly significant observation is that the relative reactivities of anisole and benzene towards acetoxylation and methylation are greater than 200 and 1.7 respectively. Such a large difference in reactivity cannot be satisfactorily explained if both reactions involve radical attack on the neutral molecule. It is consistent, however, with methylation involving attack on the neutral molecule and acetoxylation involving attack on the radical cation.

A novel example of homolytic benzoyloxylation is the reaction of 2,5-dimethylfuran with dibenzoyl peroxide. Reaction takes place in the side chain rather than in the nucleus (88 a). A possible explanation is given in Scheme 11. The key feature is again the formation of a radical

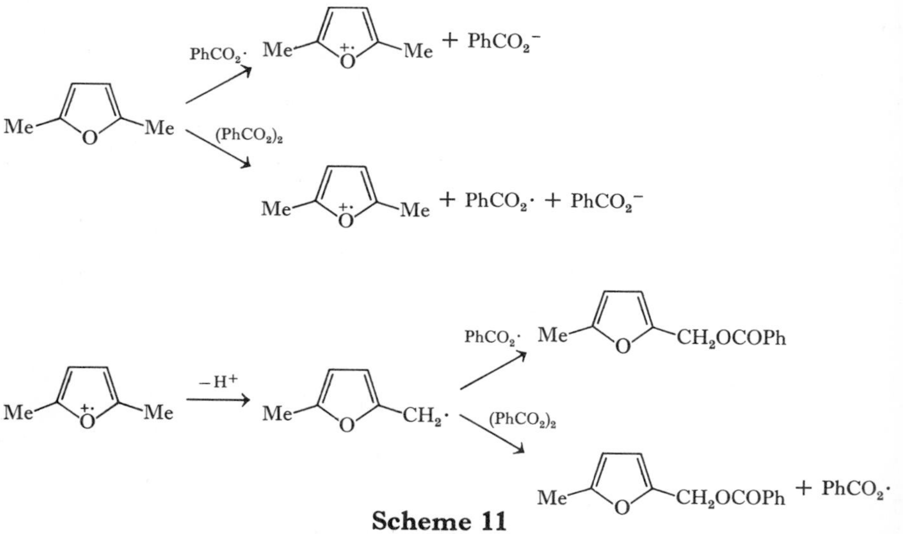

Scheme 11

cation, either in electron-transfer reaction between 2,5-dimethylfuran and benzoyloxy radicals, or in an induced decomposition of dibenzoyl peroxide by 2,5-dimethylfuran. The resultant radical cation then undergoes loss of a proton in a heterolytic fragmentation process rather than reacting with benzoyloxy radicals or with dibenzoyl peroxide.

11.4 *ALKYLATION*

Homolytic alkylation of aromatic substrates has been studied in far less detail than arylation. The methylation of a wide range of aromatic compounds has, however, been studied, but mainly from the point of view of correlating their reactivity with theoretical parameters. Much less attention has been given to accurate determination of partial rate factors. This stems, at least in part, from the absence of such convenient sources of alkyl radicals.

11.4.1 *Sources of alkyl radicals*

The sources of alkyl radicals mentioned in this section are those which have been most frequently used in homolytic aromatic alkylations.

(i) *From diacyl peroxides.* The decomposition of diacetyl peroxide provides a source of methyl radicals, decomposition occurring readily in solution at 70–80 °C. The acetoxy radicals formed by homolysis of the peroxide bond very rapidly lose carbon dioxide in an exothermic process (cf. chapter 2, p. 22):

$$(CH_3CO_2)_2 \longrightarrow 2CH_3COO\cdot \tag{74}$$

$$CH_3COO\cdot \longrightarrow CH_3\cdot + CO_2 \tag{75}$$

(ii) *From dialkyl peroxides.* Fragmentation of the alkoxy radicals produced in the photolysis or thermolysis of di-t-alkyl peroxides provides another convenient source of alkyl radicals. Thus di-t-butyl peroxide provides a good source of methyl radicals:

$$Me_3COOCMe_3 \longrightarrow 2Me_3CO\cdot \tag{76}$$

$$Me_3CO\cdot \longrightarrow Me_2CO + Me\cdot \tag{77}$$

This procedure is most useful for methylation of substrates containing no readily abstractable hydrogen atoms. A solvent containing such would also undergo hydrogen abstraction by the t-butoxy radicals:

$$PhCH_3 + Me_3CO\cdot \longrightarrow PhCH_2\cdot + Me_3COH \tag{78}$$

(iii) *Alkyl radicals by hydrogen abstraction.* Alkyl radicals can be obtained by hydrogen abstraction from an alkane by another radical. This method has been successfully used as a source of benzyl (97) and cyclohexyl radicals (90) by decomposing di-t-butyl peroxide in toluene and cyclohexane respectively:

$$PhCH_3 + Bu^tO\cdot \longrightarrow PhCH_2\cdot + Bu^tOH \qquad (79)$$

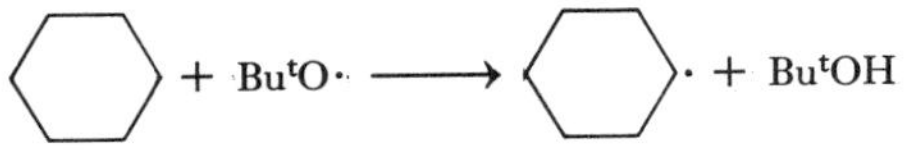

$$\langle\hexagon\rangle + Bu^tO\cdot \longrightarrow \langle\hexagon\rangle\cdot + Bu^tOH \qquad (80)$$

This method is valuable for the production of radicals from alkanes containing only one type of abstractable hydrogen.

(iv) *Photolysis of alkyl halides and organometallic compounds.* The photolysis of simple alkyl halides has been little used for the generation of alkyl radicals. One example of its use is the photolysis of 1-iodo-2-phenylacetylene, which provides the most convenient route to the 1-phenylethynyl radical (91):

$$PhC\equiv CI \xrightarrow{h\nu} PhC\equiv C\cdot + I\cdot \qquad (81)$$

Photolysis of alkylmercury(II) iodides has also been developed as a useful source of alkyl radicals (92). These compounds are readily obtainable from the corresponding Grignard reagents:

$$RMgI + HgI_2 \longrightarrow RHgI + MgI_2 \qquad (82)$$

$$RHgI \xrightarrow{h\nu} R\cdot + HgI\cdot \qquad (83)$$

The photolysis of dialkylmercury compounds provides a further source of alkyl radicals (93):

$$(PhCH_2)_2Hg \xrightarrow{h\nu} 2PhCH_2\cdot + Hg \qquad (84)$$

(v) *Pyrolysis of dialkylmercury compounds.* The pyrolysis of dibenzylmercury has been used as a source of benzyl radicals (93 a):

$$(PhCH_2)_2Hg \longrightarrow 2PhCH_2\cdot + Hg \qquad (85)$$

(vi) *Pyrolysis and photolysis of azo compounds.* 2-Cyano-2-propyl radicals have been generated by the decomposition of azoisobutyronitrile at 80 °C (98):

$$Me_2C(CN)N{=}NC(CN)Me_2 \xrightarrow{\Delta} 2Me_2\dot{C}CN + N_2 \qquad (86)$$

Szwarc and co-workers have generated trifluoromethyl radicals by photolysis of hexafluoroazomethane (93 b):

$$CF_3N{=}NCF_3 \xrightarrow{h\nu} 2CF_3\cdot + N_2 \qquad (87)$$

Trifluoromethyl radicals have also been generated by photolysis of hexafluoroacetone (93 c):

$$(CF_3)_2CO \xrightarrow{hv} 2CF_3 \cdot + CO \qquad (88)$$

11.4.2 *Mechanism of the alkylation reaction*

The mechanism of alkylation of aromatic substrates would appear to parallel that of arylation, namely reaction proceeds via an alkylcyclohexadienyl radical which dimerizes, disproportionates or suffers hydrogen abstraction. Thus, Beckwith and Waters were able to isolate 3,3'-dichloro-4,4'-dimethylbiphenyl from the methylation of chlorobenzene with di-t-butyl peroxide providing strong support for the dimerization reaction (94):

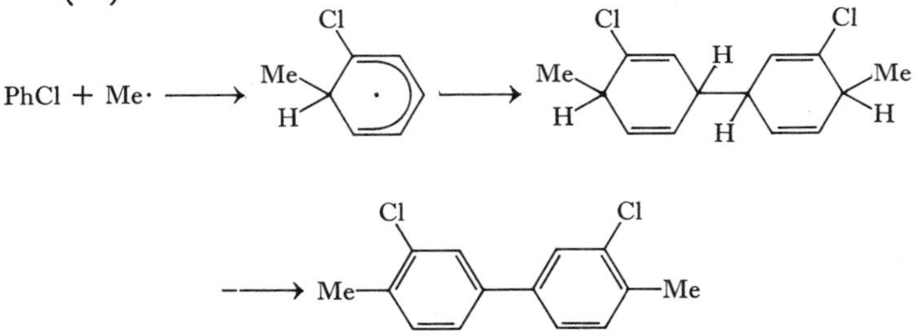

$$(89)$$

11.4.3 *Quantitative aspects of homolytic alkylation*

(i) *Isomer ratios in alkylation reactions.* The isomer distribution of alkylated products of a series of substituted benzenes has been determined. Table 11-11 records the *meta/para* ratios for the methylation, cyclohexylation, 1-phenylethynylation and, for comparison, phenylation for a range of monosubstituted benzenes. The *meta/para* ratios for the alkylations are in the order:

cyclohexylation > methylation > phenylation > 1-phenylethynylation

for the reaction of monosubstituted benzenes with +R groups and in the reverse order if the substrate contains −R groups. This is in line with the relative nucleophilicities of the radicals.

The methyl radical has a greater tendency to lose an electron than to gain an electron, and consequently can be said to be nucleophilic rather than electrophilic. The ionization potential of a secondary alkyl radical would be predicted to be less than for a methyl radical (cf. Table 3-4), and hence a cyclohexyl radical should be more nucleophilic than a methyl

TABLE 11-11 *Meta/para ratios in homolytic aromatic substitution*

Aromatic substrate	Cyclohexylation (90)	Methylation (92, 95)	Phenylation (44)	1-Phenyl- ethynylation (91)
PhOMe	5.6	1.4	1.06	—
PhMe	1.7	1.6, 2.4	0.73	—
PhBu^t	2.5	—	1.81	1.53
PhCl	2.8	2.3, 2.8	1.55	1.15
PhCN	0.09	0.21	0.33	—

radical. The electrophilic nature of an acetylenic radical follows from the *sp*-hybridization of the carbon centre.

Hammett plots for F_m and F_p for both the cyclohexylation and 1-phenylethynylation reactions have been made. These give ρ values of $+1.10$ for cyclohexylation and -1.56 for 1-phenylethynylation, consistent with the nucleophilic nature of the cyclohexyl radical and the electrophilic character of the 1-phenylethynyl radical. These compare with a ρ value of 0.05 for phenylation. In agreement with these ideas the amount of side-chain substitution decreases as the electrophilicity of the radical increases.

(ii) *Relative rates of reaction with alkyl radicals.* The complex nature of reactions of aromatic substrates with alkyl radicals means that no reliable information concerning the relative rates of attack can be obtained from competitive studies using simple alkyl radicals. Szwarc has devised an alternative method of assessing reactivity towards methyl radicals. The method consists of allowing a solution of acetyl peroxide to decompose in dilute solution in *iso*-octane and then in *iso*-octane containing a known amount of the compound under investigation. In *iso*-octane, methyl radicals only react with the solvent to give methane, apart from cage recombination reactions. Methane should, therefore, be produced in almost the same amount as carbon dioxide obtained from decarboxylation of acetoxy radicals:

$$MeCO_2\cdot \longrightarrow Me\cdot + CO_2 \qquad (90)$$

$$RH + Me\cdot \longrightarrow R\cdot + CH_4 \qquad (91)$$

In the presence of aromatic substrates some of the methyl radicals will react with the aromatic compound, and consequently the ratio of methane to carbon dioxide will be less than one. This decrease in amount of methane formed, occasioned as it is by the presence of the aromatic compound, is a measure of the reactivity of the aromatic compound towards methyl radicals. Several serious doubts can be raised as regards the assumptions on which this method is based. First, the σ-methylcyclohexadienyl radical may well undergo hydrogen abstraction by reaction with a methyl radical (reaction *92*):

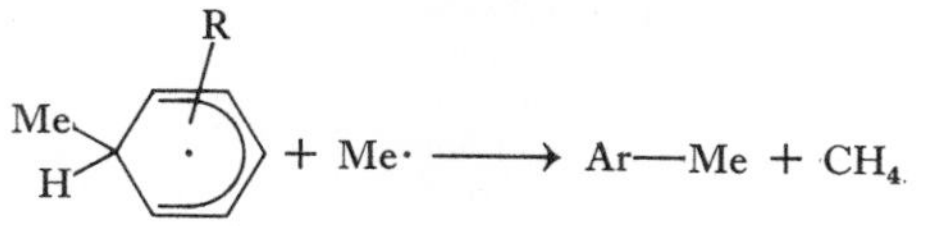

$$\text{(92)}$$

Secondly, the resultant methylated aromatic can undergo subsequent attack in the side-chain in a reaction which will again lead to the formation of methane:

$$ArCH_3 + Me\cdot \longrightarrow ArCH_2\cdot + CH_4 \qquad (93)$$

This latter is a particularly serious drawback in a determination of the methyl affinity of toluene and related compounds, though corrections have been made for this.

The absolute values of methyl affinities must be accepted with considerable reservations. Nevertheless, the methyl affinities of a whole range of compounds (Table 11-12) do closely parallel the reactivities of these same compounds towards phenyl radicals. Furthermore, Szwarc has shown that the methyl affinity is not significantly changed by changes in concentration or temperature.

Inspection of the values in Table 11-12 reveals certain points of interest. Anisole has a particularly low value, probably because it may also undergo side-chain attack in a reaction producing methane. One would not expect it to have a value below that of diphenyl ether. The relatively high value for benzonitrile can be explained on the basis of the nucleophilicity of methyl radicals. The very high value for nitrosobenzene undoubtedly arises from the radical-trap properties of the nitroso group (reaction *94*) (cf. chapter 3, p. 49):

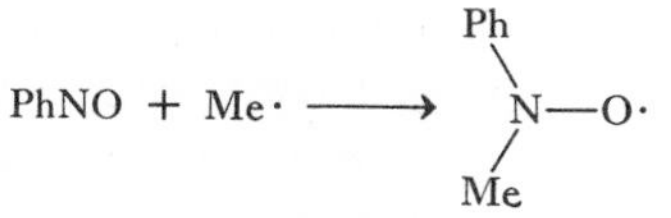

$$\text{PhNO} + Me\cdot \longrightarrow \begin{array}{c} Ph \\ \diagdown \\ N\!-\!O\cdot \\ \diagup \\ Me \end{array} \qquad (94)$$

The values for polycyclic and heterocyclic aromatic compounds have been correlated with certain theoretical parameters (see p. 456).

TABLE 11-12   *Methyl affinities for aromatic compounds* (96)

Aromatic compound	Methyl affinity
Benzene	1.0
Anisole	0.65
Diphenyl ether	2.5
Fluorobenzene	2.2
Chlorobenzene	4.2
Acetophenone	2.4
Ethyl benzoate	5.2
Benzonitrile	12.2
Nitrosobenzene	10^5
Biphenyl	5
Naphthalene	22
Phenanthrene	27
Anthracene	820
Naphthacene	9250
Pyridine	3
Quinoline	29
Isoquinoline	36
Acridine	430

The groups of both Szwarc (93 b) and Whittle (93 c) have studied the reactivity of trifluoromethyl radicals towards substituted benzenes in an analogous fashion to the study of the reactivity of methyl radicals. In contrast to the results obtained with methyl radicals, they found that the reactivity of substituted benzenes relative to benzene is reduced by electron-attracting groups but increased by electron-donating groups. This is to be expected with the fairly strongly electrophilic trifluoromethyl radicals.

11.4.4   *Alkylation of polycyclic aromatic hydrocarbons*

Detailed information about the mechanism of alkylation of aromatic compounds has been obtained from a study of the benzylation of anthracene (97). Scheme 12 indicates the products obtained and their mode of formation. Derivatives of anthracene and 9,10-dihydroanthracene were formed. This contrasts with alkylation with the less reactive 2-cyanoisopropyl radicals, which only gives derivatives of 9,10-dihydroanthracene (98) (Scheme 13). The 2-cyanoisopropyl radical is not sufficiently reactive to effect hydrogen abstraction.

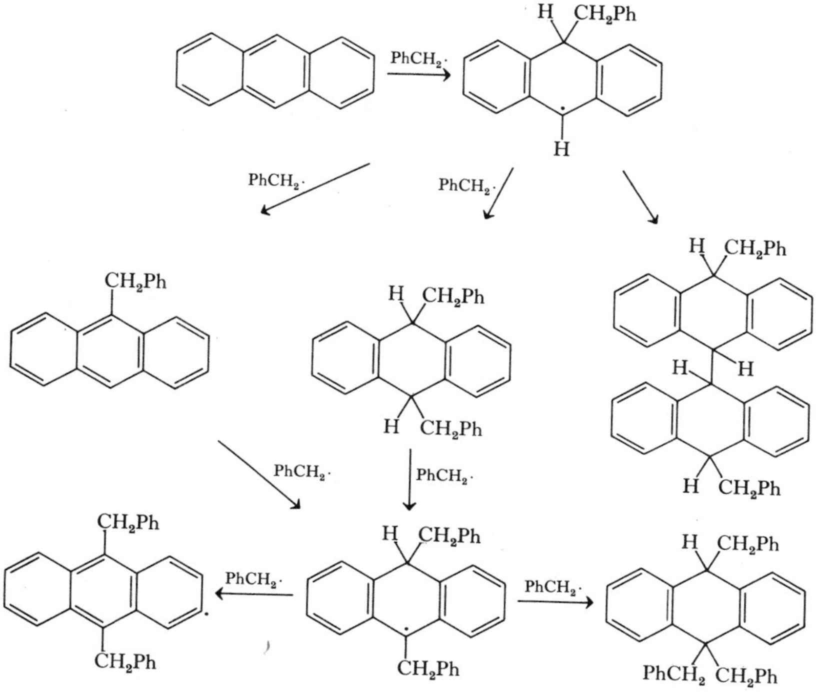

Scheme 12

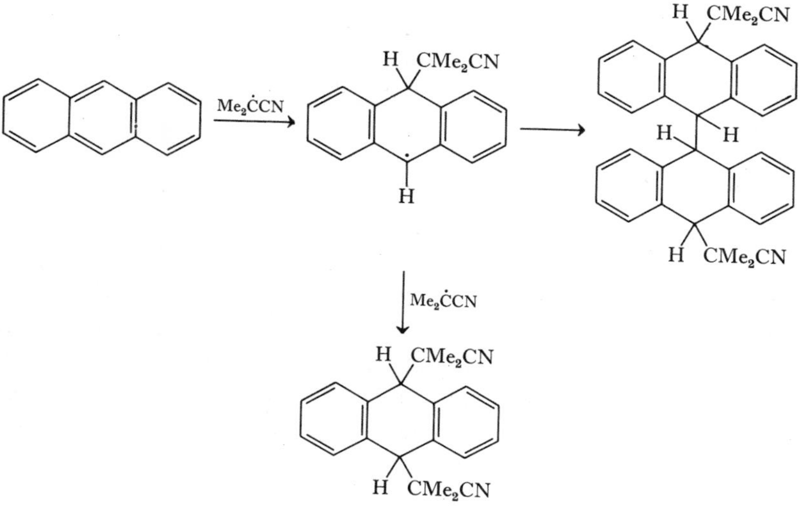

(*cis*- and *trans*-isomers)

Scheme 13

11.4.5 *Reactions of alkyl radicals with nitrogen heterocyclic compounds*

Alkylation of acridine could conceivably involve attack either on the nitrogen or the *meso*-carbon atom. Waters and Watson obtained 10-benzylacridine and 5,10-dibenzylacridan but no biacridans (99) (Scheme 14). This indicates that attack must occur at the *meso*-carbon

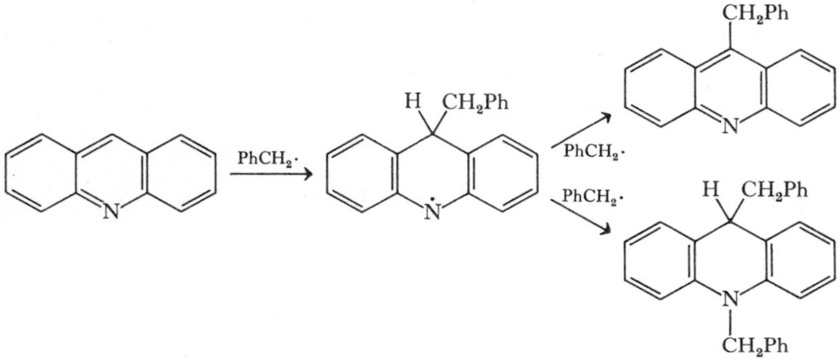

Scheme 14

atom. A similar conclusion, that attack occurs at carbon, could also be drawn from consideration of the methyl affinities for anthracene and acridine (Table 11-12). That for acridine is about half that for anthracene, indicating that attack of a methyl radical on nitrogen must be slow.

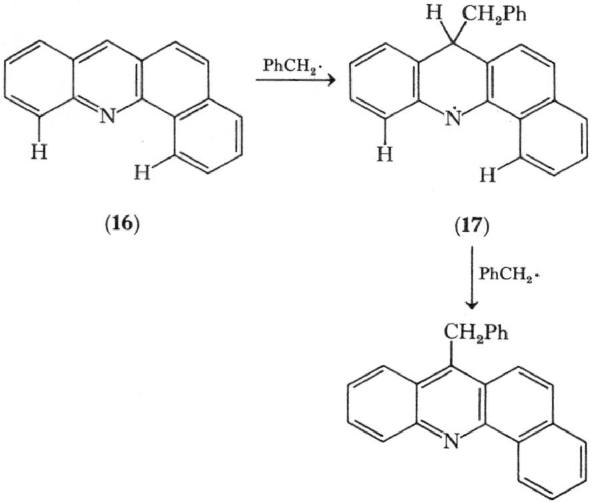

Scheme 15

Further support for initial attack occurring on the carbon comes from the benzylation of 3,4-benzacridine (**16**) and 1,2-benzacridine (**18**). The former gives a good yield of 9-benzyl-3,4-benzacridine (Scheme 15), whereas the latter is much less reactive and only 7.5 % of 9,10-dibenzyl-1,2-dibenzacridan is obtained (Scheme 16). The small amount of reac-

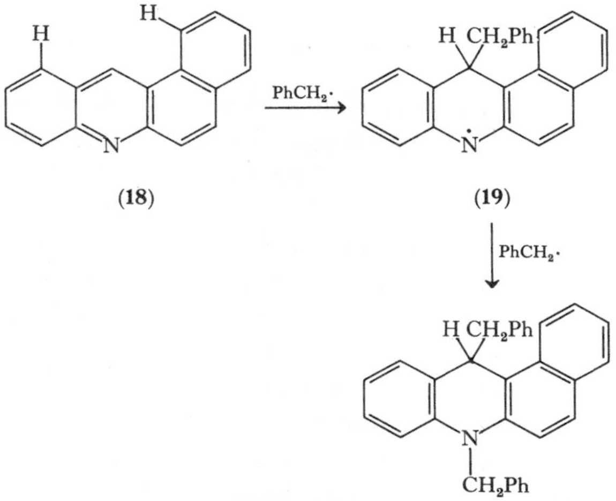

Scheme 16

tion of (**18**) is consistent with steric hindrance to attack at the *meso*-carbon atom. The radical (**19**), once formed, can then readily form the acridan by addition of a second benzyl radical. In contrast, the formation of the radical (**17**) from (**16**) is not subject to hindrance: it would then undergo hydrogen abstraction more readily than addition of a second benzyl radical to the hindered nitrogen, thus giving 9-benzyl-3,4-benzacridine.

The absence of any dimeric products, which would involve the formation of a N–N bond, is a noteworthy feature of all these reactions. The reason may be, at least in part, that such dimers might reasonably be expected to dissociate readily. Tetra-arylhydrazines are appreciably dissociated at room temperature (100).

Bass has shown that protonated heterocyclic compounds are more reactive towards homolytic aromatic substitution by alkyl radicals than the free bases (93 *a*, 101). Thus a yield of 35 % of 2- and 4-benzyl-pyridines was obtained from the benzylation of pyridine in acetic acid. This contrasts with the complete absence of any benzylated product in the reaction between pyridine and benzyl radicals in non-acidic media. The enhanced reactivity of the protonated compound may be attributed to contribution of polar forms in the transition state (20). Such effects

(20)

are obviously important only with nucleophilic radicals such as alkyl radicals. These results are in accord with predictions, using molecular orbital calculations of radical localization energies for the pyridinium ion (102).

Protonation had a similar effect on the reactivity of pyridine towards methyl (100) and phenyl radicals (103), and also on the reactivity of quinoline and isoquinoline towards benzyl and methyl radicals (100). Further, Bonnier and Court showed that the increase in reactivity of pyridine in acetic acid towards phenyl radicals is entirely attributable to the enhanced reactivities of positions 2 and 4, whilst that of position 3 was virtually unaltered. Much the same situation appears to prevail in the case of methylation of pyridine. In non-acidic media 3-methylpyridine accounts for over 20% of the reaction products, whereas none of this is obtained in acetic acid. The selective influence of the effect of protonation on reactivity towards radicals is also in agreement with atom localization energies (102).

A further aspect of alkylation of protonated heterocyclic compounds is the virtual absence of dimeric products, though they were produced, albeit in low yield, in the methylation of quinoline and isoquinoline in non-acidic media. This may possibly be explained by the reluctance of the radical cations to dimerize because of electronic repulsion. This is not entirely satisfactory as examples are known in which radical cations do dimerize.

11.4.6 *Molecular orbital treatment of reactivity*

In chapter 8 the reactivity of alkyl radicals, in their addition to olefins, was interpreted in terms of localization energy, which represents the best molecular orbital approach to the problem at present available. The localization method can be applied to situations where steric and inductive polar effects are minimal, and where conjugative stabilization effects between the radical and substrate are minor. In the alkylation of aromatics there is good reason to believe that the 'localized state' represents a good approximation to the transition state of the addition step. The intermediate alkylcyclohexadienyl radicals, in which one electron is essentially removed from the aromatic π-system and localized at the reaction site, have been detected spectroscopically. For simple alkyl radicals such as $CH_3\cdot$ and $CF_3\cdot$, the π-system of the aromatic is not appreciably extended in the transition state so that conjugative stabilization should be minor. Application of the method is limited only by the requirement that there be no steric or polar effects between radical and substrate, and by difficulties in computing localization energies for molecules containing heteroatoms.

For alkylation of unsubstituted mono- and polycyclic aromatic compounds polar effects between the radical and substrate should be minimal, and steric effects should be approximately constant from one compound to another. The transition state of this kind of reaction should approach most nearly to the ideal localized model, and a correlation between the activation energy of alkyl radical addition and the localization energy at the site of addition, can be expected. Coulson (104) was the first to point out a relationship between the methyl affinities of polycyclic aromatic hydrocarbons and the Hückel localization energies of the most reactive positions in the hydrocarbons. Szwarc and co-workers (96, 105) considerably extended the correlation and showed that trifluoromethyl radical addition reactions obeyed a similar relationship (106). Kooyman and Farenhorst (107) have measured the rates of addition of trichloromethyl radicals to polycyclic aromatic compounds, and these also are found to correlate with localization energies.

Dewar and Thompson (108) have calculated more reliable localization energies for various polycyclic aromatic hydrocarbons by the SCF–MO method, and have shown an even better correlation with the experimental rate data. In Fig. 11-1 the correlation of the relative rates of addition, per reactive centre, with atom localization energy, is shown for $CH_3\cdot$, $CF_3\cdot$ and $CCl_3\cdot$ radicals. The correlation is very good for all three radicals, so that the localization approximation comes close to the true situation, in the transition state, for this reaction.

The orientation of radical addition to polycyclic aromatic compounds is predicted reasonably well by localization theory. Naphthalene is the most thoroughly studied compound, phenylation of which occurs to the extent of 80% at the 1-position and 20% at the 2-position (109).

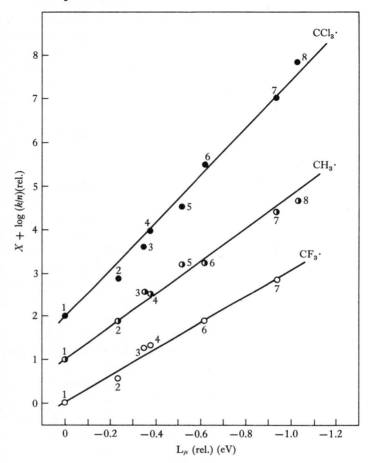

Fig. 11-1. Correlation of the relative rate of radical addition per reactive centre, log (k/n) (rel.) with the SCF–MO relative localization energies L_μ (rel.).

Experimental data for $CH_3\cdot$ and $CF_3\cdot$ radicals from Szwarc *et al.* (96, 105, 106) for $CCl_3\cdot$ from Kooyman and Farenhorst (107).

SCF–MO localization energies from Dewar and Thompson (108).

Polycyclic aromatics: 1 Benzene, 2 biphenyl, 3 phenanthrene, 4 naphthalene, 5 chrysene, 6 pyrene, 7 anthracene, 8 benzopyrene.

For $CF_3\cdot$ $X = 0$, CH_3 $X = 1.0$, CCl_3 $X = 2.0$.

Localization theory predicts that addition should occur preferentially at the 1-position. Partial rate factors, relative to benzene, have also been determined (109): 1-naphthalene 17, 2-naphthalene 4. The values calculated from localization energies are: 1-naphthalene 26, 2-naphthalene 2, in fair agreement. For the phenylation of anthracene, attack

458

is predicted to occur at the 9-position, in agreement with the experimental findings.

Some correlation of the experimental orientation and reactivity data with localization energy might also be expected for heterocyclic compounds. Polar effects will be greater than for the corresponding hydro-

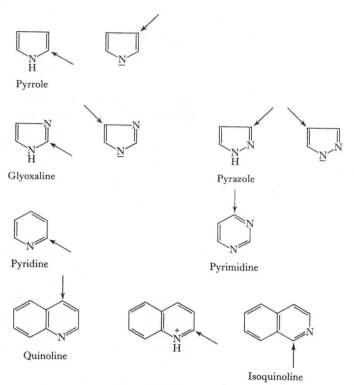

Fig. 11-2. Predictions, from localization energies, of the principal site for radical attack in nitrogen heteroaromatics (102, 110).

carbons, and also reliable localization energies are more difficult to compute because of the uncertainty as to the values of the Hückel coulomb and resonance parameters for heteroatoms. Brown and coworkers (102, 110) have derived localization energies for a number of nitrogen heterocyclic compounds, and the most reactive positions predicted for radical attack are shown in Fig. 11-2. The rates of attack of the phenyl radical at the three positions in pyridine were shown to correlate very well with localization energies which predict 2 > 4 > 3

(110). For pyrimidine, the orientation is predicted to be 4 > 2 > 5, which agrees with the results from the *p*-nitrophenylation of pyrimidine (122). The orientation of radical attack on quinoline is predicted to be 4 > 2 > 3, which agrees with the results of benzylation (101). For the quinolinium ion the orientation is predicted to be 2 > 4 > 3, which also agrees with the results of benzylation. For isoquinoline localization calculations suggest position 1 should be most reactive. Free-radical benzylation and bromination of isoquinoline occurs mainly at the 1-position (93 a).

For radical additions to substituted aromatic compounds both polar and steric effects play an important part (Table 11-2) in determining the substitution pattern. In these circumstances the localization energy approximation cannot be expected to hold so completely, nor can it be expected to explain the variation in substitution pattern found with substituted phenyl radicals (Table 11-4, 11-5 and 11-6), since the calculations assume that the rate of attack is independent of the nature of the attacking radical.

11.5 HYDROXYLATION

The radical hydroxylation of aromatic compounds, though not a reaction of synthetic utility, is of considerable importance from a biological point of view. A 'foreign' aromatic compound is usually hydroxylated when administered to animals. Thus, benzoic acid gives a mixture of *ortho*-, *meta*-, and *para*-hydroxybenzoic acids. The formation of all three isomers led to the belief that hydroxylation proceeded by a radical process. Consequently, it is of considerable interest to study the *in vitro* hydroxylation of aromatic compounds, and to compare the products with those from *in vivo* hydroxylations.

11.5.1 *Hydroxylations with Fenton's reagent*

Hydroxyl radicals are conveniently produced by the reduction of hydrogen peroxide by single-electron reductants, particularly iron(II) and titanium(III) (Fenton's reaction):

$$H_2O_2 + Fe^{2+} \longrightarrow Fe^{3+} + OH^- + HO \cdot \qquad (95)$$

Hydroxyl radicals thus produced add to benzene to give hydroxycyclo-hexadienyl radicals, which have been identified from their e.s.r. spectrum (112, 113) (cf. p. 359):

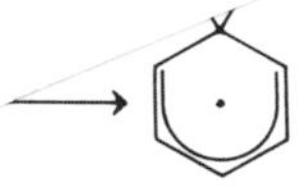

$$\longrightarrow \qquad (96)$$

The hydr_____ical has unmistakably electrophilic characteristics, as shown by the relative rates of reactivity and isomer distributions obtained from a series of monosubstituted benzenes (Table 11-13). Electron-withdrawing groups decrease the rate of reaction, and at the same time increase the amount of the *meta*-hydroxylated product. The reverse is true in the case of benzenes substituted with electron-donating groups.

TABLE 11-13 *Hydroxylation of aromatic compounds* (113, 114)

Aromatic compound PhX	Relative reactivity $\frac{PhX}{PhH}K$	Isomer distribution (%)		
		o	*m*	*p*
PhOMe	6.35	85	4	11
PhMe	—	55	15	30
PhF	—	37	18	45
PhCl	0.55	42	29	29
PhNO$_2$	0.14	24	30	46

Quantitative studies on the rate of hydroxylation of a large group of aromatic compounds have been successfully correlated with Hammett σ-values. The value of ρ is -0.41, indicative of attack by an electrophilic species (116).

Recent studies by Norman and his group have shown that the hydroxylation is much more complex than appeared at first. This is due in part to acid-catalysed fragmentations of the intermediate cyclohexadienyl radicals (e.g. reaction 97) (cf. chapter 12, p. 487). Such reactions

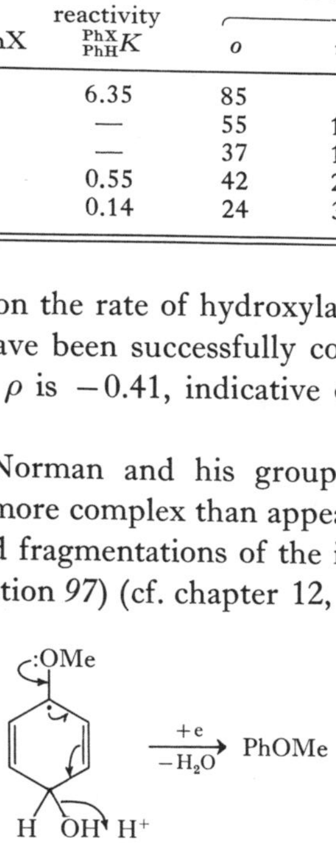

$$\xrightarrow[-H_2O]{+e} \text{PhOMe} \qquad (97)$$

decrease considerably the yield of hydroxylated products. It is also evident (cf. Table 11-14) that the yield can be increased by carrying out

Homolytic substitution

TABLE 11-14   *Effect of additives on the hydroxylation of anisole with Fenton's reagent at pH 2.5 (115)*

		Isomer distribution (%)		
Additive	Yield	*o*	*m*	*p*
—	8.4	81	0.5	18
F^-	4.6	71	9	20
F^-/O_2	11.0	57	17	26
F^-/Cu^{2+}	37.8	80	3	17

the reaction in the presence of oxygen or by adding copper(II) salts, whilst it is reduced by addition of fluoride, which complexes with iron (III) (115). The complexed iron(III) is an insufficiently powerful oxidant for the oxidation of hydroxycyclohexadienyl radicals:

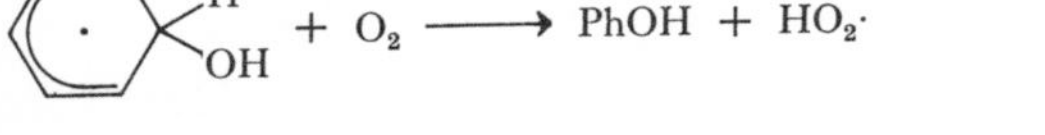

$$+ O_2 \longrightarrow PhOH + HO_2 \cdot \qquad (98)$$

$$+ Cu^{2+} \longrightarrow PhOH + Cu^+ + H^+ \qquad (99)$$

The function of these oxidants is to divert the cyclohexadienyl radicals to phenolic products by preventing their dimerization, which results in the formation of biphenyl (117):

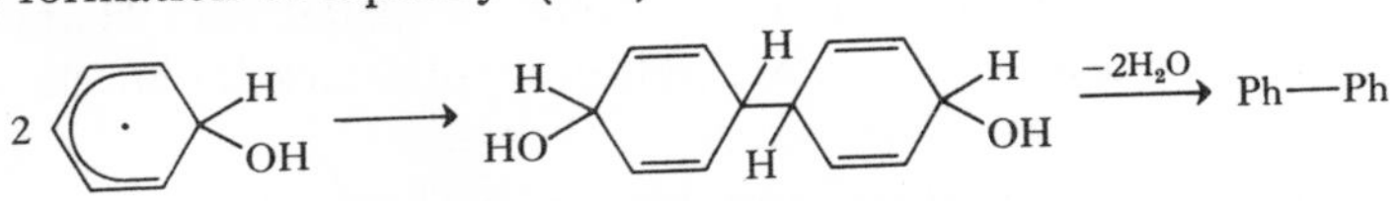

$$\xrightarrow{-2H_2O} Ph\text{---}Ph \qquad (100)$$

These oxidants also have a significant effect on the isomer distribution of phenolic products. The yield of *meta*-methoxyphenol from anisole is greatest when oxygen is present in the system. The oxidation process in this case has relatively little polar character. This is not so with metal oxidants and consequently the proportion of the *meta*-isomer decreases since the σ-radical leading to this would have a higher oxidation potential than that leading to *ortho*- or *para*-methoxyanisoles (cf. p. 431).

11.6 *AROMATIC SUBSTITUTIONS INVOLVING THE REPLACEMENT OF ATOMS OTHER THAN HYDROGEN*

All the examples which have been discussed thus far have involved reactions in which a hydrogen atom has been substituted by the incoming radical. Though this is by far the most common type of homolytic aromatic substitution, there are a number of examples scattered through the literature in which a halogen and even more exceptionally a nitro or methoxy group is replaced:

$$ArX + Ph\cdot \longrightarrow ArPh + [X\cdot] \qquad (101)$$

This type of reaction is well exemplified in arylation of hexafluorobenzene, in which any biaryl formed must involve substitution of fluorine. Williams has studied the decomposition of diaroyl peroxides in hexafluorobenzene (118). Such reactions must involve the formation of σ-complexes (**21**) which, as in previous cases, can disproportionate, dimerize or undergo oxidation (reactions *102–104*):

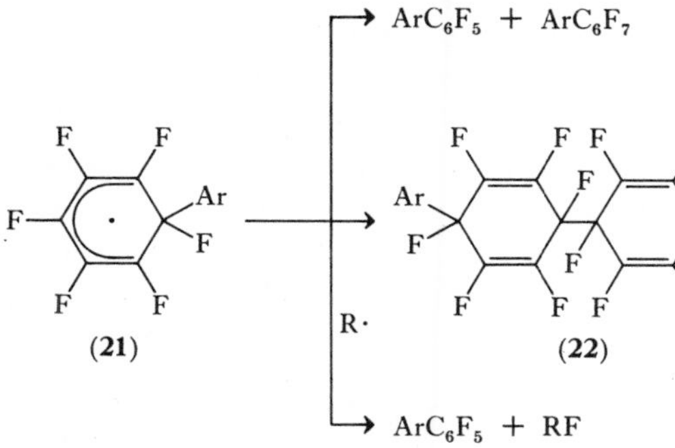

$$\longrightarrow ArC_6F_5 + ArC_6F_7 \qquad (102)$$

$$(103)$$

$$\longrightarrow ArC_6F_5 + RF \qquad (104)$$

An alternative possibility is that the radical may fragment (reaction *105*):

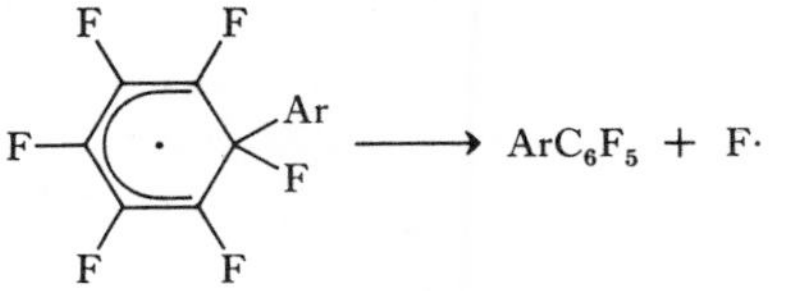

$$\longrightarrow ArC_6F_5 + F\cdot \qquad (105)$$

A significant feature of these reactions is that the yield of biaryl is considerably greater when the total reaction product is distilled than when the products are analysed directly. This is consistent with decomposition of the dimeric product (**22**) during distillation. These reactions also result in the formation of a small amount of the appropriate fluorobenzene derivative as a result of abstraction of fluorine by aryl radicals (reaction 104; R = Ar).

The phenylation of bromo- and chloro-pentafluorobenzene gave mixtures of products containing no pentafluorobiphenyl, indicative of preferential substitution of fluorine rather than chlorine or bromine (119). The same picture emerged from a study of the cyclohexylation of dihalobenzenes, from which it was found that the ease of replacement of halogens was in the order: F > I > Br > Cl (120). This establishes unequivocally that the C–X bond is not involved in the rate-determining step. The stronger electron-withdrawing effect of fluorine accounts for initial attack at that site in the case of the nucleophilic cyclohexyl radicals, but the subsequent removal of fluorine still requires considerable energy. The ease of halogen displacement is greatest in *ortho*-dihalobenzenes, probably as a result of the presence of electron-withdrawing groups on adjacent atoms facilitating attack, and also because of the relief of steric and electrostatic interaction on formation of the σ-radical. The results from the cyclohexylation of *o*-dihalobenzenes also show that replacement of halogen, particularly fluorine, can occur more readily than substitution of hydrogen. A further example of replacement of halogen is seen in the formation of both 9-chloro-10-phenylanthracene and 9-phenylanthracene in the phenylation of 9-chloroanthracene (121).

$$(106)$$

We have deliberately excluded from our discussion of homolytic reactions involving replacement of halogen the Ullmann reaction, as it is very dubious whether this reaction proceeds by a radical mechanism. The consensus of opinion is that it proceeds through an arylcopper, reaction of which with the iodoaromatic compound, gives the biaryl (122):

$$\text{ArI} + 2\text{Cu} \longrightarrow \text{ArCu} + \text{CuI} \qquad (107)$$

$$\text{ArCu} + \text{ArI} \longrightarrow \text{Ar—Ar} + \text{CuI} \qquad (108)$$

A nitro group may also on occasion be displaced, as evidenced by the formation of some pentafluorobiphenyl from the reaction of dibenzoyl peroxide with pentafluoronitrobenzene (119). This may well arise by fragmentation of the radical (23):

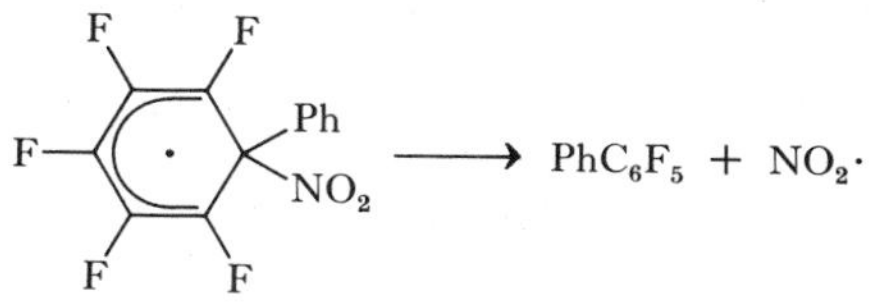

$$\longrightarrow PhC_6F_5 + NO_2 \cdot \qquad (109)$$

(23)

The probability of this type of reaction occurring is greatest with electrophilic radicals which are most likely to form stable charge-transfer complexes.

The only recorded example of replacement of a methoxy group in homolytic aromatic arylation is the replacement of the 2-methoxy group in 1,2,3-trimethoxybenzene (reaction *110*) (88).

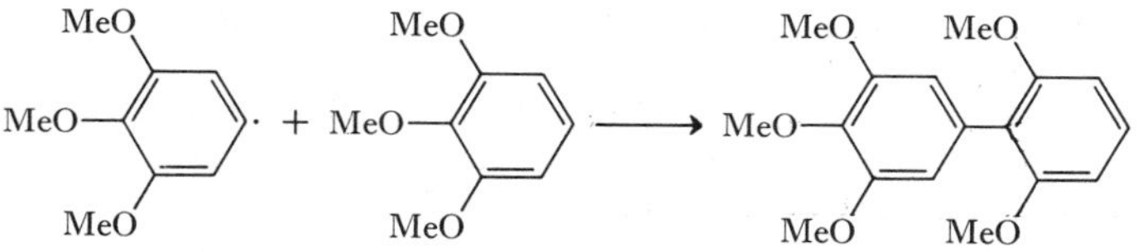

$$(110)$$

GENERAL REFERENCES

G. H. Williams, *Homolytic Aromatic Substitution*, Pergamon Press, Oxford, 1960.

G. H. Williams, *Essays in Free-Radical Chemistry*, Chem. Soc. Special Publ. no. 24, 1970, 25.

D. H. Hey, *Adv. Free-Radical Chem.* 1967, 2, 47.

R. O. C. Norman and G. K. Radda, *Advances in Heterocyclic Chemistry*, 1963, 2, 131.

K. C. Bass and P. Nababsing, *Adv. Free-Radical Chem.* 1971, 4, 1.

SPECIFIC REFERENCES

1. W. S. M. Grieve and D. H. Hey, *J. Chem. Soc.* 1934, 1797; D. H. Hey, *J. Chem. Soc.* 1934, 1966.
2. D. H. Hey and W. A. Waters, *Chem. Rev.* 1937, **21**, 169.
3. C. Shih, D. H. Hey and G. H. Williams, *J. Chem. Soc.* 1958, 2600.
4. D. R. Augood, D. H. Hey and G. H. Williams, *J. Chem. Soc.* 1953, 45.

5. D. Bryce-Smith and P. Clarke, *J. Chem. Soc.* 1956, 2264.
6. R. Huisgen and R. Crashay, *Annalen*, 1957, **607**, 46.
7. R. L. Hardie and R. H. Thomson, *J. Chem. Soc.* 1957, 2512.
8. R. L. Hardie and R. H. Thomson, *J. Chem. Soc.* 1958, 1286.
9. D. H. Hey, S. Orman and G. H. Williams, *J. Chem. Soc.* 1961, 565.
10. R. A. McClelland, R. O. C. Norman and C. B. Thomas, *J. Chem. Soc. Perkin* I, 1972, 570.
11. D. H. Hey, C. J. M. Stirling and G. H. Williams, *J. Chem. Soc.* 1956, 1475.
12. R. A. Abramovitch and O. A. Koleoso, *J. Chem. Soc.* B, 1968, 1292.
13. M. Kobayashi, H. Minato, N. Kobori and E. Yamada, *Bull. Chem. Soc. Japan*, 1970, **43**, 1131.
14. M. Kobayashi, H. Minato, N. Watanabe and N. Kobori, *Bull. Chem. Soc. Japan*, 1970, **43**, 258.
15. D. H. Hey, C. J. M. Stirling and G. H. Williams, *J. Chem. Soc.* 1955, 3963.
16. P. J. Bunyan and D. H. Hey, *J. Chem. Soc.* 1960, 3787.
17. D. H. Hey, C. J. M. Stirling and G. H. Williams, *J. Chem. Soc.* 1954, 2747.
18. O. C. Dermer and M. T. Edmison, *Chem. Rev.* 1957, **57**, 77.
19. M. Eberhardt and E. L. Eliel, *J. Org. Chem.* 1962, **27**, 2289.
20. D. H. Hey, M. J. Perkins and G. H. Williams, *Chem. and Ind.* 1963, 83.
21. G. H. Williams, *Essays in Free-Radical Chemistry*, Chem. Soc. *Special Publ.* no. 24, 1970, 25.
22. W. E. Bachmann and R. A. Hoffman, *Org. Reactions*, 1944, **2**, 244.
23. E. C. Butterworth and D. H. Hey, *J. Chem. Soc.* 1938, 116.
24. R. Huisgen and G. Horeld, *Annalen*, 1949, **562**, 137.
25. C. Rüchardt and B. Freudenberg, *Tetrahedron Letters*, 1964, 3623.
26. J. I. G. Cadogan, R. M. Paton and C. Thomson, *Chem. Comm.* 1969, 614; 1970, 229.
27. C. Rüchardt and E. Merz, *Tetrahedron Letters*, 1964, 2431.
28. J. I. G. Cadogan, *J. Chem. Soc.* 1962, 4257; J. I. G. Cadogan, D. A. Roy and D. M. Smith, *J. Chem. Soc.* 1966, 1249.
29. C. S. Rondestvedt and H. S. Blanchard, *J. Amer. Chem. Soc.* 1955, **77**, 1769.
30. J. Elks and D. H. Hey, *J. Chem. Soc.* 1943, 441.
31. J. I. G. Cadogan, *Essays in Free-Radical Chemistry*, Chem. Soc. *Special Publ.* no. 24, 1970, 71.
32. R. O. C. Norman and W. A. Waters, *J. Chem. Soc.* 1958, 167.
33. J. I. G. Cadogan, *Pure Appl. Chem.* 1967, **15**, 153.
34. S. C. Dickerman and K. Weiss, *J. Org. Chem.* 1957, **22**, 1070.
34a. A. L. J. Beckwith and R. J. Leydon, *Tetrahedron*, 1964, **20**, 791.
35. D. H. Hey, M. J. Perkins and G. H. Williams, *J. Chem. Soc.* 1965, 110.
35a. W. A. Pryor and K. Smith, *J. Amer. Chem. Soc.* 1967, **89**, 1741.
36. F. F. Gadallah and R. M. Elofson, *J. Org. Chem.* 1969, **34**, 3335.
37. J. M. Blair, D. Bryce-Smith and B. W. Pengilly, *J. Chem. Soc.* 1959, 3174.
38. R. K. Sharma and N. Kharasch, *Angew Chem. Internat. Edn.* 1968, **7**, 36.
39. W. Wolf and N. Kharasch, *J. Org. Chem.* 1965, **30**, 2493.

40. J. E. Leffler, W. J. M. Mitchell and B. C. Menon, *J. Org. Chem.* 1966, **31**, 1153.
41. D. I. Davies, D. H. Hey and M. Tiecco, *J. Chem. Soc.* 1965, 7062; D. I. Davies, J. N. Done and D. H. Hey, *J. Chem. Soc.* C, 1969, 1392.
42. D. F. DeTar and R. A. J. Long, *J. Amer. Chem. Soc.* 1958, **80**, 4742.
43. R. J. Convery and C. C. Price, *J. Amer. Chem. Soc.* 1958, **80**, 4101.
44. E. L. Eliel, S. Meyerson, Z. Welvart and S. H. Wilen, *J. Amer. Chem. Soc.* 1960, **82**, 2936.
45. J. Saltiel and H. C. Curtis, *J. Amer. Chem. Soc.* 1971, **93**, 2056.
46. D. J. Atkinson, M. J. Perkins and P. Ward, *J. Chem. Soc.* C, 1971, 3240.
47. M. Kobayashi, H. Minato and N. Kobori, *Bull. Chem. Soc. Japan,* 1969, **42**, 2738.
48. D. I. Davies, D. H. Hey and B. Summers, *J. Chem. Soc.* C, 1971, 2683.
49. C. S. Rondestvedt and H. S. Blanchard, *J. Org. Chem.* 1956, **21**, 229.
50. D. H. Hey, M. J. Perkins and G. H. Williams, *J. Chem. Soc.* 1963, 5604.
51. D. H. Hey, M. J. Perkins and G. H. Williams, *J. Chem. Soc.* 1964, 3412.
52. G. B. Gill and G. H. Williams, *J. Chem. Soc.* 1965, 995.
53. D. H. Hey, K. S. Y. Liang, M. J. Perkins and G. H. Williams, *J. Chem. Soc.* C, 1967, 1153.
54. D. H. Hey and M. J. Perkins, *Org. Synth.* 1969, **49**, 44.
55. D. H. Hey, K. S. Y. Liang and M. J. Perkins, *Tetrahedron Letters*, 1967, 1477.
56. G. R. Chalfont, D. H. Hey, K. S. Y. Liang and M. J. Perkins, *J. Chem. Soc.* B, 1971, 233.
57. G. R. Chalfont and M. J. Perkins, *J. Chem. Soc.* B, 1971, 245.
58. J. K. Kochi, *J. Amer. Chem. Soc.* 1963, **85**, 1958.
59. J. K. Kochi and H. E. Mains, *J. Org. Chem.* 1965, **30**, 1862.
60. J. K. Kochi, *Science*, 1967, **155**, 415.
61. J. I. G. Cadogan, P. Hibbert, M. N. U. Siddiqui and D. M. Smith, *J. Chem. Soc. Perkin* I, 1972, 2555.
62. R. A. McClelland, R. O. C. Norman and C. B. Thomas, *J. Chem. Soc. Perkin* I, 1972,562.
63. K. Tokumaru, K. Horie and O. Simamura, *Tetrahedron*, 1965, **21**, 867.
64. R. T. Morrison, J. Cazes, N. Samkoff and C. A. Howe, *J. Amer. Chem. Soc.* 1962, **84**, 4152.
65. C. M. Camaggi, R. Leardini, M. Tiecco and A. Tundo, *J. Chem. Soc.* B, 1969, 1251; L. Benati, N. La Barba, M. Tiecco and A. Tundo, *J. Chem. Soc.* B, 1969, 1253.
66. J. I. G. Cadogan, *Essays in Free-Radical Chemistry*, Chem. Soc. Special *Publ.* no. 24, 1970, 71.
67. R. Ito, T. Migita, N. Morikawa and O. Simamura, *Tetrahedron*, 1965, **21**, 955.
68. P. H. Oldham, G. H. Williams and B. A. Wilson, *J. Chem. Soc.* C, 1971, 1094.
69. D. H. Hey, *Adv. Free-Radical Chem.* 1967, **2**, 47.
70. D. H. Hey, B. W. Pengilly and G. H. Williams, *J. Chem. Soc.* 1956, 1463.
71. J. K. Hambling, D. H. Hey and G. H. Williams, *J. Chem. Soc.* 1962, 487.
72. D. I. Davies, D. H. Hey and B. Summers, *J. Chem. Soc.* C, 1970, 2653.

73. H. S. Blanchard and C. S. Rondestvedt, *J. Amer. Chem. Soc.* 1955, **77**, 1769.
74. A. Morgan, D. C. Nonhebel and P. L. Pauson, unpublished results.
75. R. A. Abramovitch, *Adv. Free-Radical Chem.* 1967, **2**, 87.
76. D. F. DeTar, *Org. Reactions*, 1957, **9**, 409.
77. A. Lewin and T. Cohen, *J. Org. Chem.* 1967, **32**, 3844.
78. B. Chauncy and E. Gellert, *Austral. J. Chem.* 1969, **22**, 993.
79. S. M. Kupchan and H. C. Wormser, *J. Org. Chem.* 1965, **30**, 3792.
80. D. H. Hey, C. W. Rees and A. R. Todd, *J. Chem. Soc.* C, 1967, 1101.
81. D. H. Hey, G. H. Jones and M. J. Perkins, *Chem. Comm.* 1970, 1438.
82. P. Ruggli and A. Staub, *Helv. Chim. Acta*, 1936, **19**, 1288; 1937, **20**, 37.
83. D. H. Hey and R. D. Mulley, *J. Chem. Soc.* 1952, 2276.
84. T. Nakata, K. Tokumaru and O. Simamura, *Tetrahedron Letters*, 1967, 3303.
85. D. I. Davies, D. H. Hey and G. H. Williams, *J. Chem. Soc.* 1958, 1878.
86. I. M. Roitt and W. A. Waters, *J. Chem. Soc.* 1952, 2695.
87. B. M. Lynch and R. B. Moore, *Canad. J. Chem.* 1962, **40**, 1461.
88. P. L. Pauson and B. C. Smith, *J. Org. Chem.* 1953, **18**, 1403.
88a. J. I. G. Cadogan, J. R. Mitchell and J. T. Sharp, *Chem. Comm.* 1971, 1433.
89. L. Herk, M. Feld and M. Szwarc, *J. Amer. Chem. Soc.* 1955, **77**, 1949.
90. J. R. Shelton and C. W. Uzelmeier, *J. Amer. Chem. Soc.* 1966, **88**, 5222.
91. G. Martelli, P. Spagnolo and M. Tiecco, *J. Chem. Soc.* B, 1970, 1413.
92. G. E. Corbett and G. H. Williams, *J. Chem. Soc.* 1964, 3477; *J. Chem. Soc.* B, 1966, 877.
93. D. H. Hey, D. A. Shingleton and G. H. Williams, *J. Chem. Soc.* 1963, 1958.
93a. K. C. Bass and P. Nababsing, *J. Chem. Soc.* C, 1969, 388.
93b. I. M. Whittemore, A. P. Stefani and M. Szwarc, *J. Amer. Chem. Soc.* 1962, **84**, 3799.
93c. S. W. Charles, J. T. Pearson and E. Whittle, *Trans. Faraday Soc.* 1963, **59**, 1156.
94. A. L. J. Beckwith and W. A. Waters, *J. Chem. Soc.* 1957, 1665.
95. B. R. Cowley, R. O. C. Norman and W. A. Waters, *J. Chem. Soc.* 1959, 1799.
96. M. Levy and M. Szwarc, *J. Amer. Chem. Soc.* 1955, **79**, 1949; W. J. Heilman, A. Rembaum and M. Szwarc, *J. Chem. Soc.* 1957, 1127.
97. A. L. J. Beckwith and W. A. Waters, *J. Chem. Soc.* 1957, 1001.
98. A. F. Bickel and E. C. Kooyman, *Rec. Trav. chim.* 1952, 71, 1137.
99. W. A. Waters and D. H. Watson, *J. Chem. Soc.* 1957, 253; 1959, 2082, 2085.
100. A. R. Forrester, J. M. Hay and R. H. Thomson, *Organic Chemistry of Stable Free Radicals*, Academic Press, London, 1968, Chapter 3.
101. K. C. Bass and P. Nababsing, *J. Chem. Soc.* C, 1970, 2169.
102. R. D. Brown and M. C. Hefferman, *Austral. J. Chem.* 1956, **9**, 83.
103. J. M. Bonnier and J. Court, *Compt. rend.* 1967, **265**, C, 133.
104. C. A. Coulson, *J. Chem. Soc.* 1955, 1435.
105. J. H. Binks and M. Szwarc, *J. Chem. Phys.* 1959, **30**, 1494.

106. A. P. Stefani and M. Szwarc, *J. Amer. Chem. Soc.* 1962, **84**, 3661.
107. E. C. Kooyman and E. Farenhorst, *Trans. Faraday Soc.* 1953, **49**, 58.
108. M. J. S. Dewar and C. C. Thompson, *J. Amer. Chem. Soc.* 1965, **87**, 4414.
109. R. Huisgen, F. Jacob and R. Grashey, *Chem. Ber.* 1959, 2206; D. I. Davies, D. H. Hey and G. H. Williams, *J. Chem. Soc.* 1958, 1878; B. A. Marshall and W. A. Waters, *ibid.* 1959, 381.
110. R. D. Brown, *Austral. J. Chem.* 1955, **8**, 100; R. D. Brown and R. D. Harcourt, *Tetrahedron*, 1960, **8**, 23.
111. B. Lythgoe and L. S. Rayner, *J. Chem. Soc.* 1951, 2323.
112. W. T. Dixon and R. O. C. Norman, *J. Chem. Soc.* 1964, 4857.
113. C. R. E. Jefcoate and R. O. C. Norman, *J. Chem. Soc.* B, 1968, 48.
114. R. O. C. Norman and G. K. Radda, *Proc. Chem. Soc.* 1962, 138.
115. C. R. E. Jefcoate, J. R. Lindsay Smith and R. O. C. Norman, *J. Chem. Soc.* B, 1969, 1013.
116. M. Anbar, D. Meyerstein and P. Neta, *J. Phys. Chem.* 1966, **70**, 2660.
117. J. R. Lindsay Smith and R. O. C. Norman, *J. Chem. Soc.* 1963, 2897.
118. P. A. Claret, G. H. Williams and J. Coulson, *J. Chem. Soc.* C, 1968, 341.
119. P. H. Oldham, G. H. Williams and B. A. Wilson, *J. Chem. Soc.* B, 1970, 1346.
120. J. R. Shelton and C. W. Uzelmeier, *Rec. Trav. chim.* 1968, **87**, 1211.
121. F. M. Brown and D. C. Nonhebel, unpublished results.
122. M. Nilsson and O. Wennerstrom, *Acta Chem. Scand.* 1970, **24**, 482; M. Nilsson and C. Ullenius, *Acta Chem. Scand.* 1970, **24**, 2379.
123. P. H. Oldham and G. H. Williams, *J. Chem. Soc.* C, 1970, 1260.

Radical fragmentations

12.1 *INTRODUCTION*

Radicals have been shown to undergo both homolytic and heterolytic fragmentation (reactions *1* and *2*):

$$R_3CO\cdot \longrightarrow R_2CO + R\cdot \qquad (1)$$

$$H^+ \; HO\!-\!CH_2\!-\!\dot{C}H\!-\!O\!-\!H \longrightarrow H_2O + CH_2CHO + H^+ \qquad (2)$$

Both of these classes of fragmentation will be discussed in this chapter.

Many, but by no means all, fragmentations could equally well be classified as eliminations (e.g. reactions *1* and *2*). This is not true for the fragmentation of t-alkoxy radicals in which the tertiary carbon atom is part of a cyclic system for which fragmentation leads to ring-opening rather than elimination (reaction *3*):

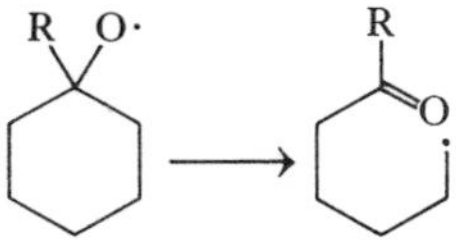

$$(3)$$

This process is distinguished from a rearrangement reaction in that there is no rearrangement of the carbon skeleton. It is possibly best classified as an intramolecular fragmentation.

12.2 *HOMOLYTIC FRAGMENTATION REACTIONS*

12.2.1 *General considerations*

The process of homolytic fragmentation is essentially the reverse of radical addition. In certain instances involving additions of weakly reactive species, such as atomic bromine or thiyl radicals, the reaction is a reversible process:

$$\text{C=C} + Br\cdot \rightleftharpoons -\overset{|}{\underset{Br}{C}}-\dot{C} \qquad (4)$$

$$\text{C=C} + RS\cdot \rightleftharpoons -\overset{|}{\underset{SR}{C}}-\dot{C} \qquad (5)$$

Equilibrium is in fact rarely obtained due to the intrusion of other reactions. However, when it is attainable it is found that temperature has a considerable effect on the position of equilibrium, largely because of entropy factors. The forward reaction, i.e. addition, results in a significant decrease in entropy and is thus favoured by low temperatures, whereas the fragmentation process entails an increase in entropy and will thus be favoured at high temperature. The fragmentation of cyclic radicals, unlike that of acyclic radicals, does not result in additional translational degrees of freedom but it does entail the creation of extra rotational degrees of freedom. Consequently, here again fragmentation should be favoured at higher temperatures.

A second general point merits discussion, namely the direction of fragmentation of differently substituted radicals. In such instances, the process proceeds so as to lead to the most stable radical (cf. reaction 6) (1):

$$\underset{\underset{Et}{|}}{\overset{\overset{Me}{|}}{Pr^i-C-OCl}} \xrightarrow{80°} \underset{\underset{Et}{|}}{\overset{\overset{Me}{|}}{Pr^i-C-O\cdot}} \xrightarrow{CCl_4} \begin{cases} \xrightarrow{(95\%)} MeCOEt + Pr^iCl \\ \xrightarrow{(3\%)} MeCOPr^i + EtCl \\ \xrightarrow{(>0.5\%)} EtCOPr^i + MeCl \end{cases} \qquad (6)$$

This aspect is dealt with in greater depth in chapter 5 in discussions on the relative stabilities of radicals. As was mentioned there, cognizance should be taken of any polar characteristics in the transition state of the fragmentation reaction. Such factors tend to accentuate the specificity of the favoured fragmentation pathway.

Discussion in this section will centre on what are essentially reversible additions, i.e. fragmentation of β-haloalkyl and β-alkylthioalkyl radicals, and also on the fragmentation of radicals not generally produced by addition reactions, e.g. the fragmentation of t-alkoxy and β-alkoxyalkyl radicals.

Radical fragmentation

12.2.2 Fragmentation of β-haloalkyl radicals

In the first instance we shall discuss the reversible addition of halogen atoms to alkenes. This is manifested in the ability of low concentrations of bromine and iodine to bring about the isomerization of alkenes in the presence of ultraviolet light (2):

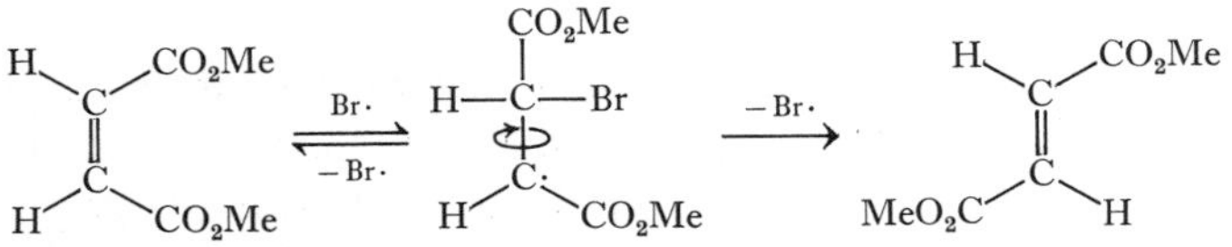

$$\tag{7}$$

Larger concentrations of halogen result in the formation of the normal addition product.

Another example of this reaction is the isomerization of cis-1,2-dibromoethylene, catalysed by radioactive bromine. The resultant trans-1,2-dibromoethylene contains radioactive bromine, as does the cis-isomer, thus providing strong support for the proposed mechanism (3) (Scheme 1).

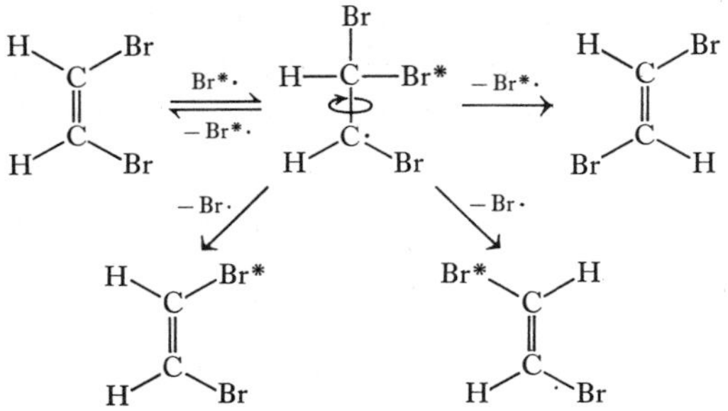

Scheme 1

The competition between elimination of a bromine atom from a β-haloalkyl radical and its reaction with molecular halogen to afford the dihalo-compound has been alluded to. This competition is of particular relevance to allylic halogenation of alkenes. Thus it has been shown that chlorination of propene at temperatures below 200 °C results in the formation of 1,2-dichloropropane, whereas at higher temperatures the major product is allyl chloride (4) (cf. chapter 7, p. 164).

The reversible nature of the addition process is more readily seen in the addition of bromine atoms. This process becomes increasingly reversible as the temperature is raised and as the concentration of bromine is decreased. Thus, bromination of cyclohexene using low concentrations of bromine at 80 °C leads to excellent yields of 3-bromo-cyclohexene (5). Similarly, ethyl crotonate formed ethyl γ-bromo-crotonate (5):

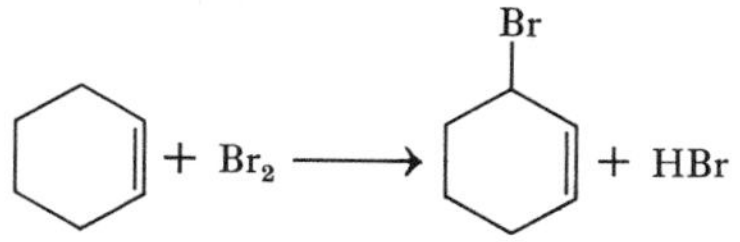

$$\text{(cyclohexene)} + Br_2 \longrightarrow \text{(3-bromocyclohexene)} + HBr \qquad (8)$$

$$CH_3CH{=}CHCO_2Et + Br_2 \longrightarrow CH_2BrCH{=}CHCO_2Et + HBr \qquad (9)$$

If the addition of bromine were to be carried out more rapidly, the predominant product in both instances would be the dibromo compound.

The reversibility of bromine addition to alkenes coupled with the irreversibility of hydrogen abstraction from allylic carbon atoms is exploited in brominations employing *N*-bromosuccinimide as the brominating agent (6): (cf. p. 190)

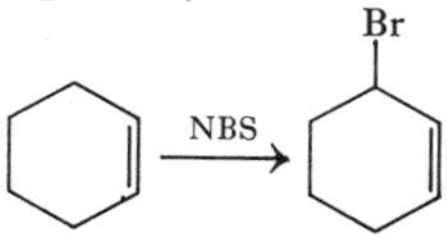

$$\text{(cyclohexene)} \xrightarrow{NBS} \text{(3-bromocyclohexene)} \qquad (10)$$

β-Haloalkyl radicals can also be generated by radical addition to haloalkenes:

$$R\cdot + CH_2{=}CHCH_2Br \longrightarrow RCH_2\dot{C}HCH_2Br \qquad (11)$$

$$R\cdot + CHCl{=}CHCl \longrightarrow RCHCl\dot{C}HCl \qquad (12)$$

The radicals thus generated can similarly undergo fragmentation to alkenes. Thus reaction of allyl bromide with excess of bromotrichloro-methane gave 3-bromo-1,1,1,5,5,5-hexachloropentane (1) together with 1,2,3-tribromopropane (2): none of the expected addition product (3) was formed:

$$CH_2{=}CHCH_2Br + CCl_3Br \nearrow \quad Cl_3CCH_2CHBrCH_2CCl_3 + CH_2BrCHBrCH_2Br$$
$$\qquad\qquad (1) \qquad\qquad\qquad (2)$$
$$\qquad\qquad\qquad\qquad\qquad\qquad (13)$$
$$\searrow\!\!\!\times \quad Cl_3CCH_2CHBrCH_2Br$$
$$\qquad\qquad (3)$$

The formation of (1) can be explained by the reaction sequence outlined in reactions (14–17):

$$\dot{C}Cl_3 + CH_2{=}CHCH_2Br \longrightarrow Cl_3CCH_2\dot{C}HCH_2Br \qquad (14)$$

$$Cl_3CCH_2\dot{C}HCH_2Br \longrightarrow Cl_3CCH_2CH{=}CH_2 + Br\cdot \qquad (15)$$
$$\mathbf{(4)} \qquad\qquad\qquad \mathbf{(5)}$$

$$Cl_3CCH_2CH{=}CH_2 + \dot{C}Cl_3 \longrightarrow Cl_3CCH_2\dot{C}HCH_2CCl_3 \qquad (16)$$

$$Cl_3CCH_2\dot{C}HCH_2CCl_3 + CCl_3Br \longrightarrow Cl_3CCH_2CHBrCH_2CCl_3 + \dot{C}Cl_3$$
$$(17)$$

Confirmation of this scheme comes from the isolation of 4,4,4-trichloro-but-1-ene **(5)** from the reaction when it is carried out using excess of allyl bromide. In this case 1,2,3-tribromopropane is again formed (reactions *18* and *19*):

$$CH_2{=}CHCH_2Br + Br\cdot \longrightarrow CH_2Br\dot{C}HCH_2Br \qquad (18)$$

$$CH_2Br\dot{C}HCH_2Br + CCl_3Br \longrightarrow CH_2BrCHBrCH_2Br + \dot{C}Cl_3 \quad (19)$$

The key step is the addition of atomic bromine generated in the fragmentation of the radical **(4)** (reaction *15*). This type of fragmentation was not observed in the corresponding reaction between allyl chloride and bromotrichloromethane which is in line with the lesser tendency of β-chloroalkyl radicals to undergo fragmentation at room temperature.

The above reaction sequence has been formulated as involving fragmentation of a β-bromoalkyl radical, whereas it could equally well be represented as involving direct bromine atom transfer from the radical to a second molecule of alkene (reaction *20*).

$$Cl_3CCH_2\dot{C}HCHBr + CH_2{=}CHCH_2Br$$
$$\longrightarrow Cl_3CCH_2CH{=}CH_2 + BrCH_2\dot{C}HCH_2Br \quad (20)$$

If this were so, then the frequency of occurrence of this type of reaction would be expected to be dependent on the concentration of alkene. One might then expect that the 'normal' addition product **(3)** would be formed in reactions carried out with a deficiency of alkene. This was not the case. It is thus concluded that these reactions are more satisfactorily formulated as fragmentations.

The fragmentation of β-chloroalkyl radicals is observed in the peroxide-catalysed reaction of hydrocarbons with 1,2-dichloroethylene (reactions *21–24*) (8):

$$RH + Bu^tO\cdot \longrightarrow R\cdot + Bu^tOH \qquad (21)$$

$$R\cdot + CHCl{=}CHCl \longrightarrow RCHCl\dot{C}HCl \qquad (22)$$
$$\mathbf{(6)}$$

$$RCHCl\dot{C}HCl \longrightarrow RCH{=}CHCl + Cl\cdot \qquad (23)$$

$$Cl\cdot + RH \longrightarrow R\cdot + HCl \qquad (24)$$

Competition with addition of chlorine is minimal because no free halogen is added to the system. There is, however, a considerable amount of higher boiling material produced as a result of reaction of the radical (**6**) with a second molecule of the olefin.

12.2.3 *Fragmentation of β-thioalkyl radicals*

The reversible nature of the addition of thiyl radicals is clearly seen by the ability of small quantities of methane thiol to bring about the isomerization of *cis*-but-2-ene (**9**):

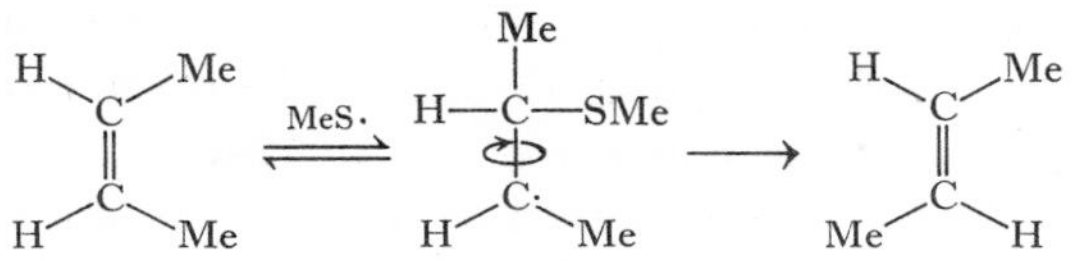

$$(25)$$

Kinetic studies showed that the decomposition of the β-thioalkyl radical (**7**) into the thiyl radical and olefin is very rapid compared to the transfer reaction, being about 80 times faster for the *trans*- and 20 times faster for the *cis*-isomer at 60 °C.

The extent of reversibility is dependent on the nature of the adduct radical, being greater the more stable the radical. Thus, there is a greater degree of reversibility in the addition of thiyl radicals to olefins than to acetylenes (**10**) as alkyl radicals are more stable than vinyl radicals. This is indicated by the higher yield of adduct obtained in the addition of methyl disulphide to propyne than to propene (reactions *26* and *27*):

$$MeC{\equiv}CH + MeS\cdot \rightleftharpoons Me\dot{C}{=}CHSMe \xrightarrow{\text{MeSSMe}} MeC(SMe)CHSMe \quad (26)$$
$$\underset{(8)}{} \qquad\qquad\qquad\qquad\qquad \underset{}{}$$

$$MeCH{=}CH_2 + MeS\cdot \rightleftharpoons Me\dot{C}HCH_2SMe$$
$$\xrightarrow{\text{MeSSMe}} MeCH(SMe)CH_2SMe \quad (27)$$
$$\underset{(9)}{}$$

The reversibility is indicated from the inverse dependence of the ratio of the yield of (**8**) and (**9**) on the disulphide concentration.

The degree of reversibility is also temperature dependent. Thus *cis*-but-2-ene underwent isomerization when it was photolysed in the presence of thiol acetic acid at room temperature but not at -78 °C (**11**).

Radical fragmentation

This is in line with predictions based on consideration of entropy factors.

The reversibility can also be suppressed to a greater or lesser degree by accelerating the hydrogen transfer by use of a thiol having a labile S–H bond in high concentration (12). This is exemplified by consideration of the reaction involving the addition of thiols, RSH, to alkenyl sulphides, CH_2=$CHCH_2SR'$. In addition to this unsymmetrical disulphide $RSCH_2CH_2CH_2SR'$, the two symmetrical disulphides, $(RSCH_2)_2CH_2$ and $(R'SCH_2)_2CH_2$, are also formed as a consequence of the reversibility of the addition of the thiyl radical. The selectivity of formation of the symmetrical compound is a measure of the efficiency of the hydrogen transfer step with respect to the reverse reaction. The results showed that this was increased by increasing the concentration of thiol and by using thiophenol rather than an alkane thiol. Alkane thiols are less prone to hydrogen abstraction than thiophenol.

The reversible addition of thiyl radicals to olefins has been successfully exploited for isomerizing caryophyllene (**10**) to isocaryophyllene (**11**) (13).

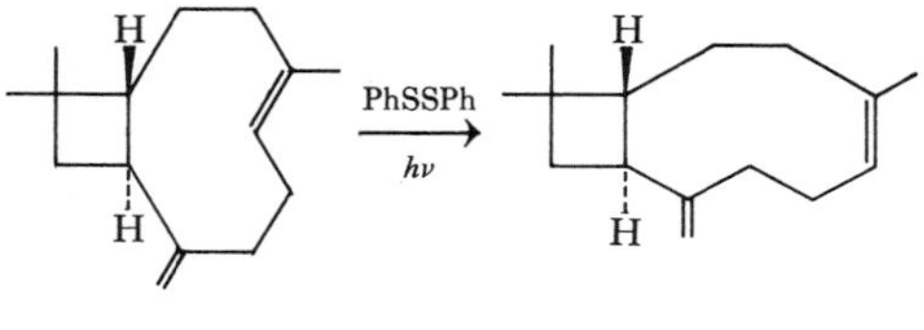

(**10**)　　　　　　　　　(**11**)

The β-elimination of thiyl radicals is seen in other reactions which proceed via β-thioalkyl radicals. Thus, the peroxide-induced reaction of β-hydroxysulphides results in the formation of carbonyl compounds and thiols (14) (reactions *28–31*):

$$CH_3CH(OH)CH_2SCH_3 + Bu^tO\cdot \longrightarrow CH_3\dot{C}(OH)CH_2SCH_3 \quad (28)$$

$$CH_3\dot{C}(OH)CH_2SCH_3 \longrightarrow CH_3C(OH)=CH_2 + CH_3S\cdot \quad (29)$$

$$CH_3C(OH)=CH_2 \longrightarrow CH_3COCH_3 \quad (30)$$

$$CH_3S\cdot + CH_3CH(OH)CH_2SCH_3 \longrightarrow CH_3\dot{C}(OH)CH_2SCH_3 + CH_3SH \quad (31)$$

The same type of approach was adopted by Kampmeier and his group who used phenyl radicals, generated from phenylazotriphenylmethane,

476

to abstract a hydrogen atom from di-t-butyl sulphide (reactions *32–33*) (15).

$$\text{Ph·} + CH_3\overset{\displaystyle CH_3}{\underset{\displaystyle CH_3}{C}}SBu^t \longrightarrow PhH + \dot{C}H_2CMe_2SBu^t \qquad (32)$$

$$\dot{C}H_2CMe_2SBu^t \longrightarrow Me_2C{=}CH_2 + Bu^tS· \qquad (33)$$

It is difficult to see why in this reaction the β-thioalkyl radicals are not scavenged rapidly by the relatively high concentration of triphenyl-methyl radicals present in solution. This could be explained by postula-ting that the radical has a bridged structure (**12**) and consequently has

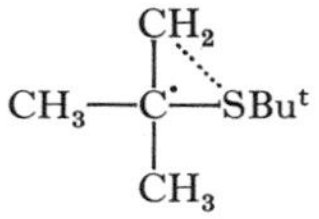

$$CH_3-\overset{\displaystyle CH_2}{\underset{\displaystyle CH_3}{C}}{\cdots}SBu^t$$

(**12**)

increased stability. This would also explain why hydrogen abstraction occurs 58 times as readily from di-t-butyl sulphide as from neopentane. Alternatively, it could be that the elimination is a concerted process (reaction *34*):

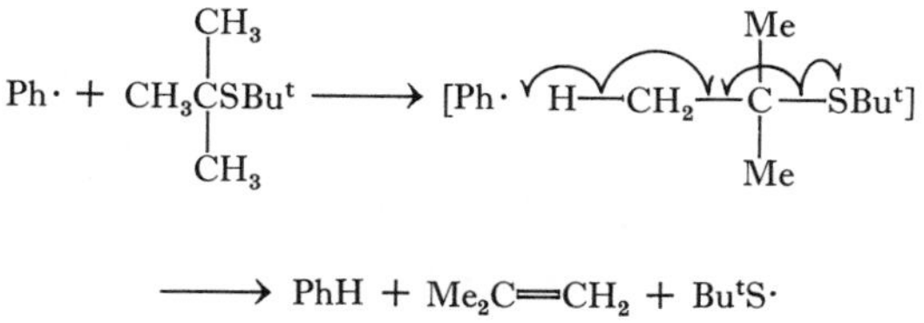

$$\text{Ph·} + CH_3\overset{\displaystyle CH_3}{\underset{\displaystyle CH_3}{C}}SBu^t \longrightarrow [Ph·\ H{-}CH_2\ \overset{\displaystyle Me}{\underset{\displaystyle Me}{C}}{-}SBu^t]$$

$$\longrightarrow PhH + Me_2C{=}CH_2 + Bu^tS· \qquad (34)$$

This would readily account for the increased reactivity of the α-hydro-gens as well as for the inability of the triphenylmethyl radical to act as a scavenger.

In reactions involving the addition of thiyl radicals to allyl halides, the resultant radical could undergo two different modes of fragmentation according to whether a halogen atom or a thiyl radical was split off (reaction *35*):

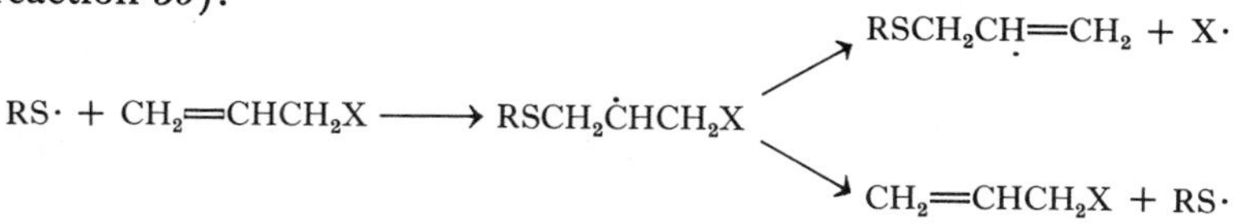

$$RS· + CH_2{=}CHCH_2X \longrightarrow RSCH_2\dot{C}HCH_2X \begin{array}{c} \nearrow RSCH_2CH{=}CH_2 + X· \\[2mm] \searrow CH_2{=}CHCH_2X + RS· \end{array} \qquad (35)$$

Radical fragmentation

Addition of methane thiol to α-methylallyl chloride gave the expected addition product (13) plus 2-chloro-2-methylpropylmethyl sulphide (14) (16):

$$CH_3SH + CH_2\!=\!\overset{\overset{\displaystyle CH_3}{|}}{C}CH_2Cl \longrightarrow CH_3SCH_2\overset{\overset{\displaystyle CH_3}{|}}{C}CH_2Cl + CH_3SCH_2\overset{\overset{\displaystyle CH_3}{|}}{\underset{\underset{\displaystyle Cl}{|}}{C}}\!-\!CH_3$$

$$\qquad\qquad\qquad\qquad\qquad\qquad\quad (13)\qquad\qquad\qquad (14)\quad (36)$$

The formation of this latter product is presumed to arise from the ionic addition of hydrogen chloride to the alkene (16) derived by fragmentation of the adduct radical (15):

$$CH_3S\cdot + CH_2\!=\!\overset{\overset{\displaystyle CH_3}{|}}{C}CH_2Cl \longrightarrow CH_3SCH_2\overset{\overset{\displaystyle CH_3}{|}}{C}CH_2Cl \qquad (37)$$
$$\qquad\qquad\qquad\qquad\qquad\qquad\qquad\qquad (15)$$

$$CH_3SCH_2\overset{\overset{\displaystyle CH_3}{|}}{\dot{C}}CH_2Cl \longrightarrow CH_3SCH_2\overset{\overset{\displaystyle CH_3}{|}}{C}\!=\!CH_2 + Cl\cdot \qquad (38)$$
$$\quad (15)\qquad\qquad\qquad\qquad (16)$$

$$CH_3SCH_2\overset{\overset{\displaystyle CH_3}{|}}{C}\!=\!CH_2 + HCl \longrightarrow CH_3SCH_2\overset{\overset{\displaystyle CH_3}{|}}{\underset{\underset{\displaystyle Cl}{|}}{C}}CH_3 \qquad (39)$$

$$\qquad (16)\qquad\qquad\qquad\qquad\qquad (14)$$

Support for this mechanism is seen in the formation of the symmetrical dialkyl sulphide (17) and by the absence of any of the rearranged compound (14) when the reaction was carried out using excess of methane thiol:

$$CH_3SCH_2\overset{\overset{\displaystyle CH_3}{|}}{C}\!=\!CH_2 + CH_3SH \longrightarrow CH_3SCH_2\overset{\overset{\displaystyle CH_3}{|}}{C}HCH_2SCH_3 \qquad (40)$$
$$\qquad\qquad\qquad\qquad\qquad\qquad\qquad\qquad (17)$$

In the analogous reaction of methane thiol with allyl bromide the only product was that resulting from addition of hydrogen bromide to allylmethyl sulphide (reactions 41–43):

$$CH_3S\cdot + CH_2\!=\!CHCH_2Br \longrightarrow CH_3SCH_2\dot{C}HCH_2Br \qquad (41)$$

$$CH_3SCH_2\dot{C}HCH_2Br \longrightarrow CH_3SCH_2CH\!=\!CH_2 + Br\cdot \qquad (42)$$

$$CH_3SCH_2CH\!=\!CH_2 + HBr \longrightarrow CH_3SCH_2CHBrCH_3 \qquad (43)$$

This is consistent with the readier loss of bromine than chlorine from *β*-haloalkyl radicals.

12.2.4 *Fragmentation of β-phosphinoalkyl radicals*

There are considerable analogies between the addition of phosphines and thiols to alkenes. Thus Pellon has shown that the isomerization of *cis*-but-2-ene occurs during the addition of various phosphines in the presence of a radical initiator (17):

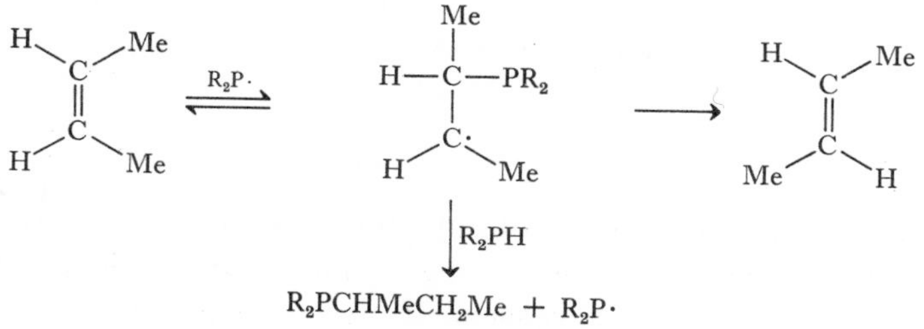
H
Me
C
C
H
Me
$R_2P\cdot$
Me
H—C—PR_2
C·
H
Me
R_2PH
$R_2PCHMeCH_2Me$ + $R_2P\cdot$
H
Me
C
C
Me
H

(44)

12.2.5 *Elimination of alkyl radicals*

The fragmentation reactions which have been discussed thus far have been the reverse of addition reactions in which either the forward or backward reaction predominated according to the reaction conditions:

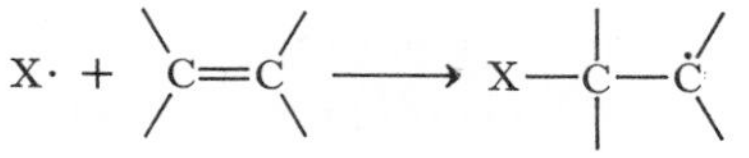
X· + C=C
X—C—C

(45)

This reversibility of the addition process is not observable in reactions of alkyl radicals with alkenes at moderate temperatures. Thus, isomerization of *cis*-but-2-ene does not occur in its reaction with trichloromethyl radicals (18). The reverse reaction can, however, sometimes be effected at high temperatures. This probably occurs in the thermal degradation of polymers.

Elimination reactions of alkyl radicals become much more feasible when such a process leads to a decrease in strain. This is exemplified in the free-radical addition of carbon tetrachloride to *β*-pinene (reactions *46–48*) (19):

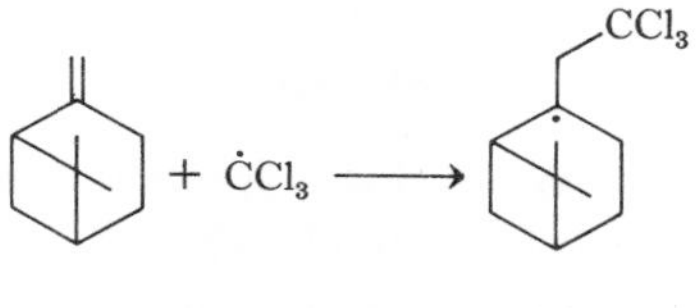
+ $\dot{C}Cl_3$
CCl_3

(46)

(18)

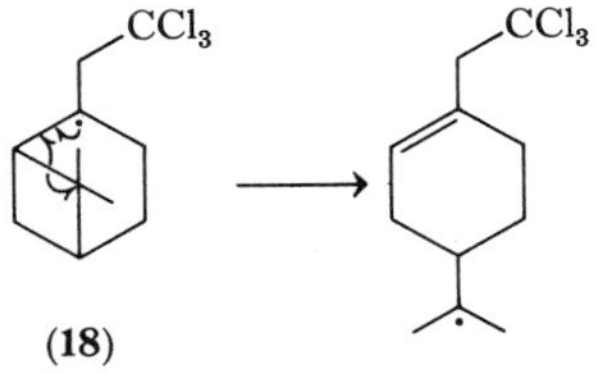

Radical fragmentation

$$(47)$$

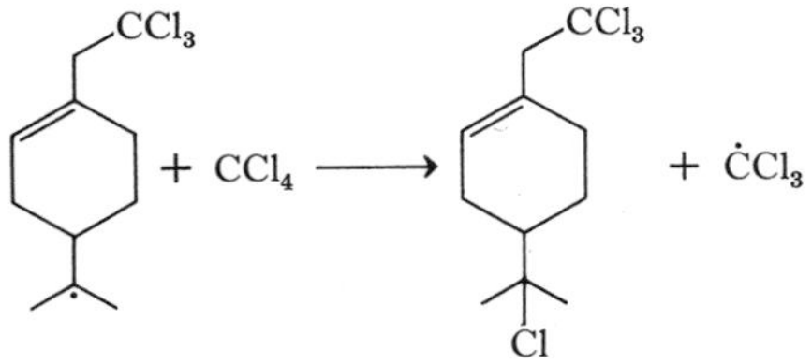

$$(48)$$

The driving force for the intramolecular fragmentation or rearrangement of the radical (18) is the relief of angle strain in the cyclobutane ring which accompanies the rearrangement, even though there is a small increase in entropy.

The same type of reasoning can be used to explain the rearrangement of cyclopropylmethyl radicals to homoallyl radicals of which there are many examples.

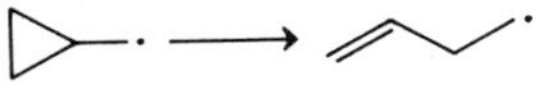

$$(49)$$

Thus, addition of thiols to 2-cyclopropylpropene gives both rearranged (22) and unrearranged products (21) (Scheme 2) (20).

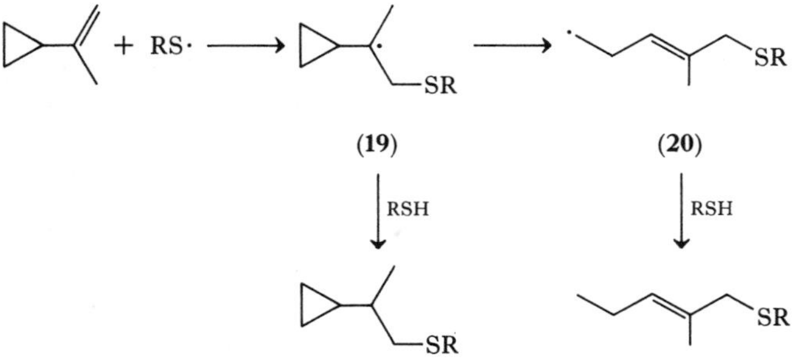

Scheme 2

Addition of thiophenol gives exclusively the unrearranged product (21) whereas reaction with methane thiol, which is a less efficient transfer agent, gives both products. It would be of some interest to obtain the

480

radical (**19**) by some other route to compare the relative ease of elimination of the thiyl radical with its ease of rearrangement to (**20**).

This type of rearrangement is only encountered with β-cycloalkylalkyl radicals when there is considerable relief of strain as a result of cleavage of the cycloalkyl ring. It is only observed with cyclopropylmethyl and cyclobutylmethyl radicals. The cyclopentylmethyl radical has no tendency to rearrange to the hex-5-en-1-yl radical. In fact the latter undergoes very ready cyclization to give the former (**21**):

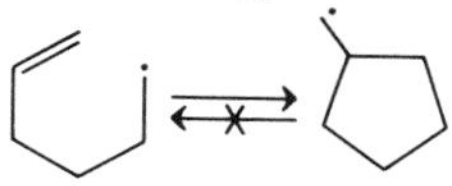

$$(50)$$

Closely related to the cyclopropylmethyl-homoallyl rearrangement is the rearrangement of the radical (**23**) to (**24**) in which an ethylene oxide ring is opened in an elimination reaction:

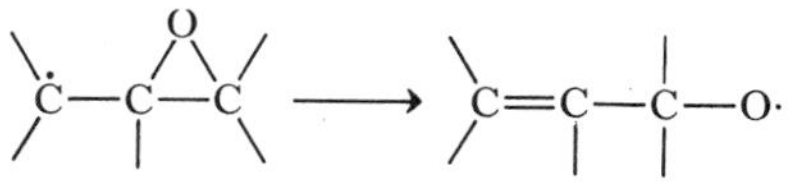

$$(51)$$

(**23**)   (**24**)

Thus, decomposition of cyclohexene oxide in the presence of di-t-butyl peroxide leads to cyclohexen-2-ol amongst other products:

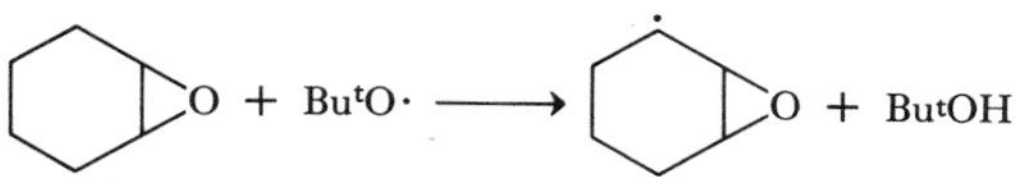

$$(52)$$

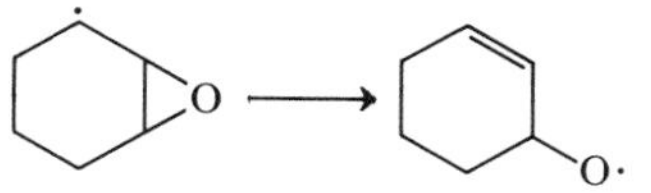

$$(53)$$

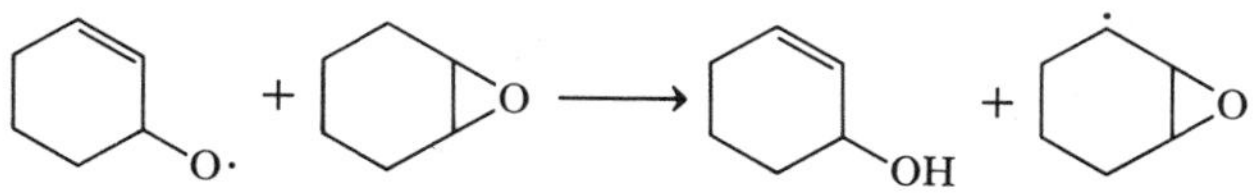

$$(54)$$

Further discussion of these reactions will be deferred to chapter 13. Mention has been made here of this type of reaction because of the close relationship to fragmentation reactions. These reactions could in fact be classified best as intramolecular fragmentations.

Radical fragmentation

12.2.6 Fragmentation of alkoxy radicals

The decomposition of t-butoxy radicals to acetone and methyl radicals is one of the best known of all radical reactions:

$$Me_3CO\cdot \longrightarrow Me_2CO + Me\cdot \qquad (55)$$

Unlike reactions involving β-thioalkyl and β-haloalkyl radicals, these reactions are not reversible. The extent to which fragmentation occurs, increases with increasing temperature, as indicated from the ratio of acetone to t-butanol produced in the decomposition of di-t-butyl peroxide in the same solvent at different temperatures (23).

t-Alkoxy radicals which can undergo fragmentation in more than one way, do so in a direction leading to the most stable alkyl radical (reactions 6, 56 and 57) (1, 24):

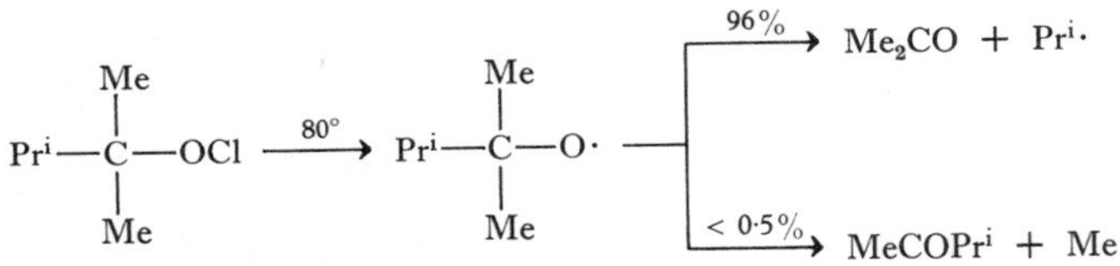

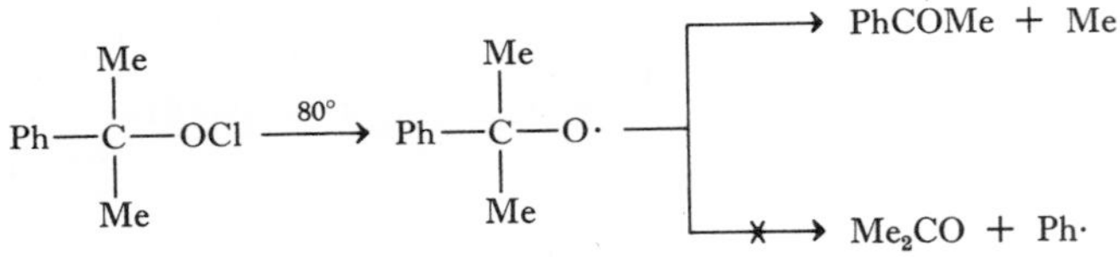

Fragmentation leading to loss of hydrogen atoms or aryl radicals does not occur. This is well illustrated by the fact that triarylmethoxy radicals undergo rearrangement rather than fragmentation (25) (cf. chapter 13, p. 505):

$$Ar_3CO\cdot \longrightarrow Ar_2\dot{C}OAr \qquad (58)$$

Intramolecular fragmentation occurs readily when two of the alkyl groups are joined as part of a cycloalkyl system (reactions 59–61) (26, 27):

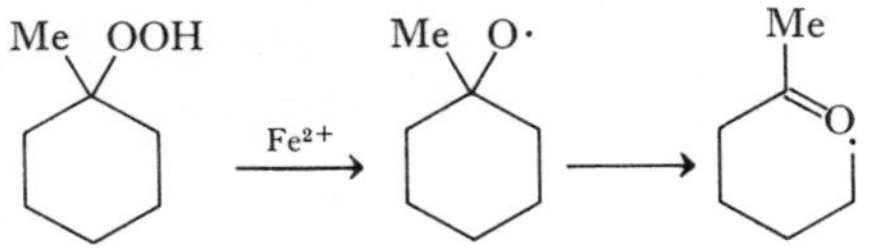

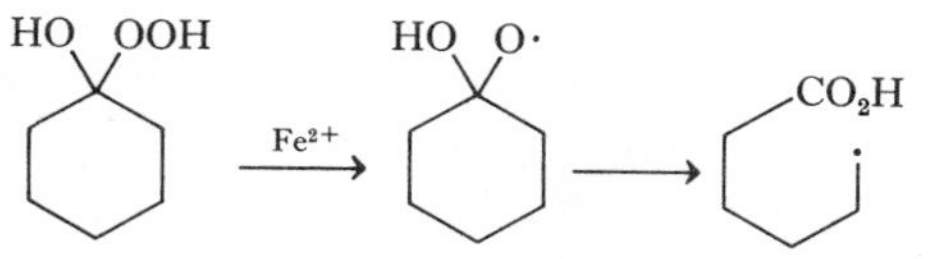

$$(60)$$

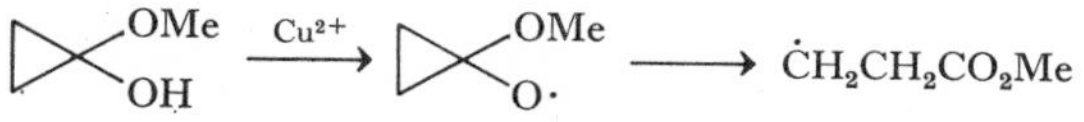

$$(61)$$

When the possibility exists of two modes of fragmentation, that leading to the more stable radical predominates (26):

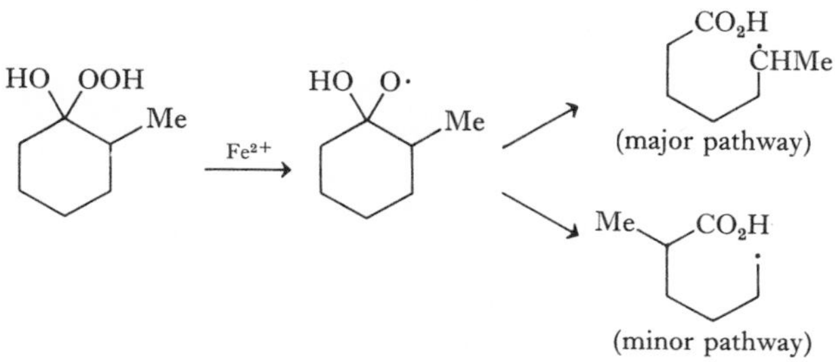

$$(62)$$

A particularly interesting intramolecular fragmentation is encountered in the thermal or photolytic decomposition of the cyclic dimeric ketone peroxide derived from cyclohexanone (27 a) (Scheme 3).

These examples illustrate the great ease with which cycloalkoxy radicals undergo fragmentation leading to acyclic radicals. The reactions occur very much more readily than the previously-mentioned fragmentation of cycloalkylmethyl radicals.

12.2.7 *Fragmentation of α-alkoxyalkyl radicals*

The fragmentation of α-alkoxyalkyl radicals resembles that of alkoxy radicals in that both lead to alkyl radicals:

$$PhĊHOR \longrightarrow PhCHO + R\cdot \qquad (63)$$

In the example quoted the fragmentation process competes with dimerization of the α-alkoxybenzyl radicals:

$$PhĊHOR \longrightarrow \begin{array}{c} PhCHOR \\ | \\ PhCHOR \end{array} \qquad (64)$$

The extent of fragmentation under given reaction conditions is determined by the stability of the alkyl radical eliminated in the fragmentation process (Table 12-1) (28).

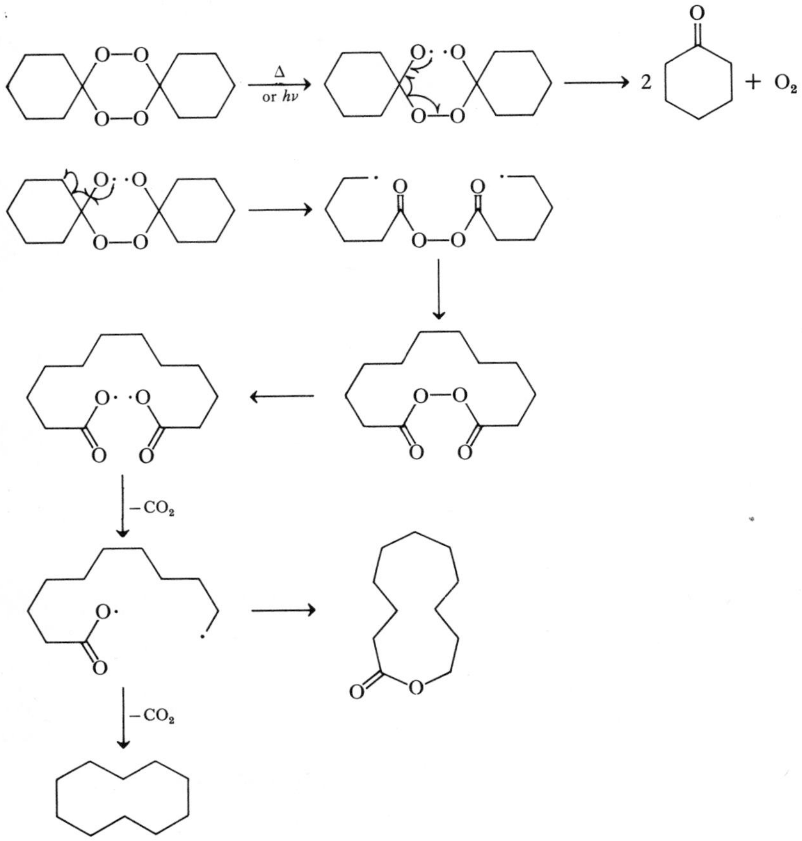

Scheme 3

TABLE 12-1   *Relative ease of fragmentation of*
α-alkoxybenzyl radicals (28)

R	Ph	Me	Et	Pr^i	Bu^t	PhCH$_2$	Ph$_2$CH
$\dfrac{[PhCHO]^a}{[(PhCHOR)_2]}$	0	0.23	0.83	2.27	3.0	∞	∞

[a] See reactions *63* and *64*.

Essentially the same process is encountered in the rearrangement of oxirane radicals (29):

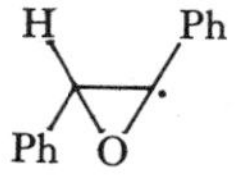

$$\text{(65)}$$

The relief of strain which accompanies this reaction results in exclusive fragmentation of the radical.

Radicals derived from both cyclic and acyclic acetals similarly undergo fragmentation to give ester radicals (30, 31):

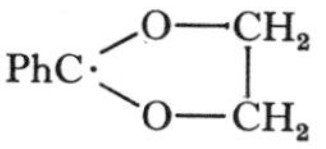

$$PhC \cdots \begin{matrix} O-CH_2 \\ | \\ O-CH_2 \end{matrix} \longrightarrow PhCO_2CH_2CH_2 \cdot \qquad \text{(66)}$$

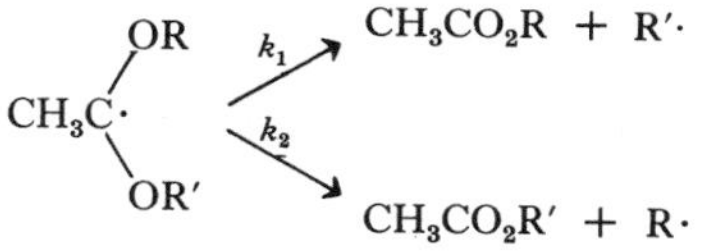

$$CH_3\overset{\cdot}{C} \begin{matrix} OR \\ \\ OR' \end{matrix} \begin{matrix} k_1 \nearrow \\ k_2 \searrow \end{matrix} \begin{matrix} CH_3CO_2R + R' \cdot \\ \\ CH_3CO_2R' + R \cdot \end{matrix} \qquad \text{(67)}$$

In this second example, the predominant pathway is again that leading to the more stable radical (Table 12-2) (31). The selectivity observed in this reaction is in fact somewhat less than that encountered in the fragmentation of alkoxy radicals. This has been attributed to the greater importance of polar effects in that process.

TABLE 12-2   *Relative rates of elimination of alkyl radicals*
from acetal radicals (31)

R'	Bu^n	Bu^{sec}	Bu^t	$CH_2=CHCH_2$	$PhCH_2$
$k_1/k_2{}^a$	1.00	4.10	18.7	25	22.4

a See reaction 67 (R=Bu^n).

The rearrangement of α-alkoxystyrenes to alkyl phenyl ketones is an example of a reaction in which the key step is the fragmentation of an α-alkoxyalkyl radical:

$$R \cdot + PhC \overset{OR}{=} CH_2 \longrightarrow Ph\overset{OR}{\underset{\cdot}{C}}CH_2R \qquad \text{(68)}$$

$$Ph\overset{OR}{\underset{\cdot}{C}}CH_2R \longrightarrow PhCOCH_2R \qquad \text{(69)}$$

485

Radical fragmentation

Evidence for this is based on the observations that (i) the reaction is catalysed by peroxides, (ii) optically active groups migrate with racemization, and (iii) neopentyl groups migrate without rearrangement (32).

12.2.8 Decarbonylation of acyl radicals

Applequist and Kaplan have shown that the ease of decarbonylation of acyl radicals is related to the stability of the derived alkyl radicals (Table 12-3) (33). The fragmentation of the acyl radical competes with its reaction with solvent:

$$
\text{RĊO}
\begin{array}{c}
\xrightarrow{k_1} \text{R}\cdot + \text{CO} \\
\xrightarrow[k_2]{\text{CCl}_4} \text{RCOCl} + \dot{\text{C}}\text{Cl}_3
\end{array}
\tag{70}
$$

TABLE 12-3   *Relative ease of decarbonylation of acyl radicals* (33)

R	Pr^n	Pr^i	Bu^t
$k_1/k_2{}^a$	0.137	1.14	12.3

[a] See reaction 70.

12.2.9 Decarboxylation of acyloxy radicals

This is an extremely facile process, particularly for acyloxy radicals derived from aliphatic acids:

$$
\text{RCO}_2\cdot \longrightarrow \text{R}\cdot + \text{CO}_2
\tag{71}
$$

The lifetime of the acetoxy radical has been estimated to be only 10^{-9} to 10^{-10} sec (34). They have in fact little if any existence outside the radical cage. Benzoyloxy radicals are considerably more stable, as evidenced by their ability to substitute reactive aromatic compounds. The decomposition of benzoyloxy radicals is a slightly endothermic process ($\Delta H = 13$ kJ mol^{-1}), in contrast to the decomposition of acetoxy radicals which is exothermic to the extent of 70 kJ mol^{-1} (35). This difference is presumably due to conjugation between the carboxy group and the aromatic nucleus.

486

12.3 *HETEROLYTIC FRAGMENTATION OF RADICALS*

It is only within the past few years that it has become recognized that fragmentation of radicals can be brought about by acids or bases. This has been largely due to the work of Norman and his colleagues.

12.3.1 *Heterolytic fragmentation of hydroxycyclohexadienyl and related radicals*

The e.s.r. spectrum obtained during the oxidation of phenylacetic acid with hydroxyl radicals is dependent on the pH of the system (36). In fairly strong acid solution the spectrum of the benzyl radical is observable, whereas at higher pH a complex spectrum is obtained which is compatible with that expected from addition of hydroxyl radicals to the aromatic system (**25**):

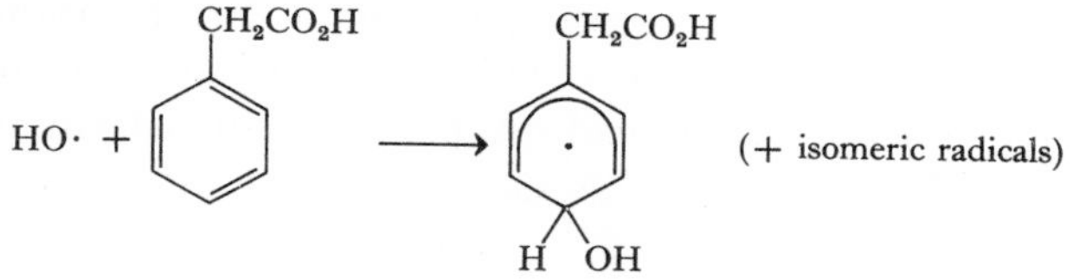

$$HO\cdot +\qquad\qquad\longrightarrow\qquad\qquad (+ \text{ isomeric radicals}) \qquad (72)$$

(**25**)

That the benzyl radicals are derived from (**25**) and not as a result of abstraction of the acidic proton with concerted cleavage of the CH_2–CO bond (reaction 73), is indicated by the pH-dependence of the reaction:

$$PhCH_2CO_2H + HO\cdot \longrightarrow PhCH_2\cdot + CO_2 + H_2O \qquad (73)$$

This type of reaction is in marked contrast to the behaviour of acetic acid, which on reaction with hydroxyl radicals gives the $\dot{C}H_2CO_2H$ radical.

A more reasonable scheme is that the intermediate hydroxycyclohexadienyl radicals are susceptible to acid cleavage (reaction 74):

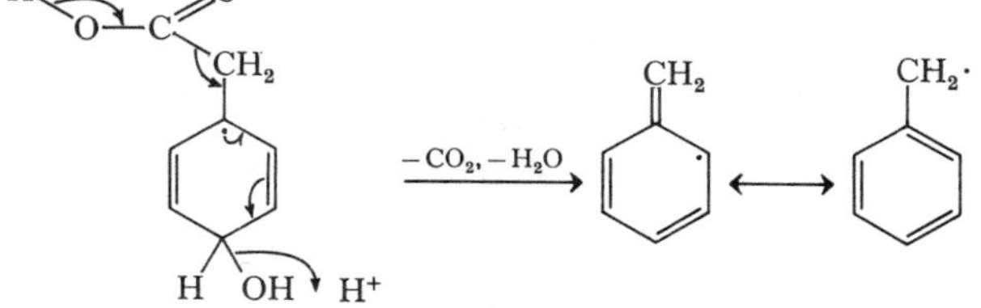

$$\xrightarrow{-CO_2, -H_2O} \qquad\qquad \longleftrightarrow \qquad\qquad (74)$$

Radical fragmentation

Product studies of the reactions of phenylacetic acid with hydroxyl radicals have complemented the e.s.r. studies (37). Thus, whereas 16% of bibenzyl was obtained from the reaction of phenylacetic acid at pH 1, only 4% was obtained at pH 6. This gives excellent support to the proposed mechanism, as the higher concentration of benzyl radicals at low pH would give an increased yield of bibenzyl.

The same mechanism accounts for the formation of benzyl radicals and formaldehyde from the oxidation of 2-phenylethanol (36):

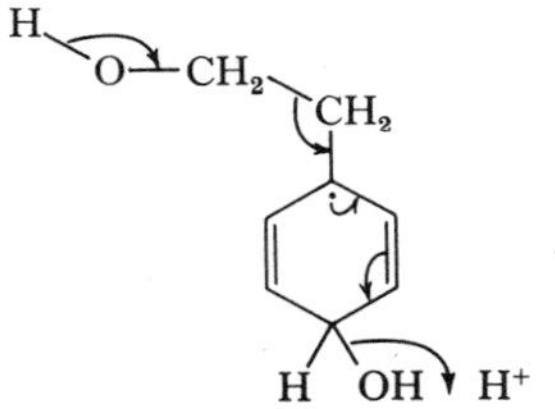

$$\longrightarrow \text{PhCH}_2\cdot + \text{CH}_2\text{O} + \text{H}_2\text{O} \qquad (75)$$

It is, however, rather less easy to account for the formation of 2-phenylethyl radicals from 3-phenylpropanoic acid or phenoxymethyl radicals from phenoxyacetic acid. A possible rationalization for the production of 2-phenylethyl radicals involves the intermediacy of the bridged radical (26) (reaction 76):

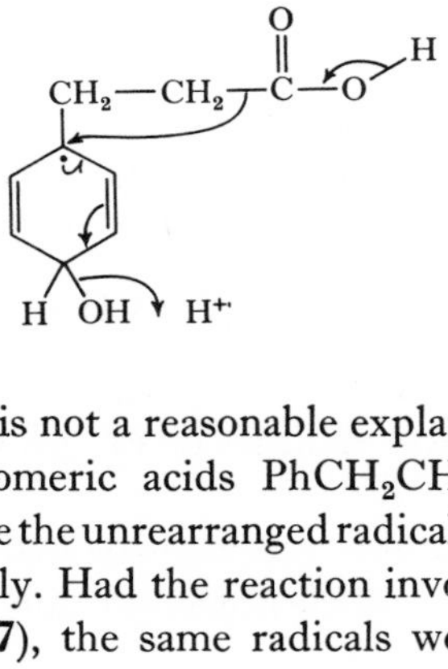

$$\longrightarrow \quad + \text{CO}_2 + \text{H}_2\text{O} \qquad (76)$$

(26)

That this is not a reasonable explanation is indicated from the oxidation of the isomeric acids $\text{PhCH}_2\text{CHMeCO}_2\text{H}$ and $\text{PhCHMeCH}_2\text{CO}_2\text{H}$ which gave the unrearranged radicals, $\text{PhCH}_2\text{CHMe}\cdot$ and $\text{PhCHMeCH}_2\cdot$, respectively. Had the reaction involved the intermediacy of the bridged radical (27), the same radicals would have been produced from both acids.

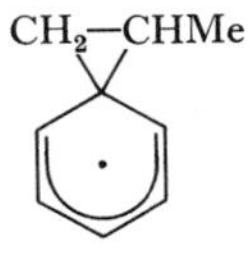

(27)

488

Consequently, Norman has proposed that the reaction proceeds as indicated in reaction (77) in which there is no formal bond between the aromatic nucleus and the carbon atom β to it (36):

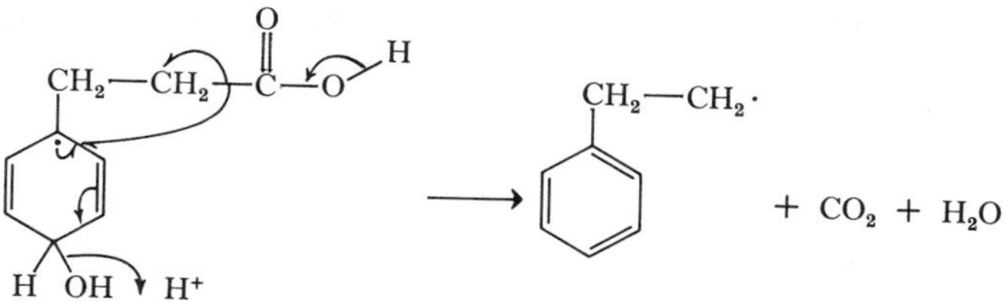

$$+ \ CO_2 + H_2O \qquad (77)$$

Whilst there are clearly many details of this mechanism which require further clarification, it does account fairly satisfactorily for the experimental observations. The same pathway can be invoked to explain the formation of phenoxymethyl radicals in the oxidation of phenoxyacetic acid. In this case the yield of 1,2-diphenoxyethane rises from 0.7% at pH 6 to 24% at pH 1.

The product composition from the oxidations of benzene, toluene, phenol and anisole with Fenton's reagent is also markedly dependent on the pH of the system (38). Thus, the ratio of bibenzyl to cresols in the oxidation of toluene changes from 0.44 to 22 as the pH is decreased from 3.6 to 1.3. From this it is inferred that the hydroxycyclohexadienyl radical (28) undergoes acid-catalysed fragmentation (reaction 78):

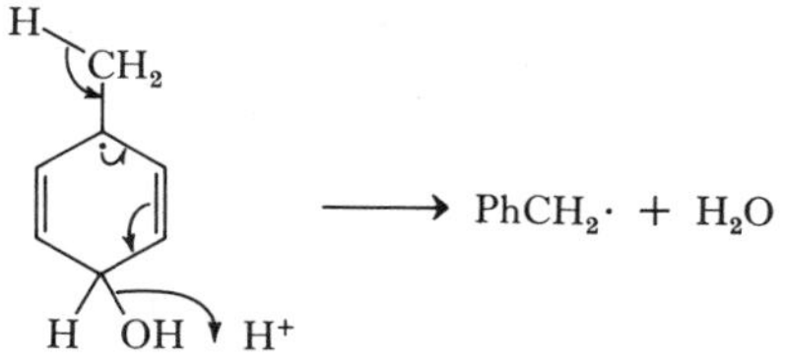

$$\longrightarrow \ PhCH_2\cdot + H_2O \qquad (78)$$

(28)

Phenol behaves similarly to toluene, and at low pH the radical (29) undergoes acid-catalysed fragmentation to give the phenoxy radical (reaction 79) (39). This latter has been identified by its e.s.r. spectrum:

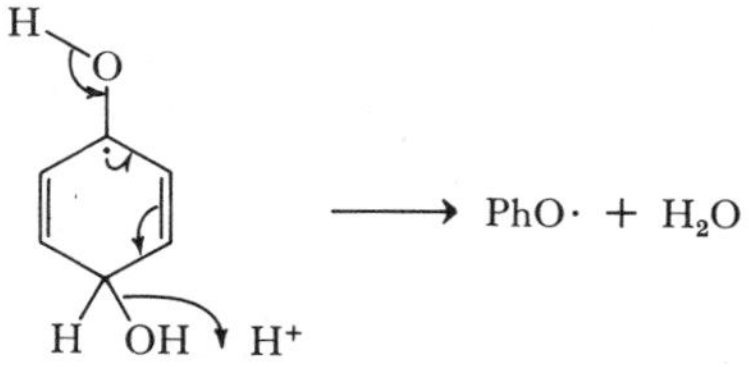

$$\longrightarrow \ PhO\cdot + H_2O \qquad (79)$$

(29)

Radical fragmentation

This type of fragmentation has also been encountered with benzoyl-oxycyclohexadienyl radicals. The yield of bibenzyl from reactions of dibenzoyl peroxide with toluene and with phenylacetic acid is increased by carrying out the reactions in the presence of trifluoroacetic acid (37):

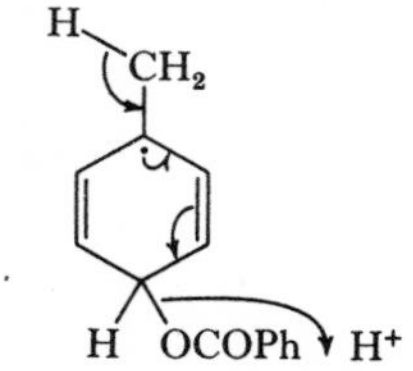

$$\longrightarrow \quad PhCH_2\cdot \ + \ PhCO_2H \qquad\qquad (80)$$

The fragmentation of the cyclohexadienyl radicals (30) produced by interaction of aromatic compounds with the sulphate radical anion is more facile than fragmentation of hydroxycyclohexadienyl radicals. This is indicated by the observation of the e.s.r. spectrum of the benzyl and 2-phenylethyl radicals in the persulphate oxidations of phenylacetic and 3-phenylpropanoic acids at a pH as high as 4 (reaction 81) (40):

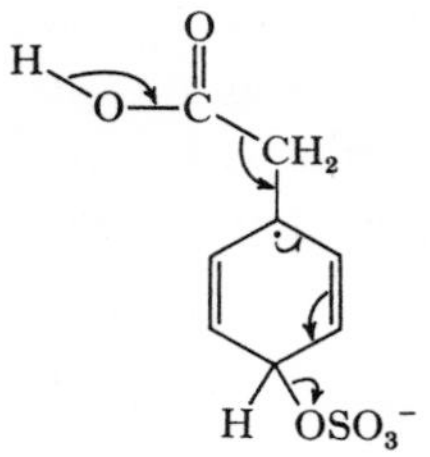

$$\longrightarrow \quad PhCH_2\cdot \ + \ CO_2 \ + \ HSO_4^- \qquad\qquad (81)$$

(30)

The efficiency of this type of reaction is also indicated by the formation of bibenzyl in a yield of 60% from the persulphate oxidation of phenylacetic acid at pH 1. This contrasts with a yield of only 16% in an oxidation using hydroxyl radicals.

12.3.2 Heterolytic fragmentation of α-hydroxyalkyl radicals

Oxidation of ethylene glycol with hydroxyl radicals in neutral or weakly acidic solution results in the formation of the expected radical (31) as a result of hydrogen abstraction from the methylene group:

$$HOCH_2CH_2OH + HO\cdot \longrightarrow HOCH_2\dot{C}HOH + H_2O \qquad (82)$$

(31)

The e.s.r. spectrum of the solution at low pH shows the formation of a new radical which has been assigned the structure (**32**). This arises as a result of an acid-catalysed fragmentation (reaction *83*) (41):

$$H^+ \quad HO—CH_2—\dot{C}H—O—H \longrightarrow H_2O + \dot{C}H_2CHO \qquad (83)$$

(**32**)

The same radical is produced in the oxidation of 2-chloroethanol under acidic conditions as a consequence of the same type of fragmentation process:

$$Cl—CH_2—\dot{C}H—O—H \longrightarrow \dot{C}H_2CHO + HCl \qquad (84)$$

It has been suggested by Norman that the driving force for this process arises, at least in part, from the stability of the resultant radical (**32**) (41):

$$\dot{C}H_2—CH{=}O \longleftrightarrow CH_2{=}CH—O\cdot$$
(**32**)

2-Cyanoethanol does not undergo this process and reaction of this with hydroxyl radicals gives only the radical $\dot{C}HOHCH_2CN$:

$$CH_2OHCH_2CN \xrightarrow{HO\cdot} \dot{C}HOHCH_2CN \xrightarrow[\times]{H^+} \dot{C}H_2CHO \qquad (85)$$

This is because cyanide is a poorer leaving group than either chloride or hydroxide in acid solution. That hydroxide is a better leaving group than cyanide suggests that the former is protonated and leaves as water.

There are a number of other examples of this process (reactions *86–88*) (41, 42):

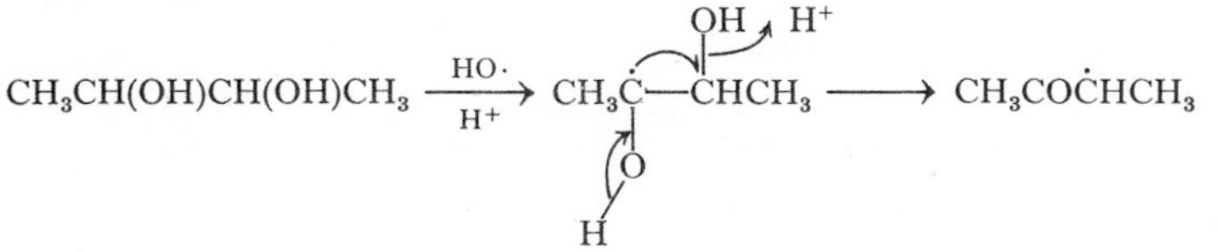

$$CH_3CH(OH)CH(OH)CH_3 \xrightarrow[H^+]{HO\cdot} CH_3\dot{C}—\overset{\overset{\displaystyle OH}{|}}{C}HCH_3 \longrightarrow CH_3CO\dot{C}HCH_3 \qquad (86)$$

$$(HO)_2CHCCl_3 \xrightarrow{HO\cdot} \overset{HO}{C}—C—Cl \longrightarrow HO_2C\dot{C}Cl_2 \qquad (87)$$

491

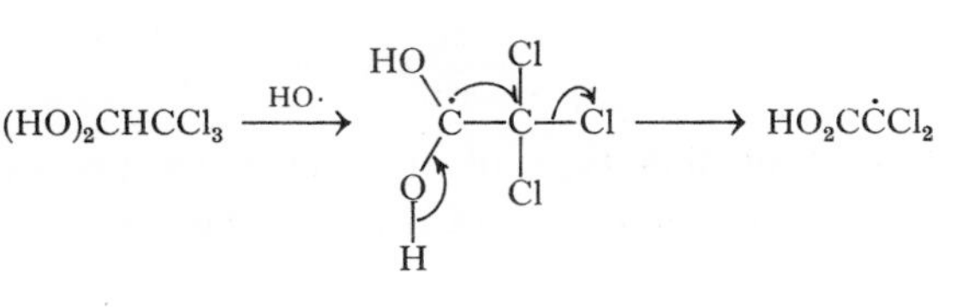

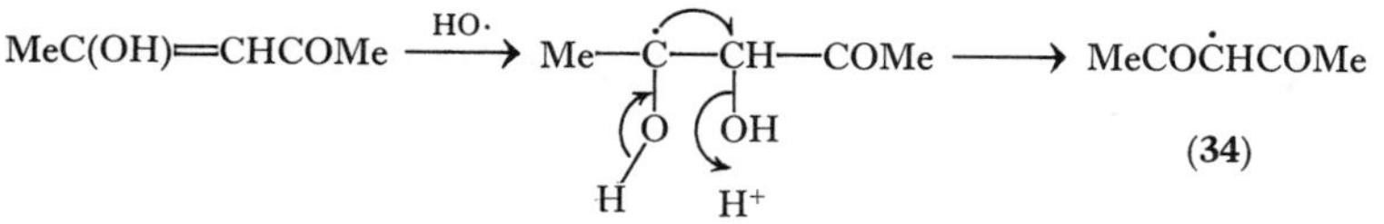

$$\text{MeC(OH)}{=}\text{CHCOMe} \xrightarrow{\text{HO·}} \text{Me}-\overset{\displaystyle}{\underset{\displaystyle}{C}}-\text{CH}-\text{COMe} \longrightarrow \text{MeCOCHCOMe} \quad (88)$$

(33)

(34)

Consistent with the proposed reaction sequences, the proportions of radicals present in the oxidation of such compounds are pH-dependent. For example, the e.s.r. spectrum of the radical (33) is observed in the oxidation of acetylacetone at pH 5, whereas at pH 1 that of the radical (34) is seen.

A somewhat similar fragmentation process occurs in the oxidation of 1,2-dimethoxyethane at low pH (reactions *89–91*) (42):

$$\text{MeOCH}_2\text{CH}_2\text{OMe} \xrightarrow{\text{HO·}} \text{MeOĊHCH}_2\text{OMe} + \text{ĊH}_2\text{OCH}_2\text{CH}_2\text{OMe} \quad (89)$$

(35)   (36)

$$\text{MeÖ}{-}\text{ĊH}{-}\text{CH}_2{-}\text{OMe} \quad \text{H}^+ \longrightarrow \text{Me}\overset{+}{\text{O}}{=}\text{CHCH}_2\cdot + \text{MeOH} \quad (90)$$

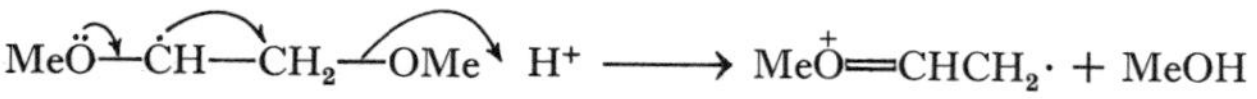

$$\text{Me}\overset{+}{\text{O}}{=}\text{CHCH}_2\cdot \xrightarrow[-\text{H}^+]{\text{H}_2\text{O}} \text{MeO}\overset{\text{OH}}{\overset{|}{\text{C}}}\text{HCH}_2\cdot \quad (91)$$

(37)

The e.s.r. spectrum of the radicals produced in this oxidation shows the presence of (35) and (36) at pH 2. As the pH of the solution is decreased, the proportion of [(35)] : [(36)] falls and the spectrum of a new radical (37) can be observed. The same type of process is also seen in the oxidation of 2-methoxyethanol:

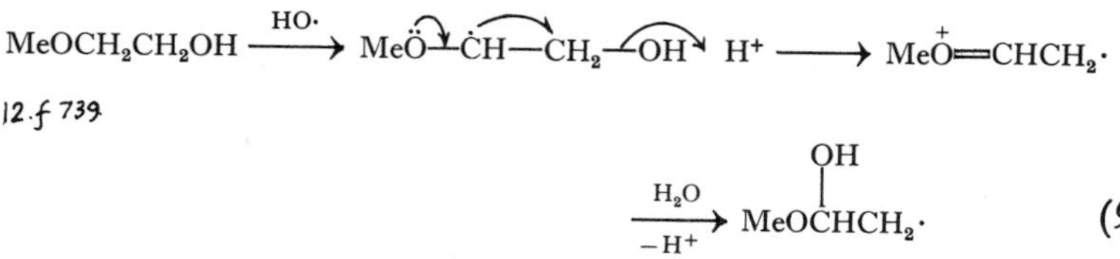

$$\text{MeOCH}_2\text{CH}_2\text{OH} \xrightarrow{\text{HO·}} \text{MeÖ}{-}\text{ĊH}{-}\text{CH}_2{-}\text{OH} \quad \text{H}^+ \longrightarrow \text{Me}\overset{+}{\text{O}}{=}\text{CHCH}_2\cdot$$

$$\xrightarrow[-\text{H}^+]{\text{H}_2\text{O}} \text{MeO}\overset{\text{OH}}{\overset{|}{\text{C}}}\text{HCH}_2\cdot \quad (92)$$

Many more examples of this type of fragmentation process will undoubtedly come to light in the course of the next few years.

12.3.3 *Fragmentation of radical cations*

Oxidation of reactive methylaromatic compounds, such as *p*-methyl-anisole, with manganese(III) acetate proceeds by electron transfer from the aromatic compound with the resultant formation of a radical cation (43). Subsequent loss of a proton in a slow step affords the benzylic radical. This step is, in effect, a further example of a heterolytic radical fragmentation (reaction *94*):

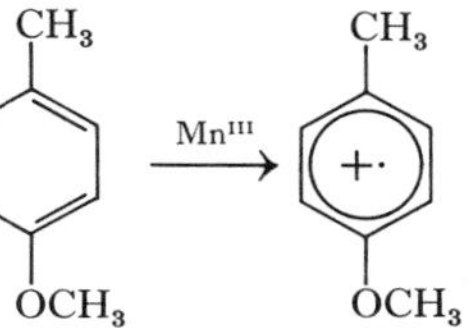

$$\text{(93)}$$

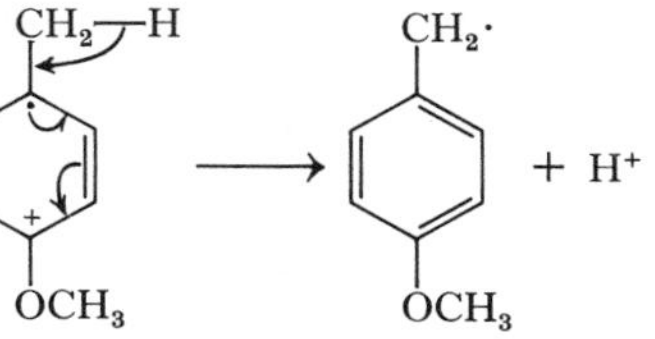

$$\text{(94)}$$

Fragmentations of this type probably occur quite frequently in oxidations with electron-transfer reagents, as well as in anodic oxidations. Thus, anodic oxidation of 9-methylanthracene-10-acetic acid involves initial formation of the radical cation (**38**) rather than discharge of the carboxylate group (44). Fragmentation of this gives the radical (**39**), dimerization of which leads to 1,2-bis(10-methylanthryl)ethane (Scheme 4).

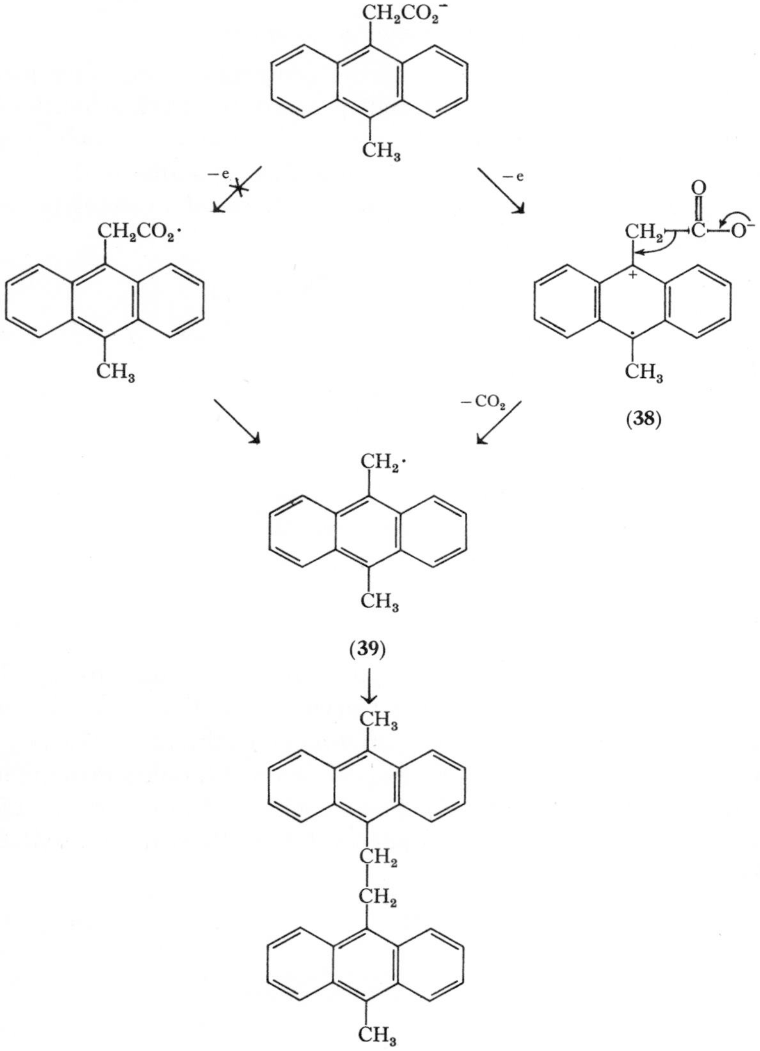

Scheme 4

There is evidence that anodic oxidation of phenols results initially in the formation of a radical cation, which subsequently gives the phenoxy radical (Scheme 5) (45).

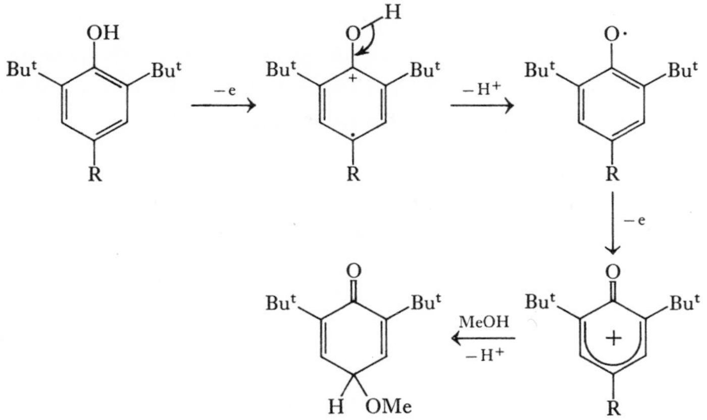

Scheme 5

This process can also be classified as a heterolytic fragmentation. In the anodic oxidation of 2,6-di-t-butyl-4-methyl phenol, products arising from the benzylic cation (**42**) are also formed. These arise as a result of further oxidation of the radical (**41**), which in turn is formed from the radical cation (**40**) in an alternative fragmentation pathway (Scheme 6).

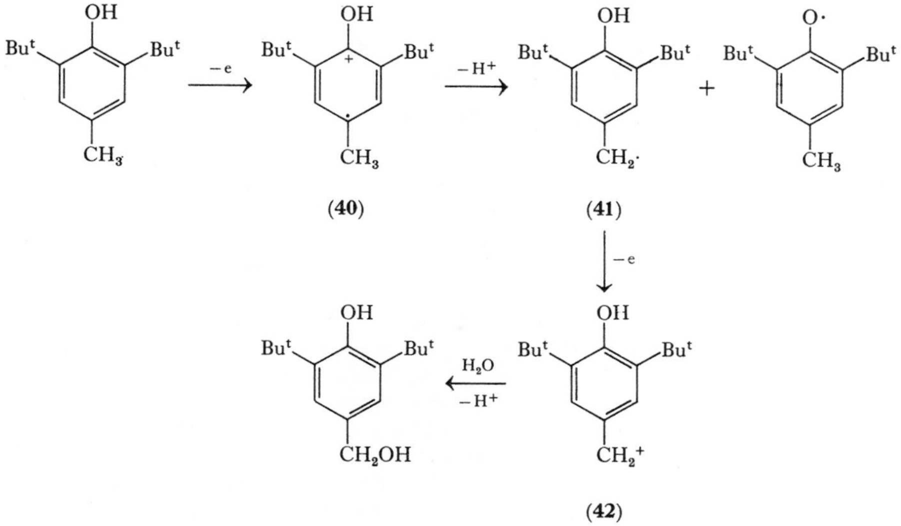

Scheme 6

495

Radical fragmentation

The same type of fragmentation process can be envisaged as occurring in the anodic oxidation of aromatic amines to arylamino radicals (43). Further oxidation of these leads to the cation (44) which reacts with nucleophiles in the reaction medium (Scheme 7) (46).

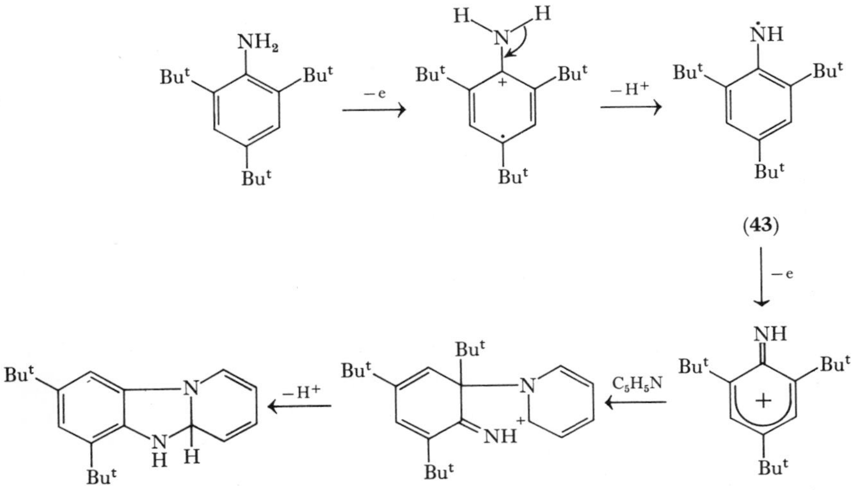

Scheme 7

GENERAL REFERENCE

E. S. Huyser, *Free-Radical Chain Reactions*, Wiley Interscience, New York, 1969, Chapter 8.

SPECIFIC REFERENCES

1. F. D. Greene, M. L. Savitz, F. D. Osterholtz, H. H. Lau, W. N. Smith and P. M. Zanet, *J. Org. Chem.* 1963, **28**, 55.
2. F. Wachholtz, *Z. phys. Chem.* 1927, **125**, 1.
3. H. Steinmutz and R. M. Noyes, *J. Amer. Chem. Soc.* 1952, **74**, 4141.
4. F. F. Rust and W. E. Vaughan, *J. Org. Chem.* 1940, **5**, 472.
5. B. P. McGrath and J. M. Tedder, *Proc. Chem. Soc.* 1967, 80.
6. J. Adam, P. A. Gosselain and P. Goldfinger, *Nature*, 1953, **171**, 523.
7. C. Djerassi, *Chem. Rev.* 1948, **43**, 271.
8. L. Schmerling and J. P. West, *J. Amer. Chem. Soc.* 1949, **71**, 2015.
9. C. Walling and W. Helmreich, *J. Amer. Chem. Soc.* 1959, **81**, 1144.
10. E. I. Heiba and R. M. Dessau, *J. Org. Chem.* 1967, **32**, 3837.
11. N. P. Neuriter and F. G. Bordwell, *J. Amer. Chem. Soc.* 1960, **82**, 5354.
12. D. N. Hall, A. A. Oswald and K. Griesbaum, *J. Org. Chem.* 1965, **30**, 3829.

13. K. H. Schulte-Elte and G. Ohloff, *Helv Chim. Acta*, 1968, **51**, 548.
14. E. S. Huyser and R. M. Kellogg, *J. Org. Chem.* 1966, **31**, 3666.
15. J. A. Kampmeier, R. P. Geer, A. J. Meskin and R. M. D'Silva, *J. Amer. Chem. Soc.* 1966, **88**, 1257.
16. D. N. Hall, *J. Org. Chem.* 1967, **32**, 2082.
17. J. Pellon, *J. Amer. Chem. Soc.* 1961, **83**, 1915.
18. P. S. Skell and R. C. Woodworth, *J. Amer. Chem. Soc.* 1955, **77**, 4638.
19. D. M. Oldroyd, G. S. Fisher and L. A. Goldblatt, *J. Amer. Chem. Soc.* 1950, **72**, 2407.
20. E. S. Huyser and J. D. Taliaferro, *J. Org. Chem.* 1963, **28**, 3442.
21. J. F. Garst and F. E. Barton, *Tetrahedron Letters*, 1969, 587.
22. E. Sabatino and R. J. Gritter, *J. Org. Chem.* 1963, **28**, 3437.
23. J. H. Raley, F. F. Rust and W. E. Vaughan, *J. Amer. Chem. Soc.* 1948, **70**, 1336.
24. M. S. Kharasch, A. Fono and W. Nudenberg, *J. Org. Chem.* 1951, **16**, 113.
25. M. S. Kharasch, A. C. Poshkus, A. Fono and W. Nudenberg, *J. Org. Chem.* 1951, **16**, 1458.
26. H. E. De La Mare, J. K. Kochi and F. F. Rust, *J. Amer. Chem. Soc.* 1963, **85**, 1437.
27. S.E. Schaafsma, H. Steinberg and Th. J. DeBoer, *Rec. Trav. chim.* 1966, **85**, 70.
27a. P. R. Story, D. D. Denson, C. E. Bishop, B. C. Clark and J.-C. Farine, *J. Amer. Chem. Soc.* 1968, **90**, 917.
28. W. H. Chick and S. H. Ong, *Chem. Comm.* 1969, 216.
29. A. Padwa and N. C. Das, *J. Org. Chem.* 1969, **34**, 816.
30. E. S. Huyser and Z. Garcia, *J. Org. Chem.* 1962, **27**, 2716.
31. E. S. Huyser and D. T. Wang, *J. Org. Chem.* 1964, **29**, 2720.
32. K. B. Wiberg and B. I. Rowland, *J. Amer. Chem. Soc.* 1955, **77**, 1159.
33. D. E. Applequist and L. Kaplan, *J. Amer. Chem. Soc.* 1965, **87**, 2194.
34. L. Herk, M. Feld and M. Szwarc, *J. Amer. Chem. Soc.* 1961, **83**, 2998.
35. L. Jaffe, E. J. Prossen and M. Szwarc, *J. Chem. Phys.* 1957, **27**, 416; M. Szwarc and L. Herk, *J. Chem. Phys.* 1958, **29**, 438.
36. R. O. C. Norman and R. J. Pritchett, *J. Chem. Soc.* B, 1967, 926.
37. R. O. C. Norman and P. M. Storey, *J. Chem. Soc.* B, 1970, 1099.
38. J. R. E. Jefcoate, J. R. Lindsay Smith and R. O. C. Norman, *J. Chem. Soc.* B, 1969, 1013.
39. J. R. E. Jefcoate and R. O. C. Norman, *J. Chem. Soc.* B, 1968, 48.
40. R. O. C. Norman, P. M. Storey and P. R. West, *J. Chem. Soc.* B, 1970, 1087.
41. A. L. Buley, R. O. C. Norman and R. J. Pritchett, *J. Chem. Soc.* B, 1966, 849.
42. D. J. Edge, B. C. Gilbert, R. O. C. Norman and P. R. West, *J. Chem. Soc.* B, 1971, 189.
43. P. J. Andrulis, M. J. S. Dewar, R. Dietz and R. L. Hunt, *J. Amer. Chem. Soc.* 1966, **88**, 5473.
44. J. P. Coleman and L. Eberson, *Chem. Comm.* 1971, 1300.
45. A. Ronlán and V. D. Parker, *J. Chem. Soc.* C, 1971, 3214.
46. G. Cauquis and J.-L. Cros, *Bull. Soc. chim. France*, 1971, 3760, 3765.

13

Radical rearrangements

13.1 *INTRODUCTION*

There are far fewer examples of rearrangements of radicals than of carbonium ions. Particularly notable is the inability of alkyl groups to undergo a 1,2-migration in radicals (reaction *1*) (1) which compares with the extreme ease of such a process with the corresponding cation (reaction *2*) (2):

$$Me_3CCH_2\cdot \xrightarrow{\quad\times\quad} Me_2\dot{C}CH_2Me \qquad (1)$$

$$Me_3CCH_2{}^+ \xrightarrow{\qquad} Me_2\overset{+}{C}CH_2Me \qquad (2)$$

No rearrangements involving the 1,2-migration of hydrogen have been unambiguously shown to occur, though they have been proposed from time to time. There are, however, numerous examples of 1,2-aryl and 1,2-halogen rearrangements, together with instances in which vinyl, acyl and acetoxy groups migrate.

The reason why alkyl groups fail to undergo simple 1,2-migrations in radical reactions has been attributed to the fact that such rearrangements would contravene the conservation of orbital symmetry (3). This is because of the somewhat sweeping generalization that radical reactions are governed by the same factors as their anionic counterparts, as they have the same highest occupied orbitals. There appears, however, to be some doubt as to how rigorously a first-order treatment of the Woodward–Hofmann rules may be applied to radical reactions (4). Clearly this is a topic which merits further examination from a theoretical aspect. Considerations of orbital symmetry do not apply to reactions proceeding through bridged intermediates. Such an intermediate (**1**) can be written for the migration of an aryl group. Similar intermediates (**2**) are feasible for reactions involving the migration of chlorine and bromine, which have vacant *d*-orbitals and may accommodate the extra electron.

498

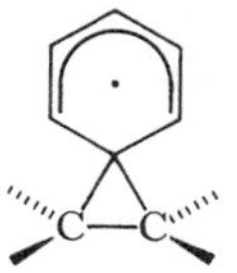

(1) 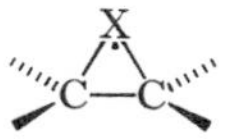   (2)

1,2-Alkyl migrations are commonly encountered in carbonium ion chemistry. The transition state (3) for such a process involves a π-com-

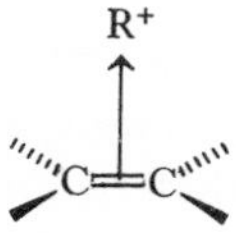

(3)

plex between the double bond and the migrating alkyl carbonium group. This is a reasonably stable arrangement. The extra electron in the case of a radical rearrangement must occupy a high-energy molecular orbital, and hence such processes do not occur, except when a bridged-radical intermediate is possible. Even when it is possible, the extent of such a migration is generally very much less than with the analogous cation.

There are a number of examples in which alkyl groups are formally observed to have migrated. In all cases examined, with one exception (reaction *3*), these proceed by a fragmentation–addition mechanism. No explanation has been offered for this particular reaction:

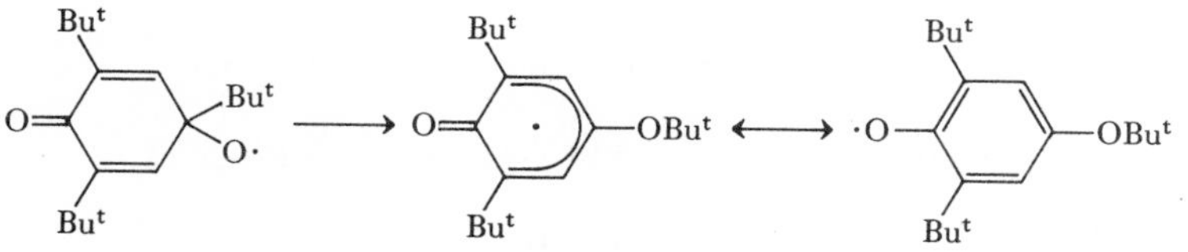

Reactions which are merely intramolecular fragmentations will not be discussed here, since they are best considered alongside other fragmentation processes (see chapter 12). Thus the fragmentation of cyclic alkoxy radicals (reaction *4*) will not be included here:

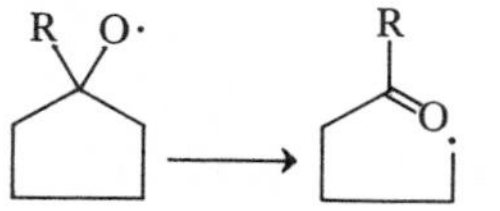

(*4*)

Radical rearrangements

Hydrogen migrations, however, will be discussed in this chapter, though these too are strictly not rearrangements since there is no alteration of the carbon skeleton.

13.2 REARRANGEMENTS INVOLVING THE MIGRATION OF AN ARYL GROUP

13.2.1 1,2-Migration of aryl groups to carbon

The tendency for a carbonium ion to undergo rearrangement is greatest in those cases in which the rearranged ion is more stable than the initial ion. The same reasoning would be expected to hold in radical systems. It was thus not unexpected that neophyl radicals (4) rearrange fairly readily to 2-methyl-1-phenyl-2-propyl radicals (5) by a 1,2-shift of the phenyl group (6):

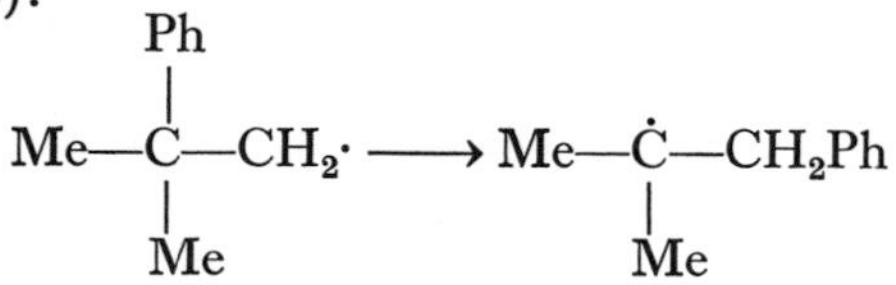

$$\underset{\overset{|}{\underset{Me}{|}}}{\overset{\overset{Ph}{|}}{Me-C-CH_2\cdot}} \longrightarrow \underset{\overset{|}{\underset{Me}{|}}}{Me-\overset{\cdot}{C}-CH_2Ph}$$

$$(4) \qquad\qquad (5)$$

The driving force for the rearrangement is the formation of a tertiary radical from a primary radical.

Experimental support for this class of rearrangement came from the isolation of isobutylbenzene as well as t-butylbenzene from the reaction of di-t-butyl peroxide with 3-methyl-3-phenylbutanal (Scheme 1).

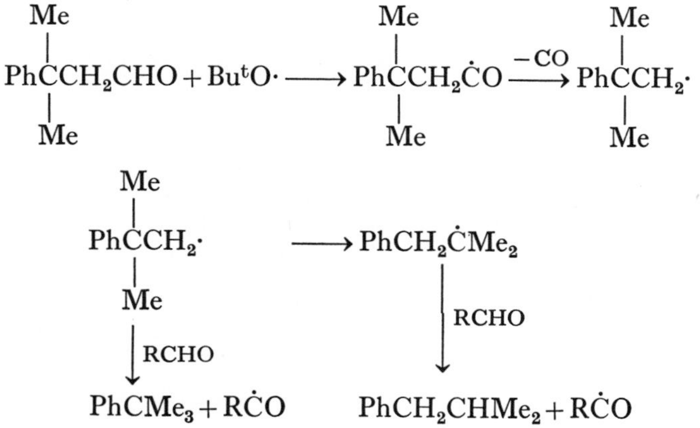

$$\underset{\overset{|}{\underset{Me}{|}}}{\overset{\overset{Me}{|}}{PhCCH_2CHO}} + Bu^tO\cdot \longrightarrow \underset{\overset{|}{\underset{Me}{|}}}{\overset{\overset{Me}{|}}{PhCCH_2\overset{\cdot}{C}O}} \xrightarrow{-CO} \underset{\overset{|}{\underset{Me}{|}}}{\overset{\overset{Me}{|}}{PhCCH_2\cdot}}$$

$$\underset{\overset{|}{\underset{Me}{|}}}{\overset{\overset{Me}{|}}{PhCCH_2\cdot}} \longrightarrow PhCH_2\overset{\cdot}{C}Me_2$$

$$\Big\downarrow RCHO \qquad\qquad \Big\downarrow RCHO$$

$$PhCMe_3 + R\overset{\cdot}{C}O \qquad PhCH_2CHMe_2 + R\overset{\cdot}{C}O$$

Scheme 1

500

The extent of rearrangement of β-arylalkyl radicals is markedly dependent on the lifetime of the radical, as well as on the activation energy for the rearrangement process. The lifetime of the radical is dependent on the intrinsic stability of the unrearranged radical, and also on the ease with which it undergoes chain transfer in the environment of the reaction. The stability of the rearranged radical plays an important part in determining the activation energy for the rearrangement process.

The importance of chain transfer to the degree of rearrangement is illustrated by the dramatic decrease in the extent of rearrangement caused by the addition of thiols to the reaction system (7):

$$PhCMe_2CH_2\cdot + PhCH_2SH \longrightarrow PhCMe_3 + PhCH_2S\cdot \qquad (5)$$

A similar, though smaller, decrease in the amount of rearrangement is seen if the reaction is carried in more concentrated solution (8). Under these conditions chain transfer with aldehyde occurs:

$$PhCMe_2CH_2\cdot + PhCMe_2CH_2CHO \longrightarrow PhCMe_3 + PhCMe_2CH_2\dot{C}O \qquad (6)$$

The importance of the relative stabilities of the unrearranged and rearranged radicals is illustrated by the data in Table 13-1. The extent of rearrangement is greatest when the unpaired electron can be delocalized by phenyl groups as in the case of the rearrangement of the $Ph_3CCH_2\cdot$ radical. Rearrangement also occurs when the rearranged radical is more highly substituted than the initial radical. This is exemplified by the ability of the neophyl radical to undergo rearrangement, whereas the reverse reaction involving the rearrangement of the isomeric $PhCH_2CMe_2\cdot$ radical does not occur.

TABLE 13-1   *Extent of rearrangement in β-phenylalkyl radicals*

Initial radical structure	Rearranged radical structure	Radical source	Extent of rearrangement (%)	Reference
$Ph_3CCH_2\cdot$	$Ph_2\dot{C}CH_2Ph$	$(RCO_2)_2$	100	9
		$RCHO/(Bu^tO)_2$	100	10
$Ph_2CMeCH_2\cdot$	$Ph\dot{C}MeCH_2Ph$	$RCHO/(Bu^tO)_2$	100	11
$Ph_2CHCH_2\cdot$	$Ph\dot{C}HCH_2Ph$	$(RCO_2)_2$	63	9
$PhCMe_2CH_2\cdot$	$Me_2\dot{C}CH_2Ph$	$(RCO_2)_2$	49	9
		$RCHO/(Bu^tO)_2$	57	6
$PhCHMeCH_2\cdot$	$Me\dot{C}HCH_2Ph$	$(RCO_2)_2$	39	9
$PhCH_2CH_2\cdot$	$PhCH_2CH_2\cdot$	$RCHO/(Bu^tO)_2$	3	11
$PhCH_2CMe_2\cdot$	$PhCMe_2CH_2\cdot$	$RCHO/(Bu^tO)_2$	0	12

Another factor which helps to promote rearrangement is the relief of steric congestion in the initial radical. This explains why $PhCMeEtCH_2\cdot$ is more extensively rearranged than $PhCMe_2CH_2\cdot$ (14). Any resonance delocalization of the unpaired electron would be similar in these two radicals, and consequently the greater degree of steric compression in $PhCMeEtCH_2\cdot$ gives a rational explanation for its greater tendency to undergo rearrangement.

It is not possible to say whether the reaction proceeds through a distinct bridged-radical intermediate or whether it merely proceeds via a bridged transition state. The available evidence tends to support the latter pathway. Thus Kochi and Krusic failed to observe any indication of the bridged species (6) in a study of the e.s.r. spectrum of the 2-phenylethyl radical (15). The spectrum consisted of a triplet of triplets with no resolvable interaction with the aromatic hydrogens. This contrasts with the 2-p-methoxyphenylethyl cation, the n.m.r. spectrum

H_2C-CH_2

(6)

of which is characterized by a single line for the methylene groups, showing that these are equivalent (16). Secondly, the extent of rearrangement of the 1-^{14}C-2-phenylethyl radical was only 3 % (11). Had a bridged intermediate been completely formed, the degree of rearrangement would have been 50 %. This is precisely the situation that occurs in the solvolysis of 2-phenylethyl-1,1-d_2 tosylate, which proceeds via a bridged carbonium ion (17). Thirdly, any bridged-radical intermediate does not lend anchimeric assistance in the formation of the radical. Thus, the rates of decomposition of $(PhCMe_2CH_2CO_2)_2$, $(PhCH_2CH_2CO_2)_2$ and $(PhCH_2CH_2CH_2CO_2)_2$ are all comparable (18). If the phenyl groups helped to stabilize the incipient alkyl radical, a considerable enhancement in the rate of decomposition should be observable. Neighbouring-group participation by aryl groups was likewise shown not to be important in the decomposition of peresters (19). This again contrasts with the large accelerating effect of phenyl groups in solvolyses of neophyl derivatives. The above evidence, whilst not indicating unequivocally that a bridged-radical intermediate, as opposed to a bridged transition state, is not formed, is certainly consistent with such a hypothesis.

502

These results do not enable one to decide whether the rearranged product is derived directly from the bridged radical (9), or whether it is formed from the rearranged radical (10). Rüchardt has successfully distinguished between these pathways by showing that the optically active aldehyde (7) undergoes decarbonylation to (11) with at least 98 % racemization (14) (Scheme 2). If reaction had involved attack by a

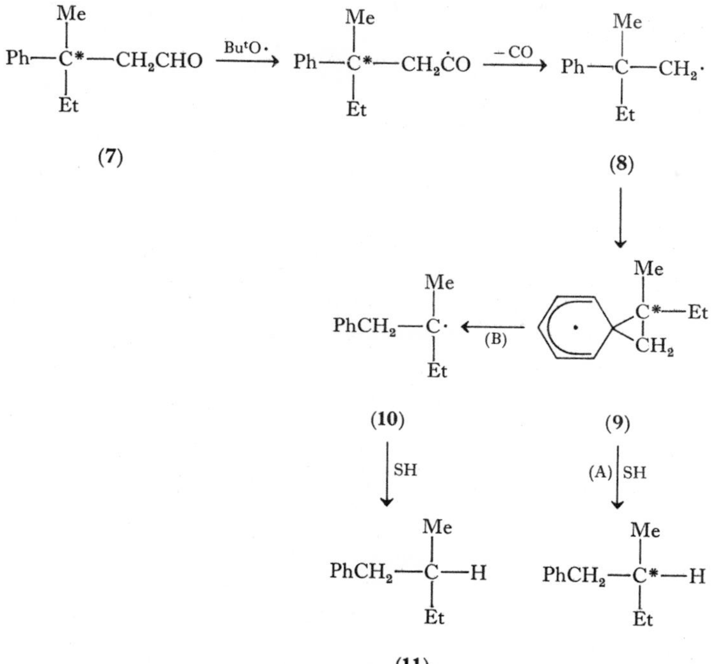

Scheme 2

bridged-radical intermediate on a second molecule of aldehyde, then retention of configuration would have been observed, as with the analogous carbonium-ion rearrangement (route A). Racemization indicates that the bridged radical (9) rearranges to the acyclic radical (10) before reaction with a second molecule of aldehyde.

Rüchardt has also examined the influence of substituents on the extent of rearrangement of substituted neophyl radicals, $ArCMe_2CH_2\cdot$. His results (Table 13-2) show that electron-withdrawing groups favour rearrangement. This is particularly noticeable for the strongly electron-withdrawing cyano and nitro groups. This may well be due to the ability

TABLE 13-2   *Relative amounts of migration (%)*
of aryl groups in ArCMe$_2$CH$_2$· radicals (20)

Substituent in aryl group	Radical source	
	Perester	Aldehyde
p-MeO	0.36	0.35
p-F	0.40	0.38
p-Me	0.72	0.65
H	1.00	1.00
o-MeO	1.12	1.10
m-Cl	1.38	1.55
m-Br	1.59	1.70
p-Cl	1.76	1.82
p-Br	1.91	1.79
p-NO$_2$	31	—
p-CN	35	19

of these groups to impart some carbonium-ion character to the transition state (12) which would be expected to be considerably stabilized by

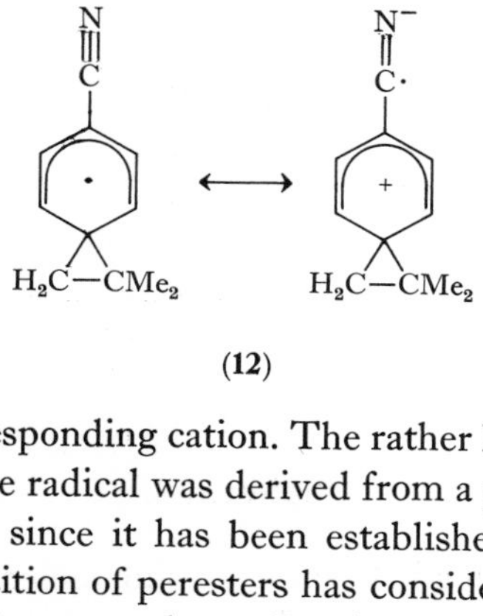

(12)

analogy with the corresponding cation. The rather larger influence of the cyano group, when the radical was derived from a perester, is consistent with such reasoning, since it has been established that the transition state in the decomposition of peresters has considerable polar character (21). It would be of interest to know if any anchimeric assistance could be observed in this instance.

13.2.2   *Migration of aryl groups to a more remote carbon radical centre*

Winstein has observed that aryl radicals can also undergo 1,4-migrations (7):

$$PhCMe_2CH_2CH_2CH_2· \longrightarrow PhCH_2CH_2CH_2\dot{C}Me_2 \qquad (7)$$

The mechanism for this process is essentially an intramolecular alkylation of the aryl ring with subsequent fragmentation of the spiro-cyclo-hexadienyl radical (13) to give the more stable radical (reaction 8):

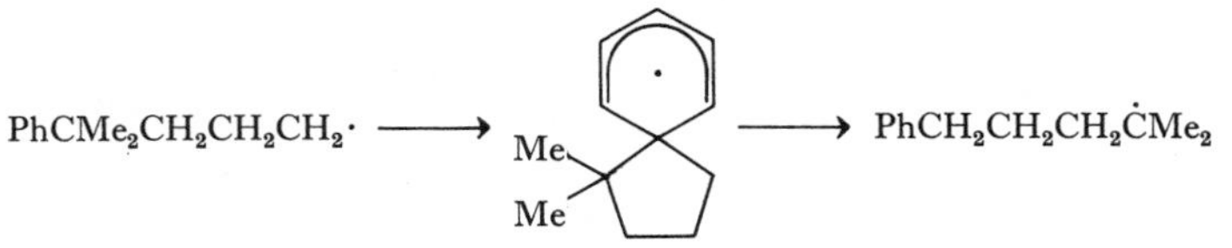

$$PhCMe_2CH_2CH_2CH_2\cdot \longrightarrow \text{(13)} \longrightarrow PhCH_2CH_2CH_2\dot{C}Me_2 \qquad (8)$$

(13)

13.2.3 *1,2-Migration of aryl groups to oxygen*

Triarylmethoxy radicals, unlike t-alkoxy radicals containing alkyl groups, do not undergo fragmentation to aryl radicals and ketones:

$$Ar_3CO\cdot \overset{\times}{\longrightarrow} Ar_2CO + Ar\cdot \qquad (9)$$

This is presumably because of the instability of the aryl radical and hence there is little driving force for the fragmentation process. Triaryl-methoxy radicals do, however, readily undergo rearrangement as evidenced by the formation of the diphenyl ether of benzopinacol (14) from the decomposition of triphenylmethyl peroxide:

$$Ph_3COOCPh_3 \longrightarrow 2Ph_3CO\cdot \longrightarrow Ph_2\dot{C}OPh \longrightarrow \begin{matrix} Ph_2COPh \\ | \\ Ph_2COPh \end{matrix} \qquad (10)$$

(14)

As with 1,2-aryl migrations to carbon, the rearrangement proceeds through a bridged transition state (15):

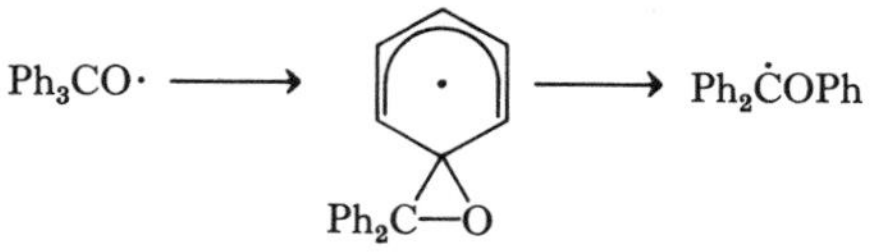

$$Ph_3CO\cdot \longrightarrow \underset{Ph_2C-O}{\text{(15)}} \longrightarrow Ph_2\dot{C}OPh \qquad (11)$$

(15)

The driving force for the rearrangement stems from the greater stability of the rearranged radical in which the unpaired electron can be delocalized onto two aromatic residues, and additionally the new C–O bond is somewhat stronger than a C–C bond. Examination of rearrangements of aryldiphenylmethoxy radicals indicates that the ability of the migrating group to delocalize an unpaired electron is probably the dominant factor in controlling the migratory aptitude of these aryl groups (22).

Radical rearrangements

The same type of behaviour is seen with triphenylsilyloxy radicals (reaction *12*) (23):

$$Ph_3SiOOSiPh_3 \longrightarrow 2Ph_3SiO\cdot \longrightarrow 2Ph_2\dot{S}iOPh \longrightarrow \begin{matrix} Ph_2SiOPh \\ | \\ Ph_2SiOPh \end{matrix} \qquad (12)$$

13.2.4 Migration of aryl groups to a more remote oxygen radical centre

Thermolysis of both bis-3,3,3-triphenylpropionyl peroxide (9), and t-butyl 3,3,3-triphenylperpropionate (24) gives, amongst other products, phenyl 2,2-diphenylacrylate, the formation of which involves rearrangement of the intermediate acyloxy radical:

$$(Ph_3CCH_2CO_2)_2 \longrightarrow Ph_2C{=}CHCO_2Ph \qquad (13)$$

$$Ph_3CCH_2CO_3Bu^t \longrightarrow Ph_2C{=}CHCO_2Ph \qquad (14)$$

The mechanism of this process involves an Ar_1-5 aryl migration (Scheme 3).

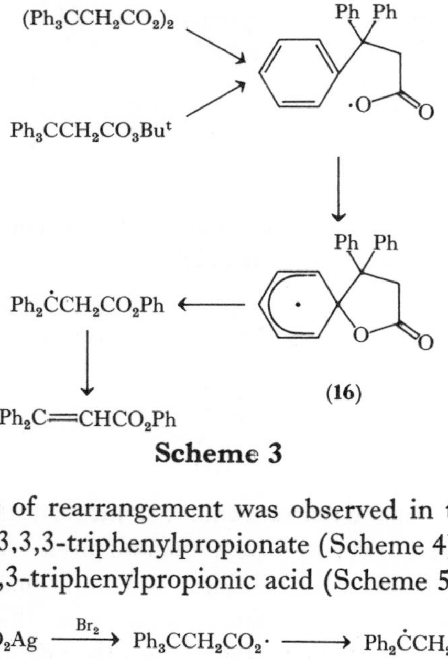

Scheme 3

The same type of rearrangement was observed in the Hunsdiecker reaction of silver 3,3,3-triphenylpropionate (Scheme 4) (25), and in the electrolysis of 3,3,3-triphenylpropionic acid (Scheme 5) (26).

$$Ph_3CCH_2CO_2Ag \xrightarrow{Br_2} Ph_3CCH_2CO_2\cdot \longrightarrow Ph_2\dot{C}CH_2CO_2Ph$$
$$\downarrow$$
$$Ph_2C{=}CBrCO_2Ph$$

Scheme 4

$$Ph_3CCH_2CO_2H \xrightarrow[-H^+]{-e} Ph_3CCH_2CO_2\cdot \longrightarrow Ph_2\overset{+}{C}CH_2CO_2Ph$$

$$\downarrow -e$$

$$Ph_2C(OMe)CH_2CO_2Ph \xleftarrow[-H^+]{MeOH} Ph_2\overset{+}{C}CH_2CO_2Ph$$

Scheme 5

No rearrangement took place in the thermolysis of bis-4,4,4-triphenyl-butyryl peroxide or in the Hunsdiecker reaction of the silver salt of 4,4,4-triphenylbutyric acid (Scheme 6).

$$Ph_3CCH_2CH_2CO_2Ag \xrightarrow{Br_2} Ph_3CCH_2CH_2CO_2\cdot$$

(17)

$$\nearrow Ph_2CCH_2CH_2CO_2Ph$$
$$\searrow_{-CO_2} Ph_3CCH_2CH_2\cdot$$

$$\downarrow Br_2$$

$$Ph_3CCH_2CH_2Br$$

Scheme 6

It is difficult to see why this should be, since models indicate that the transition state (**18**) for an Ar_1-6 migration of the radical (**17**) is com-

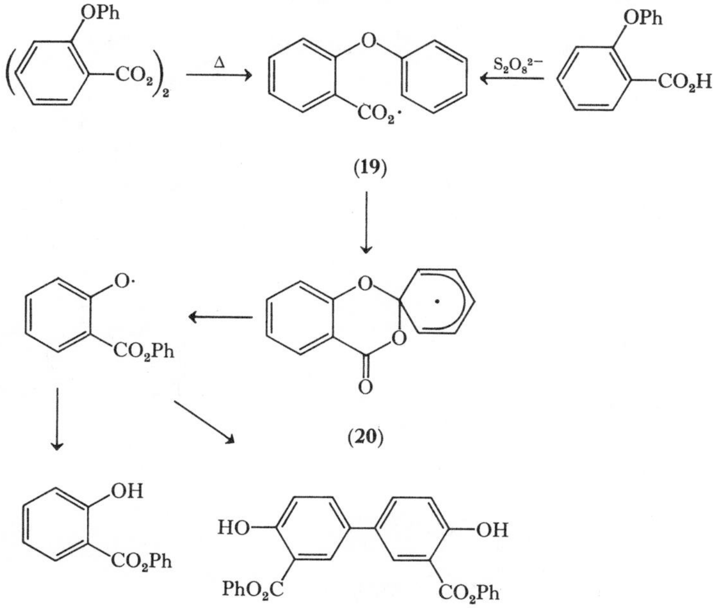

(19)

(20)

Scheme 7

parable in strain to that for the transition state **(16)** for an Ar_1-5 rearrangement.

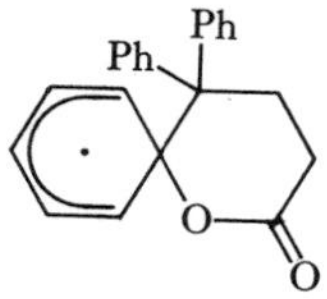

(18)

There are several examples of Ar_1-6 migrations involving migrations of aryl groups from oxygen to oxygen. Thus the main product from the decomposition of bis-2-phenoxybenzoyl peroxide is phenyl salicylate (Scheme 7) **(27)**. The transition state **(20)** in this reaction does not

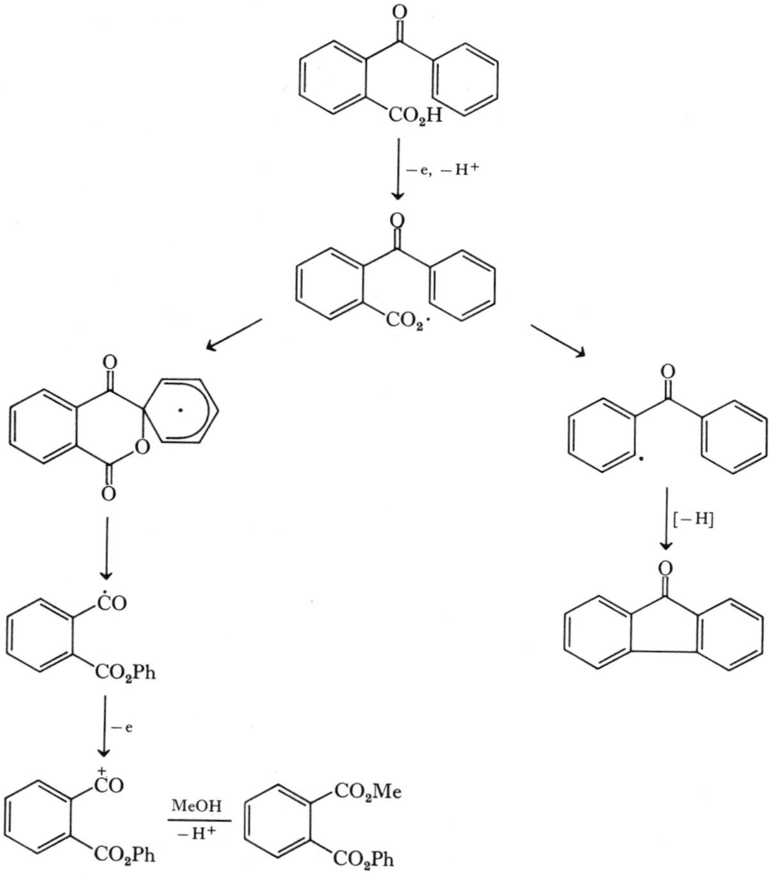

Scheme 8

appear from models to be significantly more favourable than (18). The difference in behaviour may lie more in the greater stability of (19) towards decarboxylation compared with (17), since aroyloxy radicals are much less readily decarboxylated than acyloxy radicals.

Support for the idea that the lifetime of the acyloxy radical is an important factor in determining whether rearrangement occurs is gained from the electrolysis of *o*-benzoylbenzoic acid. At low temperatures extensive rearrangement occurs, whereas at higher temperatures little or no rearrangement is observed (Scheme 8) (28). Decarboxylation of the acyloxy radical would be much faster at higher temperatures.

The rearrangement of *o*-phenoxybenzoyloxy radicals has also been observed by Thomson and Wylie in the persulphate oxidation of *o*-phenoxybenzoic acids, though under the reaction conditions employed dimeric products were isolated (Scheme 7) (29). A similar rearrangement of an aryl group from oxygen to sulphur occurs in the persulphate oxidation of *o*-thiophenoxybenzoic acids (Scheme 9) (30).

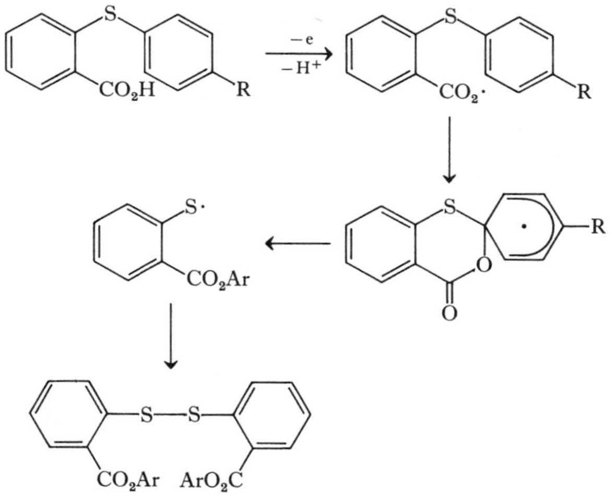

$(Ar = p\text{-}RC_6H_4)$

Scheme 9

13.2.5   *Migrations of aryl groups to silicon*

In contrast to the ease with which 2,2,2-triphenylethyl radicals rearrange, there appears to be no tendency whatsoever for the triphenylsilylmethyl radical (21) to rearrange (32):

$$Ph_3SiCH_2\cdot \xrightarrow{\quad\times\quad} Ph_2\dot{S}iCH_2Ph \qquad (15)$$
$$\text{(21)}$$

Several explanations have been advanced to account for this (32). The most cogent of these is the possibility of stabilization of the radical (21) as a result of interaction of the unpaired electron with the *d*-orbitals of silicon, and secondly that the degree of steric crowding around silicon in (21) is less than that around carbon in $Ph_3CCH_2\cdot$. A third factor is the destabilization of the transition state for rearrangement of (21) as compared to its carbon analogue. The degree of strain of the silacyclopropane ring is probably greater than that of a cyclopropane ring. These factors are either reduced or eliminated in the case of γ- and δ-silyl radicals, both of which are prone to rearrange (32).

13.3 REARRANGEMENTS INVOLVING THE MIGRATION OF A VINYL GROUP

13.3.1 *Simple 1,2-vinyl migrations*

The major product from the decarbonylation of 3-methyl-4-pentenal (22) with di-t-butyl peroxide was pent-1-ene, the formation of which involves rearrangement of the radical (23):

$$\underset{\text{(22)}}{CH_2{=}CHCHCH_2CHO} \xrightarrow{Bu^tO\cdot} CH_2{=}CHCHCH_2\dot{C}O$$

with CH_3 substituent on the CH carbons.

$$\xrightarrow{-CO} CH_2{=}CHCHCH_2\cdot \qquad (16)$$
$$\text{(23)}$$

with CH_3 substituent.

The rearrangement of the radical (23) could involve the migration of either a vinyl or a methyl group (reactions *17* and *18*) (33):

$$CH_2{=}CHCHCH_2\cdot \quad \text{(with } CH_3 \text{)}$$
$$\text{(23)}$$

$$\xrightarrow[\text{migration}]{\text{vinyl}} CH_3\dot{C}HCH_2CH{=}CH_2 \qquad (17)$$

$$\xrightarrow[\text{migration}]{\text{methyl}} CH_2{=}CH\dot{C}HCH_2CH_3 \qquad (18)$$

No *bona fide* examples of 1,2-shifts of alkyl groups at temperatures below 200 °C have been recorded in radical reactions. Consequently, it is very probable that the rearrangement involves migration of a vinyl group. This proceeds by a bridged intermediate radical (24):

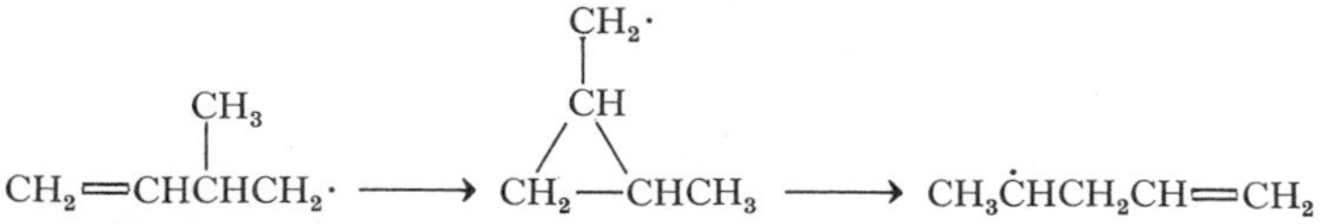

$$CH_2\!\!=\!\!CHCHCH_2 \cdot \longrightarrow CH_2\!\!-\!\!CHCH_3 \longrightarrow CH_3\dot{C}HCH_2CH\!\!=\!\!CH_2 \quad (19)$$

with structures showing CH$_3$ and the bridged CH$_2\cdot$/CH intermediate (24)

The driving force for the reaction is the formation of a secondary radical from a primary radical. Support for this scheme includes the observation that the extent of rearrangement decreased with increasing aldehyde concentration as a result of increasing competition of chain transfer with rearrangement.

The specifically labelled *cis*-deuterated aldehyde (25) gave both *cis*- and *trans*-1-deuteriobut-1-enes (Scheme 10) (34). This observation indicates that the cyclopropylmethyl radical has sufficient life-time to allow rotation about the C–C bond to occur before ring-opening. Transition-state lifetimes are generally considered to be short relative to internal rotation lifetimes, and consequently this result supports the contention that the cyclopropylmethyl radical is a reaction intermediate and not merely a transition state. At high aldehyde concentrations there is a small but significant increase in the amount of *cis*-product resulting from a chain-transfer reaction involving the initial radical.

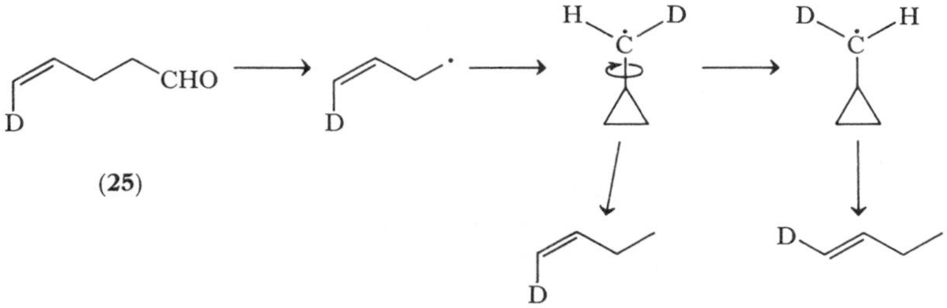

Scheme 10

Consecutive cyclization and fragmentation frequently leads to migration of vinyl groups in substituted homoallylic radicals. Reaction invariably proceeds to give the more stable radical. Schemes 11–14 provide further examples of this type of rearrangement (35–38).

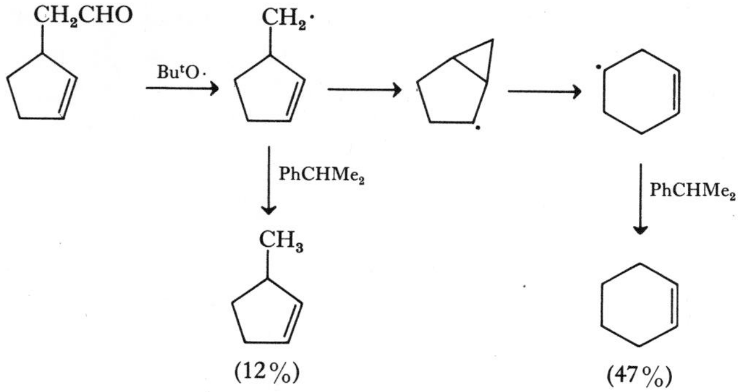

Radical rearrangements

CH₂CHO

CH₂·

Bu^tO·

PhCHMe₂

CH₃

(12%)

PhCHMe₂

(47%)

Scheme 11

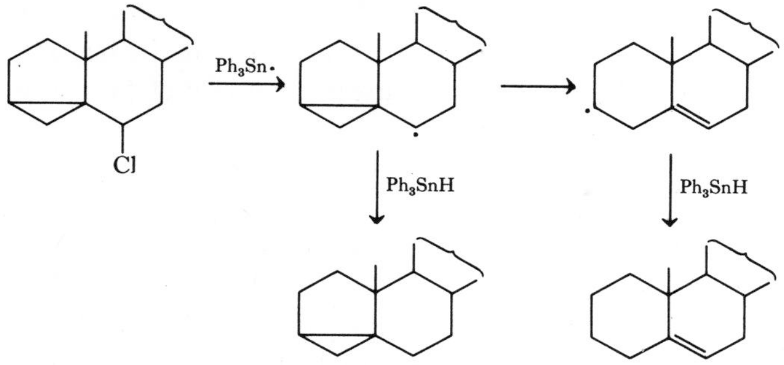

Cl

Ph₃Sn·

Ph₃SnH

Ph₃SnH

Scheme 12

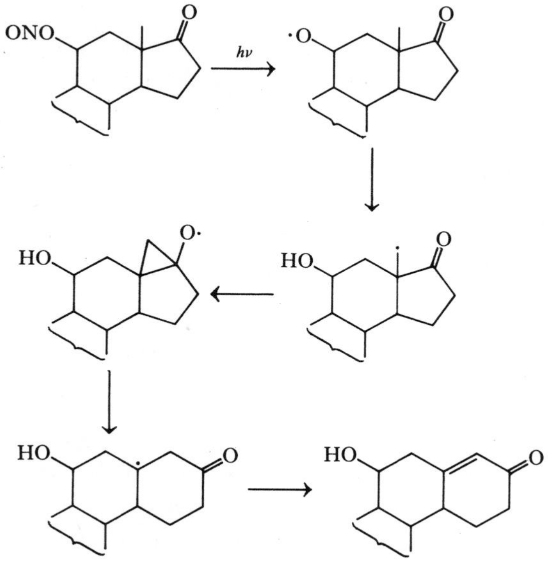

ONO

O

hν

·O

O

HO

O·

HO

O

HO

O

HO

O

Scheme 13

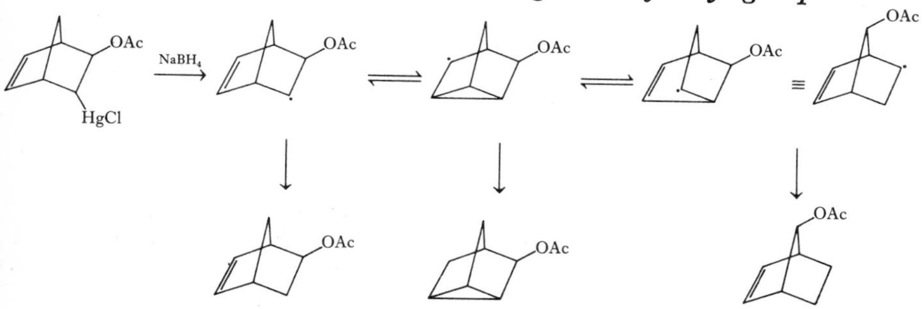

Scheme 14

13.3.2 *Cyclopropylcarbinyl-homoallyl rearrangements*

This type of rearrangement, which involves the fragmentation of cyclopropylcarbinyl radicals, has been discussed in chapter 12 (p. 480):

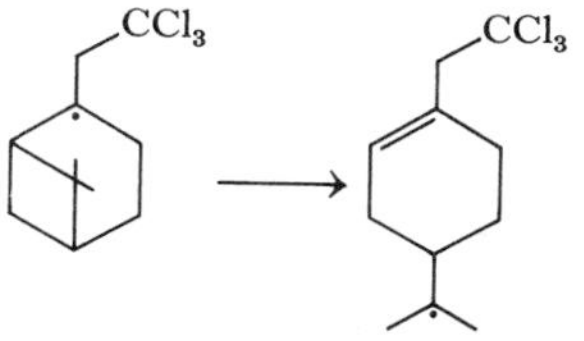

$$(20)$$

13.4 REARRANGEMENTS INVOLVING THE MIGRATION OF HALOGEN ATOMS

13.4.1 *Rearrangements in additions to haloalkenes*

Addition of hydrogen bromide to 3,3,3-trichloropropene in the presence of ultraviolet light or dibenzoyl peroxide gives exclusively 1,1,2-trichloro-3-bromopropane (39). The formation of this is a result of a 1,2-chlorine shift in the intermediate radical:

$$CCl_3CH{=}CH_2 + Br\cdot \longrightarrow CCl_3\dot{C}HCH_2Br \qquad (21)$$
$$(26)$$

$$CCl_3\dot{C}HCH_2Br \longrightarrow \dot{C}Cl_2CHClCH_2Br \qquad (22)$$
$$(26) \qquad\qquad\qquad (27)$$

$$\dot{C}Cl_2CHClCH_2Br + HBr \longrightarrow CHCl_2CHClCH_2Br + Br\cdot \qquad (23)$$

The rearrangement proceeds through the bridged radical intermediate (**28**) in which the extra electron is accommodated in the empty *d*-orbital of the chlorine. In agreement with this interpretation of the mechanism there are no reported instances of 1,2-fluorine migrations as fluorine

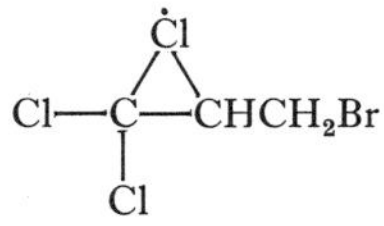

$$\underset{\underset{Cl}{|}}{\overset{\overset{\cdot}{Cl}}{\underset{}{\triangle}}}$$

$$Cl—\underset{\underset{Cl}{|}}{C}—CHCH_2Br$$

(28)

could not accommodate the extra electron. Thus, addition of hydrogen bromide to 3,3-dichloro-3-fluoropropene yields 1-fluoro-1,2-dichloro-3-bromopropane as the only rearranged product:

$$CFCl_2CH\text{=}CH_2 + Br\cdot \longrightarrow CFCl_2\overset{\cdot}{C}HCH_2Br \qquad (24)$$

$$CFCl_2\overset{\cdot}{C}HCH_2Br \nearrow\kern-1.2em\times \overset{\displaystyle \overset{\cdot}{C}Cl_2CHFCH_2Br}{\text{(29)}} \qquad\qquad (25)$$

$$\searrow \overset{\displaystyle \overset{\cdot}{C}FClCHClCH_2Br}{\text{(30)}} \qquad (26)$$

$$\overset{\cdot}{C}FClCHClCH_2Br + HBr \longrightarrow CHFClCHClCH_2Br + Br\cdot \qquad (27)$$

This is in spite of the fact that the radical **(29)** is more stable than **(30)**.

The driving force for rearrangement is, as in other instances, the formation of a more stable radical. Thus, the radical **(27)** is more stable than **(26)** because chlorine can delocalize the unpaired electron more effectively than hydrogen (cf. chapter 5, p. 106). The extent of rearrangement is less if one of the chlorines in the trichloromethyl group is replaced by either hydrogen or fluorine:

$$CCl_3CH\text{=}CH_2 + HBr \xrightarrow{\ h\nu\ } CHCl_2CHClCH_2Br \qquad (28)$$
$$(100\%)$$

$$CHCl_2CH\text{=}CH_2 + HBr \xrightarrow{\ h\nu\ } CH_2ClCHClCH_2Br \qquad (29)$$
$$(90\%)$$
$$+ CHCl_2CH_2CH_2Br$$
$$(10\%)$$

$$CFCl_2CH\text{=}CH_2 + HBr \xrightarrow{\ h\nu\ } CHFClCHClCH_2Br + CFCl_2CH_2CH_2Br$$
$$(30\%) \qquad\qquad (70\%$$
$$(30)$$

The extent of rearrangement in radical additions to alkenes is also determined by the nature of the addend. The more exothermic, and hence more rapid, is the second stage of the addition reaction, the smaller is the proportion of rearrangement which takes place. Thus the extent of rearrangement is only 53 % for the photochemical addition of bromine to 3,3,3-trichloropropene, whilst complete rearrangement occurs for the addition of hydrogen bromide (39).

There are few, if any, examples of 1,2-bromine shifts during addition reactions of alkenes. This is because the intermediate radicals tend to undergo elimination rather than rearrangement:

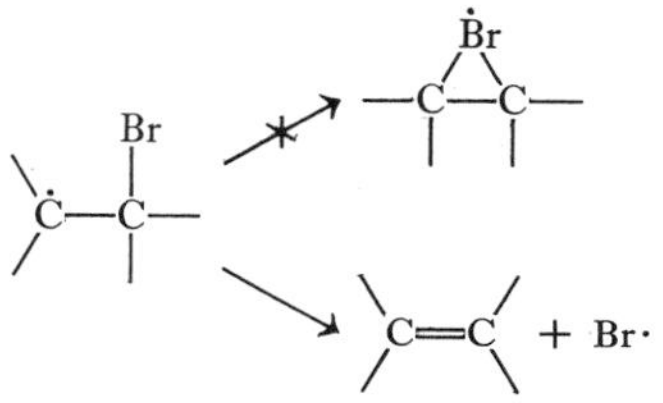

$$(31)$$

$$(32)$$

13.4.2 Rearrangements in halogenation of alkyl halides

β-Haloalkyl radicals, which can undergo 1,2-migration of halogen, are also generated in the halogenation of alkyl halides. The photochemical chlorination of t-butyl bromide and isopropyl bromide gave respectively 92 % and 15 % of the rearranged products (40) (reactions *33–35* and *36–38*):

$$Me_3CBr + Bu^tO \cdot \longrightarrow Me_2CBrCH_2 \cdot + Bu^tOH \qquad (33)$$

$$Me_2CBrCH_2 \cdot \longrightarrow Me_2\dot{C}CH_2Br \qquad (34)$$

$$Me_2\dot{C}CH_2Br + Bu^tOCl \longrightarrow Me_2CClCH_2Br + Bu^tO \cdot \qquad (35)$$

$$Me_2CHBr + Bu^tO \cdot \longrightarrow MeCHBrCH_2 \cdot + Bu^tOH \qquad (36)$$

$$MeCHBrCH_2 \cdot \longrightarrow Me\dot{C}HCH_2Br \qquad (37)$$

$$Me\dot{C}HCH_2Br + Bu^tOCl \longrightarrow MeCHClCH_2Br + Bu^tO \cdot \qquad (38)$$

The extent of rearrangement was greater for t-butyl bromide as it resulted in the formation of a tertiary radical from a primary radical.

13.5 REARRANGEMENTS INVOLVING THE MIGRATION OF ACYL GROUPS

Thermolysis of the perester (31) at 130 °C gave monomeric and dimeric products in which migration of the acetyl group had occurred (41). This occurred much more readily than migration of a phenyl group, possibly

Radical rearrangements

because of the greater stability of the resultant radical:

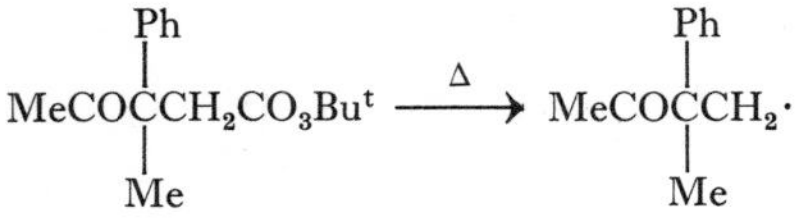

$$\underset{\underset{\text{Me}}{|}}{\overset{\overset{\text{Ph}}{|}}{\text{MeCOCCH}_2\text{CO}_3\text{Bu}^t}} \xrightarrow{\Delta} \underset{\underset{\text{Me}}{|}}{\overset{\overset{\text{Ph}}{|}}{\text{MeCOCCH}_2\cdot}} \qquad (39)$$

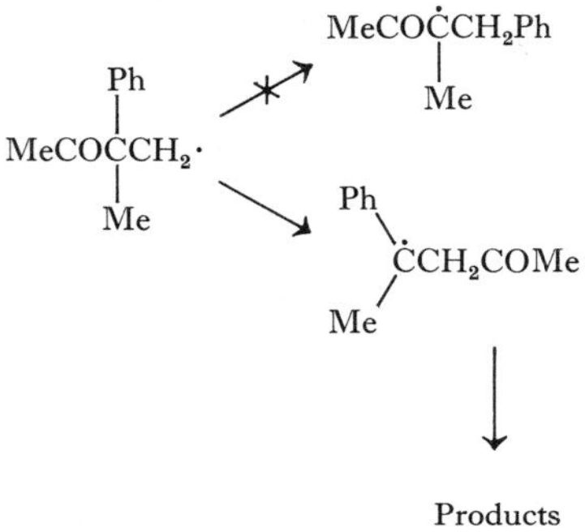

(31)

$$\underset{\underset{\text{Me}}{|}}{\overset{\overset{\text{Ph}}{|}}{\text{MeCOCCH}_2\cdot}} \qquad \qquad \overset{\overset{\text{MeCOCCH}_2\text{Ph}}{|}}{\underset{\text{Me}}{}} \qquad (40)$$

Products

As in other examples of 1,2-migrations, the rearrangement proceeds through a bridged intermediate radical (32).

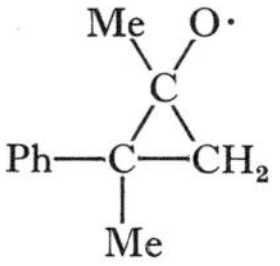

$$\text{Ph}-\underset{\underset{\text{Me}}{|}}{\overset{}{\text{C}}}-\text{CH}_2$$

(32)

13.6 REARRANGEMENTS INVOLVING THE MIGRATION OF ACYLOXY GROUPS

Rearrangement of the acetoxy group occurs during the radical-chain decarbonylation of the β-acetoxy-β-methylbutanal (33), with the resultant formation of the rearranged acetate (34) along with the un-rearranged acetate (35) (42):

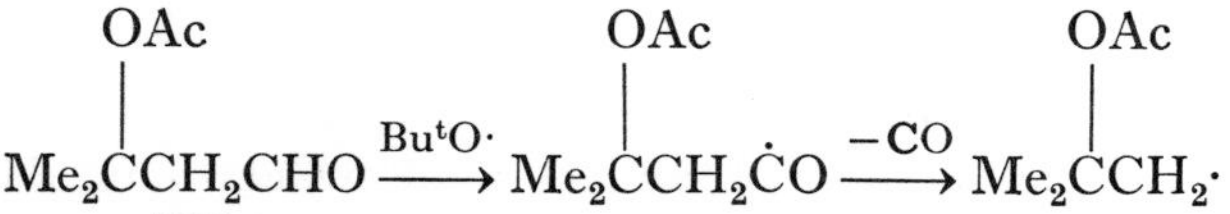

$$\underset{(33)}{\overset{\overset{\text{OAc}}{|}}{\text{Me}_2\text{CCH}_2\text{CHO}}} \xrightarrow{\text{Bu}^t\text{O}\cdot} \overset{\overset{\text{OAc}}{|}}{\text{Me}_2\text{CCH}_2\dot{\text{C}}\text{O}} \xrightarrow{-\text{CO}} \overset{\overset{\text{OAc}}{|}}{\text{Me}_2\text{CCH}_2\cdot} \qquad (41)$$

$$
\overset{\text{OAc}}{\underset{|}{\text{Me}_2\dot{\text{C}}\text{CH}_2\cdot}} \longrightarrow \text{Me}_2\dot{\text{C}}\text{CH}_2\text{OAc} \tag{42}
$$

$$
\text{Me}_2\dot{\text{C}}\text{CH}_2\text{OAc} + \text{RCHO} \longrightarrow \text{Me}_2\text{CHCH}_2\text{OAc} + \text{R}\dot{\text{C}}\text{O} \tag{43}
$$
$$
(34)
$$

$$
\overset{\text{OAc}}{\underset{|}{\text{Me}_2\text{C}}\text{CH}_2\cdot} + \text{RCHO} \longrightarrow \overset{\text{OAc}}{\underset{|}{\text{Me}_2\text{C}}\text{CH}_3} + \text{R}\dot{\text{C}}\text{O} \tag{44}
$$
$$
(35)
$$

The driving force for rearrangement is the formation of a tertiary radical from a primary radical.

Rearrangement was similarly observed during the radical additions of ethyl cyanoacetate, diethyl malonate and cyclohexane to allylic esters (43):

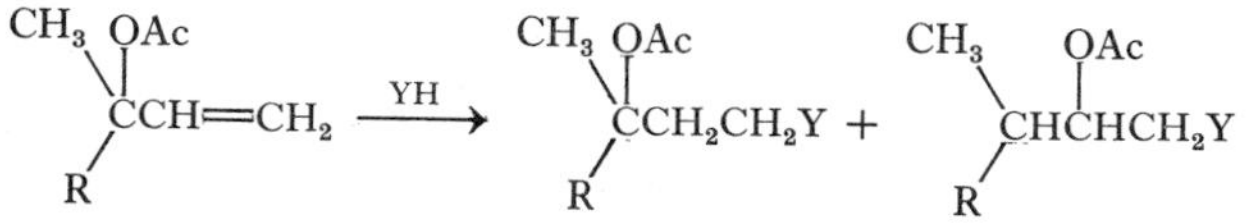

$$
\overset{\text{CH}_3}{\underset{R}{\diagup}}\overset{\text{OAc}}{\underset{|}{\text{C}}}\text{CH}{=}\text{CH}_2 \overset{\text{YH}}{\longrightarrow} \overset{\text{CH}_3}{\underset{R}{\diagup}}\overset{\text{OAc}}{\underset{|}{\text{C}}}\text{CH}_2\text{CH}_2\text{Y} + \overset{\text{CH}_3}{\underset{R}{\diagup}}\overset{\text{OAc}}{\underset{|}{\text{CH}}}\text{CHCH}_2\text{Y} \tag{45}
$$

The extent of rearrangement was much greater for the addition of diethyl malonate and cyclohexane than for ethyl cyanoacetate. This is in keeping with the much greater reactivity of ethyl cyanoacetate than diethyl malonate in radical additions, rearrangement competing effectively with hydrogen transfer for diethyl malonate but not for ethyl cyanoacetate (44).

Surzur and Teissier have considered three mechanisms for the rearrangement process (Scheme 15, on page 518) (43):

The process (*a*) corresponds to an elimination addition process and is considered most improbable because of the extreme ease with which the acetoxy radical undergoes decarboxylation, and because of the absence of any products resulting from addition of acetoxy or methyl radicals to the starting material. It is just possible that this type of pathway occurs as a cage process. The intermediate radical in route (*c*) is strictly analogous to the bridged carbonium ion which is well authenticated, and is clearly a feasible route, as is route (*b*), which is strictly analogous to that which occurs in other 1,2-migrations of radicals.

Beckwith has shown that the radicals (**36** and **37**) can be detected by e.s.r. spectroscopy when t-butyl acetate is treated with hydroxyl radicals, but that no signal could be observed which corresponded to the presence

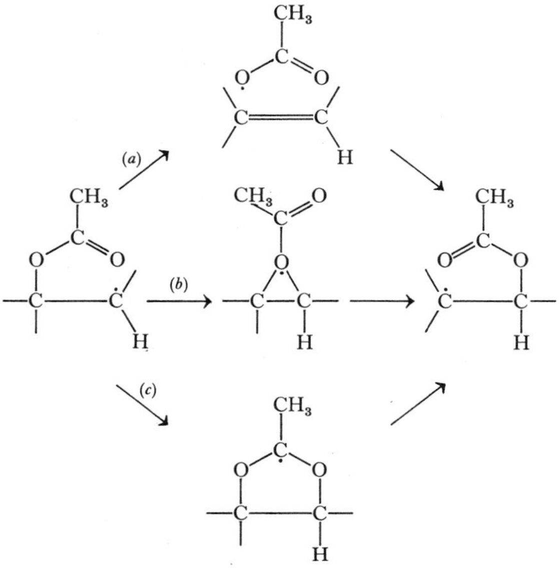

Scheme 15

of the radical (**38**) (45). Accordingly he concludes that the mechanism for rearrangement follows path (*b*):

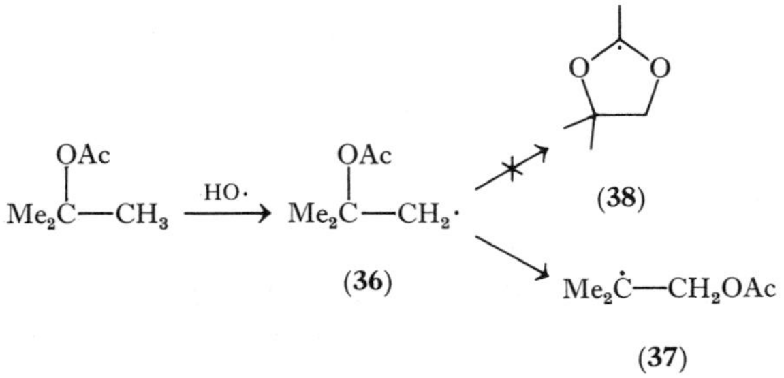

$$\text{(46)}$$

If this is the correct mechanism for the migration of an acetoxy group, it is not surprising that no examples involving 1,3-migration of acetoxy groups have been observed either in the decarbonylation of γ-acetoxy-aldehydes or in the addition of ethyl cyanoacetate to β-acetoxyalkenes (42, 43). Had the migration involved intramolecular addition to a carbonyl group followed by subsequent fragmentation (route *c*), one might reasonably expect to see examples of 1,3-migration:

$$\text{Me}_2\overset{\text{OAc}}{\underset{}{\text{C}}}\text{CH}_2\text{CH}_2\cdot \;\longrightarrow\!\!\!\times\!\!\!\longrightarrow\; \text{Me}_2\overset{\cdot}{\text{C}}\text{CH}_2\text{CH}_2\text{OAc} \qquad\qquad (47)$$

13.7   REARRANGEMENTS INVOLVING THE MIGRATION OF HYDROGEN ATOMS

Strictly, intramolecular hydrogen transfer reactions are not rearrangements in that there is no change in the carbon skeleton. They are, however, more conveniently considered along with other rearrangements.

13.7.1   *1,2-Hydrogen migrations*

There are no authenticated examples of 1,2-hydrogen migrations in radicals (46). Thus, in the photochemical chlorination of 2-deuterioisobutane the ratio of t-butyl chloride to isobutyl chloride was identical with that of the deuterium chloride to hydrogen chloride. This points to the absence of deuterium or hydrogen migration:

$$Me_3CD \xrightarrow{Cl\cdot} \begin{cases} \rightarrow DCl + Me_3C\cdot \longrightarrow Me_3CCl \\ \rightarrow HCl + Me_2CDCH_2\cdot \longrightarrow Me_2CDCH_2Cl \end{cases} \qquad (48)$$

13.7.2   *1,3- and 1,4-Hydrogen migrations*

Reutov has shown the occurrence of 1,3-hydrogen migration in propyl radicals from a study of the decomposition of 2-^{14}C-dibutyryl peroxide in carbon tetrachloride (47). The resultant propyl chloride contained the labelled carbon in both the 1- and the 3-positions:

$$(CH_3CH_2{}^{14}CH_2CO_2)_2 \xrightarrow[-2CO_2]{\Delta} 2CH_3CH_2{}^{14}CH_2\cdot \qquad (49)$$

$$CH_3CH_2{}^{14}CH_2\cdot + CCl_4 \longrightarrow CH_3CH_2{}^{14}CH_2Cl + \dot{C}Cl_3 \qquad (50)$$

$$CH_3CH_2{}^{14}CH_2\cdot \longrightarrow \cdot CH_2CH_2{}^{14}CH_3 \qquad (51)$$

$$^{14}CH_3CH_2CH_2\cdot + CCl_4 \longrightarrow {}^{14}CH_3CH_2CH_2Cl + \dot{C}Cl_3 \qquad (52)$$

The absence of 1,2-hydrogen migration indicates that the mechanism for the rearrangement must involve some sort of four-membered transition state (**39**) as opposed to consecutive 1,2-hydrogen migrations.

$$\begin{array}{c} CH_2 \\ \diagup \quad \diagdown \\ H_2C \quad\quad {}^{14}CH_2 \\ \diagdown \quad \diagup \\ H \end{array}$$

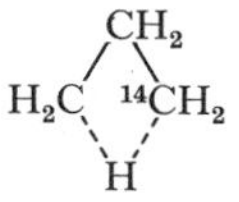

(**39**)

Reutov has suggested that 1,3-hydrogen migration only occurs when the transition state for rearrangement has a high degree of symmetry, and it is for this reason that 4-phenylbutyl radicals do not undergo rearrangement.

4-Phenylbutyl radicals also fail to undergo 1,4-hydrogen migration even though the resultant radical would be a stabilized benzylic radical (48):

$$PhCH_2CH_2CH_2CH_2\cdot \xrightarrow{\quad\times\quad} Ph\dot{C}HCH_2CH_2CH_3 \qquad (53)$$

Both 1,3- and 1,4-hydrogen shifts are, however, observed in the case of the radicals obtained by fragmentation of the radicals derived from tetrahydrofuran and tetrahydropyran (49) (reactions *54–58*) ($n = 2$ or 3). This is probably a consequence of the greater stability of the initial radical:

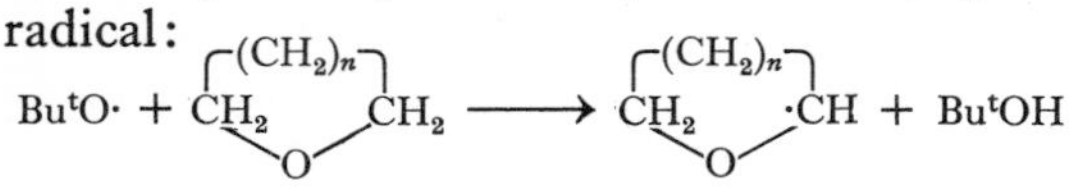

$$Bu^tO\cdot + \underset{\text{(ring: } (CH_2)_n, CH_2, CH_2, O)}{} \longrightarrow \underset{\text{(ring: } (CH_2)_n, CH_2, \cdot CH, O)}{} + Bu^tOH \qquad (54)$$

$$\underset{\text{(ring: } (CH_2)_n, CH_2, \cdot CH, O)}{} \longrightarrow \dot{C}H_2(CH_2)_nCHO \qquad (55)$$

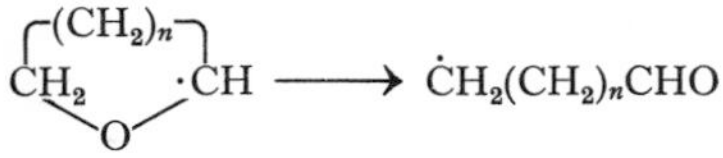

$$\dot{C}H_2(CH_2)_nCHO \longrightarrow CH_3(CH_2)_n\dot{C}O \qquad (56)$$

$$CH_3(CH_2)_n\dot{C}O + C_6H_{13}CH{=}CH_2 \longrightarrow CH_3(CH_2)_nCOCH_2\dot{C}HC_6H_{13} \qquad (57)$$

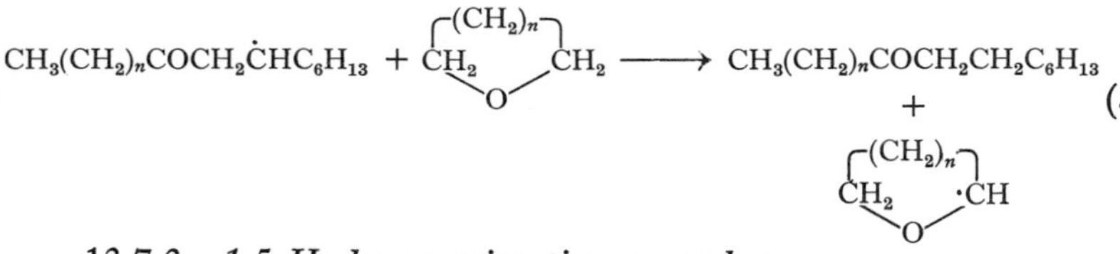

$$CH_3(CH_2)_nCOCH_2\dot{C}HC_6H_{13} + \underset{\text{(ring: } (CH_2)_n, CH_2, CH_2, O)}{} \longrightarrow CH_3(CH_2)_nCOCH_2CH_2C_6H_{13} \\ + \qquad\qquad (58) \\ \underset{\text{(ring: } (CH_2)_n, CH_2, \cdot CH, O)}{}$$

13.7.3 *1,5-Hydrogen migrations to carbon*

In contrast to the small number of examples of 1,2-, 1,3- and 1,4-hydrogen migrations, there are numerous examples of 1,5-hydrogen migration. Thus, decomposition of bis-6-phenylhexanoyl peroxide gives products which are derived from the rearranged 5-phenylpentyl radical (50, 51):

$$(PhCH_2CH_2CH_2CH_2CH_2CO_2)_2 \xrightarrow[-2CO_2]{\Delta} 2PhCH_2CH_2CH_2CH_2CH_2\cdot \qquad (59)$$

$$PhCH_2CH_2CH_2CH_2CH_2\cdot \longrightarrow Ph\dot{C}HCH_2CH_2CH_2CH_3 \qquad (60)$$

$$Ph\dot{C}HCH_2CH_2CH_2CH_3 \longrightarrow CH_3(CH_2)_3\overset{\overset{\displaystyle Ph}{|}}{C}H\overset{\overset{\displaystyle Ph}{|}}{C}H(CH_2)_3CH_3 \qquad (61)$$

$$\text{PhĊHCH}_2\text{CH}_2\text{CH}_2\text{CH}_3 \xrightarrow{\text{Cu(OAc)}_2} \text{PhCH=CHCH}_2\text{CH}_2\text{CH}_3$$

$$\begin{array}{c} \quad\quad\quad\quad\quad\quad\quad\quad\quad\quad\quad \text{OAc} \\ \quad\quad\quad\quad\quad\quad\quad\quad\quad\quad\quad | \\ +\text{PhĊHCH}_2\text{CH}_2\text{CH}_2\text{CH}_3 \quad (62) \end{array}$$

Beckwith (52), using e.s.r. spectroscopy, showed that 1,5-hydrogen migration occurs in the case of the radicals (**41** and **43**), generated by reduction of the appropriate diazonium salts (**40** and **42**). Products derived from the rearranged radicals were also isolated (Schemes 16 and 17).

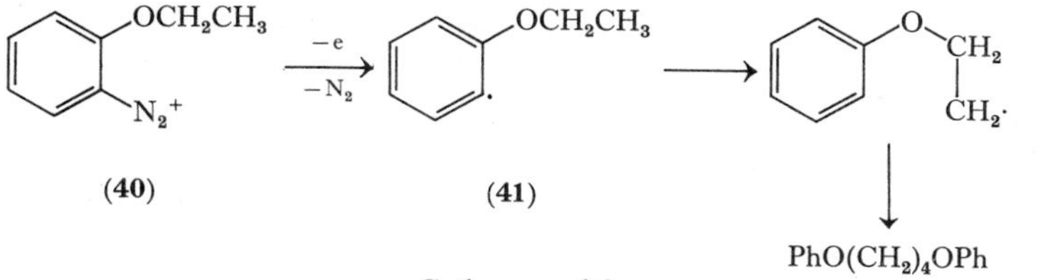

Scheme 16

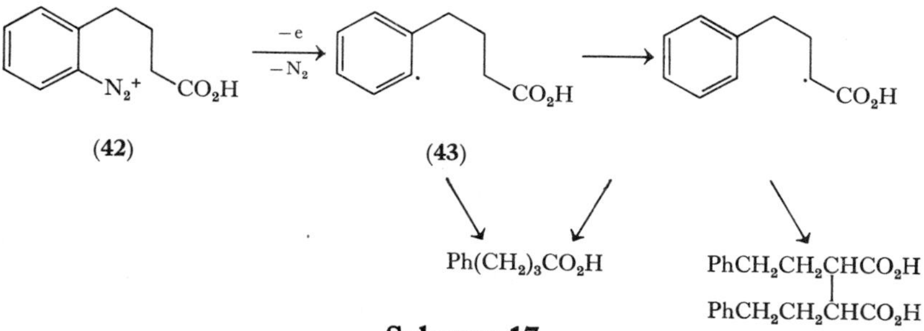

Scheme 17

Tedder has explained the anomalously high reactivity of the hydrogens attached to C-5 and the low reactivity of the hydrogens of the t-butyl group in the radical chlorination of 2,2-dimethylhexane on the basis of 1,5-hydrogen transfer (53):

$$\begin{array}{c} \text{CH}_2\text{·} \\ | \\ \text{Me}_2\text{ĊCH}_2\text{CH}_2\text{CH}_2\text{CH}_3 \longrightarrow \text{Me}_3\text{CCH}_2\text{CH}_2\text{ĊHCH}_3 \quad (63) \end{array}$$

The ease with which 1,5-hydrogen transfer occurs has been attributed to the fact that the transition state (**44**) resembles the geometry of a cyclohexane ring. Eclipsing interactions in the transition state for 1,4-hydrogen migration would militate against a facile reaction. In addition the two carbon atoms between which the hydrogen transfers would be too far apart to allow ready migration.

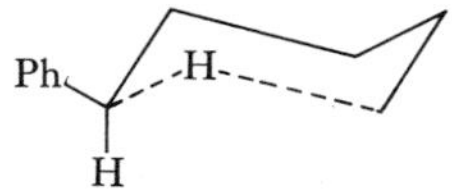

(44)

13.7.4 *1,5-Hydrogen migrations to oxygen*

The relative ease with which 1,5-hydrogen migration from carbon to carbon occurs is paralleled in the case of migrations from carbon to oxygen. Barton (65, 66), Beckwith (54), Walling and Padwa (55), Kochi (56), Greene (57), and Kabasakalian and Townley (58) have all shown that alkoxy radicals undergo facile 1,5-hydrogen migration (cf. reaction *64*):

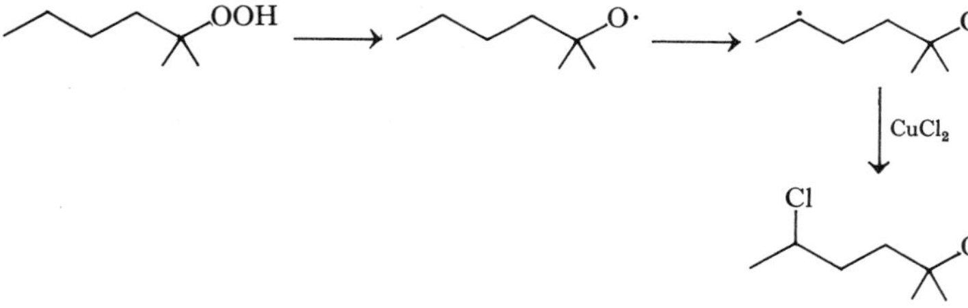

(64)

The same results were obtained independent of the source of the alkoxy radical. No instances of 1,2-, 1,3- or 1,4-hydrogen migration were encountered by these workers. Walling and Padwa did, however, come across cases in which 1,6-hydrogen migration competed with 1,5-hydrogen shift, albeit not very efficiently (Scheme 18).

$$PhCH_2CH_2CH_2CH_2CMe_2OCl \xrightarrow{h\nu} PhCH_2CH_2CH_2CH_2CMe_2O\cdot$$

PhCH₂CHClCH₂CH₂CMe₂OH   PhCH₂ĊHCH₂CH₂CMe₂OH

(9 parts) $\xrightarrow[-RO\cdot]{ROCl}$ +

PhCHClCH₂CH₂CH₂CMe₂OH   PhĊHCH₂CH₂CH₂CMe₂OH
(1 part)

Scheme 18

The driving force for 1,5-hydrogen migration from carbon to oxygen is greater than that for the analogous shift to carbon, as the new O–H

bond is stronger than the C–H bond which is broken. The greater ease
with which the reaction occurs is reflected in the very high proportion
of migration which is frequently encountered (cf. Scheme 19) (51).

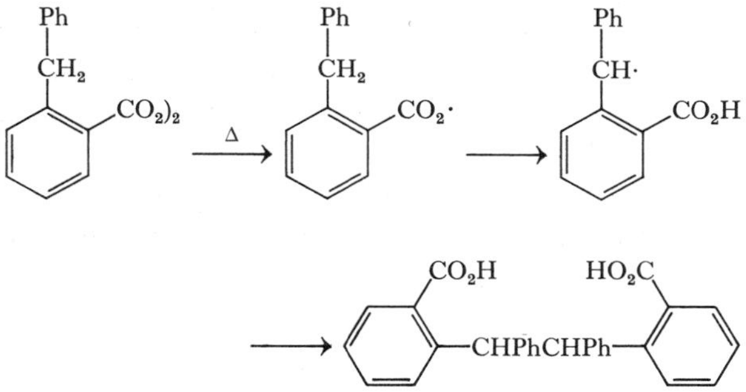

Scheme 19

Intramolecular hydrogen migration does not, however, occur when this
would involve abstraction of an aryl hydrogen as in the 2-phenylethoxy
radical (59), because an aryl C–H bond is stronger than an alkyl C–H
bond and an aryl radical is less stable than an alkyl radical:

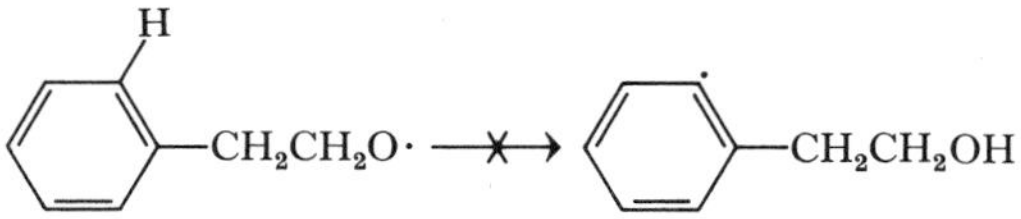

$$(65)$$

The ease with which intramolecular hydrogen migration occurs is
very sensitive to steric factors. The transition state for migration has to
have a quasi-chair form which allows favourable orbital overlap. When
this is not possible, as for cyclohexyloxy radicals, no 1,5-hydrogen shift
occurs (60). In this case the transition state would necessitate the
cyclohexane ring adopting a boat conformation (**45**). The extra energy
involved in this precludes hydrogen migration.

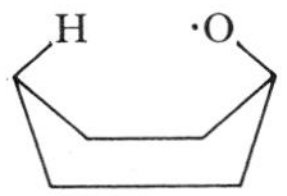

(45)

13.7.5 *The Barton reaction*

1,5-Hydrogen migration in radicals forms the basis of an exceedingly valuable synthetic procedure for the synthesis of compounds not readily accessible by other means (61–63). The essence of this reaction involves the generation of a radical centre in a steric environment such that 1,5-hydrogen migration can occur via a six-membered chair-like transition state. Scheme 20 indicates the general mechanism for such a reaction, the essential feature of which is the generation of an alkoxy radical which subsequently gives rise to a carbon radical after hydrogen migration.

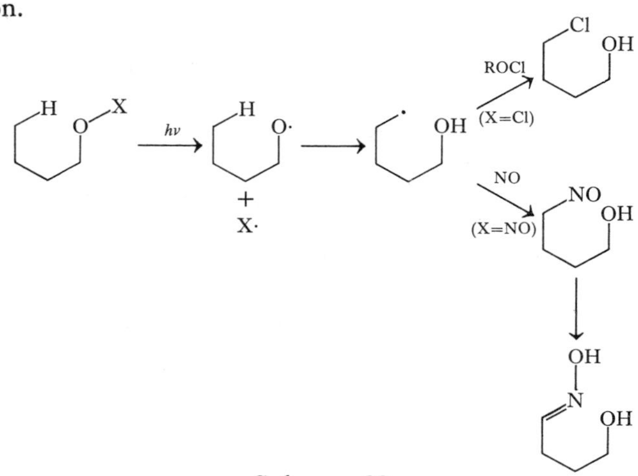

Scheme 20

The reaction has been particularly successfully applied to the synthesis of C-18 and C-19 derivatives of steroids. The formula (**46**) indicates that C-18 derivatives should be formed by generation of the alkoxy radical at C-8, C-11, C-15 or C-20, and C-19 derivatives from a radical centre at C-2, C-4, C-6, C-8 or C-11.

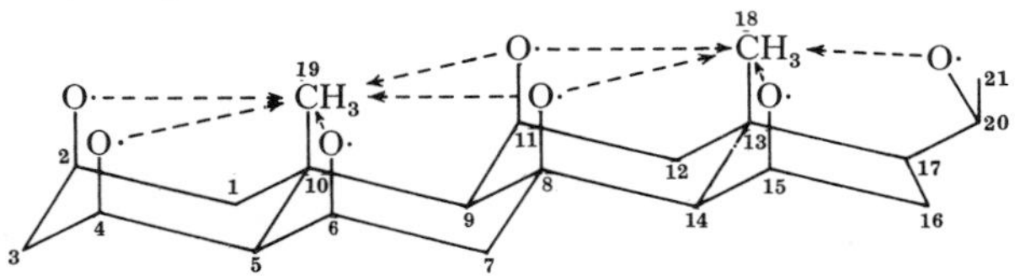

(**46**)

Schemes 21–23 give simple examples of the Barton reaction based on photolysis of nitrite esters.

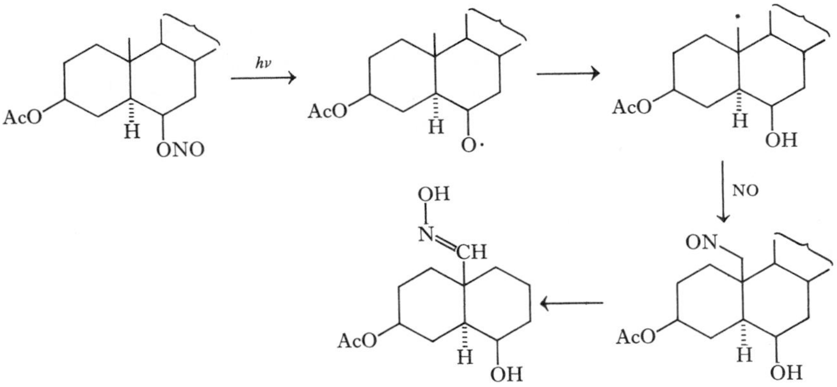

(67%)

Scheme 21

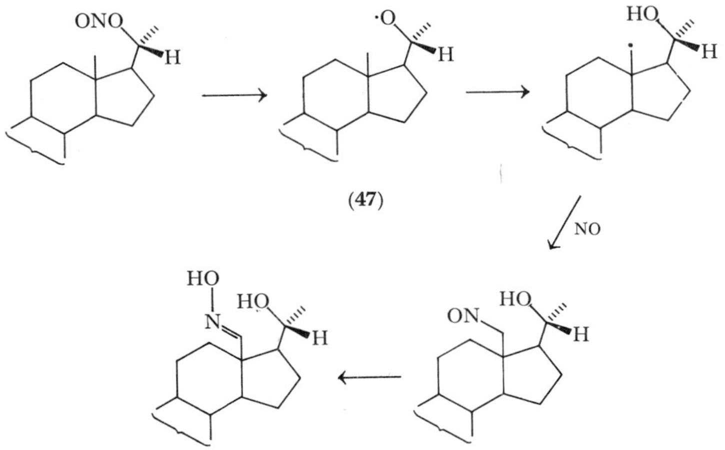

(47)

(34%)

Scheme 22

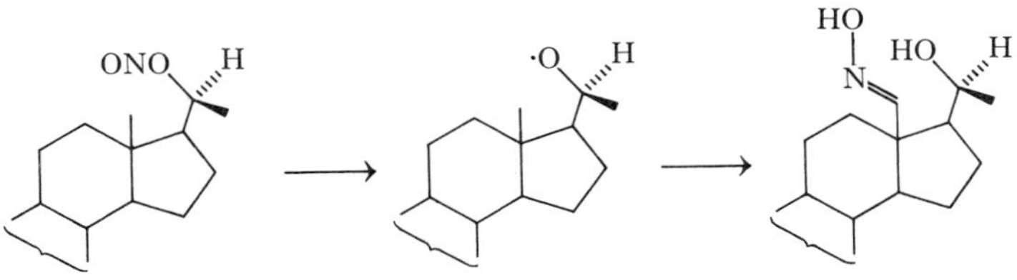

(48)   (60–65%)

Scheme 23

Radical rearrangements

It is interesting to note that the yield of product from the photolysis of the nitrite of 20α-hydroxy-4-pregnen-3-one was considerably greater than that from the 20β-isomer. The reason for this becomes clear from a consideration of the transition states (**49** and **50**) for the hydrogen migration of the radicals (**47** and **48**). The transition state (**49**) suffers a 1,3-diaxial interaction with hydrogens at C-12. This is absent in (**50**) and consequently reaction proceeds more efficiently for the radical (**48**).

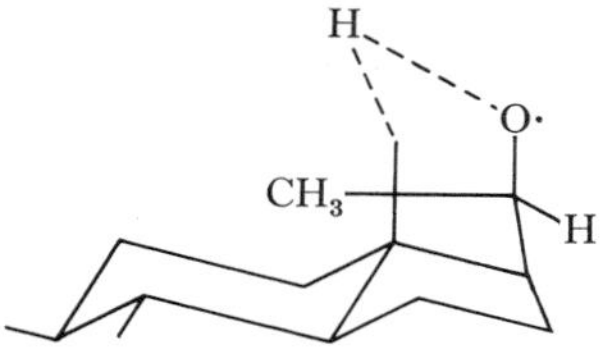

(49) 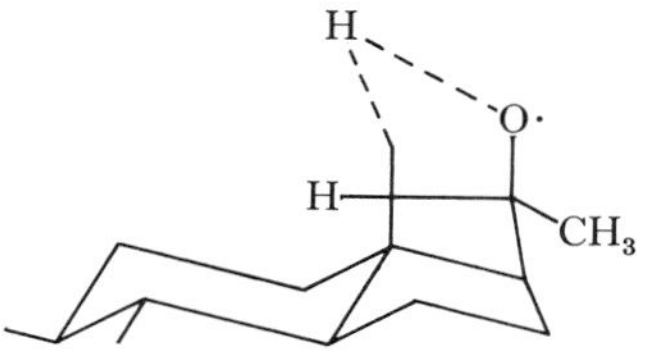   (50)

The distance separating the two atoms involved in the hydrogen migration is important. This can be varied to some extent by changes in the basic skeleton involving introduction of trigonal centres. Thus, whilst the radical from the nitrite (**51**) results in equal attack at both C-18 and C-19, introduction of a second double bond as in (**52**) results in exclusive reaction at C-18. This second double bond pushes the 19-methyl group away from the radical centre, thereby favouring attack at C-18 (62).

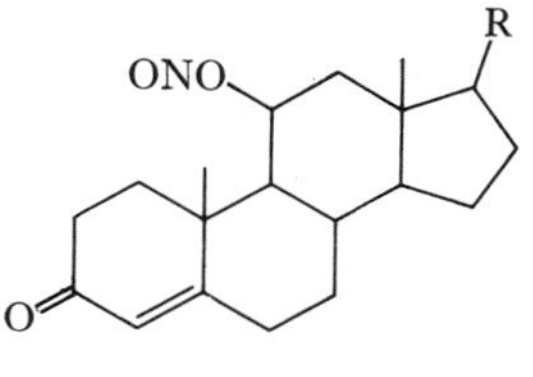

(51) 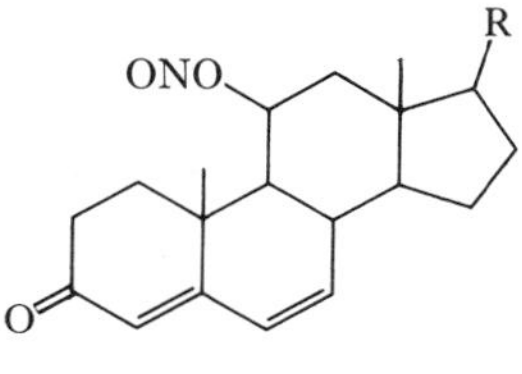   (52)

There are instances in which the rearranged radical undergoes further intramolecular rearrangement before reacting in an intermolecular reaction with nitric oxide (Scheme 24) (64).

Hypochlorites behave similarly on irradiation and provide a route to cyclic ethers (Scheme 25) (65). A related reaction which has also been used to good effect is the generation of a hypoiodite *in situ* from the alcohol, lead tetra-acetate and iodine (63).

526

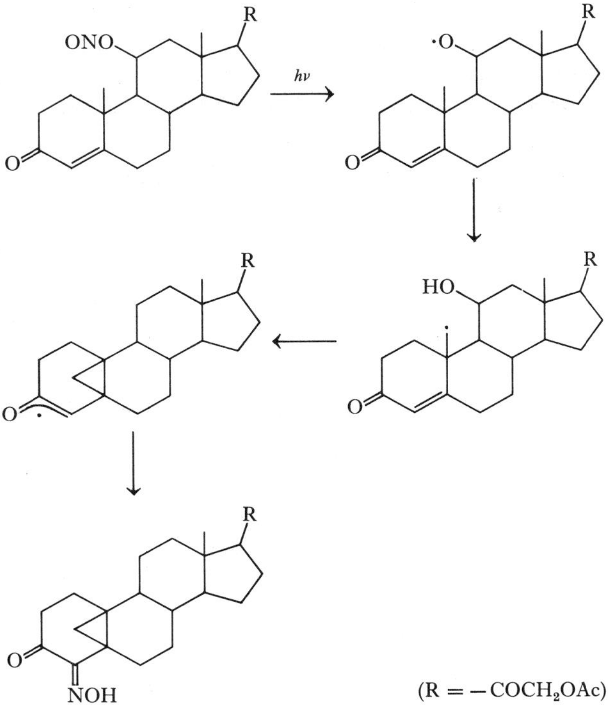

(R = − COCH$_2$OAc)

Scheme 24

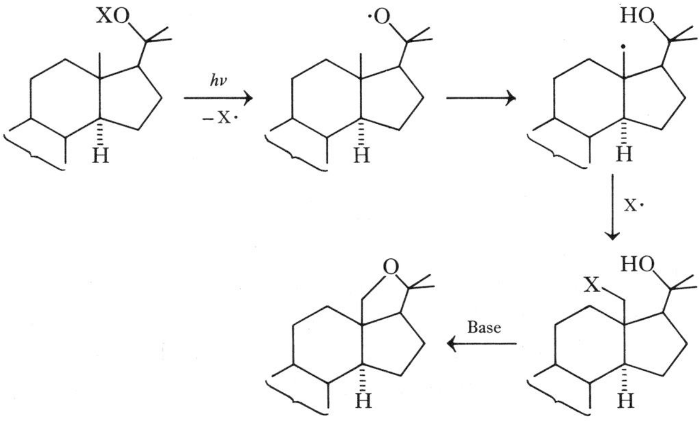

Scheme 25

13.7.6 *Hydrogen migration from carbon to nitrogen*

Photolysis or thermolysis of N-chloroamines in strongly acidic media (the Hofmann–Löffler–Freytag reaction) results in the formation of pyrrolidines and piperidines (66) after treatment of the chloramines with base. The groups of Wawzonek (67) and Corey (68) have established the homolytic nature of the reaction (Scheme 26).

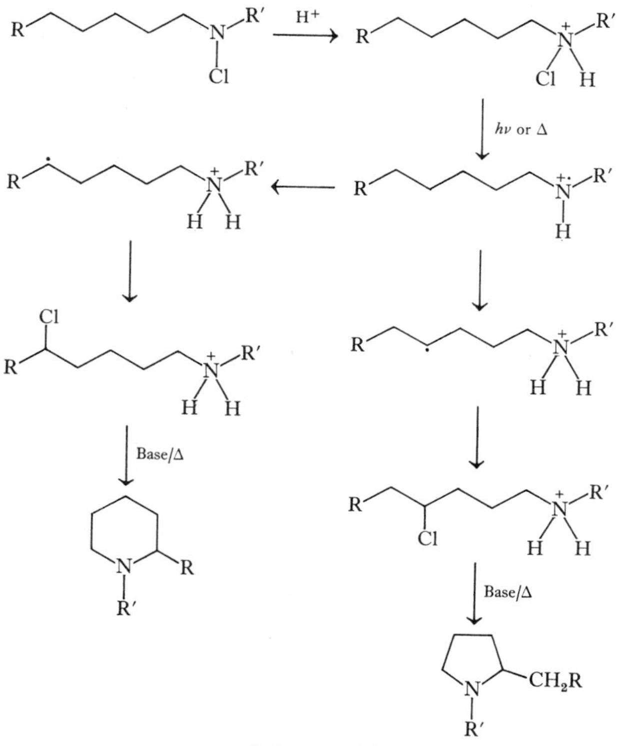

Scheme 26

The hydrogen migration in this reaction is rather less specifically 1,5-hydrogen migration than for the comparable migrations to oxygen. In all other aspects the two reactions show considerable similarity.

This reaction has been used as a valuable synthetic tool (reactions *66* and *67*) (69, 70):

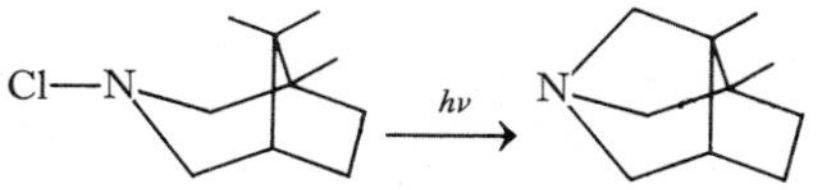

$$(66)$$

(67%)

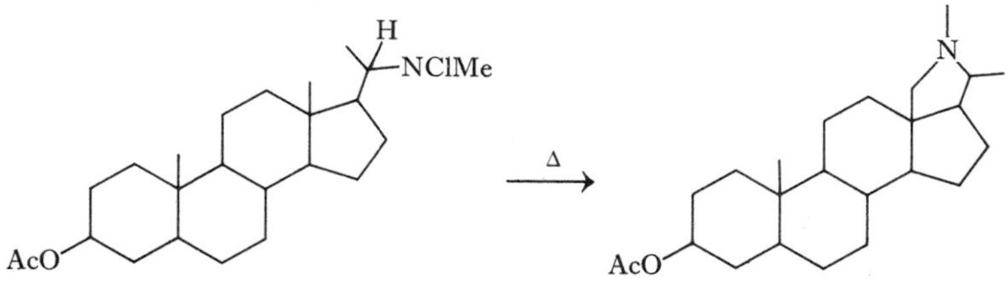

$$(67)$$

The initial radical cation may also be generated in a redox reaction:

$$R_2NHCl + Fe^{2+} \longrightarrow R_2\overset{+\cdot}{N}H + Cl^- + Fe^{3+} \qquad (68)$$

Thus, *N*-chloro-di-n-butylamine, on treatment with iron(II) sulphate and persulphate, gave an excellent yield of *N*-butylpyrrolidine (reaction *69*):

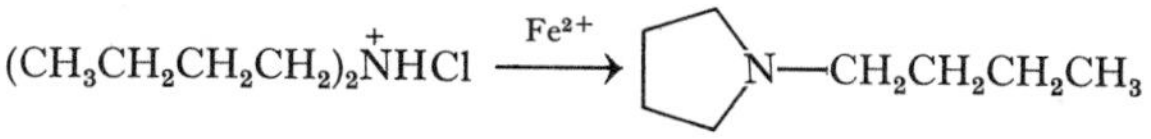

$$(CH_3CH_2CH_2CH_2)_2\overset{+}{N}HCl \xrightarrow{Fe^{2+}} \underset{}{N}\!\!-CH_2CH_2CH_2CH_3 \qquad (69)$$

(91%)

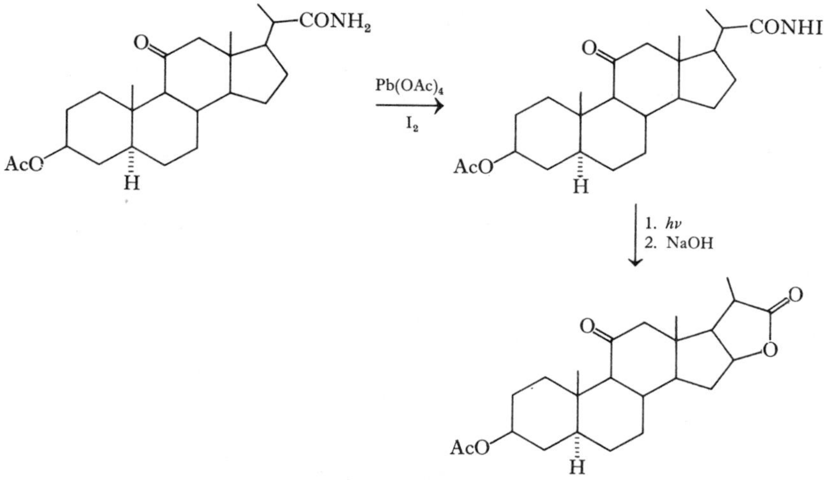

Scheme 27

Radical rearrangements

Somewhat related to the Hoffmann–Löffler–Freytag reaction is the photolysis of *N*-halogenoamides which provides a convenient and easy route to lactones. Barton and Beckwith first developed the procedure for use with steroid derivatives (Scheme 27) (71). The course of the reaction is indicated in Scheme 28.

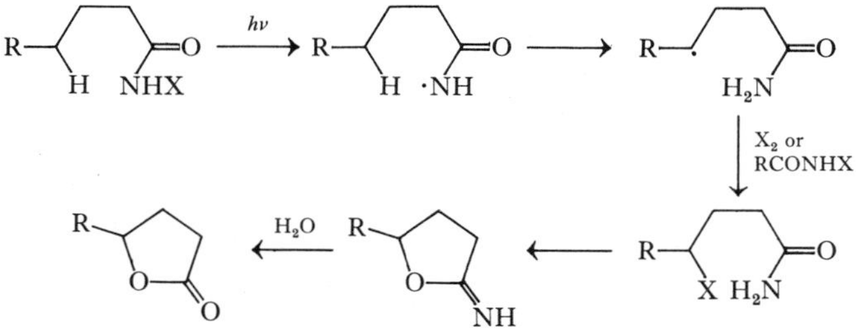

Scheme 28

GENERAL REFERENCES

R. Kh. Friedlina, *Adv. Free-Radical Chem.* 1965, **1**, 211.

A. L. J. Beckwith, *Essays in Free-Radical Chemistry, Chem. Soc. Special Publ.* no. 24, 1970, 239.

C. Walling in *Molecular Rearrangements*, ed. P. de Mayo, Interscience, New York, 1963, vol. 1, p. 407.

O. A. Reutov, *Pure Appl. Chem.* 1963, **7**, 203.

SPECIFIC REFERENCES

1. J. K. Kochi, *J. Amer. Chem. Soc.* 1963, **85**, 1958.
2. J. March, *Advanced Organic Chemistry*, McGraw-Hill, New York, 1968, p. 781.
3. M. J. Perkins, in *Organic Reaction Mechanisms*, ed. B. Capon and C. W. Rees, Interscience, London, 1968, p. 293.
4. M. J. S. Dewar and S. Kirschner, *J. Amer. Chem. Soc.* 1971, **93**, 4290, 4291.
5. C. D. Cook and M. Fraser, *J. Org. Chem.* 1964, **29**, 3716.
6. S. Winstein and F. H. Seubold, *J. Amer. Chem. Soc.* 1947, **69**, 2916.
7. S. Winstein, R. Heck, S. Lapporte and K. Baird, *Experientia*, 1956, **12**, 138.
8. C. Rüchardt, *Chem. Ber.* 1961, **94**, 2599, 2609.
9. W. Rickatson and T. S. Stevens, *J. Chem. Soc.* 1963, 3960.
10. D. Y. Curtin and J. C. Kauer, *J. Org. Chem.* 1960, **25**, 880.
11. L. Slaugh, *J. Amer. Chem. Soc.* 1959, **81**, 2262.
12. D. Y. Curtin and M. G. Hurwitz, *J. Amer. Chem. Soc.* 1952, **74**, 5381.
13. W. H. Urry and N. Nicolaides, *J. Amer. Chem. Soc.* 1952, **74**, 5163.
14. C. Rüchardt and H. Trautwein, *Chem. Ber.* 1965, **98**, 2478.
15. J. K. Kochi and P. J. Krusic, *J. Amer. Chem. Soc.* 1969, **91**, 3940.

16. G. A. Olah, M. B. Comisarow, E. Namanworth and B. Ramsey, *J. Amer. Chem. Soc.* 1967, **89**, 5259.
17. J. E. Nordlander and W. G. Deadman, *J. Amer. Chem. Soc.* 1968, **90**, 1590.
18. R. C. P. Cubbon, *Progr. Reaction Kinetics*, 1970, **5**, 29.
19. C. Rüchardt and R. Hecht, *Chem. Ber.* 1965, **98**, 2460, 2471.
20. C. Rüchardt and S. Eichler, *Chem. Ber.* 1962, **95**, 1921.
21. J. P. Lorand and P. D. Bartlett, *J. Amer. Chem. Soc.* 1966, **88**, 3294.
22. M. S. Kharasch, A. C. Poshkus, A. Fono and W. Nudenberg, *J. Org. Chem.* 1951, **16**, 1458.
23. A. K. Shubber and R. L. Dannley, *J. Org. Chem.* 1971, **36**, 3784.
24. W. H. Starnes, *J. Amer. Chem. Soc.* 1963, **85**, 3708.
25. J. W. Wilt and J. A. Lundquist, *J. Org. Chem.* 1964, **29**, 921.
26. H. Breederveld and E. C. Kooyman, *Rec. Trav. chim.* 1957, **76**, 297.
27. D. F. DeTar and A. Hlynsky, *J. Amer. Chem. Soc.* 1955, **77**, 4411.
28. P. J. Bunyan and D. H. Hey, *J. Chem. Soc.* 1962, 324, 2771.
29. R. H. Thomson and A. G. Wylie, *J. Chem. Soc.* C, 1966, 321.
30. P. M. Brown, P. S. Dewar, A. R. Forrester, A. S. Ingram and R. H. Thomson, *Chem. Comm.* 1970, 849.
31. J. W. Wilt, O. Kolewe and J. F. Kraemer, *J. Amer. Chem. Soc.* 1969, **91**, 2624.
32. J. W. Wilt and C. F. Dockus, *J. Amer. Chem. Soc.* 1970, **92**, 5813.
33. L. K. Montgomery, J. W. Matt and J. R. Webster, *J. Amer. Chem. Soc.* 1967, **89**, 923.
34. L. K. Montgomery and J. W. Matt, *J. Amer. Chem. Soc.* 1967, **89**, 3050, 6556.
35. L. Slaugh, *J. Amer. Chem. Soc.* 1965, **87**, 1522.
36. S. J. Cristol and R. V. Barbour, *J. Amer. Chem. Soc.* 1968, **90**, 2832.
37. H. Reimann, A. S. Capomaggi, T. Strauss, E. P. Oliveto and D. H. R. Barton, *J. Amer. Chem. Soc.* 1961, **83**, 4481; M. Akhtar, *Adv. Photochem.* 1966, **2**, 263.
38. G. A. Grey and W. R. Jackson, *J. Amer. Chem. Soc.* 1969, **91**, 6205.
39. R. Kh. Freidlina, *Adv. Free-Radical Chem.* 1965, **1**, 211.
40. P. S. Skell, R. G. Allen and N. D. Gilmour, *J. Amer. Chem. Soc.* 1961, **83**, 504.
41. W. Reusch and C. L. Kart, *Abstr. Amer. Chem. Soc. Meeting*, San Francisco, 1968, p. 79.
42. D. D. Tanner and F. C. P. Law, *J. Amer. Chem. Soc.* 1969, **91**, 7535.
43. J.-M. Surzur and P. Teissier, *Bull. Soc. chim. France*, 1970, 3060.
44. J. I. G. Cadogan, D. H. Hey and J. T. Sharp, *J. Chem. Soc.* B, 1967, 803.
45. A. L. J. Beckwith and P. K. Tindal, *Austral. J. Chem.* 1971, **24**, 2099.
46. H. C. Brown and G. A. Russell, *J. Amer. Chem. Soc.* 1952, **74**, 3995.
47. O. A. Reutov, *Pure Appl. Chem.* 1963, **7**, 203.
48. D. F. DeTar and C. Weiss, *J. Amer. Chem. Soc.* 1956, **78**, 4296; 1957, **79**, 3041.
49. T. J. Wallace and R. J. Gritter, *J. Org. Chem.* 1962, **27**, 3067.
50. C. A. Grob and H. Kammuller, *Helv. Chim. Acta*, 1957, **40**, 2139.
51. J. K. Kochi and R. D. Gilliom, *J. Amer. Chem. Soc.* 1964, **86**, 5251.

52. A. L. J. Beckwith and W. B. Gara, *J. Amer. Chem. Soc.* 1969, **91**, 5689, 5691.
53. V. R. Desai, A. Nechvatal and J. M. Tedder, *J. Chem. Soc.* B, 1970, 386.
54. B. Acott and A. L. J. Beckwith, *Austral. J. Chem.* 1964, **17**, 1342.
55. C. Walling and A. Padwa, *J. Amer. Chem. Soc.* 1963, **85**, 1591.
56. J. K. Kochi, *J. Amer. Chem. Soc.* 1963, **85**, 1958.
57. F. D. Greene, M. L. Savitz, H. H. Lau, F. D. Osterholtz and W. N. Smith, *J. Amer. Chem. Soc.* 1961, **83**, 2196.
58. P. Kabasakalian and E. R. Townley, *J. Amer. Chem. Soc.* 1962, **84**, 2711.
59. P. Kabasakalian, E. R. Townley and M. D. Yudis, *J. Amer. Chem. Soc.* 1962, **84**, 2716.
60. P. Kabasakalian and E. R. Townley, *J. Amer. Chem. Soc.* 1962, **84**, 2724.
61. M. Akhtar, *Adv. Photochem.* 1966, **2**, 263.
62. R. H. Hesse, *Adv. Free-Radical Chem.* 1967, **3**, 83.
63. K. Heusler and J. Kalvoda, *Angew. Chem. Internat. Edn*, 1964, **3**, 525.
64. D. H. R. Barton and J. M. Beaton, *J. Amer. Chem. Soc.* 1961, **83**, 4083.
65. M. Akhtar and D. H. R. Barton, *J. Amer. Chem. Soc.* 1961, **83**, 2213.
66. M. E. Wolff, *Chem. Rev.* 1963, **63**, 55.
67. S. Wawzonek and P. J. Thelan, *J. Amer. Chem. Soc.* 1950, **72**, 2118.
68. E. J. Corey and W. R. Hertler, *J. Amer. Chem. Soc.* 1960, **82**, 1657.
69. W. R. Hertler and E. J. Corey, *J. Org. Chem.* 1959, **24**, 572.
70. P. Buchschacher, J. Kalvoda, D. Arigoni and O. Jeger, *J. Amer. Chem. Soc.* 1958, **80**, 2905.
71. D. H. R. Barton, A. L. J. Beckwith and A. Goosen, *J. Chem. Soc.* 1965, 181.

14

Radical cyclizations

14.1 *INTRODUCTION*

Scrutiny of the chemical literature over the past few years reveals a rapidly increasing interest in radical cyclizations, particularly in the cyclization of hex-5-en-1-yl radicals. These latter can either cyclize to cyclopentylmethyl or cyclohexyl radicals:

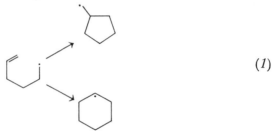

$$(1)$$

Particular attention will be devoted to cyclizations of these and related radicals in this chapter because of their widespread occurrence in radical chemistry. Examination of the factors controlling the course of cyclization can also give information pertaining to the most favourable geometry of the transition state in radical additions, cyclizations being intramolecular additions. Other types of radical cyclization involving both aliphatic and aromatic systems will also be considered.

14.2 *CYCLIZATION OF HOMOALLYL RADICALS*

Homoallyl radicals are formed reversibly by fragmentation of cyclopropylmethyl radicals (see chapter 12, p. 480):

$$\triangleright\!\!-\mathrm{CH_2}\!\cdot\ \rightleftharpoons\ \diagup\!\!\diagdown\!\!\diagdown\ \cdot \qquad (2)$$

In most simple examples the equilibrium lies far to the right, as indicated by the absence of any products containing the cyclopropane ring.

Radical cyclization

This is not the case when the cyclized radical is stabilized as is the case for the cyclization of the 4,4-diphenylbut-2-en-1-yl radical (Scheme 1) (1).

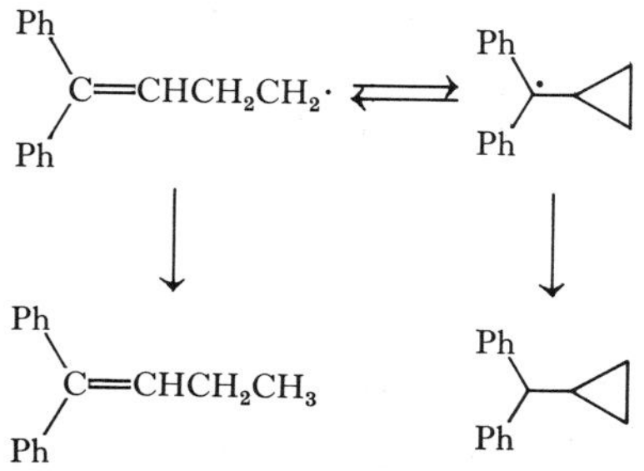

Scheme 1

The same situation also occurs when the architecture of the molecule is such as to favour cyclization (Scheme 2) (2).

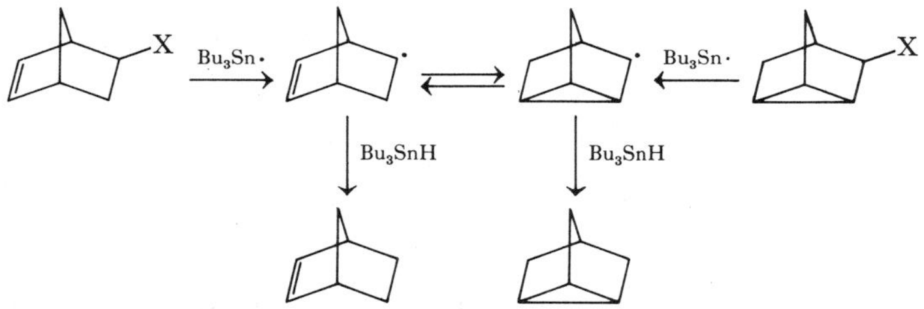

Scheme 2

The odd feature about cyclization of homoallyl radicals is that in all known instances it proceeds to give the less stable of the two possible radicals:

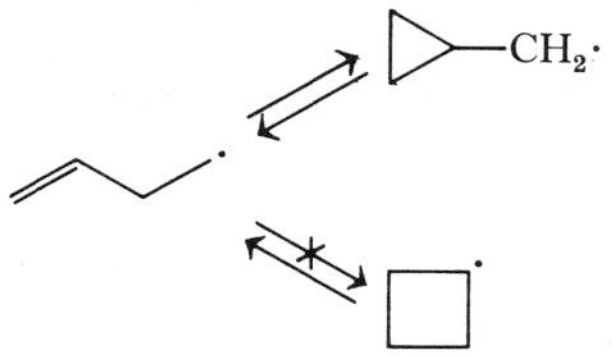

$$(3)$$

No instances are known in which a cyclobutyl radical is obtained even transitorily from a homoallyl radical. Thus, the deuteriated radical (**1**) undergoes rearrangement to the radical (**2**) via the cyclopropylmethyl radical (**3**). If cyclization to a cyclobutyl radical had occurred, then deuterium would have been seen in the products attached to C-3 (Scheme 3).

534

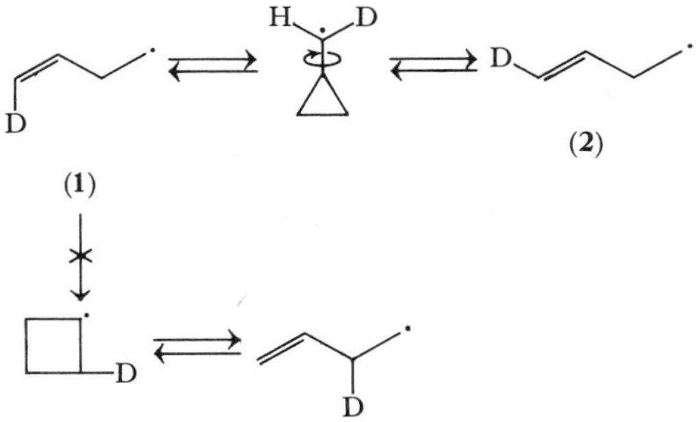

Scheme 3

Cyclobutyl radicals do, however, undergo fragmentation to homoallyl radicals, as evidenced by the formation of butadiene from the decomposition of biscyclobutanecarbonyl peroxide in the presence of copper salts (reaction *4*) (4):

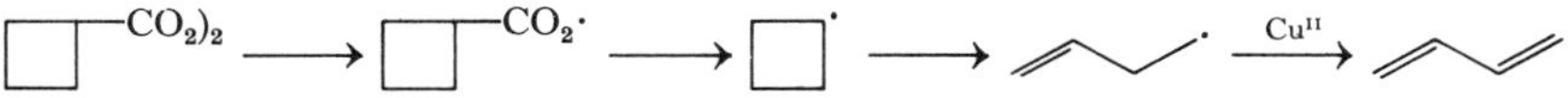

Fragmentation of cyclobutyl radicals is very much less facile than that of cyclopropylmethyl radicals, as indicated from the fact that the e.s.r. spectra of the latter can only be observed at temperatures below $-120°C$, whereas the spectra of cyclobutyl radicals can be observed at room temperature (5). The energy of activation for the fragmentation of cyclobutyl radicals is quite considerable (~ 75 kJ mol^{-1}) (6). Thus, the energy profiles for the fragmentation of cyclopropylmethyl and cyclobutyl radicals must be very different. On the basis of the principle of microscopic reversibility, this is also true for the cyclization of homoallyl radicals to cyclopropylmethyl and cyclobutyl radicals. As will be discussed later (p. 539) this difference arises from different stereoelectronic requirements for the two possible cyclization pathways.

14.3 CYCLIZATION OF PENT-4-EN-1-YL RADICALS

Cyclization of pent-4-en-1-yl radicals has been studied in much less detail than that of but-3-en-1-yl or hex-5-en-1-yl radicals, but it is apparent that when this occurs it does so to give only cyclopentyl radicals. This is to be expected, since cyclization to give cyclopentyl radicals is a more exothermic reaction to the extent of 75 kJ mol^{-1} than

Radical cyclization

is cyclization to cyclobutylmethyl radicals:

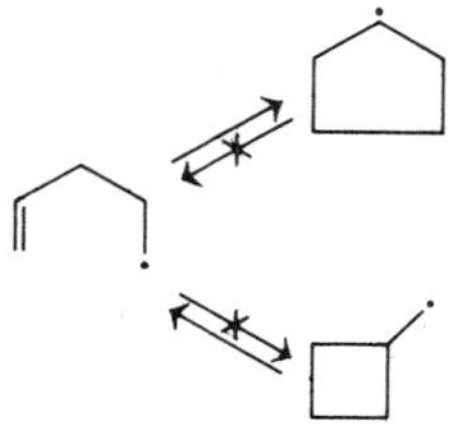

(5)

Cyclization does not, however, normally occur under conditions in which hex-5-en-1-yl radicals cyclize more or less quantitatively. For example, reaction of tributyltin hydride with 5-bromopent-1-ene gives none of the cyclized product, cyclopentane. Under the same conditions 6-bromohex-1-ene gives methylcyclopentane as the main product (Schemes 4 and 5).

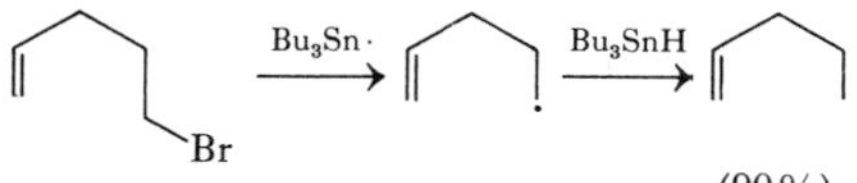

Scheme 4

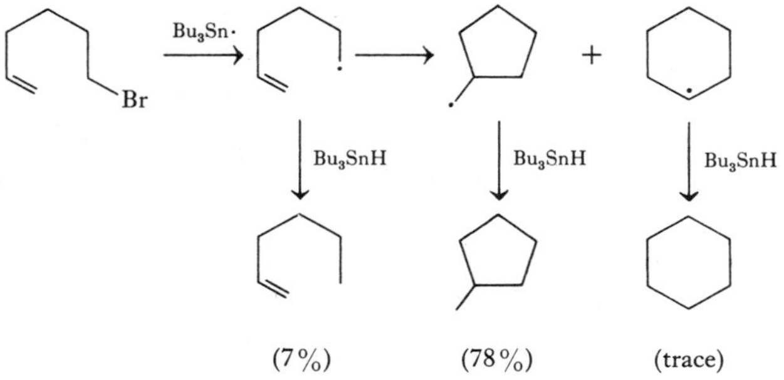

Scheme 5

This difference in behaviour is general, as exemplified by the absence of any cyclic products from the reaction of pent-4-ene-1-thiol with triethyl phosphite (8), and from the thermolysis of bis-5-hexenoyl-peroxide (9). Significant yields of cyclized products are only obtained when the molecule contains structural or electronic features which might favour this type of behaviour (reactions 6 and 7) (10, 11), or when the acyclic radical is stabilized by suitable substituents which thus give

536

it a greater lifetime (reaction *8*) (12):

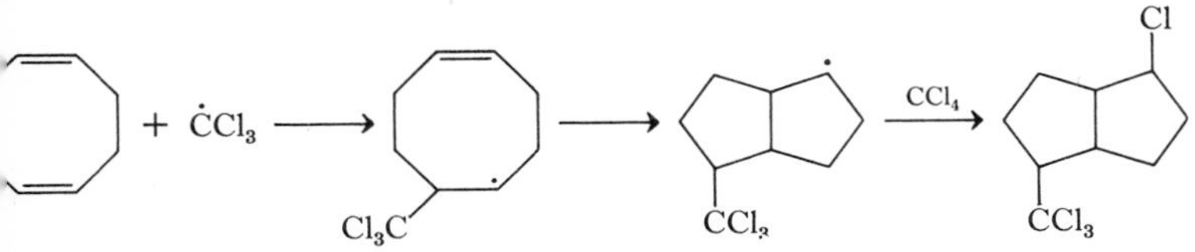

$$(6)$$

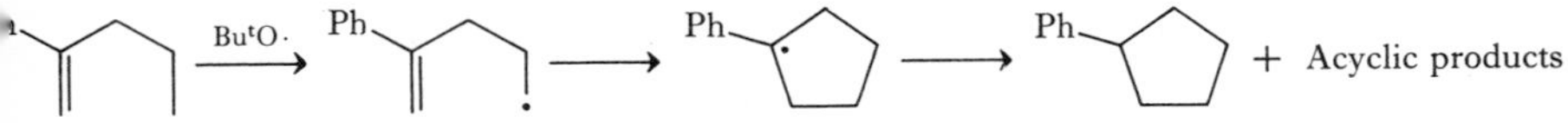

$$(7)$$

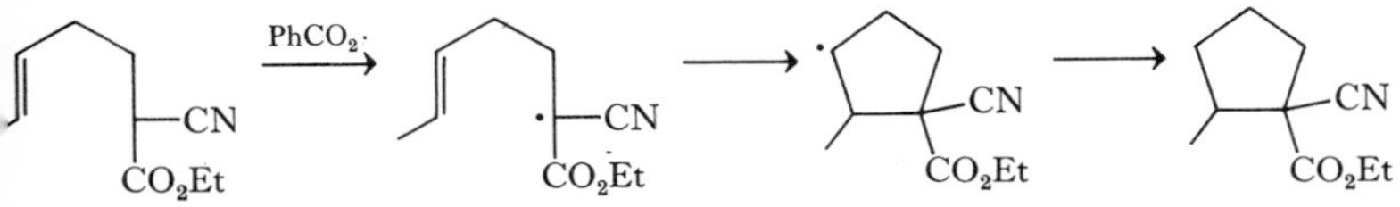

$$(8)$$

The reluctance with which pent-4-en-1-yl radicals cyclize contrasts with the ease of fragmentation of cyclobutylmethyl radicals.

14.4 CYCLIZATION OF HEX-5-EN-1-YL AND RELATED RADICALS

A large number of examples are known in which cyclization of the hex-5-en-1-yl radical occurs (Scheme 5 and reactions *9–11*) (7, 8):

$$(9)$$

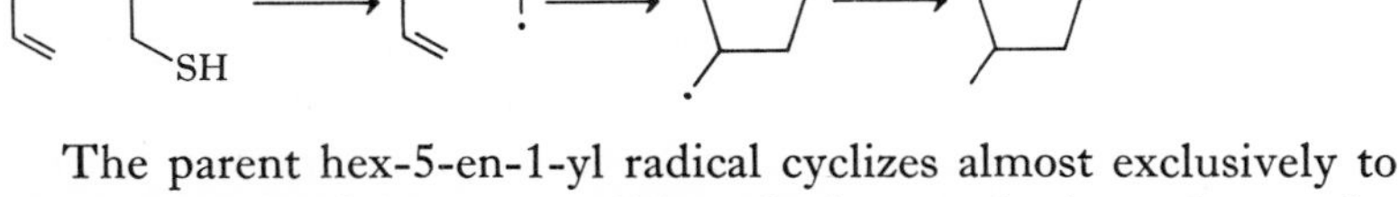

$$(10)$$

(Y = CH_2, $CHCO_2Et$, NH, NCOR; R_F = C_3F_7)

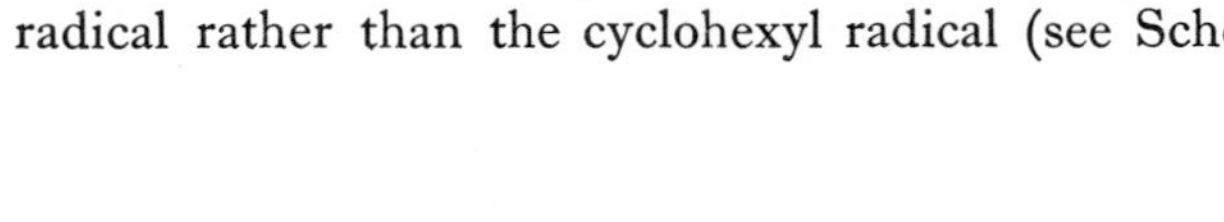

$$(11)$$

The parent hex-5-en-1-yl radical cyclizes almost exclusively to give the less stable of the two possible radicals, namely the cyclopentylmethyl radical rather than the cyclohexyl radical (see Scheme 5) (7), i.e. the reaction proceeds to give the thermodynamically less stable radical:

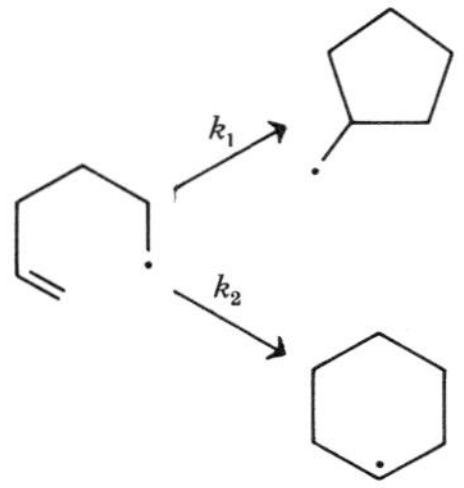

(12)

The cyclization of hex-5-en-1-yl radicals is subject to kinetic control, as both the cyclopentylmethyl and cyclohexyl radicals show no tendency to fragment (8, 12, 15), and hence one can say $k_1 \gg k_2$.

Julia and his colleagues have demonstrated that cyclization of substituted hex-5-en-1-yl radicals proceeds somewhat differently in that cyclohexane derivatives are frequently obtained and are sometimes the major cyclic product. This is particularly so when the acyclic radical is stabilized by cyano or carboethoxy groups as for the radical (3) (Scheme 6) (16).

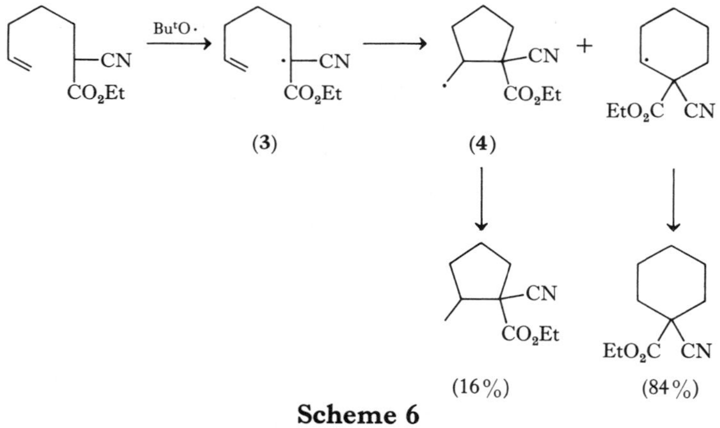

Scheme 6

The increased amount of cyclohexane derivatives appears to arise from the reversibility of the cyclization process, as evidenced by the increasing amount of cyclohexane derivatives obtained on increasing the temperature of the reaction. Julia and Maumy (16) have also shown that more of the cyclopentane derivatives are obtained when the reaction is conducted in an efficient hydrogen-donor solvent such as toluene, which can react with the radical (4) before equilibration occurs.

It thus seems that with simple hex-5-en-1-yl radicals the nature of the products is subject to kinetic control, whereas with more stable substituted radicals, such as (3), thermodynamic control is operative. This is merely a rationalization of the experimental facts, and says nothing about why cyclization of hex-5-en-1-yl radicals to cyclopentylmethyl radicals is faster than cyclization to cyclohexyl radicals. To explain this it is necessary to consider the stereoelectronic requirements of the cyclization process. Beckwith has suggested, on the basis of examination of models, that the formation of the transition state for cyclization involves interaction of the unpaired electron with the lowest unfilled orbital of the π-system. Consequently, the approach of the radical centre occurs preferably in the plane of the π-system and along a line extending almost vertically from one of the terminal carbon atoms (17). This view suggests that bond formation will occur at the terminus of the olefinic linkage most readily approached by the radical centre. Models of the hex-5-en-1-yl radical clearly show that five-membered ring formation should be favoured, though the differences between the formation of five- and six-membered rings are not large. This explanation is also compatible with the ready cyclization of hept-6-en-1-yl systems and with the reluctance of pent-4-en-1-yl radicals to cyclize. The ease of formation of a five-membered ring when the transition state for cyclization can attain the correct geometry is strikingly illustrated by the relative ease of cyclization of the 2-(Δ^3-cyclopentenyl)ethyl radical to the nornbornyl radical (18). In the transition state (5) for this process the radical centre can interact with the π-orbital of the olefinic bond along a line vertically through one of the olefinic carbon atoms.

$$(13)$$

(5)

This model also provides an excellent rationalization of the ease of fragmentation reactions which are the reverse of cyclization reactions. The transition state for fragmentation involves interaction between the radical centre and the lowest unfilled orbital of the bond undergoing fission. In the case of cyclobutyl radicals the orbital of the unpaired electron is orthogonal to the σ^*-orbital of the bond which would undergo fission. Consequently, fragmentation of cyclobutyl radicals is a process requiring high activation energy, and equally cyclization of but-3-en-1-yl radicals to cyclobutyl radicals would also have to surmount a considerable energy barrier. This is not so for the formation of cyclopropyl-methyl radicals.

Cyclization of monosubstituted hex-5-en-1-yl radicals, e.g. (6), gives rise to both cyclopentane and cyclohexane derivatives (16). It is by no means certain that this is merely a reflection of thermodynamic control on the reaction and in some instances it would appear that kinetic control is operative.

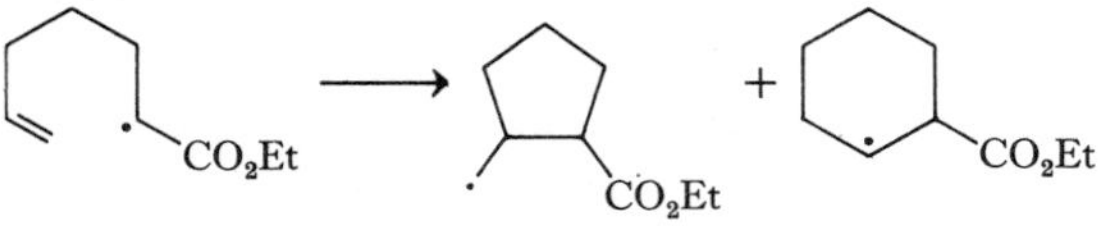

$$\qquad\qquad\qquad\qquad\qquad\qquad\qquad (14)$$

(6)

The cyclization reaction is much less exothermic in the case of more stable radicals such as (3). Consequently, the transition state will resemble rather less closely the reactants, in which the reaction centres are sp^2-hybridized, and rather more closely the products, in which the reaction centres are sp^3-hybridized. Beckwith has shown by examination of models that in this situation cyclization to the six-membered ring would be preferred.

This explanation could also be advanced to explain why cyclization of the radical (7) proceeds exclusively to give five-membered cyclic products (19), whereas the analogous thiyl radical (8) gives mainly the six-membered products (20). Cyclization of this latter would be much less strongly exothermic:

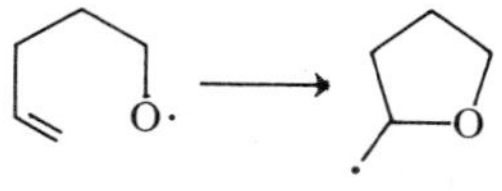

$$\qquad\qquad\qquad\qquad\qquad\qquad\qquad (15)$$

(7)

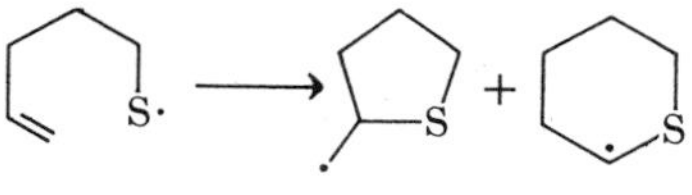

$$\qquad\qquad\qquad\qquad\qquad\qquad\qquad (16)$$

(8)

Significant amounts of products derived from the 9-decalyl radical (9) are obtained from cyclization of the cyclohexenylbutyl radical (reaction *17*) (17). This arises not so much from stabilization of the

radical centre but from steric hindrance by the hydrogens in the cyclohexene ring to formation of the spiro-radical (**10**).

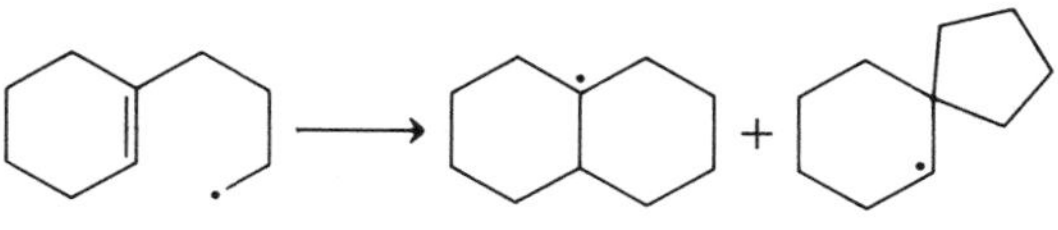

$$\text{(17)}$$

(**9**)   (**10**)

The behaviour of hex-5-en-1-yl radicals is in sharp contrast to that
of the corresponding cations, which invariably lead to the formation of
a six-membered ring (21). The reason for this difference lies in the
different stereoelectronic requirements for radical and cationic cyclizations. In the latter, the positively charged centre approaches in the nodal
plane of the π-system, thus giving a triangular arrangement of groups in
the transition state. Such a situation in which two electrons are dis-
persed over three centres is common for many electrophilic reactions. It
is not feasible for the corresponding radical reactions, as the extra
electron would necessarily occupy an antibonding orbital (cf. p. 498).

14.5 CYCLIZATION OF HEX-5-YN-1-YL AND RELATED RADICALS

Hex-5-yn-1-yl radicals undergo cyclization to form a five- rather than a
six-membered ring (22). The facile nature of the cyclization is indicated
by the absence of any acyclic products (Scheme 7). The ease of this

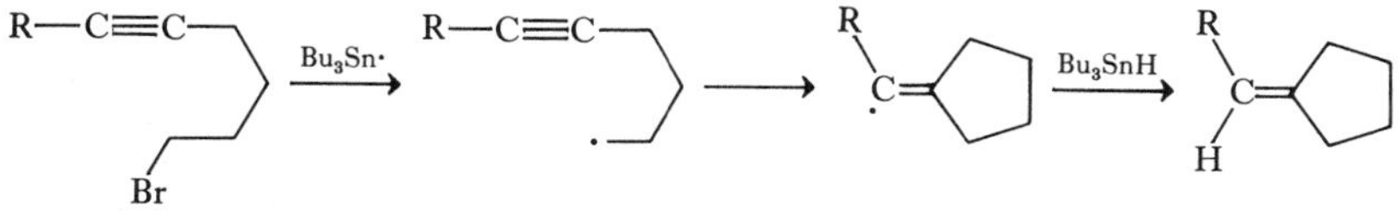

Scheme 7

cyclization is in marked contrast to the complete absence of any cyclic
product from the analogous oxy radical (**11**) (23). The reason for this is
by no means clear but it may arise as a result of electrostatic repulsion
between the electron-rich oxygen and the electrons of the acetylenic
linkage:

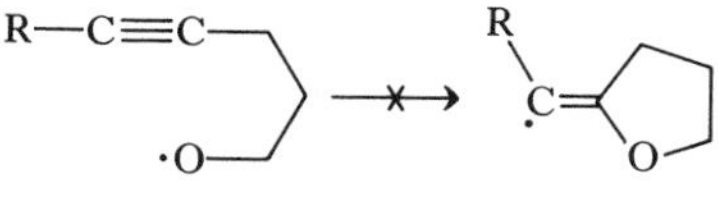

$$\text{(18)}$$

(**11**)

Radical cyclization

14.6 CYCLIZATION OF HEPT-6-EN-1-YL RADICALS

Few reactions are known in which hept-6-en-1-yl radicals are generated. Such evidence as there is seems to indicate that they do cyclize to the cyclohexane derivative, but that the cyclization process is much less facile than for hex-5-en-1-yl radicals, as is indicated by lower yields of cyclized products (11):

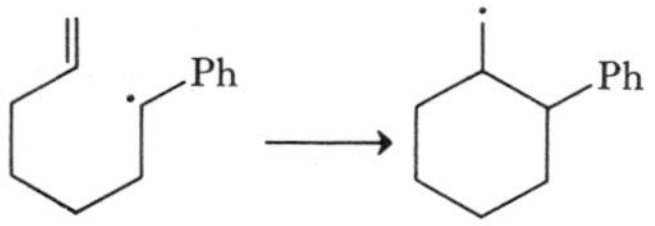

$$(19)$$

Rather better yields of cyclized product are obtained when more stable radicals are used (reaction *20*) (24):

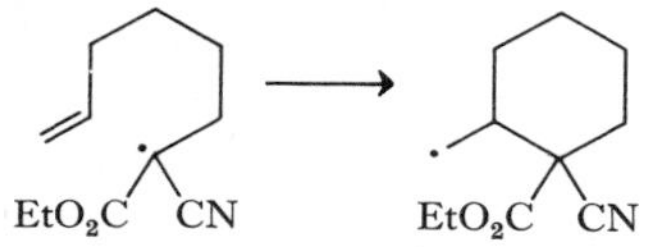

$$(20)$$

In this reaction none of the seven-membered product was detected. Models indicate that formation of a six-membered ring is likely to be favoured, and, in addition, this leads to the thermodynamically more stable product.

14.7 CYCLIZATIONS INVOLVING AROMATIC SYSTEMS

14.7.1 *Cyclizations of alkenylaryl radicals*

Beckwith has concluded from his studies of the cyclization of alkenylaryl radicals that, as in cyclizations of alkenylalkyl radicals, the direction of cyclization is controlled by the stereoelectronic requirements of the transition state rather than by the thermodynamic stability of the cyclized radical (25). Thus *o*-allyloxyphenyl and *o*-butenylphenyl radicals undergo cyclization to give only the five-membered ring. This was indicated both by e.s.r. spectroscopy and by product analysis:

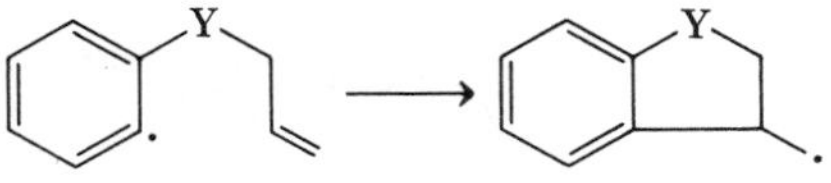

$$(21)$$

(Y = O, CH$_2$)

14.7.2 *Intramolecular homolytic aromatic substitutions by an aryl radical centre*

Extensive studies have been made on radical cyclizations of aromatic systems in which an aryl radical attacks an aromatic system. The Pschorr reaction (*22*), which is the classical example of such cyclization, has already been discussed in chapter 11, p. 439. Generally in this type of reaction cyclizations leading to the formation of six-membered rings are more favoured than those leading to five-membered rings, primarily because of steric effects.

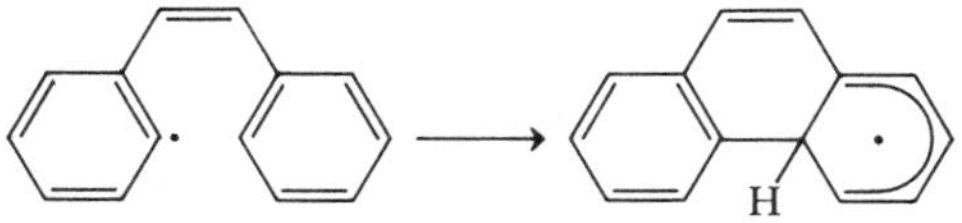

$$(22)$$

14.7.3 *Intramolecular homolytic aromatic substitutions by an alkyl radical centre*

Intramolecular alkylations of aromatic systems are comparatively widespread, though the ease with which such reactions occur is much less than for intramolecular arylations. This is reflected in the much greater tendency for intermolecular reactions to compete (Scheme 8) (26). The absence of a fairly rigid structure for the radical probably accounts for the much greater tendency of radicals, such as 4-phenylbutyl radicals, to undergo intermolecular reactions.

$$PhCH_2CH_2CH_2CH_2 \cdot \longrightarrow$$

$$\downarrow Cu^{II}$$

$$PhCH_2CH_2CH=CH_2$$

$$(86\%)$$

$$\xrightarrow{Cu^{II}} \quad (10\%)$$

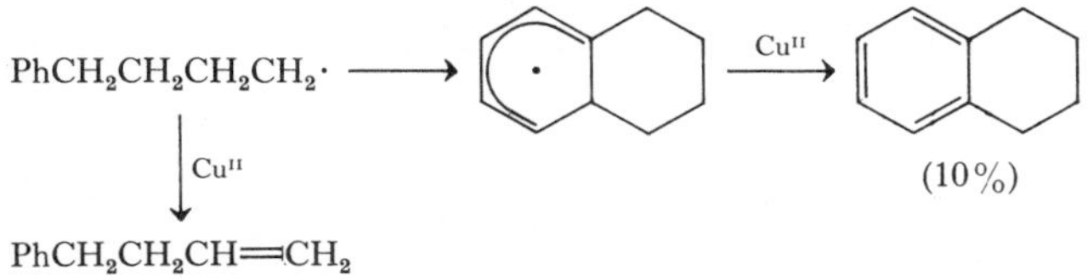

Scheme 8

This lack of reactivity with regard to cyclization is also illustrated by the failure of the stabilized radical (**12**) to give any of the cyclized radical (**13**) (27). This is in marked contrast to the analogous hex-5-en-1-yl radical. This difference arises because of the much greater reactivity of alkenes than arenes towards radical attack:

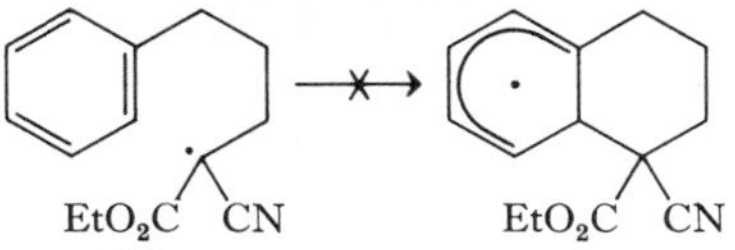

$$\xrightarrow{\quad\times\quad} \qquad (23)$$

EtO$_2$C CN   EtO$_2$C CN

(**12**)   (**13**)

Radical cyclization

GENERAL REFERENCES

A. L. J. Beckwith, *Essays in Free-Radical Chemistry, Chem. Soc. Special Publ.* no. 24, 1970, 239.

M. Julia, *Pure Appl. Chem.* 1967, **15**, 167.

SPECIFIC REFERENCES

1. T. A. Halgren, M. E. H. Howden, M. E. Medof and J. D. Roberts, *J. Amer. Chem. Soc.* 1967, **89**, 305.
2. C. R. Warner, R. J. Strunk and H. G. Kuivila, *J. Org. Chem.* 1966, **31**, 3381.
3. L. K. Montgomery and J. W. Matt, *J. Amer. Chem. Soc.* 1967, **89**, 6556.
4. J. K. Kochi and H. E. Mains, *J. Org. Chem.* 1965, **30**, 1862.
5. J. K. Kochi, P. J. Krusic and D. R. Eaton, *J. Amer. Chem. Soc.* 1969, **91**, 1879.
6. A. S. Gordon, R. S. Smith and C. M. Drew, *J. Chem. Phys.* 1962, **36**, 824.
7. C. Walling, J. H. Cooley, A. A. Ponaras and E. J. Racah, *J. Amer. Chem. Soc.* 1966, **88**, 5361.
8. C. Walling and M. S. Pearson, *J. Amer. Chem. Soc.* 1964, **86**, 2262.
9. R. C. Lamb, W. E. McNew, J. R. Sanderson and D. C. Lunney, *J. Org. Chem.* 1971, **36**, 174.
10. R. Dowbenko, *Tetrahedron*, 1964, **20**, 1843.
11. H. Pines, N. C. Sih and D. B. Rosenfield, *J. Org. Chem.* 1966, **31**, 2255.
12. R. C. Lamb, P. W. Ayers and M. K. Toney, *J. Amer. Chem. Soc.* 1963, **85**, 3483.
13. R. A. Sheldon and J. K. Kochi, *J. Amer. Chem. Soc.* 1970, **92**, 4395.
14. N. O. Brace, *J. Org. Chem.* 1966, **31**, 2879; 1969, **34**, 2441; 1971, **36**, 3187.
15. J. F. Garst, P. W. Ayers and R. C. Lamb, *J. Amer. Chem. Soc.* 1966, **88**, 4260.
16. M. Julia and M. Maumy, *Bull. Soc. chim. France*, 1969, 2415, 2427, and earlier papers.
17. D. L. Struble, A. L. J. Beckwith and G. E. Gream, *Tetrahedron Letters*, 1968, 3701.
18. J. W. Wilt, S. N. Massie and R. B. Dakek, *J. Org. Chem.* 1970, **35**, 2803.
19. R. D. Rieke and N. A. Moore, *Tetrahedron Letters*, 1969, 2035.
20. J.-M. Surzur, M.-P. Crozet and C. Dupuy, *Compt. rend.* 1967, **264**, C, 610.
21. P. D. Bartlett, W. D. Closson and T. J. Cogdell, *J. Amer. Chem. Soc.* 1965, **87**, 1308.
22. J. K. Crandall and D. J. Keyton, *J. Amer. Chem. Soc.* 1969, **91**, 1653.
23. R. D. Rieke and B. J. A. Cooke, *J. Org. Chem.* 1971, **36**, 2674.
24. J.-M. Surzur and G. Torri, *Bull. Soc. chim. France*, 1970, 3070.
25. A. L. J. Beckwith and W. B. Gara, *J. Amer. Chem. Soc.* 1969, **91**, 5689, 5691.
26. J. K. Kochi and R. D. Gilliom, *J. Amer. Chem. Soc.* 1964, **86**, 5251.
27. M. Julia, *Pure Appl. Chem.* 1967, **15**, 167.

15

Radical displacement reactions

15.1 *INTRODUCTION*

A radical displacement reaction may be defined as a homolytic process in which a group, forming part of an organic substrate, is replaced by a radical. The overall reaction may be represented by:

$$R\cdot + SA \longrightarrow RS + A\cdot \qquad (1)$$

We are concerned here with those reactions in which the displacement occurs by a bimolecular process in which the radical $R\cdot$ attacks the centre S, cleaving the S–A bond homolytically and displacing $A\cdot$:

$$R\cdot + SA \longrightarrow [R\cdots S\cdots A] \longrightarrow RS + A\cdot \qquad (2)$$

This process is given the symbol S_H2 by analogy with the similar bimolecular nucleophilic substitution process S_N2. If S is a univalent atom such as hydrogen or halogen, then process (2) is simply an atom abstraction reaction, e.g.:

$$R\cdot + H\!-\!CH_3 \longrightarrow RH + CH_3\cdot$$

Atom abstractions are by far the most common kind of radical displacement reaction, but they are dealt with in chapters 7 and 8. Our specific concern here is the case when $R\cdot$ attacks a multivalent centre S, with displacement of $A\cdot$ which might be a single atom, or a polyatomic radical, e.g.:

$$Ph\cdot + CH_3SSCH_3 \longrightarrow PhSCH_3 + CH_3S\cdot$$

This is equivalent to the abstraction of a group of atoms from AS.

One difficulty in the identification of S_H2 reactions is that the same overall result as (1) may be obtained by homolysis of AS, followed by combination of S· with the radical R·.

$$AS \longrightarrow A\cdot + S\cdot \qquad (3)$$

$$R\cdot + S\cdot \longrightarrow RS \qquad (4)$$

Observation of the product RS is not by itself sufficient evidence to establish the bimolecular displacement mechanism.

Displacement reactions

Ingold and Roberts (1) point out that the attack of R· on S should be favoured when S has vacant p- or d-orbitals available for co-ordination with R·, since a bond between R and S can then be formed before any bonds are broken. Radical displacement reactions can therefore be expected at atoms in groups II and III of the periodic table which have vacant p-orbitals, for example at boron (2):

$$\text{Me·} + \text{BEt}_3 \longrightarrow \text{MeBEt}_2 + \text{Et·}$$

Similarly, displacements can be expected at atoms in the second and third rows of the periodic table which have energetically accessible d-orbitals, for example, at phosphorus (3):

$$\text{Bu}^t\text{O·} + \text{Me}_3\text{P} \longrightarrow \text{Bu}^t\text{OPMe}_2 + \text{Me·}$$

and at sulphur (4):

$$\text{Ph·} + \text{EtSSEt} \longrightarrow \text{PhSEt} + \text{EtS·}$$

Unfortunately, homolytic reactions of compounds based on elements from these areas of the periodic table are not well known. When the reactions of radicals centred on these elements become more fully investigated, many more examples of radical displacement reactions will undoubtedly come to light.

15.2 DISPLACEMENT AT MERCURY

The displacement reaction at mercury probably occurs by addition of the attacking radical to the central mercury atom, which expands its valence to three:

$$\text{R·} + \text{HgX}_2 \longrightarrow \text{RHgX}_2 \longrightarrow \text{RHgX} + \text{X·} \tag{5}$$

The addition complex then breaks down to give the displacement products, as the carbon–mercury bond is weak and can readily be cleaved homolytically.

The reaction of iodine with alkylmercury compounds can proceed by a radical reaction which involves attack by iodine atoms at the mercury centre:

$$\text{I·} + \text{HgR}_2 \longrightarrow \text{R·} + \text{RHgI} \tag{6}$$

$$\text{R·} + \text{I}_2 \longrightarrow \text{RI} + \text{I·} \tag{7}$$

An addition complex is probably formed first, IHgR_2, but this has never been detected. This radical pathway competes with an ionic reaction, which is important in polar solvents, and with certain reactants (5, 6). Jensen and co-workers have studied the reaction of bromine with optically active s-butylmercury(II) bromide. In nonpolar solvents the radical process predominated, and complete racemization was observed (7):

$$(-)\text{-}Bu^sHgBr + Br\cdot \longrightarrow Bu^{s\cdot} + HgBr_2 \qquad (8)$$

$$Bu^{s\cdot} + Br_2 \longrightarrow (\pm)\text{-}Bu^sBr + Br\cdot \qquad (9)$$

In the same way 4-alkylcyclohexylmercury(II) halides were cleaved by radical attack of halogen atoms in non-polar solvents (8). Attack on *cis*- or *trans*-4-t-butylcyclohexylmercury(II) bromide gave a mixture of approximately equimolar amounts of the *cis*- and *trans*-4-t-butylcyclohexyl bromides. The ionic reaction which occurred in polar solvents proceeded with retention of the original configuration.

Displacement at mercury can also be brought about by alkyl radicals. Photolysis of mixtures of hexadeuterioacetone and dimethylmercury in the gas phase results in the formation of CH_3CD_3, which probably arises via the displacement shown (9):

$$CD_3COCD_3 + h\nu \longrightarrow 2CD_3\cdot + CO$$

$$CD_3\cdot + (CH_3)_2Hg \longrightarrow CD_3HgCH_3 + CH_3\cdot \qquad (10)$$

$$CH_3\cdot + CD_3\cdot \longrightarrow CH_3CD_3 \qquad (11)$$

The rates of formation of the isotopic ethanes were shown to obey the cross-combination relationship:

$$R(CH_3CD_3)/[R(C_2H_6)R(C_2D_6)]^{\frac{1}{2}} = 2$$

which strengthens the belief that they are only formed in combination reactions such as (*11*). Rebbert and Ausloos also followed the rate of reaction (*10*) by measurement of the CD_3HgCH_3 produced (10). The rate of methyl radical formation, as judged by this procedure, was slower than that observed from the rate of formation of the ethanes. Other methyl-producing reactions such as (*12*) and (*13*) were proposed to account for this:

$$R\cdot + (CH_3)_2Hg \longrightarrow RHgCH_3 + CH_3\cdot \qquad (12)$$

$$CH_3HgCH_2\cdot \longrightarrow CH_3\cdot + HgCH_2 \qquad (13)$$

Displacement at mercury by phenyl radicals has also been reported (11). Both *cis*- and *trans*-4-methylcyclohexylmercury(II) bromides, when treated with benzoyl peroxide, gave an approximately equimolar

Displacement reactions

mixture of *cis-* plus *trans-*4-methylcyclohexylmercury(II) bromides. *Trans-*2-chlorovinylmercury(II) chloride was isomerized in a similar way:

$$(PhCO_2)_2 \longrightarrow 2PhCO_2\cdot \longrightarrow 2Ph\cdot + 2CO_2$$

$$Ph\cdot + cis\text{-}RHgBr \longrightarrow PhHgBr + R\cdot$$

$$R\cdot + cis\text{-}RHgBr \longrightarrow cis\text{-} + trans\text{-}RHgBr + R\cdot$$

15.3 DISPLACEMENT AT BORON

The boron atom has a vacant p-orbital in the valence shell, so that an addition complex with an incoming radical can readily be formed. The reaction of aralkanes with trihaloboranes leads to replacement of a halogen atom by an aralkyl radical, e.g.:

$$PhCH_3 + BBr_3 \xrightarrow{AIBN} PhCH_2BBr_2$$

This probably occurs by a radical displacement reaction (1). A similar reaction is probably involved in the photolysis of aryl halides with trihaloboranes (12). The aryl haloboranes formed initially are hydrolysed in the work-up procedure to give dihydroxyboranes or boroxanes.

Kochi and Krusic have shown that alkoxyl radicals can displace alkyl radicals from boranes. The e.s.r. spectra of the displaced alkyl radicals have been observed during the photolysis of di-t-butyl peroxide and a trialkylborane (3), e.g.:

$$Bu^tO\cdot + Bu_3{}^nB \longrightarrow Bu^n\cdot + Bu_2{}^nBOBu^t \qquad (14)$$

It has been demonstrated that the rate of reaction falls with increased branching in the alkyl groups attached to boron. The relative reactivities at 30 °C, were found by Davies *et al.* to be: $Bu_3{}^nB$ (1.0), $Bu_3{}^iB$ (0.03), $Bu_3{}^sB$ (0.008) (13). The decrease in rate shows that the reaction is controlled by the increased steric congestion about the boron atom, rather than the stability of the displaced alkyl radical.

Several reactions in which an alkyl radical displaces a second alkyl radical from a trialkylborane have also been proposed. Grotewold and Lissi generated methyl radicals by photolysis of azomethane, acetone and biacetyl (14). Ethane, propane and butane were obtained as products, which suggested the reaction:

$$Me\cdot + BEt_3 \longrightarrow MeBEt_2 + Et\cdot \qquad (15)$$

The results with biacetyl suggested that the acetyl radical itself is capable of displacing an alkyl radical from the borane:

$$MeCO \cdot + Et_3B \longrightarrow MeCOBEt_2 + Et \cdot \qquad (16)$$

Bell and Platt (15) have suggested a similar displacement with trifluoromethyl radicals generated in the presence of trimethylborane:

$$CF_3 \cdot + BMe_3 \longrightarrow CF_3BMe_2 + Me \cdot \qquad (17)$$

15.4 *DISPLACEMENT AT GROUP IV ELEMENTS*

Displacement reactions at saturated carbon atoms are very rarely encountered. The only well-authenticated example of this kind of reaction is the attack of halogen atoms on cyclopropane rings:

$$X \cdot + \triangledown \longrightarrow XCH_2CH_2CH_2 \cdot \qquad (18)$$

This reaction, which is discussed in greater detail in chapter 7, p. 176 may also be considered as addition of the halogen atom to the pseudo-π-system of the cyclopropane. Displacement from sp^3-hybridized carbon is difficult because there are no vacant orbitals to form a bond with the incoming radical. This difficulty is overcome in the case of cyclopropane derivatives by the presence of the pseudo-π-system.

Displacement is also known to occur at unsaturated carbon atoms, as in ketones, e.g.:

$$CH_3 \cdot + CF_3COCF_3 \longrightarrow (CF_3)_2CCH_3 \longrightarrow CF_3 \cdot + CF_3COCH_3$$
$$\underset{O}{\overset{|}{}}$$

This also may be considered as addition of the radical to the carbonyl carbon followed by decomposition of the alkoxy radical so formed.

Few displacement reactions at silicon or germanium have been observed, in spite of the availability of vacant d-orbitals. There is little doubt, however, that in this case further reactions will soon be discovered as more work is done in this area. Trifluoromethyl radicals are known to substitute tetramethylsilane (15):

$$CF_3 \cdot + Me_4Si \longrightarrow Me_3SiCF_3 + Me \cdot \qquad (19)$$

Davidson and co-workers obtained trimethyliodosilane from the reaction of iodine with hexamethyldisilane (16). They explained this in

Displacement reactions

terms of attack by iodine atoms at silicon:

$$I\cdot + Me_3SiSiMe_3 \longrightarrow Me_3SiI + Me_3Si\cdot \qquad (20)$$

$$Me_3Si\cdot + I_2 \longrightarrow Me_3SiI + I\cdot \qquad (21)$$

Displacement at tin is well documented and can occur under much milder conditions. Clark and co-workers (17) have photolysed mixtures of hexamethylditin and various olefins in the gas phase. The products were found to be telomers of formula $Me_3SnM_xSnMe_3$, where M represents the olefin. Formation of these products undoubtedly involves attack by fluoroalkyl radicals at the tin–tin bond of hexamethylditin with displacement of a trimethyltin radical. For example, with tetrafluoro-ethylene, the main products were formed as indicated in (22) to (24).

$$Me_3SnSnMe_3 + h\nu \longrightarrow 2Me_3Sn\cdot$$

$$Me_3Sn\cdot + CF_2{=}CF_2 \longrightarrow Me_3SnCF_2CF_2\cdot \qquad (22)$$

$$Me_3SnCF_2CF_2\cdot + CF_2{=}CF_2 \longrightarrow Me_3Sn(C_2F_4)_2\cdot \qquad (23)$$

etc.

$$Me_3Sn(C_2F_4)_x\cdot + Me_3SnSnMe_3 \longrightarrow Me_3Sn(C_2F_4)_xSnMe_3 + \cdot SnMe_3 \qquad (24)$$

Similar telomers were obtained from trifluoroethylene and hexafluoro-propene. A small amount of the product $Me_3SnCHFCHF_2$ was isolated from the reaction with trifluoroethylene indicating that $=CHF$ is the main site of addition. The product $Me_3SnCF_2CHFCF_3$ was isolated in the case of hexafluoropropene, so that addition occurs mainly at $=CF_2$ in this molecule. For other olefins such as $CF_2{=}CFCl$ and $CF_2{=}CFBr$, the displacement reaction (24) could not compete with hydrogen abstraction or decomposition, and no telomers were formed.

$$Me_3SnCF_2CFCl\cdot + RH \longrightarrow Me_3SnCF_2CFClH + R\cdot \qquad (25)$$

$$Me_3Sn\cdot + CF_2{=}CFBr \longrightarrow Me_3SnCF{=}CF_2 + Br\cdot \qquad (26)$$

Jackson (18) has suggested a similar displacement reaction at tin to account for the formation of Me_3SnCF_3 in the photolysis of hexa-methylditin with trifluoromethyl iodide:

$$Me_3Sn\cdot + CF_3I \longrightarrow Me_3SnI + CF_3\cdot \qquad (27)$$

$$CF_3\cdot + Me_3Sn\overset{.}{S}nMe_3 \longrightarrow Me_3SnCF_3 + Me_3Sn\cdot \qquad (28)$$

Compounds containing tin–tin bonds are also cleaved by peroxides under very mild conditions (19):

$$PhCO_2\cdot + Et_3SnSnEt_3 \longrightarrow PhCO_2SnEt_3 + \cdot SnEt_3 \qquad (29)$$

$$Et_3Sn\cdot + (PhCO_2)_2 \longrightarrow PhCO_2SnEt_3 + PhCO_2\cdot \qquad (30)$$

Tin–carbon bonds can also be cleaved by attack of t-butoxy radicals (20). Thus tri-n-butyltin hydride and di-t-butyl peroxide yield t-butoxy-tributyltin:

$$Bu^tO\cdot + Bu_3{}^nSnH \longrightarrow Bu^tOH + Bu_3{}^nSn\cdot \qquad (31)$$

$$Bu_3{}^nSn\cdot + Bu_3{}^nSn\cdot \longrightarrow Bu_3{}^nSnSnBu_3{}^n \qquad (32)$$

$$Bu^tO\cdot + Bu_3{}^nSnSnBu_3{}^n \longrightarrow Bu^tOSnBu_3{}^n + Bu_3{}^nSn\cdot \qquad (33)$$

15.5 DISPLACEMENT AT GROUP V ELEMENTS

Displacement at nitrogen does not occur under the usual conditions of radical reactions. Phosphorus is the only element in Group V for which the displacement reaction is well established (1, 21). Alkoxyl radicals can add to tervalent phosphorous compounds to give phosphoranyl radicals, which subsequently undergo α- or β-scission.

$$RO\cdot + PX_3 \longrightarrow RO\dot{P}X_3 \begin{array}{c} \xrightarrow{\;\alpha\text{-scission}\;} ROPX_2 + X\cdot \qquad (34) \\[2ex] \xrightarrow[\beta\text{-scission}]{} O{=}PX_3 + R\cdot \qquad (35) \end{array}$$

The phophorus atom has vacant d-orbitals available which enable the valence to be increased to four. The phosphoranyl radical can then decompose by α-scisson (34) where one of the substituents on phosphorous is eliminated, or by β-scisson to produce an alkyl radical and the phosphine oxide. Only the α-scisson route can be classified as a radical displacement reaction.

The formation of phosphoranyl radicals has been postulated since 1957 (1, 21) and recent e.s.r. evidence confirms the earlier deductions from product analyses. When di-t-butyl peroxide was decomposed photolytically in the presence of a trialkylphosphine, the e.s.r. spectrum of the displaced alkyl radical was observed (3):

$$Bu^tO\cdot + PEt_3 \longrightarrow Bu^tOPEt_2 + Et\cdot \qquad (36)$$

When trimethylphosphine was used, the intensity of the methyl radical spectrum was less than expected, and a second spectrum, attributed to the phosphoranyl radical, was observed:

$$Bu^tO\cdot + PMe_3 \longrightarrow Bu^tO\dot{P}Me_3 \longrightarrow Bu^tOPMe_2 + Me\cdot \qquad (37)$$

At higher temperatures and in more polar solvents only the products of β-scisson were observed. With trialkylphosphites only β-scisson was observed:

$$Bu^tO\cdot + P(OEt)_3 \longrightarrow Bu^t\cdot + O{=}P(OEt)_3 \qquad (38)$$

In certain other phosphites, however, such as triphenyl phosphite, the α-scission route occurs. This is presumably because of the greater stability of the phenoxy radicals which are displaced:

$$Bu^tO\cdot + P(OPh)_3 \longrightarrow Bu^tOP(OPh)_2 + PhO\cdot \qquad (39)$$

It appears that carbon radicals can also add to phosphorus compounds to produce the phosphoranyl radical which subsequently decomposes. Phenyl radicals, generated from phenylazotriphenylmethane, were shown to displace alkyl radicals from several alkyl diphenyl phosphinites (22):

$$Ph\cdot + Ph_2POPr^i \longrightarrow Ph_3\dot{P}OPr^i \longrightarrow Ph_3PO + Pr^i\cdot \qquad (40)$$

Phosphoranyl radical intermediates have also been suggested for the reaction of phosphorus trihalides with olefins:

$$\cdot PBr_2 + Ol \longrightarrow \cdot OlPBr_2 \qquad (41)$$

$$\cdot OlPBr_2 + PBr_3 \longrightarrow Br_2\dot{P}OlPBr_3 \qquad (42)$$

$$Br_2\dot{P}OlPBr_3 \longrightarrow Br_2POlBr + \cdot PBr_2 \qquad (43)$$

$$Br_2\dot{P}OlBr_3 \longrightarrow Br_2POlPBr_2 + Br\cdot \qquad (44)$$

15.6 *DISPLACEMENT AT GROUP VI ELEMENTS*

Numerous displacement reactions at peroxidic oxygen by alkyl radicals have been identified. Displacement has not been observed in other oxygen compounds, presumably because the O–C and O–H bonds are strong in contrast to the weak O–O bond. The decomposition of peroxides is a complex process which is considered in greater detail on pages 15 and 429. Peroxides decompose by a straightforward unimolecular process:

$$ROOR \longrightarrow 2RO\cdot \qquad (45)$$

Frequently this is accompanied by a higher-order 'induced' decomposition, which is brought about by attack of a radical on the peroxidic oxygen:

$$R' + ROOR \longrightarrow R'OR + RO\cdot \qquad (46)$$

This displacement reaction has been observed with aroyl peroxides, but its occurrence is critically dependent on conditions such as solvent (from which the attacking radical may be derived), and substituents in the peroxide. Resonance-stabilized radicals attack the peroxidic oxygen in this way. Thus, the induced decomposition of dibenzoyl peroxide in benzene is effected by phenylcyclohexadienyl radicals:

$$(PhCO_2)_2 \longrightarrow PhCO_2\cdot \longrightarrow Ph\cdot + CO_2$$

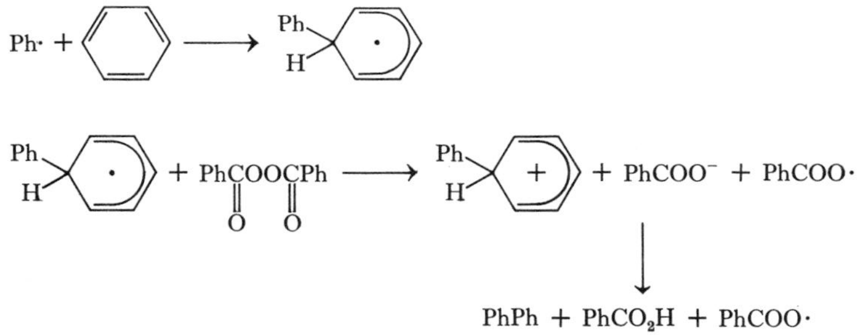

$$PhPh + PhCO_2H + PhCOO\cdot$$

Triphenylmethyl radicals attack in a similar way (25):

$$Ph_3C\cdot + (PhCO_2)_2 \longrightarrow Ph_3COCOPh + PhCO_2\cdot$$

The reaction was accelerated by electron-withdrawing substituents in the peroxide, which lends support to the idea that there must be appreciable dipolar character in the transition state (1).

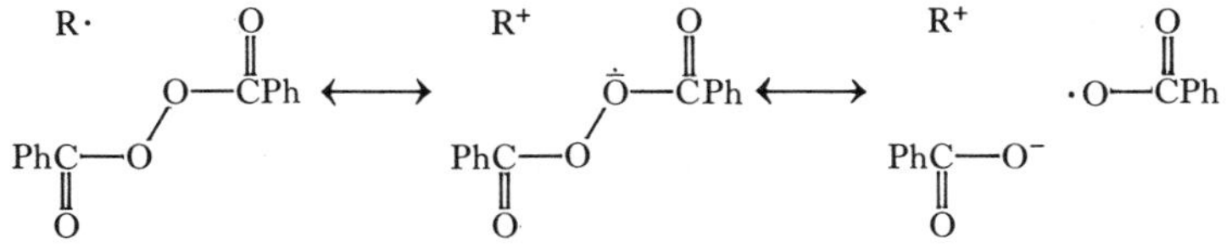

(1)

Strongly nucleophilic radicals also induce decomposition of the peroxide by a displacement process, as for example in the decomposition of dibenzoyl peroxide in ethanol or other primary and secondary

Displacement reactions

alcohols (26). The attacking radical is an α-hydroxyalkyl radical:

$$CH_3\dot{C}HOH + (PhCO_2)_2 \longrightarrow PhCO_2CH(OH)CH_3 + PhCO_2\cdot$$

$$PhCO_2CH(OH)CH_3 \longrightarrow PhCO_2H + CH_3CHO$$

Highly reactive radicals on the other hand induce the decomposition by attack at the *ortho-* or *para*-positions of the aromatic peroxide. Walling and co-workers have shown that methyl radicals generated in the decomposition of neat acetylaroyl peroxides behave in this way (23):

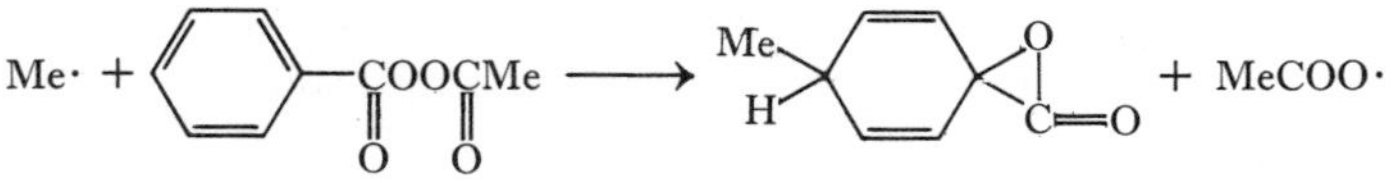

(47)

They suggested that an α-lactone would first be formed, which would then rearrange to an *ortho-* or *para*-substituted benzoic acid:

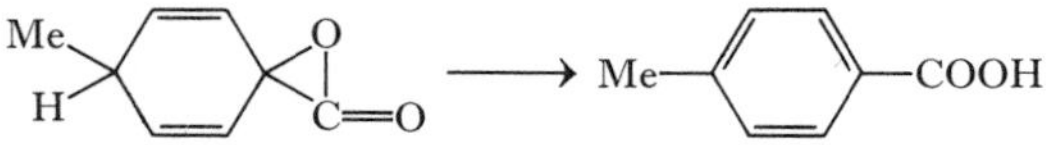

(48)

The induced decomposition of diacyl peroxides generally occurs by displacement at peroxidic oxygen, e.g.:

$$Me\cdot + (MeCO_2)_2 \longrightarrow MeCO_2Me + MeCO_2\cdot \qquad (49)$$

Dialkyl peroxides also undergo induced decomposition in certain solvents. For instance the rate of decomposition of di-t-butyl peroxide is considerably faster in primary or secondary alcohols or amines (27). The key step is believed to be attack at the peroxidic oxygen by α-hydroxy or α-amino radicals:

$$R_2\dot{C}OH + Bu^tOOBu^t \longrightarrow Bu^tOH + R_2C{=}O + Bu^tO\cdot \qquad (50)$$

Radical displacement at sulphur is, perhaps, the best established, and most well known example of homolytic substitution. Displacement at the weak S–S bond has been observed for many kinds of attacking radicals. The displacement reaction in disulphides has been observed for attack by alkyl (24), aryl (14), vinyl (28), thiyl (29) and tin radicals (30):

$$CH_3\cdot + CH_3SSCH_3 \longrightarrow CH_3SCH_3 + CH_3S\cdot \qquad (51)$$

$$Ph\cdot + RSSR \longrightarrow PhSR + \cdot SR \qquad (52)$$

$$Bu^nS\cdot + MeC{\equiv}CH \longrightarrow Bu^nSCH{=}\dot{C}Me \qquad (53)$$

$$Bu^nSCH{=}\dot{C}Me + Bu^nSSBu^n \longrightarrow Bu^nSCH{=}C(Me)SBu^n \qquad (54)$$

$$RS\cdot + R'SSR' \longrightarrow RSSR' + R'S\cdot \qquad (55)$$

$$Bu_3{}^nSn\cdot + PhCH_2SSCH_2Ph \longrightarrow Bu_3{}^nSnSCH_2Ph + PhCH_2S\cdot \qquad (56)$$

554

Displacement can also take place in sulphides, but this reaction does not occur easily unless the departing radical forms only a weak bond to the sulphur.

Quantitative estimates of the rates of reaction of disulphides towards phenyl, polystyryl and tri-n-butyltin radicals have been obtained. Some relative rate data are shown in Table 15-1.

TABLE 15-1   *Relative rate constants for displacement in disulphides*

$$R\cdot + R'SSR' \longrightarrow RSR' + R'S\cdot$$

Disulphide	Ph·[a] k(rel.) 60 °C	Polystyryl[b] k(rel.) 60 °C	$Bu_3{}^nSn$·[c] k(rel.) 30 °C
MeSSMe	86	68	—
EtSSEt	76	—	—
Pr^nSSPr^n	70	—	—
Pr^iSSPr^i	16	—	—
Bu^nSSBu^n	—	17	14
Bu^iSSBu^i	—	15	13
Bu^sSSBu^s	—	3.2	—
Bu^tSSBu^t	1.0	1.0	1.0
PhSSPh	—	435	110
$PhCH_2SSCH_2Ph$	—	70	9

[a] From refs. 4, 31.   [b] From ref. 32.   [c] From ref. 30.

The slowest rate in each case is for attack at di-t-butyl disulphide, where approach to the sulphur atom is most hindered. The rate increases as the size of the alkyl substituents on sulphur decreases, which suggests that steric effects are a major factor in determining the reactivity. The rate of attack on diphenyl disulphide is very high for both polystyryl and tri-n-butyltin radicals, presumably because of the stability of the PhS· radical which is displaced.

Pryor and co-workers (29, 4) have gained some insight into the stereochemistry of the displacement reaction from a study of the homolytic substitution at sulphur. It might be expected that the reaction would proceed in a similar manner to the S_N2 Walden inversion. The attacking radical would approach from the backside and the configuration would be inverted:

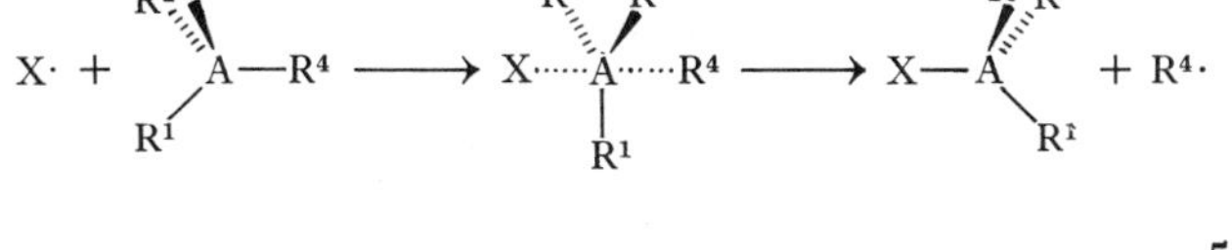

Displacement reactions

The simplest way to study this reaction would be to carry out a homolytic substitution at an optically active centre, and so determine directly whether the configuration is inverted. Unfortunately, displacement at saturated carbon atoms is extremely rare, and no example of this kind of reaction has yet been discovered. Pryor and Guard (4) attacked this problem indirectly. The rates of a series of nucleophilic substitution reactions at sulphur were compared with a series of nucleophilic substitutions at carbon:

$$Y^- + RSX \longrightarrow YSR + X^-$$
$$Y^- + RCH_2X \longrightarrow YCH_2R + X^-$$

As the alkyl substituents R were varied, a very good correlation between the rates of the two reactions was obtained. Nucleophilic substitution at carbon was known to proceed with inversion of configuration and hence it was concluded that the nucleophilic substitution at sulphur was also of this type. The rates of a series of homolytic substitutions at disulphides were next shown to correlate very well with the nucleophilic substitution rates:

$$Y\cdot + RSSR \longrightarrow YSR + \cdot SR$$

It was thus postulated that the homolytic displacement reaction would also proceed with inversion of configuration. Subsequently, Pryor and Smith (31) showed that a large number of reactions, some of which were known to proceed without inversion, could also be correlated with the nucleophilic substitution reactions. The range of alkyl substituents R was simply not great enough for any mechanistic significance to be attached to the results.

Pryor and Smith have reasoned that the homolytic substitution does, nevertheless, proceed with inversion of configuration. They suggested that the lowest energy route would involve a trigonal bipyramidal intermediate (2) with the substituent R, and the two sulphur lone pairs, coplanar:

$$Y\cdot + RSSR \longrightarrow Y\cdots \overset{\overset{\displaystyle R}{|}}{S} \cdots SR \longrightarrow YSR + \cdot SR$$

(2)

Electron repulsion between the coplanar electron pairs and the breaking and forming bonds would be minimized in this arrangement. This implies inversion of configuration at the attacked sulphur atom. The argument applies with equal force to bimolecular homolytic substitutions at atoms other than sulphur.

556

REFERENCES

1. K. U. Ingold and B. P. Roberts, *Free-Radical Substitution Reactions*, Wiley-Interscience, New York, 1971.
2. J. Grotewold and E. A. Lissi, *J. Chem. Soc.* B, 1968, 264.
3. J. K. Kochi and P. J. Krusic, *J. Amer. Chem. Soc.* 1969, **91**, 3944.
4. W. A. Pryor and H. Guard, *J. Amer. Chem. Soc.* 1964, **86**, 1150.
5. F. R. Jensen and B. Rickborn, *Electrophilic Substitution of Organomercurials*, McGraw-Hill, New York, 1968.
6. O. A. Reutov and I. P. Beletskaya, *Reaction Mechanisms of Organometallic Compounds*, North-Holland Publishing Co., Amsterdam, 1968.
7. F. R. Jensen, L. D. Whipple, P. K. Wedegaenter and J. A. Landgrebe, *J. Amer. Chem. Soc.* 1959, **81**, 1262; 1960, **82**, 2466.
8. F. R. Jensen, L. H. Gale and J. F. Rodgers, *J. Amer. Chem. Soc.* 1968, **90**, 5793.
9. R. E. Rebbert and P. Ausloos, *J. Amer. Chem. Soc.* 1963, **85**, 3086.
10. R. E. Rebbert and P. Ausloos, *J. Amer. Chem. Soc.* 1964, **86**, 2068.
11. F. R. Jensen and L. H. Gale, *J. Amer. Chem. Soc.* 1959, **81**, 6337.
12. R. A. Bowie and O. L. Musgrave, *J. Chem. Soc.* C, 1966, 566.
13. A. G. Davies and B. P. Roberts, *Chem. Comm.* 1969, 699; *J. Organometallic Chem.* 1969, **19**, 817.
14. J. Grotewold and E. A. Lissi, *J. Chem. Soc.* B, 1968, 264; *Chem. Comm.* 1968, 1367.
15. T. N. Bell and A. E. Platt, *Chem. Comm.* 1970, 325.
16. S. J. Band and I. M. T. Davidson, *Trans. Faraday Soc.* 1970, **66**, 406.
17. M. A. A. Beg and H. C. Clark, *Chem. and Ind.* 1960, 76; H. C. Clark, J. D. Cotton and J. H. Tsai, *Canad. J. Chem.* 1966, **44**, 903.
18. R. A. Jackson, *Adv. Free-Radical Chem.* 1969, **3**, 231.
19. G. A. Razuvaev, N. S. Yyazankin and O. A. Shchepetkova, *Zhur. Obshch. Khim.* 1960, **30**, 2498; *Tetrahedron*, 1962, **18**, 667.
20. K. Rubsamen, W. P. Neumann, R. Sommer and U. Frommer, *Chem. Ber.* 1969, **102**, 1290. See also *Chem. Ber.* 1967, **100**, 1063.
21. J. I. G. Cadogan, *Adv. Free-Radical Chem.* 1967, **2**, 220.
22. R. F. Bridger and G. A. Russell, *J. Amer. Chem. Soc.* 1963, **85**, 3734; R. S. Davidson, *Tetrahedron*, 1969, **25**, 3383.
23. C. Walling and F. S. Savas, *J. Amer. Chem. Soc.* 1960, **82**, 1738; C. Walling and Z. Čekovič, *ibid.* 1967, **89**, 6681.
24. C. Walling and B. Rubinowitz, *J. Amer. Chem. Soc.* 1957, **79**, 5326.
25. T. Seuhiro, A. Kanoya, H. Hara, T. Nakahama, M. Omori and T. Komori, *Bull. Chem. Soc. Japan*, 1967, **40**, 668.
26. P. D. Bartlett and K. Nozaki, *J. Amer. Chem. Soc.* 1947, **69**, 2299.
27. E. S. Huyser, C. J. Bredewegand and R. M. Van Scoy, *J. Amer. Chem. Soc.* 1964, **86**, 2401, 4148; E. S. Huyser and B. Amini, *J. Org. Chem.* 1968, **33**, 576; E. S. Huyser and R. H. C. Feng, *J. Org. Chem.* 1969, **34**, 1727; E. S. Huyser and A. A. Kahl, *Chem. Comm.* 1969, 1238.
28. E. I. Heiba and R. M. Dessau, *J. Org. Chem.* 1967, **32**, 3837.
29. W. A. Pryor, *Mechanisms of Sulphur Reactions*, McGraw-Hill, New York, 1962.
30. J. Spanswick and K. U. Ingold, *Internat. J. Chem. Kinetics*, 1970, **2**, 157.
31. W. A. Pryor and K. Smith, *J. Amer. Chem. Soc.* 1970, **92**, 2731.
32. W. A. Pryor and T. L. Pickering, *J. Amer. Chem. Soc.* 1962 **84**, 2705.

Index

559

Index

Index

Index

Index

Index

Index

Index